MECHANICS OF TURBULENCE OF MULTICOMPONENT GASES

ASTROPHYSICS AND SPACE SCIENCE LIBRARY

VOLUME 269

MECHANICS OF TURBULENCE OF MULTICOMPONENT GASES

by

MIKHAIL YA. MAROV

and

ALEKSANDER V. KOLESNICHENKO

M.V. Keldysh Institute of Applied Mathematics,
Russian Academy of Sciences,
Moscow, Russia

KLUWER ACADEMIC PUBLISHERS

DORDRECHT / BOSTON / LONDON

A C.I.P. Catalogue record for this book is available from the Library of Congress.

ISBN 1-4020-0103-7

Published by Kluwer Academic Publishers,
P.O. Box 17, 3300 AA Dordrecht, The Netherlands.

Sold and distributed in North, Central and South America
by Kluwer Academic Publishers,
101 Philip Drive, Norwell, MA 02061, U.S.A.

In all other countries, sold and distributed
by Kluwer Academic Publishers,
P.O. Box 322, 3300 AH Dordrecht, The Netherlands.

Printed on acid-free paper

Cover figure: 'The Red Spider Planetary Nebula', courtesy of ESA.
Garrelt Mellema, Vincent Icke (Leiden University, The Netherlands),
and Bruce Balick (Univ of Washington, Seattle, USA)

Printed in the Netherlands.

To unforgettable Natasha Marova

CONTENTS

PART II
SOME MODEL PROBLEMS OF MULTICOMPONENT TURBULENCE

FOREWORD

Space exploration and advanced astronomy have dramatically expanded our knowledge of outer space and made it possible to study the indepth mechanisms underlying various natural phenomena caused by complex interaction of physical-chemical and dynamical processes in the universe. Huge breakthroughs in astrophysics and the planetary sciences have led to increasingly complicated models of such media as giant molecular clouds giving birth to stars, protoplanetary accretion disks associated with the solar system's formation, planetary atmospheres and circumplanetary space. The creation of these models was promoted by the development of basic approaches in modern mechanics and physics paralleled by the great advancement in the computer sciences. As a result, numerous multidimensional non-stationary problems involving the analysis of evolutionary processes can be investigated using wide-range numerical experiments.
Turbulence belongs to the most widespread and, at the same time, the most complicated natural phenomena, related to the origin and development of organized structures (eddies of different scale) at a definite flow regime of fluids in essentially non-linear hydrodynamic systems. This is also one of the most complex and intriguing sections of the mechanics of fluids. The direct numerical modeling of turbulent flows encounters large mathematical difficulties, while the development of a general turbulence theory is hardly possible because of the complexity of interacting coherent structures.

Three-dimensional non-steady motions arise in such a system under loss of laminar flow stability defined by the critical value of the Reynolds number. Owing to the stretching of eddies, an abrupt distribution of the velocity pulsations with wavelength between certain minimum and maximum values, defined by viscous forces and boundaries of the flow, respectively, occurs. Both physical characteristics of a medium, such as the molecular viscosity responsible for dissipation of turbulent flow energy, and conditions at the boundary, responsible for unstable thin eddy layers exhibited as vortex tubes they produce, affect the conditions of the origin of vorticity and structure of the developed turbulence patterns. Turbulization gives rise to fast intermixing of medium particles and increasing the efficiency of momentum, heat and mass transfer. In a multicomponent reacting medium it also promotes the acceleration of chemical reactions. The attempt to get more insight into the nature of objects in which turbulence plays a significant or even decisive role parallels the growing attention and importance of studying and modeling this phenomenon and its relevant effects.

This book is based on the authors' study of the problems of large-scale developed turbulence in multicomponent reacting gas mixtures. The study is rooted in simulations of dynamical processes in the upper/middle atmospheres of the planets - the rarefied gas envelopes occupying the intermediate region between the denser atmospheric layers and adherent outer space, which are impacted by solar electromagnetic and corpuscular radiation responsible for numerous photolytic and accompanying direct and inverse chemical reactions. This relatively new branch of science known as planetary aeronomy, which appeared at the beginning of space research, comprises fundamental scientific fields such as kinetic theory of gases and hydromechanics, chemical kinetics and plasma physics, planetary astronomy and atmospheric sciences. We attempted to further expand the study of aeronomic problems to the planetary cosmogony.

Of specific interest is the investigation of turbulent gas motions when variable thermophysical properties of a multicomponent gaseous medium, compressibility of a flow, presence of chemical reactions, and influence of mass forces should be taken into account. These complementary effects do not allow, in general, to take advantage of the results obtained in the framework of the conventional description of homogeneous compressible fluid flows (in the Boussinesque approximation) widely used in the meteorology. On the other hand, the developed semi-empirical theory of the turbulent exchange coefficients for flows in a multicomponent boundary layer is not applicable to aeronomy or astrophysics because, in particular, gravitational effects in the structure of the respective equations are omitted. Therefore, a new mathematical approach for modeling multicomponent turbulence in such specific media should be developed with the purpose to describe adequately the combined processes of dynamics, heat and mass transfer, and chemical kinetics in a reactive gas continuum. By virtue of the complexity of the physicochemical picture of turbulent motion, theoretical approaches to solve the problem must be "semi-empirical" by their nature.

This is the goal of the present book. It is aimed at the development and justification of semi-empirical models of turbulence in multicomponent reacting gas mixtures as a mathematical basis to describe the dynamical structure and thermal regime of those media where chemical kinetic processes and turbulent mixing are equally important. Such an approach encompasses the following components:

- the development of a macroscopic diffusion theory for molecular transport in mixed gases as a basis to describe the heat and mass transfer in multicomponent gases;
- the development of semi-empirical models for large-scale turbulence in multicomponent reacting gas continua allowing, in particular, to satisfactorily describe turbulent transfer and the impact of turbulization of flows on chemical reaction rates;
- the development of complicated models for multicomponent turbulence including, as closing, the evolutionary transfer equations for the single-point second correlation moments of turbulent fluctuations of thermo-hydro-dynamical parameters, which are of key importance for the solution of various problems, involving both convective and diffusive transport of turbulence, as well as the pre-history of the flow;
- the development of the phenomenological approach, based on the methods of thermodynamics of irreversible processes, to the closure problem in the hydrodynamic equations of mean motion at the level of first order moments, enabling to obtain defining (rheological) relations (expressions for turbulent flows of diffusion, heat and the Reynolds stress tensor for a multicomponent medium) in a more general form as compared to those conventionally deduced using the mixing path concept;
- the advances of semi-empirical techniques, based on the developed multicomponent turbulence models, enabling to calculate the turbulent exchange factors for a medium stratified in a gravitational field, such as a planetary upper atmosphere or an accretion disk;
- the development of a simulation technique on an exterior turbulence scale in the free atmosphere of a planet using the data on fluctuation of the refractive index of air;

- the modeling of turbulent gas-dust media involving heat and mass transfer and coagulation relevant to the origin of planetary systems around stars.

In the study the authors pursued the idea to proceed from the general principles of multicomponent media mechanics when developing a phenomenological approach in order to advance the turbulence theory of reacting gases and modeling the turbulized mixtures, in application to some specific problems. The focus is aimed at planetary aeronomy, involving some relevant problems of atmospheric origin and evolution. The results obtained are also of particular interest to astrophysics, being intrinsically related to the modeling of various mechanisms underlying the evolutionary processes in stars, nebulae and galaxies, first of all to planetary cosmogony, with consequences for the formation and subsequent evolution of protoplanetary accretion discs and the accumulation of planets. Among other important applications of the multicomponent turbulence theory of reacting gases are problems concerning ecology and environmental protection related to the turbulent diffusion of contaminants by industry, gas-oil production, etc. These particular problems, including also damages, burning, and/or other accidental events, are increasingly cause for concern.

The book content conditionally can be divided into two parts. In the first part (chapters 1-5) the mathematical description methods of turbulent flows in multicomponent reacting gas mixtures are discussed thoroughly. In the second part (chapters 6-9) several particular examples of statement of the problem and/or numerical modeling of aeronomy and planetary cosmogony tasks are drawn, illustrating the efficiency of the developed approach.

The *first chapter*, having an introductory character, contains some general concepts of turbulence theory, with a special focus on the space environments where multicomponent turbulence plays a key role. In the *second chapter* the phenomenological theory of heat and mass transfer in a laminar multicomponent medium is considered and the defining relations for thermodynamic diffusion and heat fluxes in a multicomponent gas mixture are deduced based on the methods of thermodynamics of irreversible processes, taking into account the Onsager reciprocity principle. The *third chapter* is dedicated to the development of a turbulence model for a multicomponent reactive gas continuum. Using the weighted-mean Favre averaging, the differential equations of mass, momentum and energy balance, as basis of the model, are obtained, which are pertinent to the description of the mean motion of turbulent multicomponent mixtures of reacting gases. Also the rheological relations for turbulent flows of diffusion, heat and the Reynolds stress tensor are deduced. In the *fourth chapter* a complicated turbulence model for multicomponent continua with variable density (rested, as closing, on the differential transfer equations for the second correlation moments of turbulent fluctuations of thermo-hydrodynamic parameters) is developed. A thermodynamic approach to the simulation of turbulized multicomponent media is addressed in the *fifth chapter*. Finally, in the *last four chapters*, based on the developed theoretical approaches, some particular examples of the modeling problems are discussed. They include the treatment of diffusion processes in the planetary lower thermosphere; the evaluation of the turbulent exchange coefficients in the homopause; the modeling of dynamical properties of the Earth's middle atmosphere using space monitoring techniques to reveal the statistical structure of the turbulent field from the structure characteristics of the refractive index of air; and the modeling of transfer and coagulation processes in the turbulent gas-dust medium of protoplanetary nebula.

Although, because of the great complexity of the phenomena under study, these examples are of a rather limited nature, they allow us to create the basic ideas regarding the problems of the models and their possible treatment. They reflect the real advancement in the mechanics of multicomponent turbulent media and serve as a guideline to use the developed approaches for other important applications.

The book comprises the results of investigations in the field of multicomponent turbulence the authors carried out in the Keldysh Institute of Applied Mathematics of Russian Academy of Sciences for the last nearly twenty years. The basic concepts and approaches were discussed in numerous publications, most comprehensive in the two authors' monographs "Introduction to Planetary Aeronomy" (Nauka, Moscow, 1987) and "Turbulence of Multicomponent Media" (Nauka, Moscow, 1998), both in Russian.

We acknowledge many valuable discussions with our colleagues from the Keldysh Institute and other academic organizations, including workshops on hydromechanic problems, planetary physics and cosmogony, which served as an incentive to summarize our study in the book. We are grateful to academician V.S. Avduevsky, who has read the original manuscript and made some important comments. Our thanks go to K.K. Manuilov for the assistance in preparing the electronic version of the book, and to Kluwer Academic Publishers for the initiative and efforts that made this publication possible.

September, 2000

This book is dedicated to Dr. Natasha Marova, who passed away untimely during the final editing of the manuscript. Natasha was a beautiful and intelligent woman, loyal, sincere, positive, and kind to all who knew her.

Born in 1933, Natasha graduated from Moscow University of Geodesy, Aeromapping and Cartography in 1957. She earned her Ph.D. in 1968 from the P.P. Shirshov Institute of Oceanology of Russian Academy of Sciences where she worked for 45 years. She pioneered the study of bottom topography and sea mounts and was a world expert in this field. She participated in many expeditions on board scientific ocean-going vessels and made significant scientific contributions to the research programs producing major maps. She has authored approximately 150 publications in scientific refereed journals.

Natasha was dedicated to both her professional interests and her family. She was the beloved wife and devoted friend of Mikhail Marov for nearly half a century. She was a happy mother and grandmother. Her caring assistance, wise advice and uncanny ability to resolve family problems will remain an invaluable legacy for her daughter and grandsons.

Natasha was also a good professional colleague and objective critic for us both. Her expertise, experience, and patience promoted our collaboration over 30 years and numerous joint publications, including this book. It is our great fortune that she shared with us this time and place on the tiny planet Earth travelling in infinite space. The dedication of this book to the memory of Natasha Marova comes of our respect and love for her.

Mikhail Ya. Marov
Aleksander V. Kolesnichenko

April, 2001

PART 1

SEMI-EMPIRICAL MODELING OF TURBULENT MULTICOMPONENT GASES

The study of major physical-chemical mechanisms responsible for spatial-temporal distribution and variations of the basic macroparameters (density, velocity, temperature, pressure, composition, etc.) in a turbulent flow of a multicomponent reacting gas mixture is especially fruitful in conjunction with the development of turbulence models describing the most essential features of physical phenomena. Turbulent motion in a multicomponent natural medium is different from that of an incompressible homogeneous fluid in a number of features. The first difference is variability of properties of a flow when mass-averaged density, various thermophysical parameters, all transport factors, etc. depend on temperature, composition, and pressure of a medium. The spatial heterogeneity of temperature, composition, and velocity of turbulent continua leads to transferring their properties by turbulent eddies (turbulent transport of heat and mass). The latter becomes particularly complex in a multicomponent mixture. In the presence of specific chemical and photochemical transformation processes occurring under condition of turbulent mixing, the pattern of a flow is additionally complicated.

Geophysical applications frequently require to take into account some other factors, such as the influence of a planetary magnetic field on a weakly ionized mixture of atmospheric gases, effects of radiation on temperature fluctuation and turbulent transport of radiation energy, etc. Accordingly, when simulating, for example, the structure, dynamics, and thermal state of rarefied gaseous envelopes of celestial bodies, the theoretical results from a conventional model of turbulence in a homogeneous compressible fluid appear to be unacceptable. In this regard, the problem of a mathematical description and developing an adequate model for turbulence of multicomponent chemically interacting gaseous mixtures arises. Such a model must account for compressibility of the flow, variability of thermophysical properties of the medium, heat and mass transfer, effects of a gravitational field, etc. These problems will be addressed in this part of the book.

CHAPTER 1

TURBULENCE IN NATURAL MEDIA

1.1. TURBULENT MOTION OF A FLUID. GENERAL CONSIDERATIONS

Turbulent motion is a chaotic unsteady motion superimposed on basic motions of a liquid (gaseous) medium, which are referred to as statistically averaged motions. Under conditions of turbulent flow, the velocity, temperature, pressure, mass density, concentration of chemical components, index of refraction of a medium, and other hydrodynamic and thermodynamic properties of a fluid exhibit random fluctuations. Because of these irregularities they vary both in space and time. Due to formation of numerous differently sized eddies, turbulent flow features enhance the ability to transfer momentum, energy and mass of elementary liquid volumes. This results in stronger action on the flow around solid bodies and in intense heat exchange and mixing between layers, as well as in acceleration of chemical reactions. Such conditions of fluid motion arise due to loss of stability in an ordered laminar flow, when the dimensionless Reynolds number $\boldsymbol{Re} = VL/\nu$ (where V and L represent the characteristic velocity and the linear scale of the flow, respectively, and ν is the kinematic viscosity) exceeds some critical value $\boldsymbol{Re}_{cr}$. In a more general sense, the turbulence serves as one of the manifestations of diverse motions in hydrodynamic systems having a very large number of degrees of freedom and a high degree of non-linearity. In such a system, new macroscopic relationships form and the internal small scale eddy current structure becomes absolutely random with growth of $\boldsymbol{Re}$. At the same time, large scale coherent (almost ordered) eddy structures arise on the turbulent background. So, while the onset of turbulence characterizes a transfer from order to chaos, birth of order from chaos occurs within a developed turbulent flow when $\boldsymbol{Re} \gg \boldsymbol{Re}_{cr}$ (*Townsend, 1956; Cantwell, 1984; Belotserkovskii, 1997*).

Turbulence is a specific feature of numerous natural phenomena, when dynamic processes are accompanied by transfer of momentum, energy, and mass. Turbulent effects are observed at spatial scales ranging from centimeters up to megaparsecs. For example, they manifest as diverse dynamic phenomena in the terrestrial atmosphere and hydrosphere, in the atmospheres and interiors of stars and planets, in interstellar gaseous dust clouds (like planetary nebulae and proto-planetary disks), in the galactic and intergalactic medium, and in plasmas (magnetohydrodynamic, or plasma turbulence). Meteorological processes are predominantly turbulent and involve interaction of the oceans with the atmosphere, evaporation from water surfaces, vertical and horizontal heat transport, strong mixing of admixtures (including pollutants), viscous dissipation of small-scale eddy kinetic energy. Turbulence arises in many engineering devices where moving fluids, gases, or plasmas are utilized. Among others, it occurs in boundary layers and wakes of submerged solid bodies, in jet streams and in mixing layers, in channel and pipe currents by vortex excitation of oscillations of mechanical and acoustic oscillatory systems (so-called aeolian tones), in plasma beams, etc.

1.1.1. Geophysical Turbulence

Turbulent motions are always dissipative, therefore they can not be self-supporting and need energy supply from the environment. Turbulence sets up due to the growth of small perturbations in a laminar flow or owing to convective instability of motions. In the first case the turbulence energy is extracted from the kinetic energy of shear currents while in the second case it comes from the potential energy of a non-uniformly heated fluid in a gravitational field. Stratification of the atmosphere (distribution of the mass density ρ and other thermohydrodynamical parameters in the direction of gravity) and the rotation of Earth (with angular velocity $\Omega = 7.29\times10^{-5}\ \mathrm{c}^{-1}$) exert specific effects on the character of geophysical turbulence. Besides, occurrence of multiple components in the real atmosphere frequently gives rise to mixture *baroclinicity* due to dependence of ρ not only on the pressure p (as in barotropic media), but also on the temperature T and/or the concentration Z_α of its individual components. The baroclinicity is dynamically important as it results in the appearance of a source term in *Friedman's* equation for vorticity (see (1.2.1)).

In an unstable atmosphere, turbulent convection develops due to the accelerating action of the buoyancy force. Earth's rotation leads to building up turbulent boundary (*Eckman's*) layers in the atmosphere near the land surface and the bottom floor in the oceans. Because of global meridional changes in the Coriolis parameter $f = 2\Omega\cos\theta$ (where θ completes the latitude ϕ to $\pi/2$; $f = 10^{-4}c^{-1}$ everywhere except of areas close to the equator), the *Rossby-Blinova* waves occur. They generate cyclones and anticyclones in the atmosphere and synoptic eddies in the ocean, exemplifying two-dimensional macroturbulence. In turn, macroturbulence, which originated from large-scale heterogeneities in heat influx into the atmosphere from the underlying surface, generates strong microturbulence due to hydrodynamic instability of vertical wind velocity gradients. Development of rather weak microturbulence (featuring small $\boldsymbol{Re}$) in the oceans relates to the occurrence of thin quasi-homogeneous layers separated by surfaces of discontinuities in temperature and salinity (*Monin*, 1988). In studies of geophysical turbulence and in some other cases, the intensity of turbulent transport can be estimated from turbulent viscosity ν^T (or turbulent diffusivity D^T).

Insolation, or the solar heat flux into the atmosphere and on the surface of the Earth (as well as on the other Earth-like planets), is the main energy source of atmospheric physicochemical processes. A relatively small fraction of the influx transforms into kinetic energy of eddy motions at all scales. The kinetic energy of turbulent motion of a system is generated by large energy carrying eddies due to a gradient in the hydrodynamic velocity V (shear flow). The characteristic size of such eddies determine the external (integral) turbulence scale L. As an indication, eddies greater than L are non-isotropic, while eddies of smaller scales are approximately isotropic. The scale ranges from millimeters to thousands of kilometers, $l << r << L$ (where l is the so-called internal turbulence scale, comparable with the *Kolmogorov microscale* l_k) and thus virtually comprises the entire spectrum of dynamic processes in the atmosphere. In this range, a

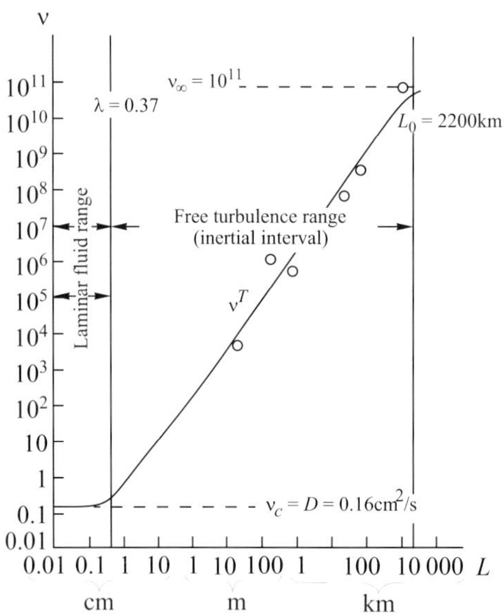

Fig. 1.1.1. Factor of vertical diffusion depending on the turbulence scale, L. The regions of laminar motion and free turbulence (inertial interval) are distinguished. The empirical points corresponding to the Richardson-Obuchov law are shown. According to *Monin, 1969*.

cascade process occurs of energy transfer from large scale eddy motions to small scale ones. The quasi-stationary conditions for occurrence of such a cascade mechanism in the Earth's turbulent atmosphere are characterized by an approximately constant quantity $\varepsilon_e \approx 3$ cm²/s³ , independent of the eddy scale r. On the one hand, this quantity represents the transport rate of rotational kinetic energy from large atmospheric eddies to smaller ones. On the other hand, it characterizes the specific velocity of dissipation of turbulent (*kinetic*) energy into heat due to molecular viscosity which occurs in eddies of the minimum *Kolmogorov scale* $l_k = \nu^{3/4}\varepsilon_e^{-1/4}$.

The described cascade process in the atmosphere takes place in the inertial interval of scales where turbulent viscosity ν^T has the form: $\nu^T(r) \propto \varepsilon_e^{1/3} r^{4/3}$ (*Richardson, 1926; Obukhov, 1941*) and matches the *Richardson-Obukhov* empirical "law of four thirds". The latter also follows from the dimensionality and similarity theory. The dependence of the effective vertical diffusivity $\nu(L)$ on the local turbulence scale $L(z)$ is shown in Figure 1.1.1.

The area of the laminar motion with small L and molecular viscosity $\nu = 0.16$ cm²/s and the area of free turbulence (inertial interval) at large L are emphasized in the figure. As it was mentioned above, for such large values of L, there is no generation or dissipation of kinetic energy but only its transfer to decreasingly smaller scales r (down to the scale $r = l_k$). Decrease in scale fits the growth of the wave number κ in the curve of the energy spectral density $E(\kappa)$ of the velocity field (see Figure 1.1.2).

Maximum turbulent viscosity ν^T in the atmosphere corresponds to the typical scale length for synoptic processes $L_0 = a / f$ (*Obukhov, 1949*), where $a = \sqrt{gH}$ is the isothermal sound velocity. At moderate Earth latitudes, L_0~3000 km and $\nu^T \approx 10^{11}$ cm²/s. Obviously, in this case the degeneration time of atmospheric kinetic energy due to turbulent viscosity $\tau(L) \sim L^2 / \nu^T(L) \sim \varepsilon_e^{-1/3} L^{2/3}$ matches the typical transformation time of the potential energy of synoptic processes $\tau = (E^{-1}\partial E/\partial t)^{-1} \cong 5 \times 10^5$ s, i.e. it approximately equals a week. Here $E \approx 10^{21}$J is the

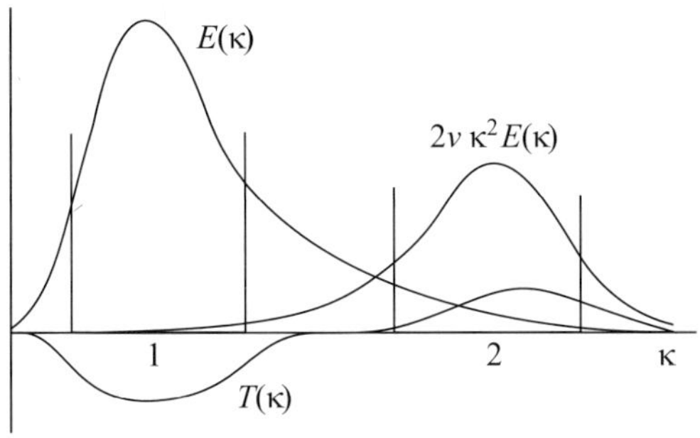

Fig. 1.1.2. Energy spectra, $E(\kappa)$, and energy dissipation, $2\kappa^2 E(\kappa)$, and the reallocation function for the energy, $T(\kappa)$. The areas 1 and 2 on the horizontal axis designate the energy and dissipation intervals, respectively. According to *Monin and Yaglom, 1975*.

total kinetic energy of the atmosphere; $\partial E / \partial t \approx (1-2)\times 10^{12}$ kW is the transformation rate of potential energy into kinetic energy (*Monin, 1969*).

As one can see, turbulence with turbulent viscosity ν^T can be considered as an efficient mechanism for dissipating kinetic energy of synoptic scale motions (large-scale eddies with characteristic dimension L). Thus, the minimum scale of synoptic motions capable to overcome viscosity is estimated to be $L_{\min} \propto \varepsilon_e^{-1/4} (\nu^T)^{3/4}$ (*Monin, 1969*). Such an approach agrees with the concept that there are universal relations between features of large- and small-scale fluctuations of turbulent flows, these relations being determined for $\boldsymbol{Re} \to \infty$ by small-scale fluctuations rather than at large-scale velocity fluctuations. The latter determines average values of all thermohydrodynamical quantities and, basically, are independent of $\boldsymbol{Re}$. This embodies the Batchelor (*1955*) hypothesis about statistical independence of large- and small-scale motions.

On the Spectrum of Atmospheric Processes. In the theoretical study of turbulent atmospheric fluctuations of wind velocity (or other thermohydrodynamical parameters fluctuating in a flow), the methods of mathematical statistics relevant to quasistationary stochastic processes, together with correlation and spectral analysis, are usually employed (see Chapter 8). The most commonly used statistical properties of a field of a random quantity $A(\boldsymbol{r},t)$ involve its average value (in what follows, the latter is designated with an overscribed bar) and variance, various correlation and structural functions, spectral density functions, etc. Because motion systems of diverse spatial-temporal scales co-exist in the atmosphere, the thermohydrodynamical parameters describing the atmosphere must be averaged. In this approach it is assumed that individual realization of turbulent flows can be described with equations of multicomponent hydrodynamics for instantaneous motion. The averaging can be performed with respect to temporal (and/or spatial) interval Δt (for instance, to fluctuation values, recorded at a fixed point during period Δt), or to an ensemble of physically admissible realizations. To obtain representative statistical estimates of the parameters governing the fluctuating flow, the interval should be properly chosen. The larger the fluctuation scales, the longer this interval must be.

Thus, to choose the size of the interval, it is necessary to know an order of magnitude (of the spectrum) of the fluctuations in wind velocity, pressure, temperature, and other thermohydrodynamical parameters of a turbulent mixture flow. The energy spectrum of the fluctuation field of any physical quantity $A(\boldsymbol{r},t)$ represents a series of curves describing dependence of $\overline{A'^2}$, the squared of the random variable amplitude A, on angular frequency ω ($\omega = 2\pi \mathrm{f}$, f is the frequency of oscillations in s^{-1} or Hz) and/or on linear dimensions κ ($\kappa = \mathrm{f}/V$ is the wave number, V is the average flow velocity at the atmosphere level under study) of multiple scale eddies.

The broad spectrum of fluctuations of the random variables mentioned above, whose periods range from a fraction of a second to thousands of years, is characteristic of the atmosphere (see *Monin, 1969*). Within this spectrum, micrometeorological oscillations with periods from a fraction of a second to a few minutes deserve particular attention. Specifically, these oscillations just arise in the atmospheric layer above the surface and represent small-scale isotropic turbulence serving as the most important mechanism for viscous dissipation. The maximum of the energy spectrum $\omega E(\omega)$ (where $E(\omega)$ is *the spectral kinetic energy density of the flow*) fits a period $\tau_m = 1/\omega \approx 1$ minute. This suits the scale of horizontal turbulent heterogeneities $L = V\,\tau_m \approx 600$ m in case of a typical air velocity in synoptic processes $V = 10$ m/s. At $\omega > 1/\tau_m$, the wind velocity spectra fit the Kolmogorov-Obukhov "law of five thirds" (*Obukhov, 1941*) $E(\omega) \sim (\varepsilon_e^{2/3}/V)(\omega/V)^{-5/3}$, and the turbulence spectrum sharply breaks in the range of maximum turbulent fluctuations frequencies $\omega \sim V\varepsilon_e^{1/4}\nu^{-3/4}$ (*Monin, 1988*).

For an isotropic turbulent flow, the cascade process of kinetic energy transfer from large scale vortex motions to small scale eddies is convenient to analyze in the space of wave numbers κ using the spatial spectral energy density $E(\kappa)$, which is defined as $E(\kappa) = \int\limits_{|\kappa'|=k} \Phi_{jj}(\kappa')d\kappa'$. Here $\Phi_{ij}(\kappa)$ is the Fourier transform of correlation function of the random velocity field $R_{ij}(\boldsymbol{r}) = \overline{V_i'(\boldsymbol{x}+\boldsymbol{r})V_j'(\boldsymbol{x})}$, where $\kappa = |\boldsymbol{\kappa}|$ is the wave number module, and summation is performed over repetitive indices. Note that $R_{jj}(0) = \overline{V_j'V_j'}$ is the kinetic energy of the fluctuating motion of a unit mass. Using the definition of $E(\kappa)$, it follows that turbulent energy equals $e \equiv \frac{1}{2}\overline{V_j'V_j'} = \frac{1}{2}\int_0^\infty E(\kappa)\,d\kappa$. Figure 1.1.2 shows the pattern of turbulent motion energy distribution at various scales, i.e. the spatial spectrum of the quantity e.

Great insight, characterizing fluctuative motion of the velocity field, follows from the "spectral form" of the *Karman-Howart* equations (*Monin, Yaglom, 1975*) describing the time variability of the spectral distribution of the turbulence energy $E(\kappa,t)$:

$$\partial E(\kappa,t)/\partial t = T(\kappa,t) - 2\nu\kappa^2 E(\kappa,t).$$

Here the first addend, $T(\kappa,t)$, describes the redistribution of energy over the turbulence spectrum due to interactions of its spectral components of wave numbers κ with all other spectral components, the redistribution being created by nonlinear "inertial terms" of the basic equations of hydrodynamics. It is important to emphasize that the redistribution of energy among spectral components occurs without a change in the total turbulent motion energy ($\int_0^\infty T d\kappa = 0$). The second addend, $2\nu\kappa^2 E(\kappa)$, describes the energy dissipation due to viscosity and characterizes the decrease in the kinetic energy of perturbation fluctuations with wave number κ, being equal to the intensity of these perturbations multiplied by $2\nu\kappa^2$. This means that, due to viscosity effects, the energy of long wavelength perturbations with smaller κ decreases much slower than the energy of short-wave perturbations. This reflects the proportionality of friction force to the velocity gradient.

The spectra with energy $E(\kappa)$ and energy dissipation $2\nu\kappa^2 E(\kappa)$ are sketched in Figure 1.1.2 together with the function $T(\kappa,t)$ which determines the redistribution of energy over the spectrum. As can be seen, negative values of function $T(\kappa)$ in the range of small κ, which fit to the maximum of large-scale motion energy in the curve $E(\kappa)$, are replaced by positive values in the range of large κ. The latter correspond to the maximum energy dissipation in the curve $2\nu\kappa^2 E(\kappa)$. This affirms the idea of the cascade energy transfer from large-scale components of motion to small-scale components. Large local gradients are characteristic of the latter and, hence, it is here where viscosity plays an important role. After Monin and Yaglom (*1975*), the first of these intervals, accumulating up to 80-90% of total turbulence energy $e = \frac{1}{2}\int_0^\infty E(\kappa)d\kappa \cong \int_0^{\kappa_0} E(\kappa)d\kappa$, is referred to as the energy interval, while the second one, where complete dissipation of energy occurs $2\nu\int_0^\infty \kappa^2 E(\kappa)d\kappa \cong 2\nu\int_{\kappa_0}^\infty \kappa^2 E(\kappa)d\kappa$, bears the name interval of viscous dissipation of the spectrum (here κ_0 is intermediate wave number value in the inertial interval between the energy and dissipation intervals).

Such an approach highlights the problem of *the rate of energy redistribution over the spectrum* $T(\kappa)$, proceeding from the basic spectral representation concept of the velocity field itself, which was developed by Batchelor (*1955*). Altogether, this brings additional support to the idea that the initial energy of large eddies, determining the dynamic and kinematic properties of a flow, is being expended on their fragmentation which is expressed by turbulent viscosity. Ultimately, this energy determines the properties of viscous dissipation when the conditions of total energy balance are met.

It is of interest to mention that based on the dimensions/similarity approach the so-called *fastest response principle* was suggested for frequency-energy spectra of various geophysical/astrophysical phenomena described as hydrodynamic motions, including the developed turbulence (*Golitsyn, 1997*). It states that the kinetic energy scale equals (by order of magnitude) the product of the energy supply rate to the system (from outside in the form of insolation or heat from the planetary interior, or generated inside as ε_e or inhomogenities in the internal field) and the least time scale pertinent to the system. This principle is intimately related with stochastics of events depending on their intensity and mode of energy transformation within the system.

1.1.2. Some Methods of Turbulence Simulation

In addition to the statistical approach to turbulence modeling, the phenomenological (semi-empirical) approach and methods of direct numerical simulation of turbulence find an expanding application. They are based on solutions of special kinetic equations or non-stationary system of the three-dimensional *Navier-Stokes equations*, although only averaged properties of motion can be obtained because of the stochasticity of this phenomenon. Nevertheless, sometimes this approach makes it possible to trace not only time evolution of the formation of various spatial structures but also to study the general dynamics and nature of turbulence development. For instance, the outcome of numerical simulations of the "overthrow" phenomenon in a hydrodynamic system (designed as a multilayer model of coupling of the simplest elements, or triplets) illustrate the cascade energy transfer in a flow in which turbulence developed. This is in agreement with the known Kolmogorov-Obukhov law (*Gledser et al., 1961*) and corroborates the general properties of the behavior of the dynamic systems.

It is interesting to note that examination of the stochastization process of dynamic systems and scenarios of passing to chaos in numerical turbulence simulations serves as an analog to the solution of incorrect (ill-posed) problems, which uses an averaging operator *of the parametrical extension* (*Tikhonov and Arsenin, 1986*). With such an approach, the ordered structure of a turbulent current, which is determined as an attractor of an asymptotically steady solution for averaged magnitudes, represents its regularized description (*Belotserkovskii, 1997*). However, it should be noticed that using the techniques of direct numerical turbulence simulation for solving practically important problems (especially those related to calculations of turbulent heat and mass transport in multicomponent chemically active mixtures) is often inconvenient or too bulky. Therefore, such problems are more suitable to be solved with the use of less sophisticated *semi-empirical* theories.

The classical approach to simulation of multicomponent turbulence is based on Reynolds' idea about averaging the hydrodynamic equations of a mixture over an ensemble of identical currents (probable realizations), or by means of another equivalent procedure. Thus, the equations obtained for the scale of averaged motion, due to nonlinearity of the initial Navier-Stokes equations, involve uncertain correlation terms (such as turbulent diffusion vectors and heat, and turbulent Reynolds stress tensors) and consequently they happen to be non-closed.

The closure of the hydrodynamic equations of a mixture, averaged according to Reynolds, is usually carried out with the help of some semi-empirical turbulence models (this just being the subject of the present study). At the same time, it is important to indicate in advance a basic drawback of such an approach. The problem is that the Reynolds averaging is carried out over all scales of turbulence. In other words, the simulations based on semi-empirical hypotheses of closure are simultaneously performed over the entire spectrum of multi-scale eddy structures. In contrast to a virtually universal spectrum of small-scale fluctuations (for diverse patterns of currents), large-scale structures substantially differ from each other in different currents (see Figure 1.1.3).

Hence, it is obvious that creation of universal semi-empirical turbulence models suitable for description of the diverse turbulent mixture currents is hopeless and therefore, that the problem is focussed on the assessment of applicability limits for the most relevant turbulence model. Nevertheless, one may assume that the involving multiparameter approximations based on evolutionary transfer equations for the higher mo-

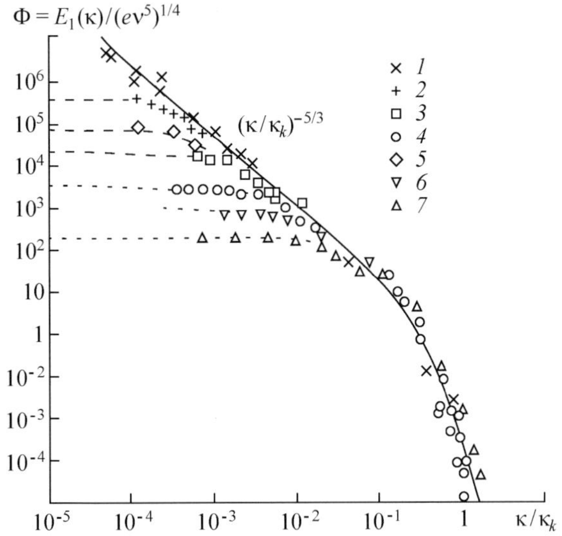

Fig. 1.1.3. Spectrum of the longitudinal turbulent pulsation component for different flow patterns: 1 – tide-ebb channel; 2 – circular stream; 3 – tube flow; 4 – permanent shear flow; 5 – trace behind a cylinder; 6 – trace behind a grid; 7 – boundary layer. According to *Chapman, 1980*.

ments of the thermohydrodynamical parameters fluctuating in a multicomponent flow will allow us to advance the development of universal models for turbulent mixture, capable to describe a fairly large number of diverse turbulent currents.

1.1.3. Turbulent Diffusion

In the complex problems related to the theoretical examination of heat and mass transfer processes in a natural turbulent multicomponent medium, the transport modeling of minor admixtures is of considerable importance (including air mass mixing regarding their chemical activity). In the atmosphere, in addition to gases, there also are aerosols of various types and sizes. They are partly involved in chemical transformations and phase transitions and include radioactive admixtures both of natural (radon, thoron, and products of their decay) and artificial (due to production and testing of nuclear weapons, nuclear power station accidents, etc.) origin. The transfer process of these admixtures and their mixing is conditioned by turbulent diffusion whose nature depends on the structure of the fluctuation velocity field and the turbulence energy distribution over fluctuations of different spatial scales.

When describing diffusion processes in a turbulent atmosphere, it is possible to distinguish the average values of the admixture concentration $\overline{Z}_\alpha$ ($Z_\alpha \equiv n_\alpha / \rho$, see Chapter 2) and their fluctuative deviations Z'_α, as well as the average magnitude $\overline{V}$ and fluctuations V' of air motion velocity. This makes it possible to depart from the diffusion equations for instantaneous concentration values to the diffusion equations for the averaged motion scale using common averaging techniques (see equation (3.1.24)). Depending on the scale, the theory of turbulent admixture diffusion distinguishes gradient-type diffusion generated by rather small eddies, and non-gradient-type diffusion created by large eddies (*Monin and Yaglom, 1971*). The phenomenological approach is applicable to the gradient-type turbulent diffusion (see section. 3.3.1).

As turbulence is a flow pattern property of a fluid rather than a property of the fluid itself, the mechanism of exchange of momentum (as well as energy and substance) implies only an implicit identity to molecular exchange of such a quantity. Nevertheless, there exists a definite analogy between mixture diffusion in the field of small-scale turbulence and molecular diffusion, which is also valid in the case of contamination of a medium by a fine-grain admixture, provided the turbulence scale is small compared to the scale of a *contamination cloud.* Such analogy is based on an assumed proportionality between the turbulent flow of some diffusing substance and the gradient of its averaged concentration. Indeed, likewise a random molecular motion is characterized by some average velocity of the molecules v_m and the free path length l_m (so that the molecular diffusivity $D \sim v_m\, l_m$), random turbulent mixing can be described by the turbulent diffusivity $D^T \sim v_t\, l_t$, where v_t is the characteristic magnitude of turbulent velocity fluctuations and l_t is local turbulence scale (the so-called mixing length). However, in contrast to v_m and l_m, the parameters v_t and l_t are flow properties rather than fluid properties. Accordingly, D and D^T are addressed as constants of proportionality; D between the molecular diffusion flow $\boldsymbol{J}_\alpha$ of some substance α and the gradient of its concentration ∇Z_α ($\boldsymbol{J}_\alpha = -D\nabla Z_\alpha$), and D^T between the turbulent flow of the substance in the fluctuating velocity field $\boldsymbol{J}_\alpha^T = \rho\overline{Z_\alpha' \boldsymbol{V}'}$ and the gradient of its average concentration $\nabla\overline{Z}_\alpha$, i.e. $\boldsymbol{J}_\alpha^T = -\overline{\rho} D^T \cdot \nabla\overline{Z}_\alpha$ (*Monin, 1962*). Note that since turbulent (or eddy) diffusion, unlike its molecular counterpart, usually is an anisotropic entity, in general, D^T and $\boldsymbol{l}_t$ must be treated as tensors.

In case of a conservative admixture, when individual properties of particles of a diffusing substance preserve while moving in a fluctuating flow, the Lagrangian concept (see formula (3.3.1)) must be used as the most natural turbulent mixing model. This allows us not to focus on details of the mixing mechanism, but to consider such a process as "common motion" of particles in a fluid where diffusion occurs. The introduction of the *Lagrangian turbulent fluctuation* $Z'_{\alpha L}$ $(\equiv Z'_\alpha + \mathbf{l}_t \cdot \nabla\overline{Z}_\alpha \cong 0)$ of the conservative property Z_α makes it possible to define a certain length $l_t = |\boldsymbol{l}_t|$ (roughly proportional to the average linear scale of the velocity fluctuations), in which the correlation between the initial and final velocities of the given Lagrangian particle disappears. This allows us to simulate the turbulent diffusivity $D^T = \overline{\boldsymbol{l}_t \cdot \boldsymbol{V}'}$ during the entire time period of the fluctuating motion (*Priestly, 1962*). The possibility of some direct influence of molecular diffusion on eddy diffusion transport must not be ruled out, provided that the amount of the diffusing substance within a large-scale vortex noticeably change within a time frame, during which the correlation between the velocity values in the vortex remains fairly high. Apparently, it also must be assumed that these multi-period diffusion processes have superposition features.

In general, qualitative distinction between turbulent and molecular diffusion in the atmosphere still lacks at moderately small vortex velocities. This gives applicability limits to the parabolic diffusion equation (see equation (3.2.5)). In particular, under conditions present in the near-land layer of the atmosphere, the inequality $D^T >> D$ is valid, which is conditioned by the fact that $l_t >> l_m$, although $v_t << v_m$ is fair. In other words, the invoked analogy to the molecular diffusion is acceptable only for limited volumes of air parcels, as the velocity of admixture transport in the turbulent atmosphere is limited by the magnitude of the wind velocity fluctuations responsible for turbulent mixing.

Moreover, the analogy to molecular diffusion is wrong in the large-scale turbulence range, when the scale l_t becomes commensurable to or exceeds the size L of contamination clouds and the distances between particles remain essentially unchanged as they are transported with large-scale motions, including large eddies. In this case, a more general statistical approach to large-scale turbulence (discussed in section 1.1.1) is suited for simulating diffusion transport. Two physical key quantities of a dissipative hydrodynamic system, that determine energy transfer from large eddies to smaller ones, are used here: the viscous dissipation velocity of the turbulent energy ε_e and the turbulence scale L (*Monin and Yaglom, 1971*). In this case, unlike in case of gradient diffusion, the specificity of a multicomponent medium is of no importance.

When simulating the turbulent diffusivity D^T for any possible atmospheric components (including pollutants), it is necessary to account for factors most strongly affecting their mixing (dispersion). In particular, of primary importance is *a priori* evaluation of the mixing length (scale of turbulence) or its approximation. Herewith thermal (and/or concentrational) stratification of the atmosphere, which virtually determines the character of diffusion processes (see section 3.3.2.), must be taken into account. As it was already mentioned, in the general case of an anisotropic fluctuating velocity (concentration) field, the diffusion transport of minor components is characterized by the turbulent diffusion tensor D^T (or the viscosity tensor ν^T). When currents feature a sharply determined direction of inhomogeneity (as is, for example, the case for an atmosphere stratified in the gravitational field), the turbulent velocity fluctuations are superimposed on its average (wind) component in the horizontal direction, the latter being enhanced in the presence of shear.

For the currents of this kind, various simplified approximations for vertical turbulent viscosity ν^T, such as the *Prandtl* mixing length model (see formula (3.3.20)) or the *Heisenberg* hypothesis on the role of small eddies in evaluating ν^T when the gradient approximation is admissible, are successfully used. These approximations are fair only, however, if there is local equilibrium in a turbulent field (see section 4.3.9) ensuring the balance between generation and dissipation of turbulent energy at any point in the turbulent medium. Nevertheless, generally speaking, such approaches are inapplicable to multicomponent chemically active mixtures because of the essential role played by chemical kinetics which breaks down the Lagrangian invariance of transferable substances. Here again, we face the necessity to develop new methods to simulate turbulent mixture transfer factors that allows one to take into account the peculiarities of multicomponent media.

1.1.4. The Dynamic Nature of Turbulence

Let us now make some general remarks concerning the dynamic nature of turbulence in a non-linear dissipative gas-fluid system, that can exchange both energy and substance with surrounding bodies. By virtue of such exchanges, the formation of various spatial-temporal structures is possible whose succession makes up the process of self-organization.

In the presence of turbulence each individual particle in a medium moves randomly, so its coordinates and direction of motion vary in time by Markov's random process law. A complete statistical description of turbulent flow is reducible to the definition of a probability measure within its phase space ($\boldsymbol{r}, \boldsymbol{p}$), involving all possible individual realizations of random thermodynamical fields which characterizes this space. Therefore, turbulence can be examined by means of the statistical mechanics of multiple particles (see, for instance, *Obukhov, 1962*). Or it can be described with the kinetic equation, which is an analogue of the Boltzmann equation in a phase space for some conditional density function of the probability distribution $f_{turb}(\boldsymbol{r}, \boldsymbol{p}, t)$, serving as the basic statistical feature of pulsating motion (*Klimantovich, 1991*).

A characteristic property of an open system with a large number ($N \to \infty$) of independent dynamic variables ($\boldsymbol{r}, \boldsymbol{p}$) is its dynamic instability due to mixing that can be imagined as exponential divergence of initially close phase trajectories. So any initial distribution of a probability density function in a phase space aims towards the limiting equilibrium distribution, that is, the most chaotic state with maximum entropy in the *Boltzmann-Gibbs-Shannon* sense. Turbulization of fluid or gas motion may be represented as a result of changing phase trajectory topologies, leading to rearrangement of the attractors and to qualitative modification (*bifurcation*) of the state of motion. Velocity correlation at any point of a flow is limited to small time intervals depending on the initial conditions. It is impossible to assess a causal dependence between velocity fields outside these intervals as well as correlation with an antecedent motion.

All this substantiates the idea of the stochastic nature of velocity fluctuations in a turbulent flow that arise due to loss of stability of laminar motion of a hydrodynamic system when changes in external controlling parameters occur (for example, the number $\boldsymbol{Re}$). From this point of view, turbulent motion is more random, than laminar motion, and turbulence is identified with chaos (or noise). The stochastic nature of turbulence is reflected by dense interlacing of phase trajectories with different asymptotic behavior (*topology*) and structure of attraction areas they enclose (*attractors*). Such behavior of trajectories in phase space means that the system possesses ergodicity. In other words, for almost all realizations of a random field, its time average is equal to its statistical average, the system's time correlation functions rapidly decay, and the frequency spectra are continuous. The ergodic property is, apparently, one of the characteristic features of a steady-state homogeneous small-scale turbulent field (see, for example, *Kampe de Feriet, 1962*).

On the other hand, in passing to the maximum shear turbulence in an open hydrodynamic system, new macro dependencies between individual areas set up due to collective interaction of the involved subsystems. This enhances the internal ordering of the system as compared to arbitrary fluctuations at the molecular level. In this case, multiple spatial-temporal turbulence development scales correspond to the coherent behavior of a large number of particles. Among other things, this relates to the emergence of clearly ordered dissipative structures with a certain degree of organization and formation of areas of increased concentration of vorticity in the form of eddy pipes and eddy layers. These structures, mentioned at the beginning of the present section, appear on the background of small scale turbulent motion. Granulae in the solar photosphere are another example of the extensive family of coherent structures in turbulent flows (*Petrovay, 2001).* All this supports the seemingly paradoxical inference that the state of mature turbulent motion, despite of its enormous complexity, is more ordered than the state of more symmetrical laminar motion. Such a phenomenon, "indicating, how difficult it is to distinguish order from chaos in case of complex motions" (*Klimantovich, 1986*), makes up a part of the general problem of self-organizing (*synergetics*).

Interest in this problem, emerging in the most diverse natural phenomena and engineering areas, has recently grown. According to modern concepts (see, for instance, *Nikolis and Prigogine, 1977*; *Haken, 1980*), the self-organizing dissipative structures of this kind occur as a result of stabilization of spatially-inhomogeneous instabilities in an open non-linear dynamic system due to coherence accompanying the process of bifurcation and, thus, in arriving at "order through fluctuations". For such systems, according to Prigogine and Stengers *(1984*) "the probability becomes an objective property generated, as it were, inside the dynamics and reflecting fundamental structure of a dynamic system". Unfortunately, it should be noted, that although understanding of the synergetic nature of turbulence as a process of self-organizing in open hydrodynamic nonlinear systems is about thirty years old, up to now the concepts of emerging coherent spatial-temporal dissipative structures have not been embodied in usable engineering calculation methods for large-scale turbulence.

1.2. TURBULENCE IN ATMOSPHERES OF PLANETS

The atmosphere of an inner planet is its gas envelope limited from the bottom by a solid underlying surface. For an outer gas-liquid giant planet, the notion of its atmosphere is synonymous to its exterior layers. It may be considered as a free flowing fluid, for which gravity is furnished by the planet, and the energy to regularly lift this envelope and to ensure that gas flows is furnished by the Sun.

As a rule, the Reynolds number $\boldsymbol{Re}$ exceeds $\boldsymbol{Re}_{cr}$ in atmospheric gas flows and consequently, the currents are turbulent. Turbulization of atmospheric currents arises because of their deformation while flowing around irregularities of the underlying surface, or when a large-scale flow becomes unstable because of increased temperature and wind velocity gradients. The loss of stability of internal gravity shear waves is the basic cause of turbulence generation in a free atmosphere. Destruction of such waves may be due to "primary" or "secondary" instabilities. A "primary" instability (a Kelvin-Helmholtz instability) develops in a shear layer between flows with different velocities, if the condition $\boldsymbol{Re} < \boldsymbol{Re}_{cr}$ is valid within a major portion of the wave layer. In case of a "secondary" instability, the flow is steady on the average but unstable volumes are localized near wave crests. Instability of this kind is characteristic mainly of layers with strongly curved vertical temperature and wind velocity profiles.

The presence of multiple components and chemical activity of atmospheric gases is one of the special features of a planetary atmosphere. Turbulence in a homogeneous flow and in a reacting multicomponent flow does not manifest itself in the same way. Changes in density, temperature, and the mixture structures due to chemical processes can result in flow turbulization. The arising density gradients generate additional vorticity through interaction with surrounding gradients or fluctuations in the pressure field. It is related to the emerging of a source term $\rho^{-1}\nabla\rho^{-1}\times\nabla p$ (equal to zero in barotropic media) in the Friedman vorticity equation

$$d(\boldsymbol{\Omega}_a/\rho)/dt = (\boldsymbol{\Omega}_a/\rho)\cdot\nabla\boldsymbol{V} - \rho^{-1}\nabla\rho^{-1}\times\nabla p + \rho^{-1}\nabla\times\boldsymbol{F}, \tag{1.2.1}$$

where $\boldsymbol{\Omega}_a = \boldsymbol{\omega} + 2\boldsymbol{\Omega}$ is the absolute velocity vortex, $\boldsymbol{\omega} = \nabla\times\boldsymbol{V}$ is the vorticity vector, $\boldsymbol{\Omega}$ is the angular velocity vector of the planet's rotation; $\boldsymbol{F}$ is the sum of accelerations due to the viscosity force and external mass forces (see, for example, *Monin et al.,1989*). When the mass density ρ is constant, all dynamic effects potentially depending on this "vorticity" term are absent.

Thus, the occurrence of local mass density inhomogeneities (gradients) constitutes a major feature of reacting multicomponent currents which is usually neglected in classical turbulence models for homogeneous fluids. Local energy generated by chemical reactions is another complicating aspect of modeling multicomponent turbulence. Local heat release in gas flows accelerates the expansion of a medium and can induce Rayleigh-Taylor instabilities (in currents stratified in a gravitational field in the presence of a buoyancy force), thus embodying a feedback to hydrodynamics. In what follows we will consider some examples of turbulent natural multicomponent media.

1.2.1. Planetary Atmospheres

Properties of planetary atmospheres strongly differ from each other even within the relatively small region of the solar system occupied by the terrestrial planets. In addition to Earth, this group comprises Mercury, Venus, and Mars. Mercury, like the Moon, is virtually deprived of an atmosphere: the density of its gas envelope does not exceed 10^{-17} g/cm^3 at the surface, though a number of dynamic effects arise by interaction of solar wind plasma with the magnetosphere and the exosphere of Mercury. Venus and Mars, as two extreme models of evolution of Earth-like planet, represent, from this viewpoint, the most interesting cases for comparative planetology in terms of specific features in their meteorology, climate formation and interior dynamics. Their atmospheres of secondary origin are oxidizing in character and mainly consist of carbon dioxide. The latter creates pressure at Venus' surface two orders of magnitude larger and at Mars' surface two orders of magnitude smaller than the pressure at Earth's surface.

At the same time, the atmospheres of the giant planets (Jupiter, Saturn, Uranus, and Neptune), are reducing in their character and mainly consist of hydrogen, helium, and hydrogen-bearing compounds (water, ammonia, methane and other hydrocarbons). It is assumed that the primary atmosphere of these gas-fluid planets having no solid surfaces have essentially remained unchanged since the formation of the solar system (*Marov, 1986*). Chemical reactions in the atmospheres and clouds of Jupiter and Saturn, exhibited as very rich color palettes for an outside observer, may be considered as an analog of the processes that occurred during the earliest evolution phases of the terrestrial planets.

Unlike stars, emitting radiation at the expense of nuclear energy generated in their interiors, planets are cold bodies reflecting and re-radiating absorbed energy from the Sun. Winds in their atmospheres are principally driven by an atmospheric thermodynamic heat engine. From the viewpoint of energy exchange, both the underlying solid (or liquid) and gas layers represent a unified open thermodynamic system. It is controlled either by incoming solar radiation, or (as in the case of Jupiter, Saturn and Neptune) also by energy from the interior, which is two-three times greater than the energy these planets receive from the Sun.

The energy absorbed by the atmosphere and transformed into internal and kinetic energy of gaseous media, eventually is radiated back to outer space as IR-radiation, ensuring total energy balance. The heat and mass transfer occur on the background of global (at scales $L >> H$, where $H = p/\rho g$ is the scale height), mesoscale ($L \approx H$) and local ($L << H$) atmospheric dynamics. The latter is responsible for the diversity of meteorological processes occurring in the atmosphere where molecular thermal conductivity plays a negligible role in the overall heat exchange. Advection, determined by planetary circulation, and convection are characteristic ways of regular heat and mass transfer. Convection is due to the fact that, as the temperature at a given pressure level systematically changes with latitude ϕ on a rotating planet, the barotropic atmosphere becomes baroclinic (see equation of state (3.2.1)), where turbulence appears as a consequence of nonlinear evolution of *Rayleigh-Taylor* instabilities. Strong motional perturbations of the medium also manifest themselves as cyclonic winds accompanied by turbulent heat and mass transfer (*Chamberlain and Hunten, 1987*). Some general quantitative criteria (*similarity parameters*) characterizing planetary dynamics, based on the energetic and thermodynamic approach, were found (*Golitsyn, 1973*).

Four types of forces, namely gravity and viscous friction ($\boldsymbol{F}$) force, the Coriolis force ($2\rho\boldsymbol{\Omega}\times\boldsymbol{V}$), and the pressure gradient force (∇p), effect an elementary air medium volume. In their presence, specific realizations of dynamic matter transfer on different planets mostly depend on the period correlation between their rotation and the thermal relaxation of their atmosphere. The relative contribution of the Coriolis force to atmospheric dynamics is determined by the *Kiebel-Rossby* number $\boldsymbol{Ro} = U/f\,L$, where U and L are the typical horizontal of velocity and length scales of synoptic processes (for the terrestrial atmosphere, $\boldsymbol{Ro} \approx 10^{-1}$).

At scales smaller than that of synoptic processes, the Coriolis force, counterbalancing horizontal pressure gradients, imparts geostrophicity to currents that relates to specific oscillations in the atmosphere with the inertial frequency $\tau = 2\Omega_z$, where Ω_z is the vertical component of the angular rotational velocity of the planet. Joint effects of planetary rotation and sphericity (variation of frequency τ (or f) with latitude ϕ, so-called "β-effect") result in the formation of the above mentioned planetary Rossby-Blinova waves. In the troughs and crests of these waves, the major generators of weather, i.e. cyclones and anticyclones, are forming. These planetary waves, horizontally traveling predominantly westward in the Earth's atmosphere, have periods exceeding both the rotational period of the planet and the periods of three other types of wave motions in the atmosphere: tidal, acoustic, and internal gravity waves (*IGW*). Short-period *IGW*s with rather small amplitudes are classed as micrometeorological os-

Fig. 1.2.1. Venus' southern hemisphere in UV. The dark horizontal Y-like pattern represents a planetary-scale wave drifting slowly with respect to atmospheric superrotation. Reappearance of such a feature every 4 days is a measure of the rotation period of the atmosphere and it is characteristic of Venus' "merry-go-round" circulation at the upper level of the clouds (approximately 65–70 km above the surface). Ultraviolet contrasts for which absorption by sulfur acid droplets and submicron sulfur particles in the clouds appear to be responsible, distinguish eddy patterns of various spatial scales in the structure of wind motions. An image made by the *Pioneer-Venus* spacecraft, courtesy of *NASA*.

cillations in the atmosphere and, together with convection, serve as a source of small-scale turbulence. Through dissipation they thus exert an energy effect on the formation of large-scale weather processes.

For Rossby-Blinova waves, *Ertel's* condition for preservation of potential vorticity Ω_* of a main current holds as:

$$d\Omega_* / dt = 0, \qquad (\Omega_* \equiv \rho^{-1}\mathbf{\Omega}_a \cdot \nabla S), \qquad (1.2.2)$$

where S is the entropy of a medium per unit mass, with an approximately vertical direction of $\nabla S/|\nabla S|$ in a stratified atmosphere – this is the so-called thermodynamic vertical; Ω_* is the potential vortex: an adiabatic Lagrangian invariant which, by definition of adiabatic processes, is also the entropy S ($dS/dt = 0$). Condition (1.2.2) follows from the Friedman vorticity equation (1.2.1) for adiabatic processes, when $\boldsymbol{F} = 0$. The oscillatory exchange between of the global vorticity gradient β, defined as $\beta = r^{-1}\partial f / \partial\phi$, and the relative vorticity of the wave itself, also is a characteristic feature (*Charney and Stern, 1962; Monin, 1988; Ingersoll et al., 1995*).

1.2.2. Dynamics of the Atmospheres of Earth and Venus

In the atmosphere of the Earth, the forces conditioned by pressure gradients, are virtually balanced by the Coriolis forces (the Rossby number $\boldsymbol{Ro} \ll 1$). Therefore, geostrophic wind is a typical synoptic feature and its zonal and meridional components ($u = -(\rho f)^{-1} \partial p / \partial y,\ \ v = (\rho f)^{-1} \partial p / \partial x$) can be assessed from the known pressure distribution. Meanwhile, the Coriolis force effect is insignificant on the very slowly rotating Venus (one revolution for 243 terrestrial days), and the condition of cyclostrophic balance turns out to be justified (*Haltiner and Martin, 1957*; *Marov and Grinspoon*, 1998). It is characteristic of the latter that the zonal (latitudinal) component, growing with altitude, is superimposed by a circulation cell (the Hadley cell counterpart) whose originates from the temperature gradient between the equator and pole. As a result, the super-rotation of the atmosphere (or the so-called carrousel circulation) arises, such that on Venus the wind velocity grows from about 0.5 m/s at the surface up to ~100 m/s at 65 km. There, close to the tropopause, the upper cloud boundary is located. These winds are clearly traceable by the traveling of characteristic ultraviolet contrasts, supposedly being due to absorption of solar radiation by sulfur allotropes (Figure 1.2.1). Although significant progress was accomplished in the study and the modeling of these specific patterns, the mechanism by which this vigorous circulation is maintained is poorly understood yet.

On the background of relatively steady stratification of Venus' troposphere, there are the convective instability zone in the middle cloud layer between 52 and 57 km and

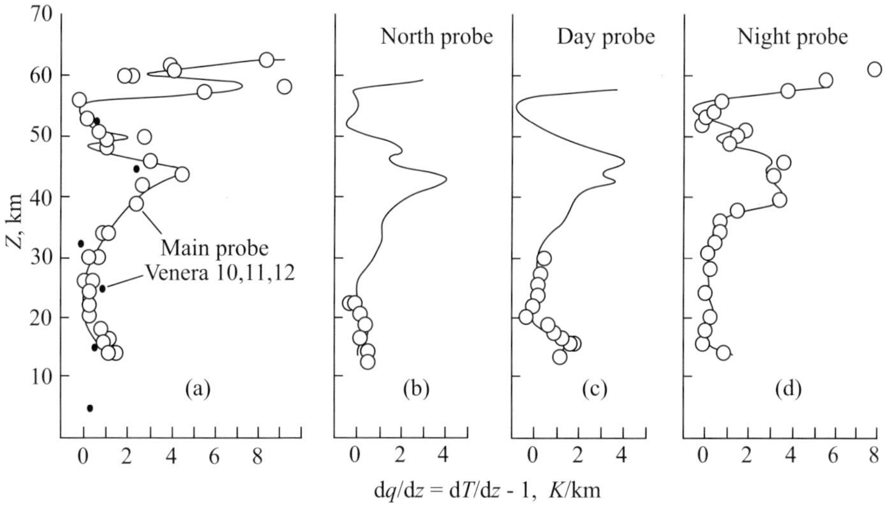

Fig. 1.2.2. Variation of the stability parameter, $d\theta / dz$, depending on height, z, in Venus' atmosphere based on the data from temperature measurements on the space probes *Venera 10, 11, 12* and *Pioneer -Venus (PV)*: (**a**) *Venera 10 to 12* (•) and *PV Large probe* (o); (**b**) *PV Northern probe*; (**c**) *PV Day probe*; (**d**) *PV Night probe*. The convective instability domains at altitudes of 52–57 km and shear flows between 45 and 50 km emerge clearly. The observed data consistency in different regions and at different times during the day argues for the rather uniform nature of atmospheric dynamics on Venus. According to *Sieff, 1983*.

the zones of shear currents, the strongest of which is observed in the altitude wind profile at 45–48 km (Figure 1.2.2.). It is noteworthy that the static stability parameter of the atmosphere here is

$$d\theta/dz = (\theta/T)(dT/dz - \gamma_a) > 0 \tag{1.2.3}$$

(where $\theta = T(p_0/p)^{\gamma-1/\gamma}$ is the potential temperature, p_0 is the standard pressure, $\gamma_a = g/C_p$ is the adiabatic gradient) and that the Richardson gradient number is

$$\boldsymbol{Ri} = (g/\theta)(\partial\theta/\partial z)(\partial V_h/\partial z)^{-2} > \boldsymbol{Ri}_{cr} = 1/4. \tag{1.2.4}$$

Therefore, theoretically, generation of turbulence is impossible, because the necessary instability condition for small perturbations to exist in the inviscous shear current is **Ri** < 0.25. Nevertheless it does occur (as it also does in the terrestrial atmosphere with similar values of the Richardson gradient number, $\boldsymbol{Ri} \approx 2$–6) and agrees with temperature fluctuations of order $0.1\,K$ at scales of about 100 m, typical of shear turbulence (*Woo et al., 1982; Seiff, 1983*). This is verified by measurements of these temperature fluctuations at 45–50 km and above 60 km, derived from radio occultation experiments using spacecraft.

Apparently, turbulence is also present in the underlying layers of the troposphere, at least in layers located higher than 15–20 km, where vertical velocity fluctuations of 0.2–0.3 m/s were detected during the parachute descent of the *Venera* landers from measurements of the vertical component of the Doppler frequency shift of the spacecraft master oscillators. A respective example is shown in Figure 1.2.3. (*Kerzhanovich and Marov, 1983*).

In turn, the atmospheric stability increases near the upper cloud boundary (~ 60–

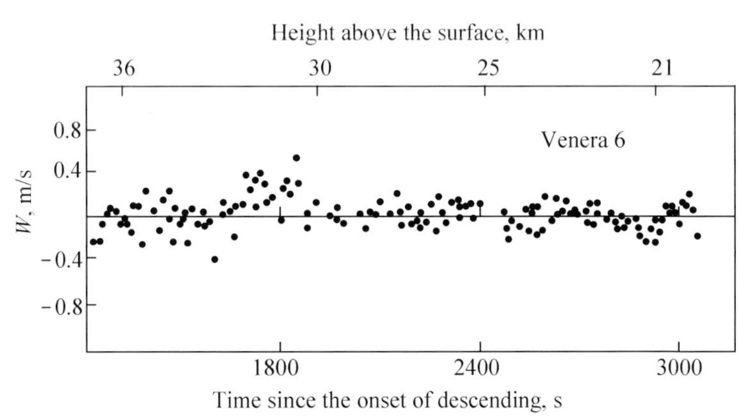

Fig. 1.2.3. Variations in vertical wind velocity components, *w*, during parachuting of the *Venera 6* lander in Venus' atmosphere. Time along the horizontal axis corresponds to the height descent from 51 to 18 km. The root-mean-square deviation of the measurement values (averaged over 10 s) does not exceed 0.2 m/s. According to *Kerzhanovich and Marov, 1983*.

65 km), where the parameter $\partial\theta / \partial z$ (Figure 1.2.2). However, as can be seen in Figure 1.2.1, apart from shear currents, there also are eddy structures at various spatial and temporal scales, associated with global circulation processes. All arguments mentioned above, suggest that turbulence is an important element of the atmospheric dynamics on Venus. Note that it could have been equally important during early evolutionary phases of Venus' atmosphere, which relates to the hypothetical primary ocean that could have been accumulated as water outgassed from the interior. This intriguing hypothesis assumes that such an ocean was lost later on due to the development of a runaway greenhouse effect (*Marov and Grinspoon, 1998*), which is corroborated by the data on enrichment of Venus' atmosphere by deuterium, a heavy hydrogen isotope. Indeed, its fraction of the total hydrogen content $N(D)/N(H) = (1.6 \pm 0.2)\times 10^{-2}$ turned out almost two orders of magnitude larger than the terrestrial fraction. This could be explained by the isotope separation process, caused by thermal dissipation of hydrogen from the atmosphere. In other words, deuterium could have been accumulated in the atmosphere as the water evaporated from the primary ocean and its molecules were dissociated by solar ultraviolet radiation.

However, the ratio $N(D)/N(H)$ mentioned above implies that the mass of the evaporated ocean would make up only a few percents of the water storage in the terrestrial oceans, which is hardly possible to assume, since Venus is a neighboring planet with a similar mass. More likely, the thermal dissipation mechanism only operated at the final phase of the primary ocean loss when the total hydrogen abundance in the Venus atmosphere *N(H)* had decreased to 2–3 % of the initial level, while during earlier phases the hydrodynamic dissipation mechanism, known as *blow-off-assisted escape*, appeared to be effective (*Hunten, 1982; Hunten et al., 1988*). According to this mechanism, heavier atmospheric components are carried away jointly with hydrogen because of the high velocity of the hydrogen flux and hence no hydrogen-deuterium fractionation occurs. One may assume that multicomponent turbulence significantly contributed to the implementation of such a process.

1.2.3. Dynamics of the Martian Atmosphere

The rarefied atmosphere of Mars possesses very small thermal inertia and its dynamics substantially differs from those of Earth and Venus. The model of global circulation based on the geostrophic balance condition ($\boldsymbol{Ro} \ll 1$) predicts similar a motion topology in the troposphere and in the stratosphere of Mars with dominance of eastward winds at high latitudes in winter and in the subtropics in summer, and with westward winds at other latitudes.

At the same time, because of the very low temperature in the polar regions in the winter hemisphere, the carbon dioxide is partially freezing out off the atmosphere and forms ice deposits. Such seasonal exchange of carbon dioxide between the atmosphere and the polar caps is the main driving mechanism of gas transfer in the meridional direction. This results in Hadley cells like configurations with ascending and descending flows and restructuring wind system near the surface and at high altitudes in the summer and winter hemispheres (*Zurek et al., 1992; Marov, 1992; 1994*). The character of the circulation is strongly influenced by surface relief (areography), on which both observable wind patterns and generation of horizontal waves of various spatial scales depend. In turn, planetary waves, conditioned by baroclinic instabilities, and interior gravitational waves manifest themselves as irregularities in temperature and vertical motion profiles in the stratosphere. They also relate to observable wave motions in cloud structures at the lee side during flowing around an obstacle, which provides evidence for the occurrence of strong shear flows in the Martian atmosphere (*Briggs and Leovy, 1974*).

The pattern of temperature, pressure and horizontal wind velocity variations at the Mars' surface, derived from direct measurements of the *Viking-2* lander during approximately three terrestrial (1.5 Martian) years, is shown in Figure 1.2.4. Apart from the characteristic seasonal mode due to CO_2 - condensation at the polar caps, there are irregular variations at smaller temporal scales related to diurnal changes, local fluctuations of the atmospheric parameters, and wave processes. The most pronounced variations occur in the records of the wind field, whose basic component is superimposed by effects of relief, surface roughness, and turbulence in the boundary layer.

Proceeding from the available experimental data, it is possible to suppose that the Martian boundary layer is generally similar to that of Earth. Its structure is schematically shown in Figure 1.2.5 as altitude profiles of the temperature T and the potential temperature θ. It is arbitrarily subdivided in a very thin (≤ 1 cm) layer of molecular diffusion heat exchange (A) adjacent to the surface (note that in the terrestrial atmosphere its thickness does not exceed ~1 mm), a relatively narrow layer (from several meters to several hundreds of meters) of the largest temperature and wind velocity variations (B), and a convective layer (C) (several kilometers thick) for which $\theta \approx$ const (*Zurek et al., 1992*). The layers (B) and (C) are characterized by the strongest turbulent convection, which the intensity substantially drops of outside the boundary layer (i.e. for layers higher than layer (C)).

The convection compensates for high static instability of the Martian atmosphere, which is close to saturation even for the very low relative content of water vapor. Interestingly enough, the efficiency of the excitation mechanism for the day time convection on Mars approximately is an order of magnitude higher than in the terrestrial atmos-

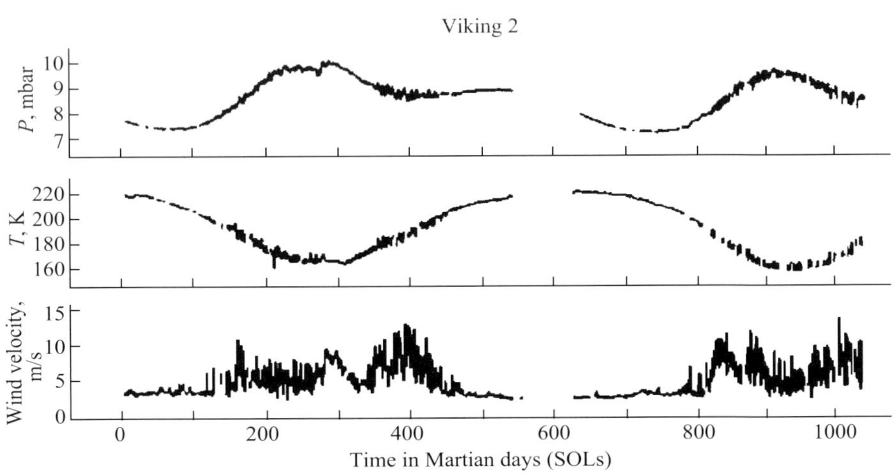

Fig. 1.2.4. Pressure, *P*, temperature, *T*, and wind velocity in the near-surface atmosphere of Mars measured on the *Viking* lander in the *Utopia* area averaged over one Martian day (SOL). Along the horizontal axis – time in SOL (1 SOL = $24^h\ 37^m\ 23^s$). According to *Zurek et al., 1992*.

phere, while at night it completely locks owing to the formation of an inversion layer (D) with a positive temperature gradient at the surface (see Figure 1.2.5). At the same time, the thickness of the nocturnal boundary layer is much greater than that of the day-time layer and, like in the terrestrial atmosphere, the strongest winds (*night-time jet*) probably blow here without being decelerated by surface friction in the presence of the inversion layer (*Andre et al., 1978*). Thus, the entire near-surface atmosphere appears turbulent at the expense of shear currents in the night jet, even under conditions of steady stratification.

Convection is also responsible for the large dust content constantly suspended in the Martian troposphere by virtue of which its optical depth usually does not happen to be less than $\tau \approx 0.2$. This creates an additional dynamic effect superimposed on the global wind system, which is caused by positive feedback between the dust content and the heating of atmospheric gas. The effect, that manifests itself in thermally generated diurnal and semi-diurnal tides, was confirmed by *Viking*'s measurements (see Figure 1.2.4). The most dramatic phenomena, however, are periodically observed global dust storms when small-sized dust particles lifts higher than 20–30 km due to turbulent mixing. Owing to the high opacity of the troposphere ($\tau \geq 5$), an anti-greenhouse effect develops and strongly damps circulation transport. Local dust storms (tornadoes, or *dust devils*) can arise even more often in some regions of the planet. Such local storms also are observed during the declining phase of a global dust storm lasting for several months.

Turbulence of a dispersed medium is intimately related to the processes responsible for triggering, maintenance, and decay of dust storms. The pattern of this kind of turbulence substantially depends on the dynamics and energy interaction of gaseous and dust phases, though details of these mechanisms are not quite clear. As we know, for Earth it is possible to evaluate turbulent momentum and heat flows from measured av-

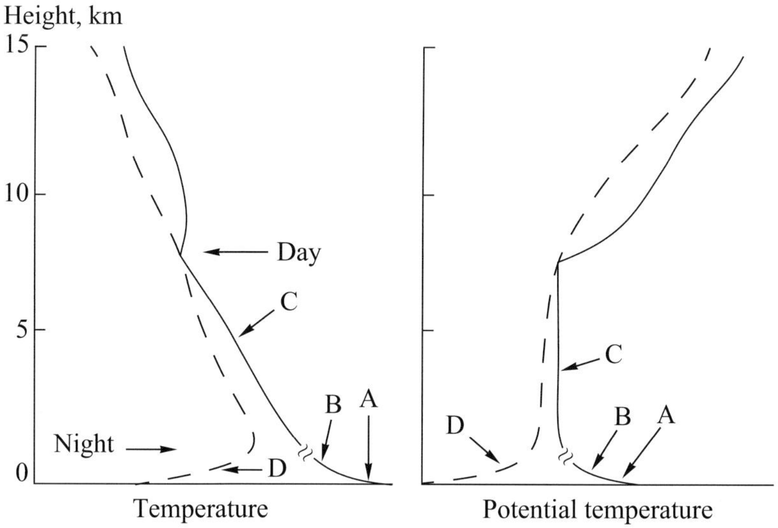

Fig. 1.2.5. Temperature, *T*, (left) and potential temperature, θ, (right) depending on height in the Martian boundary layer (solid curve - day, dashed curve – night). Three characteristic layers A (D), B and C are selected. The break between B and C means that for the layers A (D) and B the vertical scale is exaggerated. According to *Leovy, 1982* model.

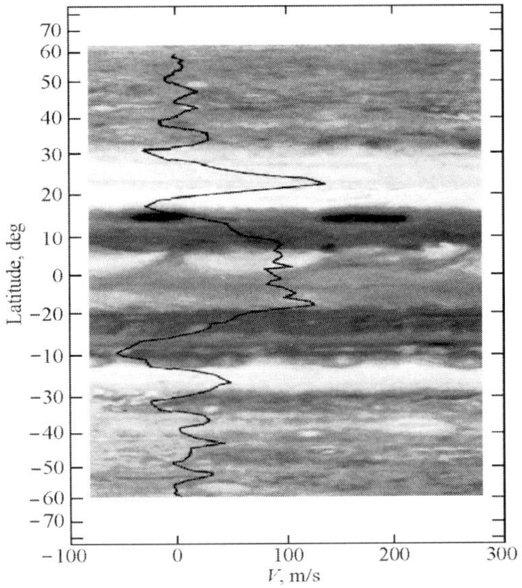

Fig. 1.2.6. Structure of zones and belts on Jupiter. Light zones correspond to ascendant flows, darker belts – to descending flows. The curve running from top to bottom depicts the highly variable wind velocities, coincident with the direction of proper planetary rotation in the zones, and opposite in the belts, thus resulting in the strong shear flows in transient regions. Inhomogeneous motions at different spatial scales observed in the cloud patterns are superimposed on this general picture. Two dark regions in the belt located at the latitudes 8–18°N are the so-called 5μ hot spots. The image transmitted by the space vehicle *Voyager*, courtesy of *NASA*.

erage velocity $u(z)$ and temperature $T(z)$ altitude profiles (*Monin and Yaglom, 1971*).

Admitting that the same approach is also applicable to the atmospheres of other planets, it is possible to reverse the problem and to estimate the respective profiles for the Martian boundary layer, these estimates being modified for the case of turbulent flows comprising heavy admixture (*Barenblatt, 1955; Golitsyn, 1973*). It was shown that the average velocity profile for neutral stratification reduces to the form:

$$u(z) = (u_* / \kappa\omega)\ln(z / z_0)\,, \tag{1.2.5}$$

where $u_* = \sqrt{\tilde{\tau} / \rho}$ is the dynamic velocity (friction velocity); $\tilde{\tau} = -\rho\overline{u''v''}$ is the turbulent momentum flux; κ is the *Karman constant*; ρ is the density of the gaseous phase. The dimensionless parameter $\omega = \nu / \xi\kappa u_*$ is determined by settling the velocity of dust particles ν, the ratio ξ of coefficients of turbulent exchange for admixture and momentum ($\xi \approx 1$), and the quantity u_*. Thus, it follows from (1.2.5) that if particles are relatively small ($\nu < \xi\kappa u_*$, or $\omega < 1$), the presence of dust in the flow results in diminishing the Karman constant κ. In other words, an increase of the velocity gradient near the surface contributes to the *saltation* effect that is, tearing large amounts of dust particles off the ground and rising them into the atmosphere.

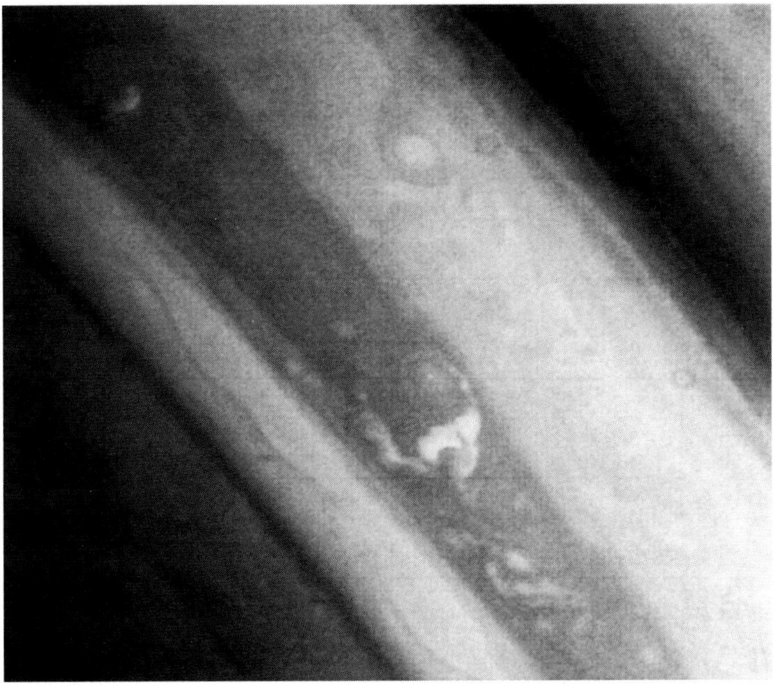

Fig. 1.2.7. Structure of zones and belts on Saturn characterized as a whole by the smaller contrasts in their coloration but the higher wind velocities as compared to Jupiter. The winds near the equator reach nearly u =500 m/s. Inhomogeneties and eddies similar to wave-like motions in the dark belt downward from the center of the image are observed in the flow patterns. The image transmitted by the space vehicle *Voyager*, courtesy of *NASA*.

1.2.4. General Circulation of the Atmospheres of Giant Planets

The structure and dynamics of the atmospheres of giant planets, whose typical representative is Jupiter, are distinguished by a number of specific features which differs them from the terrestrial planets. Their tropospheres are of great extent (~120–200 km) and the thermal structure reasonably well fits an adiabatic temperature gradient. In Jupiter's hydrogen and helium dominated atmosphere, at the pressure level of about 1 bar, a layer of visible clouds composed of ammonia crystals is located, while clouds beneath that level appear to be composed of ammonium hydrosulphide (NH_4SH) and water. However, such a composition does not explain the presence of subtle color shades in discrete cloud systems, because simple condensation of gases and vapors in the jovian atmosphere would produce white clouds only. One should therefore admit the existence of a specific atmospheric chemistry responsible for the formation of compounds with greater complexity, probably involving organic polymers. Their production is assumed to be driven by solar ultraviolet light, lightning discharges, and charged particles raining down from the magnetosphere in the polar regions.

A similar cloud structure is assumed to exist in the atmosphere of Saturn, though the appearance of its globe is more subdued than that of Jupiter. At the same time, on Uranus and Neptune, at lower effective temperatures, high cloud layers involve methane, probably under which layers of condensed ammonia and sulfur-containing compounds are located. Intense absorption of the red part of the solar spectrum by atmospheric methane explains the characteristic aquamarine color of these planets.

The main property of the atmospheric circulation on Jupiter and Saturn is the presence of an ordered system of zones and belts at low and moderate latitudes and of strong jet streams traveling in the intrinsic rotation direction of the planets in the equatorial area (Figures 1.2.6 and 1.2.7). On Saturn, it is as fast as 500 m/s, compared to 150 m/s on Jupiter. The largest temperature gradients also are observed here, while temperature differences between equator and poles are hardly noticeable (Figure 1.2.8). The higher located light zones are areas of ascending currents while the dark belts represent descending ones. For these rapidly rotating planets, $\boldsymbol{Ro} \ll 1$, therefore, owing to the Coriolis interaction of meridional currents, strong zonal flows arise between zones and belts. The strongest alternating winds with velocities more than 100 m/s occur in these transitional areas where shear currents form.

At the same time, natural convection driven by the interior heat source, together with unordered motions and numerous coherent eddy structures (convective cells) observable at high latitudes, is the main mechanism of planetary dynamics. Obviously this mechanism is rather similar to the classical Rayleigh-Taylor hydrodynamic instability problem of a horizontal layer of fluid heated up from below, i.e. to the Benard problem. However, in the case under consideration, the convective interiors are in tight dynamic interaction with the upper gaseous layer where solar energy absorption occurs. This results in extremely complicated current structures with numerous eddy formations and high turbulization patterns, as observed most clearly on the Jupiter's disk (Figure 1.2.9). Analytical and numerical studies of the convection mechanism in rapidly and uniformly rotating fluid spheres (*Busse,1970;1976; Gilman, 1977;1979*) have shown that when an internal heat source is present in a viscous heat conducting fluid, a periodic array of

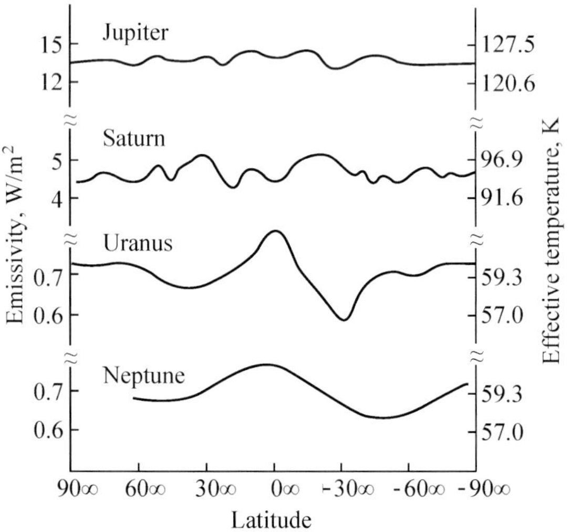

Fig. 1.2.8. Emitted power and equivalent brightness temperature vs latitude for the giant planets for 0.3–0.5 bar atmospheric pressure. The equator-to-pole temperature differences are small, while the temperature gradients mirror the changes in the zonal velocity profile. According to *Ingersoll, 1990*.

convection cells (cylinders) sets up, the latter being aligned in parallel to the rotation axis. Simultaneously, at the expense of declination of these cells due to rotation, a weak second-order mean flow arises consisting of differentially rotating coaxial cylinders, as shown in Figure 1.2.10. Such patterns of columns and cylinders, also obtained in experiments with barotropic fluids in rotated axisymmetrical containers, were associated with zones and belts in the atmospheres of Jupiter and Saturn.

It was shown that the occurrence of a secondary flow is caused by the nonlinear transformation of eddy energy into kinetic energy of the mean flow. Basically, the mean flow draws energy from convection cells by tilting them, causing a radial transfer $\overline{\rho u'v'}$ of eastward momentum against the mean momentum gradient $\partial u / \partial r$, where u' and v' are the departures from the longitudinal means of eastward and radial velocities u and v, respectively, and r is the radial direction (*Ingersoll et al., 1984*). The idea that eddies are adding kinetic energy to the mean flow is confirmed by the same sign of $\overline{\rho u'v'}$ and $\partial u / \partial r$. Unfortunately, this model yields no complete analogy to the observable currents. Also, there is no answer to the question concerning the layer thickness, within which zonal flows occur. In other words, it is unclear till how deep the temperature gradient remains close to adiabatic (γ_a) and how energy dissipation occurs that depends on the magnitude of the turbulent viscosity ν^T and the mean square velocity of vortex motions $|\partial u / \partial r|^2$.

Fig. 1.2.9. Atmospheric motions of various configurations observed on Jupiter's disk. This mosaic map is composed of nine TV snapshots made by the *Voyager* spacecraft through the violet filter from a distance of 4.7 million km. The maximum resolution on the disk is 140 km. The ordered zonal flows reflect the system of planetary circulation in the equatorial and middle latitudes, whereas fully irregular patterns are evident at higher latitudes, the latter being associated with turbulent convection driven by the interior heat source. Eddy motions at different spatial scales are clearly visible. The prominent vortex feature with anticyclonic nature – the Great Red Spot (*GRS*) is located in the southern hemisphere close to the equator. Courtesy of *NASA*.

Ingersoll and Pollard (*1982*) have attempted to solve the problem, proceeding from the assumption that turbulent viscosity values equal thermal conductivity. The latter was defined as the ratio between the turbulent heat flux and the potential temperature gradient $|\partial\theta / \partial r|$. Then the criterion for penetrating zonal currents into the deep of a gas-fluid planet may be $\boldsymbol{Ri} = N^2 |\partial u / \partial r|^{-2} \geq 1$, where $N^2 = (g / \theta)\partial\theta / \partial r$ is the *Brent-Vaisala* frequency; g is the acceleration due to gravity; and T is the temperature.

This idea agrees well with the theoretical prediction that inclined convective cells must occur in shear currents if the Richardson gradient number, $\boldsymbol{Ri} \approx 1$ (*Lipps*,1971). Besides, according to direct measurements of the atmospheric *Galileo Jupiter Probe*, that reached a pressure level of 21 bars, the wind velocity really increased during its descent (*Ingersoll et al., 1998*), though this limited depth is hardly sufficient to confirm or refute the model. On the contrary, in the upper troposphere and stratosphere wind velocities rapidly decrease with altitude.

The *Galileo* probe also experienced small and essentially random accelerations during its descent measured by the Doppler frequency shift technique generally similar to that used in the above mentioned experiments with the *Venera* probes. These accelerations were caused jointly by turbulence along the descent path and buffeting (an aerodynamic effect). Analysis of the frequency residuals in the vertical profile of Jupiter's dominating deep zonal winds may give some indication as to the scale size of turbulent cells encountered by the probe during its descent (*Atkinson et al., 1998*).

The importance of the eddy stress term $\overline{\rho u'v'}$ as a way of feeding energy into

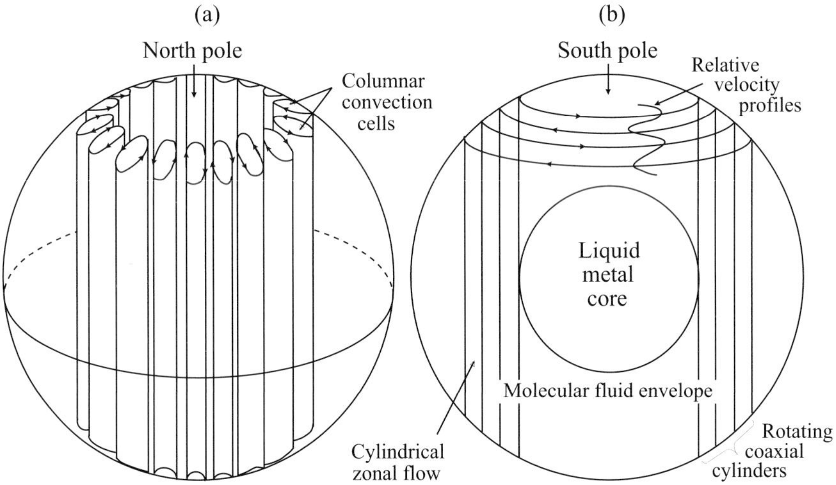

Fig. 1.2.10. Columnar convection cells **(a)** and cylindrical zonal flow **(b)** on a fast rotating liquid sphere. The columnar mode is the preferred form of convective instability in a uniformly rotating, viscous, conducting fluid, whereas the cylindrical mode is the most general form of steady zonal motion in an inviscous adiabatic fluid (*Busse, 1976*). The interaction between these two modes is analogue to the behavior of transverse convective disturbances in a sheared horizontal layer and implies power transmission of inclined convective cells to zonal flow. According to *Ingersoll and Pollard, 1982; Ingersol et al., 1984).*

zonal jets and the maintenance of planetary circulation also was studied using numerical simulations of motions of thin fluid layers on a rotating planet (*Williams, 1978; 1979*). In the barotropic model, vorticity was introduced as a small-scale checkerboard pattern and was removed by eddy viscosity. It was found that an initially small mean flow was amplified by the tendency of the $\overline{\rho u'v'}$ term to have the same sign as the meridional average velocity component $\partial u / \partial y$ (Figure 1.2.11). In turn, in the baroclinic model, eddies arose spontaneously and were supported by instabilities that drew their energy from the meridional temperature gradient. The pattern of eddy mean flow interactions was oscillatory due to the energy exchange between eddies and the mean flow, though initially it was similar to the pattern in the barotropic model.

The positive correlation between $\overline{\rho u'v'}$ and $\partial u / \partial y$ was confirmed on Jupiter by measurements on the *Voyager* spacecraft. For the atmospheric height range comprising the whole cloud sheet, the implied rate of eddy energy transfer was estimated as more than 10% of the emitted heat flux of the planet (*Ingersoll et al., 1981; 1984*). Since similar estimates for the Earth do not exceed ~ 0.1% (with total the dissipation level of all forms of kinetic energy into heat being no more than ~1%), this brings evidence that

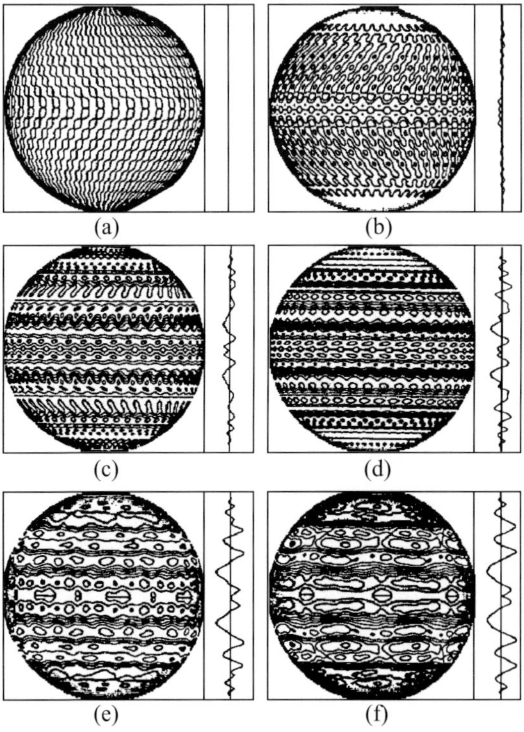

Fig. 1.2.11. The development of zonal jets from eddies in a thin atmosphere on a rotating planet. In this barotropic model (no horizontal density contrasts) the initial stream-function pattern **(a)** reflects the pattern of mechanical forcing; later stream-function patterns **(b–f)** show the emerging zonal structure. The longitudinal mean zonal velocity is shown to the right of each pattern. According to the numerical model of *Williams, 1978.*

the thermomechanical cycles of these two planets are substantially different.

Apart from Jupiter and Saturn, convection as a thermal mechanism to transport energy from the interior also should play an important role in the dynamics of Neptune's atmosphere, but not in the atmosphere of Uranus, because Uranus has no internal heat source. The most interesting feature determining the thermal conditions and dynamics of Uranus' atmosphere is the unique orientation of its rotation axis, lying almost in the plane of its circumsolar orbit. However, despite significant distinction in inclination and energetics, both planets feature qualitatively identical meridional temperature and zonal wind profiles at cloud level, although on Uranus winds are approximately twice as weak as on Neptune. Besides, it is important to emphasize that, although the energy sources capacity per unit area for Neptune is smaller by a factor of about 20 compared to Jupiter, wind velocities in Neptune's atmosphere are almost 2.5 times higher, reaching 400 m/s on the equator (Figure 1.2.12).

This is due the to very low turbulent viscosity of Neptune's atmosphere and, respectively, to the low energy dissipation level of wind motions and turbulization of shear currents (*Ingersoll et al.,1995;1998*). It is interesting to note that winds on Neptune and Uranus blow opposite to their rotational direction, which distinguishes these planets from Jupiter, Saturn, and Venus (and also from the Sun and Titan), which feature equatorial superrotation, as follows from Figure 1.2.12. When addressing Earth, one may conclude that, in contrast to Neptune, the terrestrial atmosphere has the highest dissipation level. This is mainly caused by processes related to hydrological cycles, as well as by small-scale convection and surface friction. Therefore, though Earth receives much more solar energy, the terrestrial wind velocities are almost an order of magnitude smaller than those on Neptune.

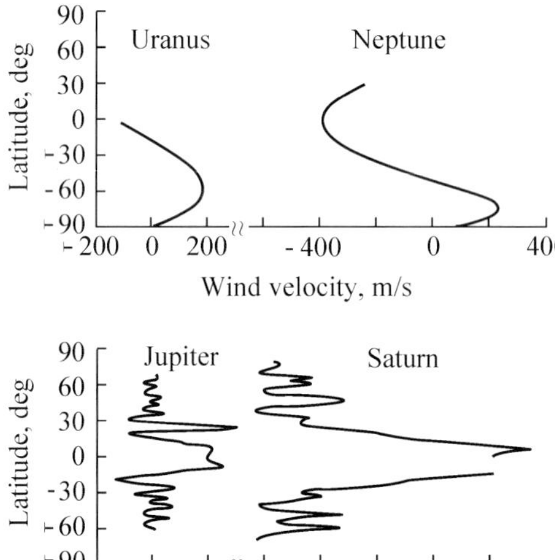

Fig. 1.2.12. Zonal velocities vs latitude for the giant planets based on a compilation of data published by different authors. In contrast to Jupiter, Saturn and Uranus, for which the superrotation is characteristic, the subrotation predominates on Neptune, where likewise on Uranus, the shear flows are also weak. According to *Ingersoll, 1995.*

The phenomenon on Uranus and Neptune returns us to the problem of the relation between planetary interiors and the observable thin upper gaseous layers, that is, to the relationship between kinetic and potential energy in the formation of atmospheric dynamics. The comparison between the measured zonal wind velocity profile u on Neptune with the results of calculations in the shallow water approximation (*Allison and Lumetta, 1990; Ingersoll et al., 1995*) is shown in Figure 1.2.13. A thin steady fluid layer of variable thickness h on a deep adiabatic sublayer was considered. The vertical relative vorticity component $\varsigma = (\nabla \times V)_z$ was determined as a function of $\partial h / \partial \phi$ under condition of potential vorticity preservation of the main current, $q(\phi)$ = const, where $q \equiv (\varsigma + f)/h$ (see condition (1.2.2)). The so-called deformation radius $L_d = \sqrt{(gh\Delta\rho/\rho)}/f$, which describes eddy effects on rapidly rotating planets, was used as a free parameter. Here $\Delta\rho/\rho$ is the ratio between the layer densities; all other designations were given earlier. The best agreement with measurements was obtained under the condition that $q(\phi)$ keeps constant not in the entire hemisphere but only within limited intervals, $\Delta\phi$. The parameter L_d, or in other words, the horizontal perturbation dimension whose kinetic energy, E_K, is comparable to the potential energy E_P, plays a key role in obtaining results from the modeling. Obviously, $E_K > E_P$ when perturbations are smaller than L_d, and vise versa. Unfortunately, the results available do not allow to put specific constraints on this parameter.

Particularly on giant planets having an internal heat source, numerous large-scale eddy formations stand out on the various cloud structure patterns (see Figure 1.2.9). The most characteristic formations are the Great Red Spot (*GRS*) on Jupiter, measuring

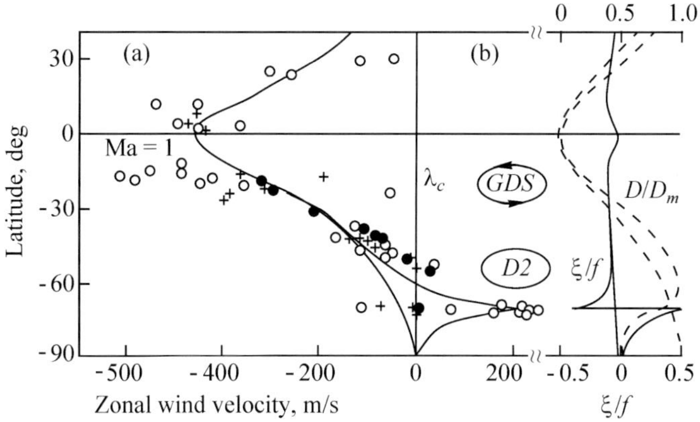

Fig. 1.2.13. Comparison of Neptune's zonal velocity profiles calculated from two shallow-water models with cloud-tracked wind data: **(a)** Cloud-tracked wind measurements in three series (○, *, +) during *Voyager* flyby; **(b)** The thickness, *D*, of the upper stable layer normalized to its maximum value, D_m, depending on latitude ($D/D_m = h(\phi)$) (dashed line) and the ratio of relative (ς) and planetary (f) vorticities (solid line) calculated from shallow-water models of *Allison and Lumetta, 1990*. The model assuming piecewise constant values for the potential vorticity, $q = (\varsigma + f)h$, in both hemispheres fits to the measurements best. The positions of the Great Dark Spot (*GDS*) and the smaller size spot (*D2*) are shown. According to *Ingersoll, 1995*.

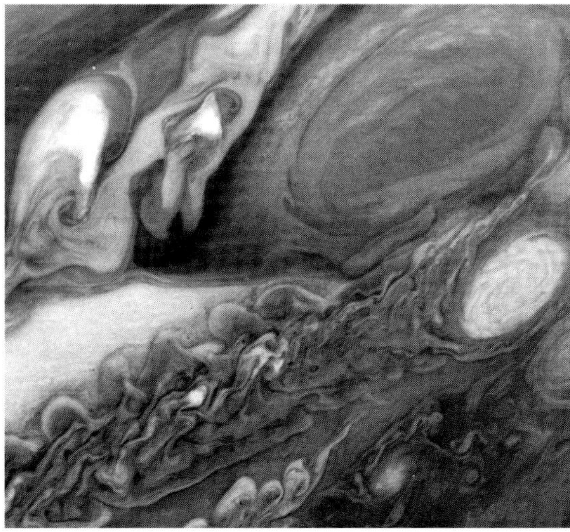

Fig. 1.2.14. The Great Red Spot (*GRS*) on Jupiter (size 25000 x 12000 km) enclosed by strongly turbulized streams to the west and south. The maximum resolution is 95 km. At the bottom right – vortices of smaller size ("white ovals"). An image made by the space vehicle *Voyager*, courtesy of *NASA*.

~25000 km (Figure 1.2.14), and the Great Dark Spot (*GDS*) on Neptune, measuring ~15000 km (Figure 1.2.15). The *GRS* represents a giant anticyclone type vortex with an estimated lifetime of several thousand years. It is located above the surrounding cloud layer due to ascending motions from beneath and the very complicated morphology of internal eddy currents.

At periphery of the *GRS* where current velocities exceed 100 m/s, particularly strong flow turbulization and exchange of gas and cloud particles between the vortex and neighboring zones occur. As yet, there is no satisfying explanation for the formation and maintenance of such steady structures in the atmospheres of Jupiter, Saturn, and Neptune on the background of chaotic small-scale activities in the form of relatively small clouds appearing and disappearing in a few hours. In addition, areas of descending motions where temperatures are higher than in surrounding clouds (the so called 5-micron hot spots, see Figure 1.2.4) have been discovered on Jupiter. These areas are associated with some local changes in the chemical composition of the atmosphere. Accidentally, the atmospheric probe jettisoned from the *Galileo* spacecraft parachuted in such a spot, which seems to explain the extremely low water vapor abundance recorded in Jupiter's atmosphere, not being characteristic of the planet as a whole (*Mahaffy et al., 1998; Carlson et al., 1998*).

The observed longitudinal-latitudinal oscillations of spots, including *GRS* and *GDS*, resemble motion patterns of the upper parts of eddies in a steadily stratified shear flow. Like ordered zonal currents, it seems natural to treat them from the point of view of hydrological cycle formation in a stratified gas-fluid medium, taking into account its chemical composition, energetics, and its matching to the stability criterion considering the diverse relationships between the internal and solar energy sources. Understanding the complete collection of hydrometeorological elements of such a system, including the interrelation between convective motions in the interiors and atmospheres, and plane-

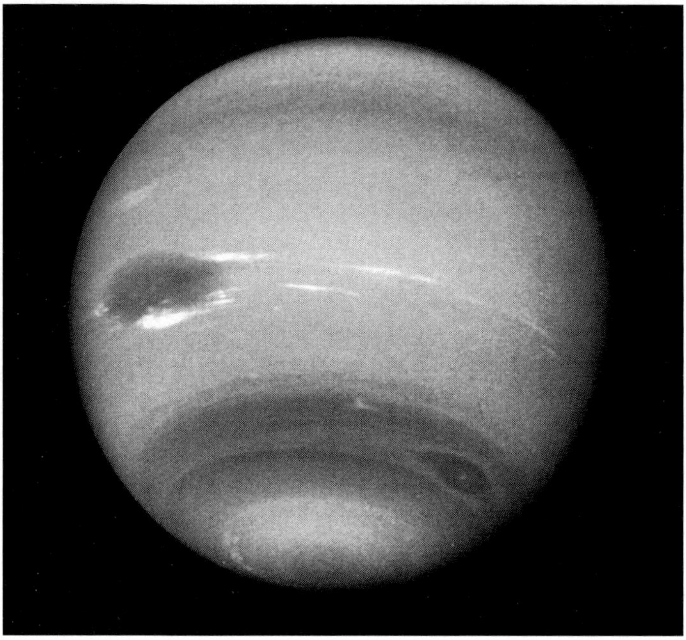

Fig. 1.2.15. The Great Dark Spot (*GDS*) on Neptune (size 15 000 x 6 000 km) and the smaller spot at the bottom right. Their periods over the planetary disc is 18.3 and 16.1 hours, respectively, while the wind velocity in the vicinity of *GDS* reaches 550 m/s. An image made by the space vehicle *Voyager*, courtesy of *NASA*.

tary circulation peculiarities, involving extremely sophisticated turbulent features, is a challenge for geophysical hydrodynamics.

1.3. TURBULENCE IN THE UPPER ATMOSPHERE OF PLANETS

The upper atmosphere of a planet is defined as the outer part of its gas envelope, which is very rarefied and frequently associated with the circumplanetary outer space. A significant fraction of it is a characteristic example of a multicomponent turbulent medium. It is subjected to direct absorption of incident solar radiation and numerous chemical transformations together with the processes of heat and mass transport. Intense solar electromagnetic radiation is principally responsible for various photochemical processes such as photoionization, photodissociation, and excitation of inner degrees of freedom of atoms and molecules. These processes are accompanied by inverse association reactions of atoms into molecules, ion recombinations, spontaneous photon emissions, and collisional deactivations. Properties of gas under influence of gravitational and electromagnetic fields are strongly dependent on the efficiency of molecular and eddy diffusion and heat transport at different height levels. Due to temperature, concentration, and pressure gradients, the multi-scale turbulent hydrodynamic motions are developed from below up to the lower thermosphere. In addition, solar corpuscular radiation and some auxiliary energy sources (such as tidal oscillations, viscous energy dissipation of internal gravity waves, etc.) affect the composition, dynamics, and energetics of the upper atmosphere.

1.3.1. The Upper Atmospheres of the Terrestrial Planets

Atmospheric regions situated above the stratosphere (~50 km) are usually referred to as the upper atmosphere of the Earth. During recent years the complete height interval from the tropopause (12–15 km) to the lower thermosphere (~110 km) was also ascribed to the middle atmosphere. It is just in the upper part of the middle atmosphere and the overlying thermosphere (defined as a region with a positive temperature gradient), where the main energy exchanges occur. These exchanges are caused by direct absorption of solar extreme ultraviolet radiation (*EUV*) and soft X-rays (approximately from 2000 $\overset{o}{A}$ to 10 $\overset{o}{A}$), and additionally by interactions with energetic particles originating from solar plasma after their acceleration in the terrestrial magnetosphere. Longer wavelength *UV* radiation (2000–4000 $\overset{o}{A}$), in particular responsible for the formation of the ozonosphere, and solar particles with higher energies (protons up to 15–30 MeV) generated in solar flares, are absorbed in the lower regions of the middle atmosphere (for details see *Akasofu and Chapman, 1972*; *Marov and Kolesnichenko, 1987*).

In addition to the absorption of electromagnetic and corpuscular energy from the Sun, the values of important upper atmosphere parameters, such as wind velocity, mass density, temperature, and both chemical and ionic composition, are determined by mass, energy, and momentum transfer from underlying regions of the middle atmosphere and

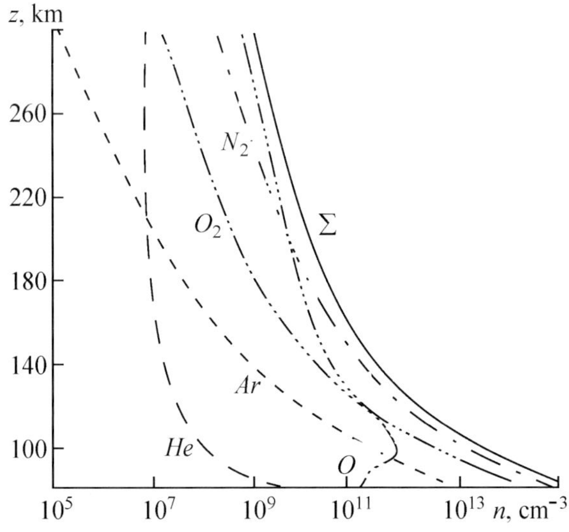

Fig. 1.3.1. Height profiles of atomic and molecular components in the terrestrial thermosphere at moderate level of solar activity; Σ – total number density in cm^{-3}.

troposphere. The middle atmosphere itself hosts a complex interplay between radiative processes, chemistry, wave dissipation and turbulence, involving non-linear dynamics and electrodynamics. The latter is mainly promoted by magnetospheric-ionospheric interaction processes which, in turn, exert the most pronounced effect on the distribution of thermohydrodynamical parameters of the high latitude thermosphere and ionosphere, due to energy dissipation of particles precipitated from the magnetosphere and excited current systems. These processes results in powerful heating of the medium accompanied by the growth of large local mass density gradients and dynamic transport.

Up to about 120 km (the height at which gravitational diffusion separation of gases occurs), the atmosphere remains well mixed with mean molecular weight, $M = 28{,}9$. So, this level serves as a boundary between the *homosphere* and the *heterosphere*. It is also called the *turbopause* (or *homopause*) and characterizes the height at which turbulent mixing becomes inefficient and molecular diffusion begins to dominate, and the height distribution of atmospheric species is controlled by their particular scale heights. However, despite a constant M in the middle atmosphere, its actual composition is subject to large variations caused by minor admixtures. This is due to the extreme complexity of chemical and dynamic processes occurring in the stratosphere and, of minor importance, in the mesosphere and the lower thermosphere.

Nitrogen (N_2), the main atmosphere component at the Earth's surface, dominates up to about 180 km. Above that level, in the thermosphere, atomic oxygen (O), originating from dissociated O_2-molecules, starts to dominate, eventually being replaced by helium (He) and hydrogen (H_2) at even higher levels (Figure 1.3.1). Depending on the thermospheric temperature, mainly conditioned by the solar activity during the 11-year

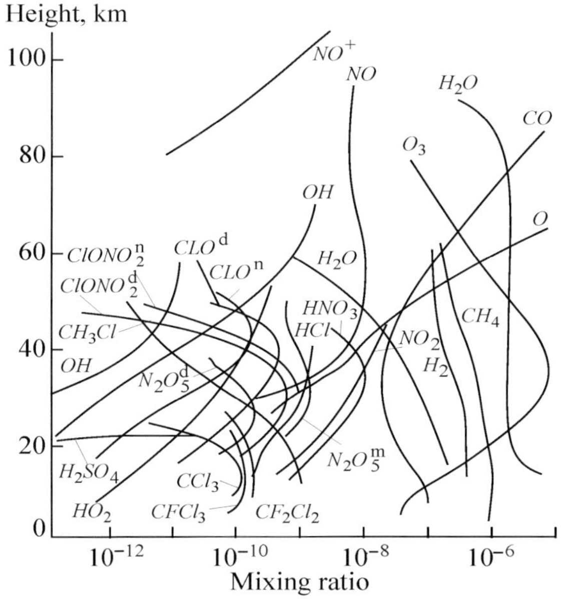

Fig. 1.3.2. Height distribution of minor species in the terrestrial middle atmosphere shown as volume mixing ratio relative to the total of main components ($N_2 + O_2 + Ar = 1$). Day (d), night (n), morning (m), and evening (e) conditions are distinguished. According to *Solar Terrestrial Research for 1980s, US National Academy 1981.*

cycle and the local time, the concentrations of *He* and *O* appear to be approximately equal to each other in the altitude range from 500 to 700 km, while the *He* and H_2 abundances are nearly the same between 900 and 1600 km. While N_2, O_2 and *O* concentrations grow with temperature, the content of H_2, on the contrary, drops because of its increasing dissociation rate.

As far as *He* is concerned, its variations are even more complicated, because they exhibit a strong latitudinal dependence, related to the season and solar cycle phase. In turn, hydrogen-, carbon-, and nitrogen-bearing compounds such as CH_4, H_2O, *OH*, CO_2, NO_x and O_3, play a noticeable role in the mechanisms of chemical transformations and radiative heat exchange. The same is true when addressing some of their more complex derivatives (including those of anthropogenic origin), and a number of other minor components, including metastable ones (Figure 1.3.2.). Aeronomic processes, especially in the presence of particles with *superthermal* velocities, responsible for non-equilibrium chemical kinetic processes, significantly influence the structure, energetics, and dynamics of these and overlying layers of the Earth's atmosphere (*Marov and Kolesnichenko, 1987; Marov et al., 1996; 1997*).

Various dynamic processes, wave motions included, are important features of the middle atmosphere and thermosphere. Dynamics, related to general circulation, causes redistribution of matter and energy at global scales. By means of mass, momentum, and heat exchange, this determines many aspects of the energy balance, implying an intimate coupling between the processes in near-planetary space. Dynamic variations of the pressure field, first of all atmospheric tides, planetary waves, and *IGW*s, play an important role in this balance being observed as spatial-temporal variations of structural parameters in various areas. Wave-wave and wave-mean flow interactions affect minor species distributions and emission variations.

Solar heating and the gravitational attraction of the Sun (as well as the attraction of

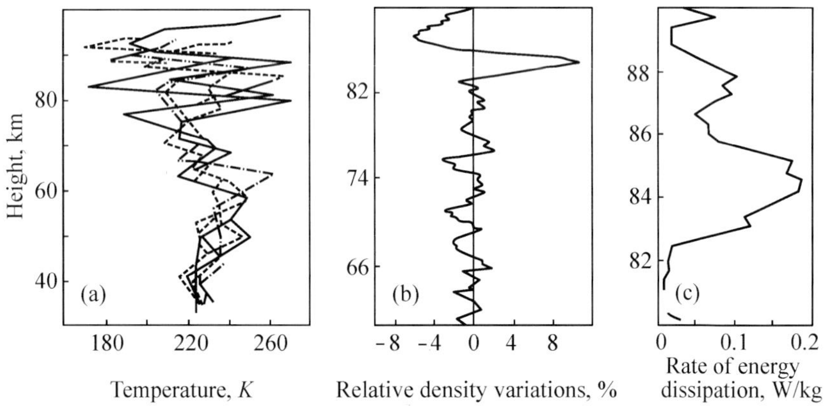

Fig. 1.3.3. An example of temperature *T* **(a)** and relative density ρ **(b)** variations depending on height in the terrestrial middle atmosphere associated with the propagation of internal gravity waves, based on high-altitude rocket and lidar measurements. It is contrasted to the rate of internal gravity wave energy dissipation extracted from the data from radar power spectra measurements **(c)**. According to *Middle Atmosphere Program/Handbook for MAP, 1985, and Geophysical Institute University of Alaska Biennial Report, 1983-1984.*

the Moon, in case of the Earth) serve as the main source of tides in the atmosphere of terrestrial planets. Emergence of the above mentioned thermal tides is characteristic of Mars, while tidal effects in the dense Venus atmosphere probably facilitate its capture into resonance rotation with Earth (see, e.g., *Kuzmin and Marov, 1974*; *Marov and Grinspoon, 1988*).

IGW dissipation is an important energy source in the upper atmosphere. Though there are many diverse sources of waves with various phase velocities, the presence of stationary irregularities at the base of an atmosphere creates a peak in the wave spectrum at zero horizontal velocity. As wave absorption results in acceleration of a medium in the direction of the wave propagation, the total effect commonly displays itself as a braking of the atmosphere (*Fels and Lindzen, 1974; Lindzen, 1981; Holton, 1982; Andrews et al., 1987*).

Diverse perturbations serve as a source of *IGW*s. They are generated due to restructuring of meteorological processes, flow of air streams around ridges, *wind shear* instabilities, heating of auroral zones, etc. In a stratified medium like an atmosphere, such waves usually propagate both in vertical and horizontal directions, though the horizontal component in an initially vertical perturbation may become predominant with increasing altitude. The heat release through energy dissipation of *IGW*s in the lower thermosphere turns out to be comparable to other energy sources related to the incident solar radiation at these altitudes (Figure 1.3.3.).

Turbulence, whose time and space morphology is not yet fully clarified, is an important factor in atmospheric dynamics and energetics, particularly in the lower thermosphere. Its origin is mainly due to convective instability, wind shears, tidal oscillations, instability and/or breaking of *IGW*s, and other perturbations. Therefore, when analyzing thermohydrodynamic processes in the middle atmosphere, it is often necessary to simultaneously consider equations describing averaged concentration, temperature, and wind velocity fields, together with turbulent motion intensity features such as the eddy diffusion, D^T. The latter sometimes serves as a free parameter of the problem. Numerous field experiments have led to the conclusion that turbulence almost permanently occurs up to ~100 km, and that within the 60–100 km layer, individual turbulent sublayers of several kilometers thick often neighbor fairly quiet sublayers. At the same altitudes, a fluctuation wind component, originating from wave motions, also occurs.

1.3.2. Turbulent Diffusion in the Atmosphere of Terrestrial Planets

The turbulent processes in the upper atmosphere of a planet are responsible for high-altitude redistribution of components (due to eddy diffusion), variability in rate of chemical reactions (due to turbulent mixing), and turbulent energy exchange (heating through viscous dissipation of turbulent energy and cooling through turbulent thermal conductivity). As a simplification already discussed in section 1.1.3, eddy diffusion D^T which determines inertial interpenetration of elementary volumes of separate components of multicomponent fluid (mixing), may be considered as analog of molecular diffusion D in a medium with averaged macroscopic properties, which assumes applicability of corresponding forms of the diffusion equation.

Accordingly, the time of smoothing out the composition heterogeneities when passing to equilibrium $t \propto H^2/d$ (where H is the characteristic scale of the order of scale height) is determined depending on the relation of magnitudes of molecular ($d = D$) or turbulent ($d = D^T$) diffusion coefficients. Respectively, as was earlier mentioned, position of the transitional area from homosphere to heterosphere, that is of homopause, corresponds to the beginning of dominance of magnitude D over D^T. Obviously, the maximum diffusion rate is achieved by complete mixing, while outside the homopause the processes of molecular diffusion are determined by molecular masses of individual components.

Unlike heterosphere where the basic mechanism of matter transport is molecular diffusion in a rarefied gas medium, the structural properties of homosphere are effected by eddy diffusion. It ensures preservation of relative abundance of atmospheric gas composition in height, except for chemically active minor species. The competing processes of molecular and turbulent transport substantially determine the patterns of the structure, dynamics, and energetics of the upper atmosphere in the turbopause region. Turbulent mixing in the homosphere also appreciably controls supply of hydrogen atoms to the *exobase* level, and thus the rate of escape of light atoms from the planetary atmosphere (*Chamberlain and Hunten, 1987*). For Earth atmosphere it is mainly hydrogen and partially helium.

Modeling the structure, dynamics, and energetics of the upper atmosphere is very complicated problem. In order to ease the problem, in some cases one-dimensional approach with the use of averaged equations of hydrodynamics of a multicomponent mixture accounted for various processes of heat and mass transfer, radiation exchange, and chemical kinetics can be efficiently used (*Marov and Kolesnichenko, 1987*). The averaging is performed in latitude, longitude, and, to various degrees, in time in such a way that these models give a simplified representation of the global average. In such a model, vertical transport of atoms and molecules is represented as globally averaged vertical diffusion described by eddy diffusion coefficient D^T. The latter serves as a fitting parameter, its value being obtained as a best fit the results of calculations to the measured high-altitude profiles of atmospheric components.

As an example, Figure 1.3.4 shows variations of several odd nitrogen species (NO, $N(^4S)$) between 90 and 120 km for different height dependencies of $D^T(z)$ that were calculated based on the diffusion-photochemistry model of the terrestrial lower thermosphere (*Kulikov and Pavlyukov, 1987*). The curves 3–5 in the figure correspond to the average, minimum, and maximum values of D^T, respectively. They vary, within the considered altitude range, between $10^6 - 10^4$ cm^2/s, the average value corresponding to $D^T = 5\times10^5$ cm^2/s. These estimates, serving for illustration purposes, were obtained proceeding from the comparison between calculated and measured thermospheric temperatures (curve 1), analysis of spectrometric mass measurements of N_2, O_2, O, and Ar concentrations (curve 2), and theoretical and measured concentration distributions of atomic oxygen (curve 3). According to other experimental evidence, D^T is closer to 10^6 cm^2/s (*Hunten, 1975*). As we see, the differences of the NO and N concentrations due to uncertainty in D^T can reach ~30%. This stresses the importance of using the most accurate estimates of this factor when evaluating turbulence influences on thermospheric processes. Let us note that they depend, in their turn, on the correctness of the calculation of odd nitrogen production rates accounted for formation of superthermal or exited particles, as well as for the ways of their thermalization and desactivation due to collisional relaxation or spontaneous radiation (*Marov et al., 1996; 1997*).

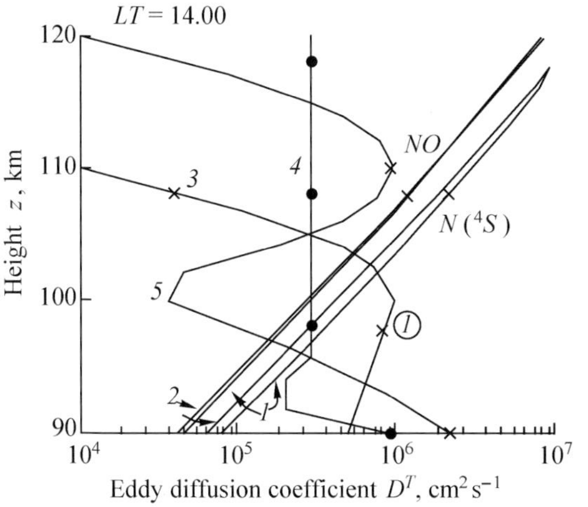

Fig. 1.3.4. Variations in odd nitrogen (*NO, $N(^4S)$*) at height 90–120 km (curves 1-2) according to the diffusive-photochemical model with different eddy diffusion factors, D^T, (curves 3-5) used as fitting parameter.

The considered patterns of the upper and middle atmospheres of the Earth have many similarities in the atmospheres of other planets in the solar system and their satellites possessing gas envelopes. Our knowledge of these unique natural multicomponent media, having a great diversity of physicochemical processes, dramatically advanced for the last three decades, mainly due to space exploration. Space vehicles enabled detailed studies of these frontiers on a comparative-planetological basis and allowed us to broaden former concepts about evolution paths of celestial bodies. When developing mathematical models of planetary atmospheres, the basic source of information are the data on chemical structure and spatial-temporal variations of structural parameters with due allowance for the crucial role of aeronomic processes in energetics and atmospheric dynamics.

In contrast to the Earth where main agents of chemical transformations are products of oxygen photolysis and nitrogen, photolysis of CO_2 determines the atmospheric chemistry on other inner planets. Sulfur- and halogen-bearing compounds are also of great importance in the middle atmosphere of Venus. They absorb solar ultraviolet radiation in the wavelength range from 4000 to 2000 $\overset{o}{A}$, while carbon dioxide absorption occurs from 2300 up to 10 $\overset{o}{A}$. Atmospheric chemistry features and related cycles, including cycles of sulfur-bearing compounds on Venus that relate to the formation of sulfuric acid clouds, depend on the specificity of aeronomical processes in the middle and upper atmosphere (*Esposito et al., 1983; Volkov et al., 1989; Marov et al., 1989; Marov and Grinspoon, 1998*). Together with diffusion and convective transport, these processes determine the redistribution of components in altitude and the overall thermal balance of the atmosphere. In addition, as there is no intrinsic magnetic field on Venus, a noticeable contribution into energetics is brought by solar wind particles and induced electrical currents in the ionospheric plasma of the planet. Even more pronounced effects occur on Mars having *localized crustal paleomagnetic anomalies* in its southern hemisphere as the *Mars Global Surveyor* mission revealed (*Acuna et al, 1999*). This significantly complicates the patterns of its interaction with solar plasma and accompanying dynamical processes resulting in the set up of "small magnetospheres" formed by the crustal magnetic fields. The distinguishing feature of the magnetic field pile up boundary at Mars most likely varies from Venus-like to Earth-like above the crustal magnetic field regions (*Krymskii et al., 2000*).

Internal gravity waves also play substantial role in the dynamics of Venus and Mars. On Venus, the periodic density perturbations, owing to wave motions with horizontal dimensions from 100 up to 600 km, were recorded on the *Pioneer Venus* spacecraft. Their source can be vertically traveling gravity waves originating in the cloud deck 50-70 km high (*Zhang et al., 1996; Bougher et al., 1997*). Braking (disintegration) of these waves in the thermosphere, similar to their dissipation in the mesosphere and lower thermosphere of the Earth (*Lindzen, 1981*), apparently represents an important energy source and momentum supply which facilitates zonal asymmetry in thermospheric circulation known as a superrotation mechanism. When modeling the middle atmosphere of Mars a link was established between the vertical structure of the eddy diffusion coefficient and the shape of the ozone profile; it was related to turbulence induced by breaking gravity waves and hence, if such an assumption is correct, ozone could serve as a tracer of turbulence (*Chassefiere et al. 1994*).

Likewise the Earth's upper atmosphere, the heat sink on other inner planets is determined by spontaneous radiation of molecules and atoms at visible and infrared wavelengths and by turbulent thermal conductivity. The intensity of atmospheric emission in lines and bands observed as dayglow and nightglow, depends on degree of non-equilibrium state of a medium and on the collisional relaxation efficiency of the excited states of atmospheric components (*Marov et al., 1997*).

The region of turbulent mixing on Venus and Mars extends higher than on Earth

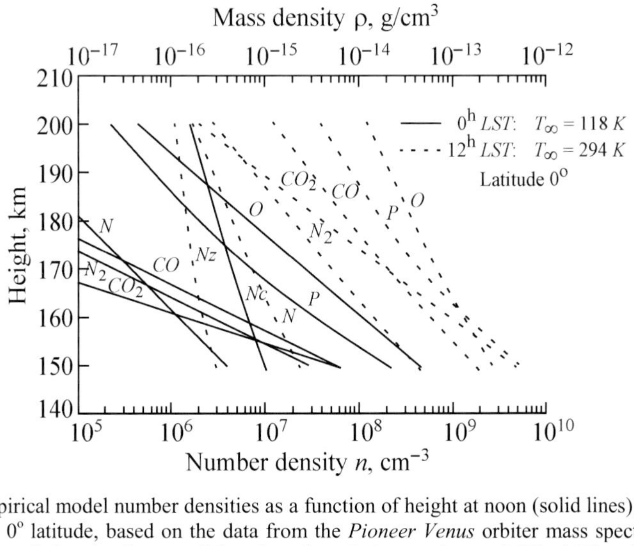

Fig. 1.3.5. Empirical model number densities as a function of height at noon (solid lines) and midnight (dashed lines), 0° latitude, based on the data from the *Pioneer Venus* orbiter mass spectrometer. It is evident that the relative dominance of the photodissociation products of CO_2 (*CO* and *O*) begins at 150 km at night time and at 170 km at day time. The total mass densities, ρ_{day} and ρ_{night}, referred to the at the upper horizontal scale are also shown. According to *Niemann et al., 1980*.

and the tropopause occurs at ~130 km high. This promotes to the increased efficiency of the recombination processes of CO and O – the products of CO_2 dissociation. Therefore, the relative dominance of these species begins only at 150 km at night and at 170 km at day time. This is shown in Figure 1.3.5, plotted from mass spectrometric measurements aboard the *Pioneer-Venus* spacecraft (*Niemann et al., 1980*). The total concentration of components substantially grows during day time due to thermal "swelling" of the atmosphere. Such behavior of Venus' thermosphere is in agreement with modeling results assuming proper values of the eddy diffusion coefficients, similar to what was discussed for the terrestrial thermosphere. Apart from the carbon dioxide and products of its photolysis, the model is also able to predict the altitude dependence of a series of minor components (*HCl, Cl,* H_2O, O_3), which were not covered by the measurements. An example of respective calculations together with the employed altitude profile, $D^T(z)$, is shown in Figure 1.3.6 (*Krasnopolsky and Parshev, 1983*).

It was found that the value $D^T = 4\times10^6$ cm^2/s fits the measurements best at altitudes of 110 km and higher. This value is larger than the average value matching the model of the lower terrestrial thermosphere. An even greater value of this parameter (10^7 cm^2/s) was obtained from the data analysis of *Pioneer-Venus* measurements during the morning (*Von Zahn et al., 1980*). However, the highest turbulent mixing efficiency is supposed to exist in the Martian thermosphere, for which an average value $D^T \approx 2\times10^8$ cm^2/s has been obtained (*Nier and McElroy, 1977*). Obviously, apart from the intense emissivity of atmospheric gas in the infrared bands of CO_2, the higher eddy diffusion efficiencies also relate to *exospheric temperatures* on Venus and Mars, which do not exceed 200–350 K, compared to the average value of 800–1000 K for Earth.

Besides, it was revealed that the entire night hemisphere of Venus, starting at the usual position of the thermosphere, is extraordinary cold: its temperature measures about 100 K, which is even much lower than the temperature of the terrestrial mesopause (~170 K). As a consequence, this zone has come to be known as *criosphere* (see

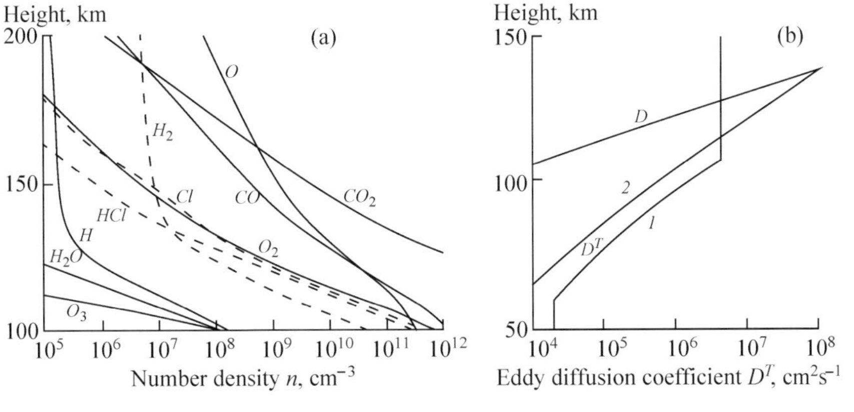

Fig. 1.3.6. Composition model for Venus' upper atmosphere **(a)** calculated for the two different eddy diffusion factor height profiles, D^T (z), 1 and 2 **(b)**. The best agreement with the data from measurements above 110 km is observed for D^T = 4 x 10^6 cm^2/s. D - the coefficient of molecular diffusion. According to *Krasnopolsky and Parshev, 1983*.

Keating et al., 1980; *Izakov and Marov, 1989; Marov and Grinspoon, 1998*). Its formation is explained by the long duration of Venus' night because of the slow intrinsic rotation of the planet and, accordingly, long isolation from the day time side. There is no similar isolation on Mars, therefore there are no so sharp differences between the thermospheric temperatures on the day and night sides.

Currently, the thermospheric general circulation model, or *TGCM* (*Dickinson et al., 1984*) provides the most adequate description of the structure and dynamics of upper atmospheres of the terrestrial planets involving temperature and wind fields, and partial

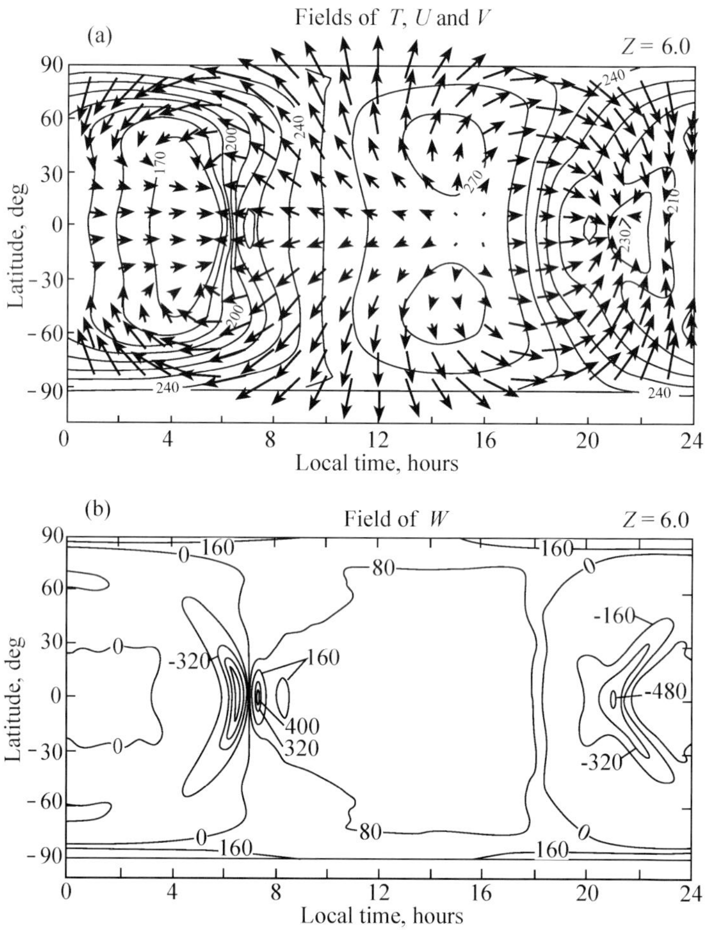

Fig. 1.3.7. Temperature, *T*, and horizontal winds, *u, v*, in the thermosphere of Mars at the pressure level slice corresponding to the average height of 210 km **(a)** and the respective vertical flow velocities **(b)** based on the *TGCM* model. Temperature contours from 170 to 270 *K* with 10 *K* intervals; the maximum temperature variation within a day is 110 *K*. Maximum wind vector (230 m/s) is at terminators and poles; maximum wind divergence and convergence is at 15^h *LT* and at 5^h *LT*, respectively. Contours of vertical velocity from 960 cm/s in the early morning to 400 cm/s in the late evening, with 160 cm/s intervals. According to *Barth et al., 1992*.

concentration. The model was successfully developed and applied for Venus and Mars (*Bougher et al., 1988*). The results of these calculations have shown that thermospheres of these planets have a number of common features, determined, as in the case of Earth, by sources and sinks of heat and by the efficiency of its redistribution depending on the rotation rate of the planet. Figure 1.3.7 (*Barth et al., 1992*) shows an example of a simulation of the temperature *T*, and horizontal (u,v) and vertical (w) winds in the Martian thermosphere at 210 km high. As we see, the diurnal variations of temperature make up $110\,K$, while winds, having regions of divergence and convergence close to the temperature maximum (~15^h *LT*) and minimum (~5^h *LT*), respectively, become as strong as 230 m/s at terminators and poles.

Velocities of the strongest downward and upward flows, related to the winds, measure from –9.6 m/s to 4 m/s in the morning and late evening, respectively. They result in significant adiabatic gas heating on the night side and cooling on the day side. In particular, because of day time *upwelling,* the maximum of thermospheric heating displaces from noon on the equator to ~15^h *LT* at $\sim 30^o$ latitude. Such a dynamic pattern manifests substantial departure of the thermal state of Mars' and Venus' thermospheres from the condition of radiative equilibrium and, at the same time, supplies evidence on the importance of both large-scale winds and small-scale processes in the distribution of neutral components.

The simulation results also revealed another interesting fact. It turned out that, when large-scale dynamics is taken into account, the value of the eddy diffusion coefficient beneath the homopause may be taken as low as $\sim 2 \times 10^3$ cm^2/s in order to ensure the best matching of the calculated ratio $n(O)/n(CO_2)$ and measurements obtained at different phases of the solar cycle. This makes the very idea of defining the homopause on Venus and Mars as a fairly sharp interface between the regions of dominance of tur-

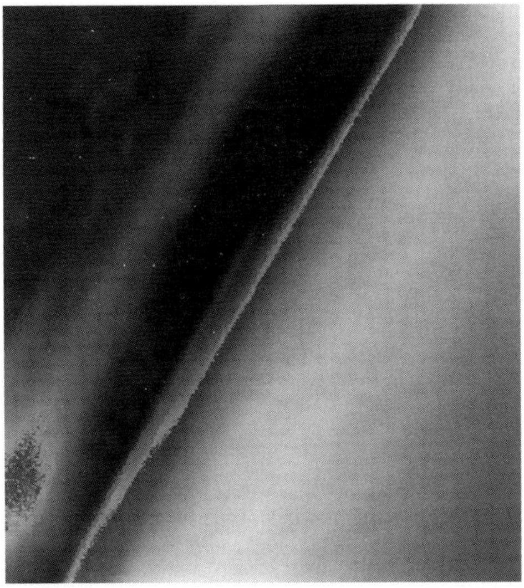

Fig. 1.3.8. Haze at the limb of Titan, the largest satellite of Saturn, observed from 22 000 km. Three cloud layers, with different optical properties apparently formed by aerosols of hydrocarbon ices, are distinguished. An image made by space vehicle *Voyager*, courtesy of *NASA*.

bulent and molecular diffusion doubtful (*Bougher et al., 1988*). Similarly, it poses the question whether a simplified approach is adequate to model a planetary upper atmosphere using the D^T coefficient as a fitting parameter.

1.3.3. The Upper Atmospheres of the Giant Planets

The upper atmosphere of a giant planet, comprising the middle atmosphere which overlies the tropopause, is also mainly formed due to direct absorption by gases and aerosols of solar ultraviolet and *EUV* radiation. This provokes numerous photochemical and chemical reactions which influence the thermal structure above the cloud deck, as well as the formation of ionospheres. In addition to various hydrogen-bearing compounds, these reactions probably involve some hydrocarbons and their derivatives responsible, in particular, for dense hazes in the upper atmospheres of Uranus and Neptune. Similar hazes are assumed to exist on the Saturnian satellite Titan (Figure 1.3.8), having a very dense atmosphere and attracting growing attention as a celestial body with a possible existence of pre-biological forms of complex organic compounds (*Hunten et al., 1990; Owen, 1994*).

Meanwhile, thermospheres of giant planets turned out to be much hotter than could be explained by solar energy absorption models. As an example, the temperature and density altitude profiles in the upper atmosphere of Saturn are shown in Figure 1.3.9. In the middle atmosphere, the temperature is virtually the same as on Jupiter and measures about $140\,K$. At the same time, the mean exospheric temperature on Saturn

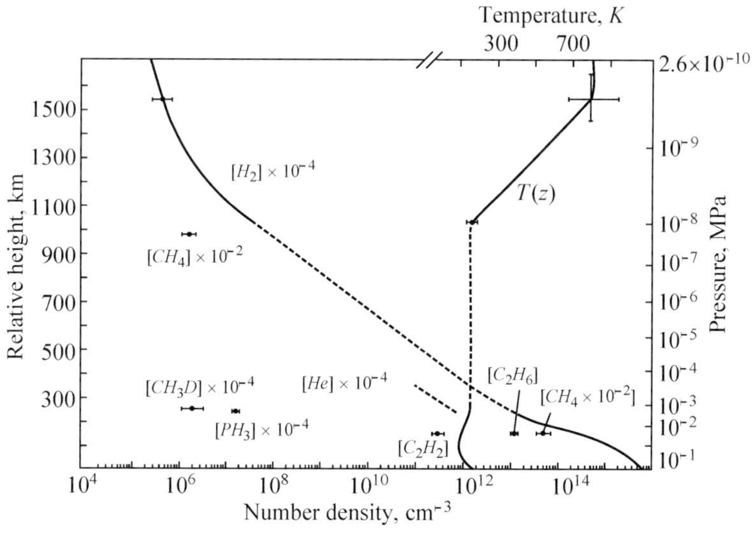

Fig. 1.3.9. Temperature and density profiles of the upper atmosphere of Saturn above 1 bar pressure level. The pressure-height relation above this level is shown in the left and right ordinates. Solid lines – measurements, dotted lines – interpolation. The horizontal bars mark the inaccuracy of the abundances of species (shown in square brackets) measured lower in the atmosphere of northern hemisphere. According to *Atreya et al., 1990*.

(600–800 K) is only slightly lower than on Jupiter where it is approximately 1100 K, whereas the exospheric temperature on Uranus is almost the same as on Saturn, and is as high as 500 K, even on distant Neptune.

As is known, other energy sources are also of importance on Jupiter. First of all, there are protons and electrons injected from the magnetosphere. They are responsible for the intense ultraviolet emissions and polar aurorae even observable from Earth and substantially contributing to the heating of the Jovian thermosphere (*Marov et al., 1997*). However, a similar mechanism hardly is assumed to be operational on other giant planets because the injection of auroral particles from their weaker magnetospheres is unable to supply enough energy to heat their thermospheres and exospheres up to the rather high necessary temperatures. Also the amount of ultraviolet photons rapidly declines with heliocentric distance. Anyway, there is no definite correlation between the observed temperatures and the heliocentric distance, or with the strength of intrinsic magnetic field of the planet. Instead, a correlation between the thermospheric temperature and turbulent exchange efficiency seems a plausible explanation. Saturn, indeed having a relatively low thermospheric temperature, possesses the highest value for the eddy diffusion coefficient, the inverse situation occuring on Uranus, while Neptune occupies an intermediate position (*Bishop et al., 1995*).

This supports the earlier assertion that the height distribution of minor components and the energy exchange at the homopause level of a planetary atmosphere radically depend on the intensity of vertical mixing. The most appropriate values of the eddy diffusion coefficient for different planets were determined proceeding from the experimental data available. This approach was based on three methods: the calculation of atomic hy-

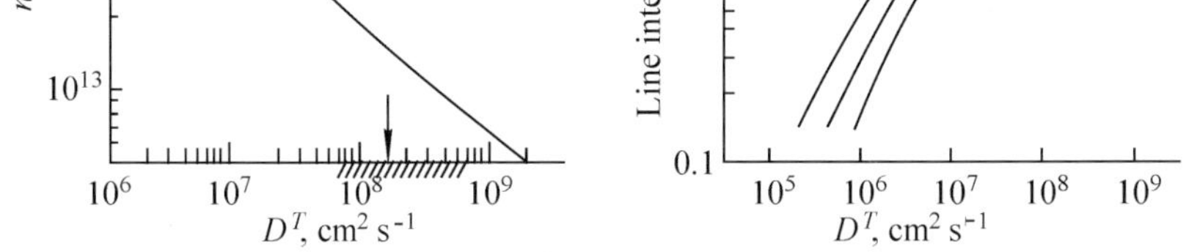

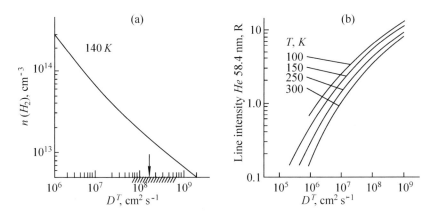

Fig. 1.3.10. Estimates of the eddy diffusion factor, D^T, in the atmosphere of Saturn based on the data from measurements of the H_2-CH_4 and He abundances: **(a)** Dependence of molecular hydrogen density on D^T at the level of Saturn's homopause corresponding to a vertical optical depth in methane of $\tau_{CH4} = 1$ in the $L\alpha$ line; $D^T \approx 2\times10^8$ cm^2/s (arrow) is the best fit to the measured gas density distribution; the shaded area along the horizontal axis takes into account a possible height deviation of the level at which $\tau_{CH4} = 1$; **(b)** Dependence of the emission power in the *HeI* 584 $\overset{o}{A}$ line (in Rayleighs) on D^T for relative helium abundance $X_{He} = 0.06$ and different temperatures at the scattering level. $D^T \approx 10^8$ cm^2/s at the Saturn homopause (~ 1110 km) for $T \approx$ 250 K is the best fit to the measured line intensity. According to *Atreya et al., 1990*.

Table 1.3.1 Turbulent diffusivity in the upper atmosphere of a planet at the homopause level.

Planet	Homopause height, km	Atmosperic pressure, bar	D^T, cm^2/s	Reference
Earth	110 – 115	10^{-7}	10^6	Marov and Kolesnichenko, 1987
Venus	130 – 135	2×10^{-8}	10^7	Von Zahn et al., 1980
Mars	135	2×10^{-10}	$(1.3-4.4)\times10^8$	Nier and McElroy, 1977
Jupiter	440	10^{-6}	$(0.7-2.2)\times10^6$	Atreya et al., 1981; McConnell et al., 1982
Saturn			$(0.7-6.0)\times10^8$ $(4.0-12)\times10^7$	Atreya et al., 1982; Sandel et al.,1982
Titan	3500	6×10^{-10}	$(0.3-3.0)\times10^8$	Smith et al., 1982
Uranus	354 – 390	$(3.7-2.0)\times10^{-5}$	$(0.5-1.0)\times10^4$	Atreya et al., 1991; Encrenaz et al., 1998
Neptune	550	10^{-6}	$(1.6-16)\times10^7$ 5×10^6	Bishop et al., 1995 Yelle et al., 1993;

drogen column abundance above the level of unit optical depth in methane from the observed planetary $L\alpha$ emission; a comparison of the measured height profiles of molecular hydrogen and methane as a heavier gas, whose density sharply starts to decrease at the homopause level, with allowance for photochemical and transport processes; and analysis of the intensity of solar radiation scattering at the $HeI\ 584\ \overset{o}{A}$ line with regard of temperature dependence of the eddy diffusion coefficient D^T and, therefore, of the relative He -content above the base of scattering atmospheric layers. All these methods have given fairly similar estimates of D^T.

Figure 1.3.10 shows an example of the H_2-density dependence upon D^T at the altitude of unit vertical optical depth in methane at $L\alpha$ in the homopause of Saturn, under the assumption $D^T \sim n^{1/2}$, where n is atmospheric number density. Evidently, the best fit to the measured gas distribution corresponds to the very large value of $D^T \approx 2\times10^8$ cm^2/s (*Atreya et al., 1990*), which means that turbulent transport in the lower thermosphere of Saturn is almost 100 times more efficient than on Jupiter. Such a difference could be caused by different thermal structures in their stratospheres and mesospheres or more vigorous tropospheric dynamics on Saturn. The latter could be driven by helium condensation in the process of its separation from hydrogen in deeper layers of the planet, which falls down to the core, which relates to the presence of the large heat flux from Saturn's interior.

The best estimates of the eddy diffusion coefficient D^T at the homopause level for different planets relating to the period of maximum solar activity are summarized in Table 1.3.1. The table is based on the data systematically treated in Atreya et al., (*1990*) and complemented with the updated information about Uranus and Neptune.

Evidently, the lowest values of D^T belong to Earth, Jupiter, and Uranus, while the highest ones are attributed to Mars, Saturn, and Titan. Venus and Neptune feature intermediate values. The foregoing discussion supports an exceptional importance of D^T for assessing the structure and thermal regime of the planetary upper atmosphere. It implies that the problem of more complete and physically reasoned definition of this parameter must be addressed, which is directly related to a detailed analysis of turbulent transfer in the relevant natural media.

As was already mentioned, a number of features inherent to the upper planetary atmosphere, such as the multicomponent mixture it is composed of, variations of the mean molecular weight, the presence of gravity and multiple chemical reactions, influence turbulence in the homosphere and the transitional region. This gives rise to specific effects which should be incorporated into the modeling of aeronomic processes, and prevents one from utilization of the theoretical approach developed for the treatment of turbulent flows in homogeneous incompressible fluids (see, e.g. *Monin and Yaglom, 1971*; *Levellen*, *1980*). The semi-empirical theory of turbulent transfer coefficients for flows in a multicomponent boundary layer (*Ievlev, 1975*) also is not quite applicable. Therefore, the problem of modeling these particular media requires the development of an adequate turbulence theory in multicomponent chemically reacting gas mixtures.

1.4. ASTROPHYSICAL AND COSMOGONIC MODELS

The process of self-organization on the background of turbulent motion is thought to be the major mechanism determining the properties of astrophysical objects in various stages of their evolution, including the origin of galaxies and galactic clusters, the birth of stars from the diffuse medium of gas-dust clouds, and the formation of protoplanetary disks which is followed by the accumulation of planetary systems. These basic concepts and models, developed on this basis, make up a foundation of the stellar and planetary cosmogony and are also referred to as an important cosmological element of the Universe. Unfortunately, many intriguing problems in this challenging field still have remained unresolved.

1.4.1. The Birth and Evolution of Stars

The birth of stars is driven by the mechanism of gravitational condensation in a giant stellar system, that is, in a galaxy involving billions of stars with various ages, masses, luminosities, and chemical compositions. Such systems represent the basic state of substance in the Universe.

A star forms from the collapse of clouds of molecular gas and dust in the interstellar medium, in other words, when a cloud of matter reaching some critical mass begins to accumulate (see, e.g., *Shu et al., 1987; Boss and Vanhala, 2000*) . Stars primarily form in the spiral arms of galaxies; the centers of their formations are genetically related to the massive, rather flat galactic disks. The interstellar gas and dust mixture with a non-uniform density distribution is concentrated in its plane of symmetry. The vast majority of stars form in giant molecular clouds and also in smaller diffuse molecular clouds which bound up the bulk of the material in the interstellar medium.

According to the theory of gravitational instability of matter in rest (*Jeans, 1969; Cassen and Woolum, 1999*), gas condensations form from the interstellar medium because of instability due to random fluctuations. This process is accompanied by the compression, chaotic motions of gas and dust particles, and intense heat release owing to conversion of potential into kinetic energy, as well as by momentum transfer from the forming condensation to peripheral areas (Figure 1.4.1). Significant amounts of dust in a molecular cloud (up to 1% by mass) plays a very important role in the process of star formation, enabling cloud collapse by disposing unwanted energy and providing a formation surface for increasingly complicated molecules from original matter, with the potential to form larger solid bodies around the *protostar*.

When a temperature of several million degrees is reached in the interior of such a protostar, the process of thermonuclear synthesis begins, and the gravity is being counterbalanced by interior gas pressure (gravitational equilibrium). Ultimately, when the compression ceases, the star occupies a definite position at the *Main Sequence* of *Hertzsprung-Russel (HR)* diagram, depending on its luminosity and thus, its mass. It rests in this position preserving its size, mass, and luminosity until the nuclear fuel is exhausted. Release of huge amounts of energy owing to nuclear reactions in the central region of the star is accompanied by radiative energy transfer and convective motions,

Fig. 1.4.1. Orion nebula located at about 1500 light years from Earth, one of the nearest regions of star formation. Its brightness is caused by light emission of new born, hot stars.

stars with small masses being entirely convective, leading to active mixing of matter in the stellar interior. Prominent regions in the *HR* diagram besides the Main Sequence, reflecting the processes of high/low mass star formation and evolution, include blue supergiants, red supergiants, red giants, hot subdwarfs, and white dwarfs.

As follows from the theory of stellar evolution, after completion of the hydrogen burn-out, a star with mass $M \approx (1.4-3)\,M_{\odot}$, where $M_{\odot}$ is mass of the Sun, leaves the Main Sequence and first transforms into a red giant when its luminosity and radius increase. Afterwards, in due course of nuclear fusion in the stellar core (successive helium, carbon, oxygen and silicon burn-out, involving the intermediate formation of respective envelopes and cores and the eventual build up of an iron core), such a star loses its mass and eventually evolves into a white dwarf. Such an object represents a gradually cooling down, degenerated star with mass $M \approx M_{\odot}$. Collapse of a star after formation of the iron core can not be balanced any more by the energy release at the expense of thermonuclear reactions in its central part. The energy of compression is consumed for the decomposition of iron nuclei down to the formation of a neutron core. The pressure in the center is determined by the degeneration of the electron gas, while the gas of atomic residuals determines the density. According to the concept of modern statistical physics, the maximum mass which can be kept by cold electrons, is equal to the *Chandrasekhar limit mass* $M = 5.75\,M_{\odot}/\mu_e$, where μ_e is the number of nucleons per electron (see, e.g., *Bethe, 1966; 1982*).

Fig. 1.4.2. *Crab* nebula, the remnant of the supernova observed by Chinese astronomers in 1054. The nebula structure is a bizarre net of filaments forming an outer shell which expands with a velocity close to $0.1c$ and interacts with the rarefied gas of the interstellar medium. An image obtained in the red line of hydrogen, *Hα*, at the *Kitt Peak* observatory.

The loss of mass during stellar evolution described by the exposed theoretical concepts is supported by observations. The formation of planetary nebulae at slow outflow of matter from red giants and flashes of novae due to sudden release of huge amounts of energy accompanying processes in stellar interiors serve as examples. The upper mass limit of stars evolving into white dwarfs is derived from observations of stellar associations.

More massive stars with $M > (3-5)M_{\odot}$, whose helium nuclei are not in a degenerated state, experience fast collapse at the final stage of evolution and drop their outer envelope owing to the formation of a strong shock wave in the center. In other words, these massive stars, which may only last for a few million years, explode. This is observed as a dazzling supernovae, throwing much of the interior into the neighboring interstellar space. The energy release at such a dramatic event exceeds that of novae thousands of times, and measures more than 10^{40} erg/s. This phenomenon can be described in the framework of the theory of strong explosions, and its propagation in the interstellar gas can be associated with a detonation wave (*Sedov, 1965; Zeldovich and Reiser, 1966; Zeldovich, 1983*).

Extremely complicated configurations form during supernovae explosions due to the interaction processes of gas expanding with sub-relativistic velocities with interstellar medium (Figures 1.4.2, 1.4.3). They are referred to as supernovae remnants. It ap-

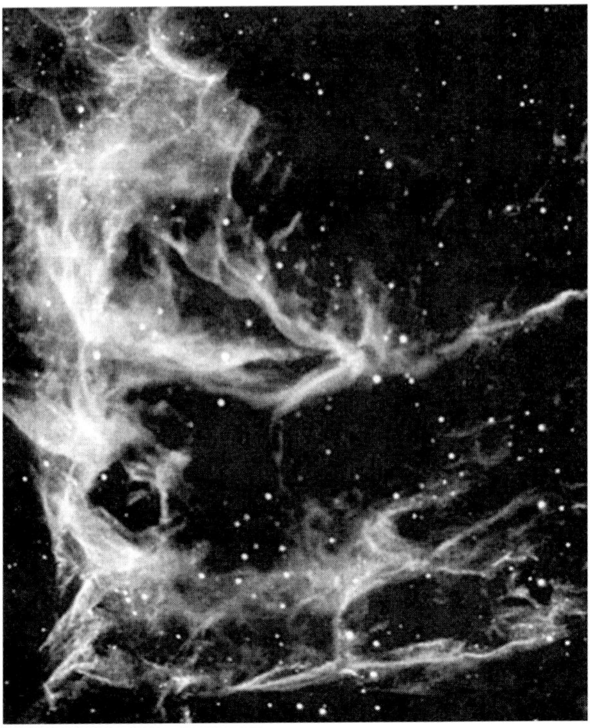

Fig. 1.4.3. *Cygnus* loop, a part of the *Vela* nebula in the *Cygnus* constellation, which is the remnant of a supernova explosion more than 150 000 years ago. The loop patterns formed in the process of expanding gas evolution. An image obtained in the *OIII* λ 5007 $\overset{o}{A}$ line at the *Kitt Peak* observatory.

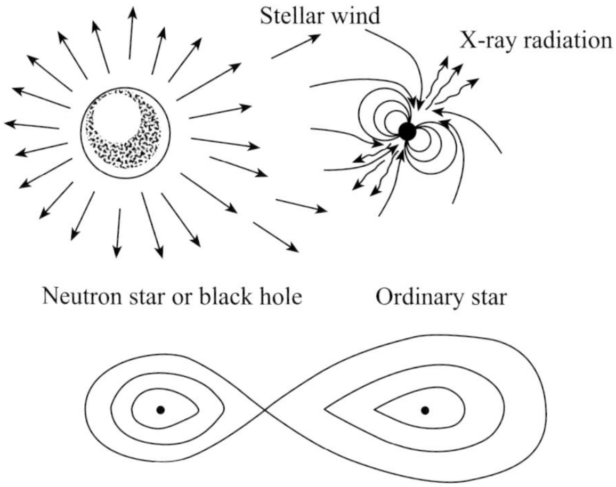

Fig. 1.4.4. The formation scheme of a simply connected region in a close binary system in the vicinity of a black hole and the assumed mechanism of X-ray generation.

pears that *nucleosynthesis* processes both occur in the stellar interiors and during the explosion. These processes are responsible for the enrichment of the interstellar medium with heavy elements and the gradual growth of the mean atomic number Z. The heavy elements predominately concentrate in cosmic dust particles, probably forming in atmospheres of cold giant stars, though numerous polyatomic molecules, including complicated organic compounds, have been detected by radio astronomical methods in the gaseous medium as well. Such a multiphase and multicomponent medium especially is characteristic of the massive, cold, and dense clouds, where the processes of gravitational collapse occur, which are assumed to be an early phase of the formation of stellar clusters and associations (*Pikelner et al, 1976*). Depending on the abundance of heavy elements in the interstellar galactic medium, one distinguishes the conditions during the formation of first and successive generations of stars.

An adequate description of the collapse phenomenon is possible only in terms of the relativistic theory of gravity based on the *Einstein's general theory of relativity*. This theory leads to basically new situations in the relativistic collapse when new phenomena arising from combining the quantum theory of matter and the theory of gravity are taken into account (*Zeldovich and Novikov, 1975*). The core of a supernovae transforms into a neutron star or a black hole, that is, a region of a peculiar state of matter having an infinitely large density and representing a spatially-temporal singularity. Experimental detection of neutron stars and black holes became possible due to radiation emitted by their interaction with a companion (for example, when a normal star loses matter owing to the powerful gravitational attraction towards the neighboring neutron star or black hole). The strongest loss of matter occurs when a star expands during its evolution and the surface reaches the *Roche limit*, i.e. the equipotential surface which forms in a tight binary system with a single focus (Figure 1.4.4).

This case is characteristic of complicated dynamic structures of mass exchange, as follows from numerical gas-dynamic models (*Bisikalo et al.,1997*). The most prominent features of such a structure include the flow of matter from a donor star with generation

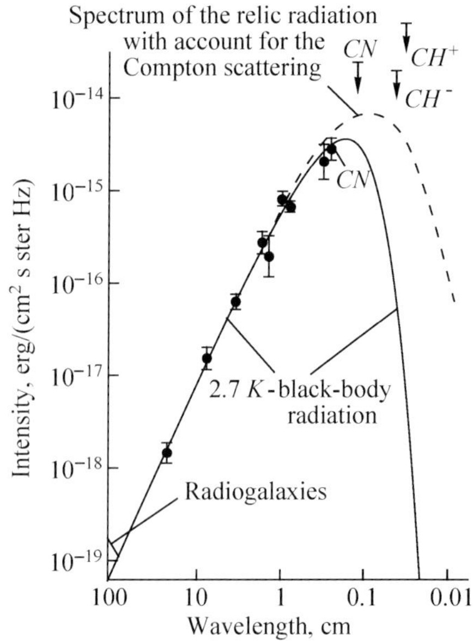

Fig. 1.4.5. The relict radiation filling up the Universe isotropically, which follows the spectrum of blackbody radiation having a temperature of 2.7*K*. Black circles on the curve – experimental points, vertical bars – measurement errors. The dotted line takes into account a possible spectrum distortion for the Compton scattering on hot electrons. Strokes with arrows correspond to the upper limit of temperature of the relict radiation deduced from the levels of population of the interstellar molecules *CN*, CH^-, and CH^+ . According to *Zeldovich and Novikov, 1975.*

of shock waves and tangential discontinuities, as well as the formation of an *accretion disk* with changing stellar wind parameters during the evolution of the system. Strong turbulization of matter and its heating up to the temperature of tens and even hundreds of million degrees accompany either the accretion process towards the surface of a neutron star possessing a strong magnetic field which directs the flow towards the poles, or the formation of a gaseous disk in a rotating binary system with a black hole inside (disk accretion). In every scenario, the directed Bremstrahlung of hot plasma at X-ray wavelengths modulated by the rotational effects of both the neutron star itself and the system as a whole sets up.

Radiation of the galactic interstellar gas, occurring mainly in the state of neutral hydrogen atoms with temperatures of tens to thousands of degrees, is observed at radio wavelengths. Modeling the structure and evolution of galaxies and the entire Universe is closely related to studying the nature of radio lines from neutral hydrogen and excited diatomic molecules in sources of ultra high-frequency radiowaves, that is, in cosmic masers concentrated in gas-dust nebulae. Also of primary interest is the nature of relict radiation (Figure 1.4.5). Detection of this radiation uniformly filling up the whole Universe stimulated the development of the hot Universe concept and the *Big Bang* theory which implicates that the Universe in the past got through a stage of being an enormously dense, hot plasma in complete thermodynamic equilibrium with Planck's radiation spectrum. Gradual cooling of the plasma in due course of its expansion from the moment of singularity perfectly corresponds to the equilibrium spectrum of the contemporary observed radiation with temperature $T = 2.7K$ (*Zeldovich and Novikov, 1975; Doroshkevich et al., 1976*).

The relativistic theory of the homogeneous isotropic Universe, based on the Einstein's gravity equations and Friedman's cosmological equations, is now generally acknowledged. Meanwhile, there is no unambiguous answer to the question whether this expansion will proceed infinitely, or it will be replaced by contraction after a certain amount of time (the theory of an *oscillating* Universe). The highlights for solving the problem are studying the relatively small spatial intensity variations of the relict radiation together with the data on radio emission of hot gas clouds in the regions of galaxy accumulations, as well as investigating the storage and properties of *dark matter*. The latter is most irritating to astronomers because we can not even see the major part (at least 90%) of the Universe's ordinary, or baryonic, matter left behind by the Big Bang. Computer modeling suggests (*Cen and Ostriker, 1998*) that the missing baryons (protons and neutrons) made in this event formed sprawling complexes of hydrogen plasma clouds. These primordial clouds condensed over time into a vast, filamentary network of plasma clouds, or a sort of cosmic cobweb that now presumably links galaxies and galaxy clusters. The challenge is to find out from observations how much matter this plasma web really contains. This would enable the answering of the intriguing question: which of the models of the Universe, opened or closed, is the most plausible.

Fig. 1.4.6. The spiral galaxy *M31* (*NGC 224*), the *Andromeda* nebula, located at 2 millions light years from the Earth. This giant vortex of gas-dust clouds comprises nearly 300 billions of stars and resembles by its size and shape our Galaxy, the *Milky Way*. It has two rather small satellites with an elliptical shape. The image is made in blue light. According to *The Hubble Atlas of Galaxies, 1961*.

1.4.2. The Role of Turbulence in the Evolution of the Universe

The above discussion is intimately related with the problem of formation of galaxies (Figure 1.4.6) that is considered as a composed element of evolution of the hot Universe. The process is focused on density variations and eddy motions in the primary plasma involving electrons, protons, positively charged helium nuclei, and photons. After the completion of recombination, dense clouds of neutral gas are eventually formed with emerging shock waves and turbulence in the compressing gas. The role of turbulence during these events is thought to be decisive in the framework of eddy theory of the origin of galaxies which is historically rooted in the solar system cosmonogy. According to this theory, the expanding early Universe was both anisotropically and dynamically structured and had a significant eddy component (*Ozernoy and Chibisov, 1970; Ozernoy, 1976*). Such a pre-galactic (or cosmological) turbulence, which is referred to as photonic, should result in density perturbations which are eventually amplified at the expense of gravitational instability. This process is thought to be responsible for the formation of *protogalaxies*. However, the eddy theory, including its versions of strong and weak turbulence, meets a number of difficulties because strong turbulence contradicts experimental evidence, while primary eddies in the weak turbulence scenario cannot ensure the observed rotation of galaxies.

The adiabatic theory of galaxy formation (*Syunyaev and Zeldovich, 1972; Doroshkevich et al., 1976*) seems more justified. It claims that turbulence emerges in a natural way and is an important component of galactic and intergalactic gas hydrodynamics. Successive processes involve the origin of dense, flattened gas clouds. Characteristic features of gas are perturbation anisotropy in the stress tensor at the expense of tidal forces; adiabatic gas compression in the vicinity of some point along one of the gas flow directions; and subsequent accretion of a major amount matter on the already compressed gas. As a result, a peculiar matter distribution emerges in the formed disk with a

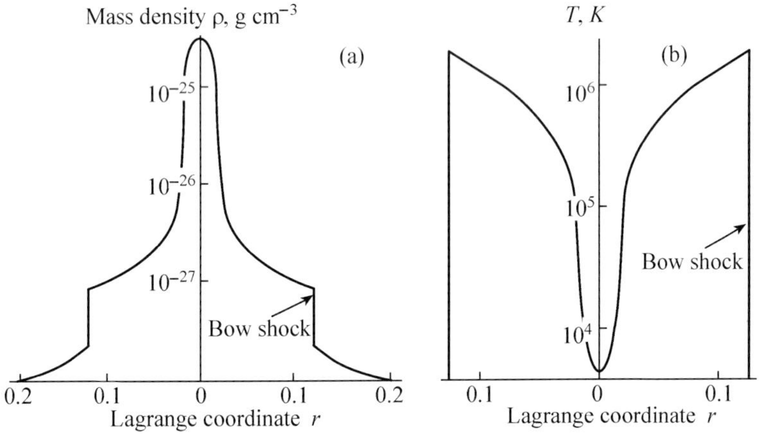

Fig. 1.4.7. Model origin of dense flattening gas clouds in the primordial matter of the Universe – "pancakes" presumably giving birth to clusters of galaxies. The matter distribution in such a disk in the phase of fast adiabatic compression in small vicinity of the point of instability resulted in the formation of a shock wave, is shown as density **(a)** and temperature **(b)** profiles in Lagrangian coordinates orthogonal to the plane of the pancake. According to *Doroshkevich et al., 1976.*

sharp density maximum in the center that steeply falls down outwards (Figure 1.4.7). The configurations, resembling pancakes, determine successive evolution of matter in galaxies and stars. The *pancake model* assumes that powerful eddy motions and turbulization of both hot and cooling down gas occurs. In the hot gas, these processes are maintained by the flow of eddy velocity through the front of a shock wave, while in the cooling gas by thermal instability and non-homogeneity in density. The latter results in desintegration of the pancake into individual clouds. Therefore, these notions imply that turbulent motions strongly influence the parameters of emerging galaxies, especially in the pancake's outer regions.

Turbulence seems to be important to the dynamical structure of the shaped galaxies. In particular, giant eddies of anticyclone types were discovered recently in the spiral galaxies, which were earlier suggested based on laboratory experiments with shallow water (*Fridman and Choruzhiy, 1998*). They are caused by centrifugal instability developed in gaseous disks and are responsible for the formation of both density waves in the galactic spiral arms and eddies. Such a mechanism is relevant to disks having appreciable (~10–15%) relative leaps in the rotational velocity curve; then hydrodynamic instability prevails over that due to gravity, which is the case for spiral galaxies. These leaps are associated with the centers of eddies. Evidently, the observed patterns are an example of shear current instabilities comprising, as a particular case, the centrifugal instability.

1.4.3. The Origin of Planetary Systems

The problems of galaxy and star evolution directly relate to the problem of origin of the solar system and planetary systems around other stars. Detection of such systems having a large variety of possible configurations, most recently received experimental support (*Major and Queloz., 1995; Major et al, 1997; Butler and Marcy, 1997; Marcy and Butler,1998; Schneider, 1999*) and greatly advanced human's exploration of the Universe. However, the masses (ranging from that of Saturn up to more than ten times that of Jupiter), the very short orbital periods at close distances to the star they orbit, and the very eccentric orbits of the majority of the companions detected so far, are not consistent with the principal characteristics of our own solar system.

All this seems to point to the anomalous nature of these presumably gaseous objects which are yet to be distinguished from brown dwarfs – failed stars unable to sustain hydrogen fusion in their cores. Anyway, these peculiar planets do not fall into our traditional scenario of the origin of the solar system, involving gaseous giants formating at relatively great distances from the Sun at low temperatures allowing ice and frozen gases to accumulate. Instead, because of the proximity to the mother star, the predicted effective temperatures of the extrasolar planets exceed 1000–1500K. Thus some extraordinary features, including potentially important chemical species to condense out on their surfaces and in their atmospheres, are expected assuming solar composition and reasonable constraints on albedo, radii, and age (*Guillot, 1999*). The latter is focused on the concurrence of migration and formation timescales based on the standard α-disk accretion model involving gaseous disk – protoplanet tidal interaction (*Terquem et al., 2000*).

As an initial concept of planetary cosmogony, one used to assume that in the process of stellar collapse a substantial portion of the cloud material, possessing noticeable angular momentum, keeps to orbit around the central condensation and leaves behind as a protoplanetary disk. Basically, these ideas rely on the *Kant-Laplace* hypothesis about simultaneous formation of the Sun and protoplanetary cloud and its rotational instability responsible for the successive separation of flat concentric rings at the periphery of the cloud. The planets later on condensed from the rings' matter. It could be that matter from outer areas of the cloud continued to accrete onto the disk, which resulted in strong turbulization of the gas-dust medium because of a mismatch between the specific angular momentum of the incident substance and that of the substance of the disk experiencing Keppler rotation (*Lissauer, 1975*). Thus the velocity gradient associated with differential angular rotation of cosmic matter around the Sun is assumed to be the main source of turbulence in the protoplanetary disk system.

While the model notions about the origin of protoplanetary nebulae are supported by observations, the mechanical and cosmochemical features of the solar system allows to highlight the problem of solid bodies' origin. Basically, the present patterns in the system of planets and satellites indicate a uniform formation process. Experimental evidence on the properties and/or composition of surfaces of the planets and small bodies,

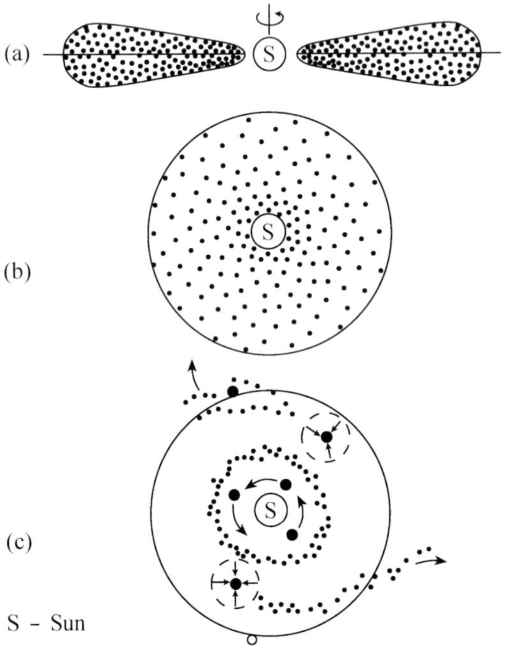

Fig.1.4.8. Model representation of the sequence of accretion of planetary size bodies from planetesimals: **(a)** Condensation of dust particles from the gas matter of the protoplanetary nebula and disk formation due to their sedimentation to the central plane; **(b)** Origin of primary clusters – embryos due to gravitational instability and their growth to planetesimals – comet-like icy bodies of a few kilometers in size; **(c)** Accumulation of planets in the process of accretion of planetesimals, followed by their partial ejection from the domain of the giant planets to the periphery of the solar system where the Kuiper belt and the Oort cloud were formed. According to *Greenberg, 1989*.

in comparison with the samples of their germs and debris – meteorites, serve as a clue, in attempt to reconstruct probable scenario and mechanisms of this formation process.

The idea of accumulating planets from a primordial cold gas-dust disk after its separation from the central condensation goes back to the 1950–1960 s (*Schmidt, 1957; Safronov, 1969*). It involves rather delicate dynamics of gravitating bodies after development of perturbations in the rotating dust subdisk near the central plane; its desintegration owing to gravitational instability partially damped by shear-generated turbulence and differential motions induced by gas drag; settling down of dust aggregates toward the disk central plane and collisional coagulation; the successive dust accretion into germ clots and then into bodies with intermediate dimensions (planetesimals); and gradually sweeping up smaller bodies by larger ones during the evolution of a swarm (Figure 1.4.8). As this process progresses, the rotation velocity of the forming bodies becomes slower than the Kepler velocity because of gained centrifugal acceleration and hence, a decrease in the gravitational attraction. This enhances their braking in the ambient gas and promotes escalation of this process. Exchange of initial matter in the radial direction should also play an important role. Its efficiency places some constraints on various evolution scenarios for the disk and the degree of chaotization when the original planetary germs form (*Greenberg, 1989*).

Further, it is assumed that the original planetesimals proceeded to grow fast enough (within thousands of years) through the concurrence of the collisional and gas drag processes, their relative efficiency being constrained by body sizes: for relatively small bodies impact velocities decreased with size and motions were dominated by gas drag. Gravitational interactions of larger bodies promoted the velocity increase due to mutual perturbations, which ultimately resulted in their "runaway" growth and the formation of numerous (up to hundreds) Moon to Mars-sized planetary embryos in the zone of the inner planets (*Weidenschilling, 1980;2000; Weidenschilling and Cuzzi, 1993; Artymowicz, 2000*). These bodies finally accreted into a few planets through catastrophic collisional processes (on a time scale of 10^7 to 10^8 years) giving rise to the contemporary configuration of the solar system.

Such a scenario follows a certain sequence of processes which underlie the origin of the solar system. Another approach, challenged by new discoveries of enigmatic planetary-mass companions around other stars, pursues the idea that chaotic dynamics played a key role in the formation, evolution and survival of planetary systems. The basic argument emerged from computer simulations is the extreme sensitivity of the model (involving a gravitational instability in the protoplanetary disk, disk-planet interactions and numerous collisions) to the initial conditions (*Malhorta,1999; Armitage and Hansen, 1999*). Thus the system evolves chaotically into a rather random configuration which might rather resemble some of the putative extrasolar planetary systems besides our solar system which appears to be completely unique.

The significance of collisional processes in every scenario of planetary system formation should be emphasized. There is progressively collected evidence of their importance in the following evolution of major planets and small bodies throughout the whole solar system history. It is supported by well established migration of asteroids and comets which scared the planetary surfaces and shaped the atmospheres of the Earth and the other inner planets by deposition of cometary-like volatiles through impact processes. These impacts, at later times, probably have repeatedly interfered with the Earth's biological evolution and may once have been essential for starting this evolution. At the same time, comets provide the link of the solar system with the Galaxy by probing the material of the protosolar cloud and thus providing a coupling of planetary cosmogony and Galactic evolution (*Marov and Rickman, 2001*).

There is some evidence in support of the idea that the formation of the protosolar nebula was triggered by a supernova explosion. This event might have occured in the neighbourhood of a compact gas cloud that had been set up due to fragmentation of a more massive gas accumulation. Short-lived radionuclides found in the *Allende* meteorite and, in particular, the correlation between the exhausted isotope ^{26}Al (whose presence in the early solar system is established reliably) and its daughter isotope ^{26}Mg served as a key argument (*Lee et al., 1976; Wasserburg and Papanastasiu, 1982*). With regard to theoretical concepts on mutual relations between the isotopes ^{24}Mg and ^{26}Mg, and ^{26}Al and ^{27}Al, the time lapse between an injection into the protosolar nebula of the products of nucleosynthesis accompanying such an explosion, and the formation of solid bodies in the solar system was deduced.

This idea is favored also by the results of modeling indicating that a pressure excess is necessary to cause gravitational collapse of a diffuse cloud similar to the parent cloud of the solar system and separation of the disk. Such an excess could be provided at the expense of shock waves generated by the supernova explosion. It is interesting that long before the studying of the Allende meteorite, famous russian astronomer Fesenkov (*1973*) has directed attention to the possibility of a similar relation. He wrote: "... The presence of desintegration products of various short-lived isotopes in meteorites supports the supposition that shortly before formation of planets the explosion of a Supernova took place which could promote the arising of various non-uniformities in the medium of the protoplanetary nebula." In view of the observable frequency of emerging supernovas in the Galaxy, such an event should not be considered as extraordinary. In particular, an assumption about quite recent (~10^5 years ago) supernova explosion in a fairly close (some tens of parsecs) neighbourhood of the Sun was put forward proceeding from the intensity analysis of ^{26}Al gamma-line and the increased background noise of space rays at soft X-ray wavelengths. Hence, it is possible that the solar system is now inside a gas cloud of increased density which formed as a result of such an explosion (*Clayton et al., 1986*).

The problem of planetary system formation directly relates to the mechanism of angular momentum ($J_\odot$) transfer from a collapsing Sun-like star to its protoplanetary disk. For a protosolar nebula with uniform density the angular momentum value is in the range of $10^{52} < J_\odot < 10^{53}$ g cm^2 s^{-1} and can be an order of magnitude higher (an upper limit) if mass concentrates in the disk's center (*Ruzmaikina and Makalkin, 1991*). Most probably, such transfer was conditioned by turbulent viscosity in the rotating, convectively unstable gas disk which has determined the time scale of its expansion (*Safronov, 1969; 1991; Ruden and Pollack, 1991; Sterzik and Morfill, 1994*). In what has followed upon the accumulation of germs, no less important role could have been played by turbulent eddies which promoted particles to accelerate and integrate more easily into "rings" of matter.

Another way to loose the angular momentum of the Sun at this early stage of its evolution relates to shear motions in the differentially rotating Keplerian disk responsible for non-linear instabilities and turbulence set up (*Shakura and Syunyaev, 1973; Zeldovich, 1981; Dulbrulle, 1993*), though such enhanced hydrodynamical turbulence in the presence of stabilizing Coriolis forces, is doubted (*Balbus and Hawley, 2000*). In a partially ionized medium, the situation mainly could be controlled by electromagnetic forces or local shear instabilities arising in a polhode magnetic field (*Alfven, 1978; Balbus and Hawley, 1991*). In other words, gas magnetic coupling and, in particular the ionization fraction, may significantly influence the dynamical state of a protoplanetary nebula at different distances from the Sun and maintain turbulence.

Which also must not be ruled out is the possibility of carry-away of surplus rotational moment during the compression phase, which is promoted by a magnetic field, though it is a rather weak in stars (*Pikelner and Kaplan, 1976*). In the following evolutionary phases, angular momentum can be carried away by the stellar wind, and as the sum of angular momentum of plasma per unit mass and the momentum linked to magnetic stresses keeps constant, the rotational moment is transferred to the surface through magnetic stresses. This results in a gradual decrease of the rotational velocity of the star (*Baranov and Krasnobaev, 1977*). However, in none of the above models, including turbulent convection, it is not yet possible to find an adequate mechanism for the emergence of effective viscosity in order to explain transfer of angular momentum $J_{\odot}$ in the radial direction from the central condensation.

Basically, conditions to set up and maintain shear turbulence in various stages of the Keplerian disk evolution should be considered for a two-phase (gas-dust) medium having differential rotational velocity ω, varying the relative abundance of dust particles η (so that the densities of gas ρ_g and dust ρ_d are related as $\rho_d = \eta\, \rho_g$) and their size distribution taking into account probable coagulation influencing the optical properties (opacity) of the medium*). Here we address only the general concept; a more detailed analysis is given in Chapter 9.

Proceeding from the reasons of dimensionality for a turbulent shear layer of thickness z, rotating with an angular velocity ω, it may be supposed that an eddy, originating from a vertical gradient of the rotational velocity $dv/dz = \omega = \ell\, \boldsymbol{Ro}$ (where $\boldsymbol{Ro}$ is the Rossby number) at $\boldsymbol{Ro} = const$, adapts to the angular velocity of the entire disk. This, however, can not be the case in the presence of dust particles (*Safronov, 1996; Cuzzi et al., 1994*). Indeed, the gradient dv/dz leads to settling down of particles near the plane $z = 0$ and the formation of a dust subdisk. This is accompanied by the arising of radial drift in the disk and by changes in thermal conditions at the expense of dissipation of turbulent energy. Obviously, turbulence, being essentially absent under conditions of complete mixing at $dv/dz = 0$, should increase while growing particles are settling down. As dv/dz increases up to $\ell\, \boldsymbol{Ro}$, the settling velocity and the time scale of these competing processes will change, which, in its turn, would affect conditions of gravitational instability determining the growth of larger bodies in the subdisk.

Attention must be paid to the fact that turbulence, generated at the boundaries of layers in the protoplanetary disk, conforms in its character to the parameters of the Eckman boundary layer with thickness $\delta \sim (\nu^T/\omega)^{1/2} \sim \Delta v / \omega\, \boldsymbol{Re}^T$, where, as before, ν^T is the turbulent viscosity, Δv is the difference between the gas disk rotational ve-

---

*) The opacity is determined mainly by micron size particles. Therefore, the heat exchange conditions vary with particle growth. This would result in a reduced convective instability and intensity of turbulent motions. It is possible, however, to assume that the duration of this process was significant and exceeded the time of particle shedding from the interstellar medium on the disk outer envelope because, apart from agglomeration, it was also controlled by transfer and evaporation in its internal regions subjected to convection (see, for details, *Nakamura et al.,1994,* and *Makalkin and Dorofeeva, 1996*).

locity and the Keplerian velocity of dust particles, and $\boldsymbol{Re}^T = \Delta v\delta / \nu^T$ is the turbulent Reynolds number (*Safronov, 1991; Cassen, 1994*). In this case, the mixing length is $\sim\delta$ and the rate of turbulent mixing is $\nu^T \sim \delta^2\omega$. In framework of the majority of the discussed evolutionary accretion disk models for the young Sun, the factor ν^T conforms to a greater velocity than the relative particle velocity, v_r. This means that the probability of integration (cohesion) of particles is higher than their fragmentation during collisionary processes, which supports the idea of gradual growth of planetary germs (planetesimals) in the dust subdisk.

Invoking the mechanism of shear turbulence supports the idea on feasibility of annular compression of the flat protoplanetary cloud followed by coagulation of planets from originally loose gas-dust clots filling a significant part of their gravisphere and slowly contracting due to inner gravitational forces (*Eneev and Kozlov, 1981a; 1981b; Myasnikov and Titorenko, 1994*). As was found from numerical experiments based on this alternative model, the very effect of annular compression of self-gravitating clots is independent of dimensions and masses of initial cloud bodies and an assumption about significant particle eccentricities in exhaustion process is not necessary. Also it is interesting to emphasize that tidal evolution at this early evolutionary phase has determined, in a natural way, the rotational motion of all planets including the inverse rotation of Venus and Uranus. During the following compression, this evolution has changed the periods and axial inclinations of their intrinsic rotation to their now-a-day values. However, the stability problem of the postulated loose, extended bodies being a rationale of the model, requires additional treatment to be justified.

1.4.4. Plasma Turbulence

The nature and evolution of astrophysical objects we have addressed are closely related to problems of plasma turbulence in space environment. Such turbulence is excited under unstable conditions within the plasma itself, for example, due to instabilities in counter configurations of magnetic fields and rapidly varying processes on the Sun, especially in photospheric plasma, flare and coronal mass ejection (*CME*) areas. Other examples are: instabilities of fast particles in the solar wind when traveling through the heliosphere and in the vicinity of shock waves (shock crossings) under interaction with planets or comets; anisotropic particle distributions by velocities in the magnetosphere of a planet, within relativistic streams in the solar corona or in the magnetosphere of a pulsar; plasma transfer through the magnetopause, accompanied by generation of hydromagnetic waves by the shock and/or ion pick up processes in the magnetosheath region where the solar wind interacts with a planet, followed by downstream turbulence and a magnetic field lines reconnection mechanism directly driven by this turbulence; magnetosphere-ionosphere coupling feedback instabilities related to substorm onset and auroral arc generation; turbulence accompany triggering mechanisms and an entire sequence of events for a substorm, in particular, turbulence in current disruption (*CD*) magnetotail regions causing efficient plasma and magnetic field transport and energy conversion from the magnetic field to the particles and *vise versa*; various turbulence in ion-acoustic solitary waves in plasmas; etc. (see, e.g., *Pudovkin and Semenov, 1985; Brinkmann et al., 1990; Kallendrode, 1998*)

It is worthwhile to note that plasma turbulence is distinguished from usual hydrodynamic turbulence by a broader spectrum of mutually interacting plasma waves (such as *Langmuir, ion-acoustic, Alfven,* and *spiral* waves) and by resonant wave and particle interaction determined by induced radiation, absorption, and scattering of waves by particles. Such interactions cause collisionless wave dissipation in case of equilibrium and the emergence of instabilities in non-equilibrium plasmas which lead to the development of turbulent processes. As this takes place, the dispersion properties of waves can vary and new modes of collective motions in the plasma occur. These can be described in terms of a phenomenological theory relying on the results of numerical simulations.

The development scales of plasma instabilities and concurrent processes of wave and particle interactions in a strong turbulence medium can be judged from the processes observed on the photosphere surface and in the atmosphere of the Sun, as a typical *G2*-star on the Main Sequence in the *HR* diagram. Both macro- and microturbulence parameters are constrained by the effects exerted on spectral lines by granular (convective) motions and by the value of the integrated disk flux, respectively, inferred from observations of the Sun and other solar-type stars and dwarfs (*Gustafsson, 1998*).

In Figure 1.4.9 a photograph of the Sun obtained in X-rays, is shown. Here one can see the dark areas of coronal holes, from which hot, solar wind plasma flows out

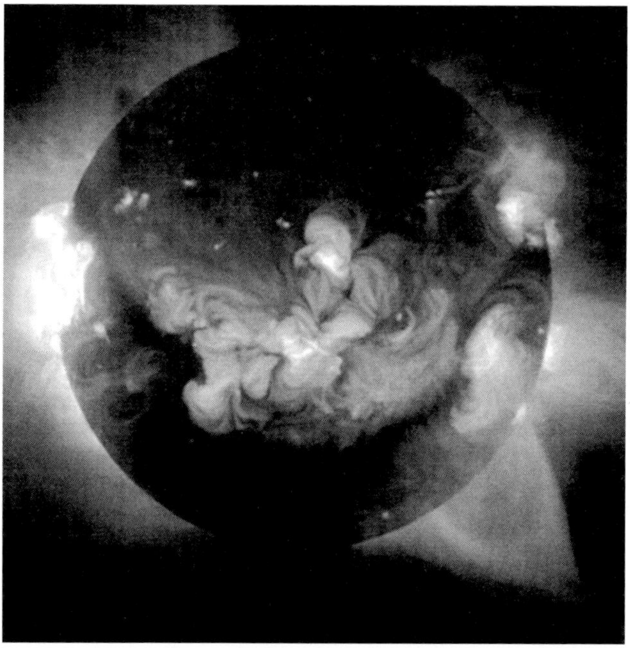

Fig. 1.4.9. The Sun in X-rays. Large bright structures with irregular shapes represent the activity regions where the closed magnetic force lines are capable to hold a hot highly turbulized gas having a temperature of $\sim 10^6 K$. Dark regions are associated with the coronal holes with disconnected topology of magnetic force lines where solar wind plasma outflow occurs. At the left side of the disk the development of a powerful solar burst is seen. An image of the space vehicle *Yukon* (England, USA, Japan), courtesy of *NASA* .

continuously, and numerous active areas related to the diverse magnetic fields topology. A strong concentration of magnetic energy and origin of electric current systems in the chromosphere result in powerful perturbations (flares), together with the formation of shock waves. As electric current layers are destroyed and field structures change, explosive releases of electromagnetic energy occur due to the non-uniformity of motions in strongly magnetized collisionless plasma; this leads to acceleration of particles, generation of powerful fluxes at short wavelength, and ejection of hot plasma clots (*Akasofu and Chapman, 1974*). The matter with frozen-in magnetic fields "pulled out" of the solar photosphere as giant coronal loops, dramatically impacts the state and structure of interplanetary space, physical processes in close vicinity to the planets, and geophysical phenomena.

Summary

- *Turbulence represents a wide-spread and sophisticated physical phenomenon occurring in various natural media and engineering systems. Atmospheres of the planets of the solar system, and in particular, their rarefied gas envelopes which are referred to as an interface between the dense atmospheres and outer space, where numerous aeronomic processes occur, are characteristic examples of such turbulized natural media.*

- *Multicomponent turbulence plays an important role in forming the structure and properties of various astrophysical objects, such as galaxies and stars in their different evolutionary phases, planetary and cometary atmospheres, protoplanetary clouds and accretion disks, and other cosmogony scenarios and models. Diffusion of minor atmospheric species depending on the structure and energy distribution of turbulent motions at different spatial scales, is a key problem in protecting the environment.*

- *Detailed study of multicomponent turbulent media, being addressed to specific features of natural and space environments and their adequate description, is focused on refined theories of such media, as a basis for the development of advanced mathematical models describing the phenomena and processes involved.*

CHAPTER 2

REGULAR MOTION OF GASEOUS MIXTURES INVOLVING PHYSICOCHEMICAL INTERACTIONS OF THE COMPONENTS

In-depth study of the correlated physical, chemical, dynamical, and radiative processes in the turbulent multicomponent medium represents an extremely sophisticated problem. The most complete and rigorous examination of such a medium can be carried out based on the kinetic theory of polyatomic ionized gaseous mixtures. The system of Boltzmann's generalized integro-differential equations for distribution functions describing particles of each sort in a mixture with retained integrals for both elastic and inelastic collisions in the right part, being supplemented by the radiative transfer equation and the Maxwell equations for electromagnetic fields, is most appropriate for the study. Such an approach was elaborately developed in a monograph of the authors (*Marov and Kolesnichenko, 1987*). They applied Chapman-Enskog's generalized method for solving the system of gas-kinetic equations for reacting mixtures and investigated in detail the efficiency of this approach. It was admitted, however, that a number of simplifications usually introduced in order to solve complex aeronomic problems practically (such as accounting for paired collisions of interacting molecules only, assuming a lack of internal structure in colliding particles when assessing molecular exchange coefficients, etc.), substantially reduce the advantages inherent to these kinetic equations.

At the same time, from a macroscopic viewpoint, a rarefied gas mixture (such as a planetary outer gas envelope) can be considered as a continuous chemically active medium and then, for its adequate mathematical description, the techniques for multicomponent mixture mechanics can be applied. This permits one to obtain a system of hydrodynamical equations with all necessary closing relationships in a phenomenological way, based on the principles of non-equilibrium thermodynamics. Such a phenomenological approach allows us to model both laminar and averaged turbulent flows (*Kolesnichenko, 1980*). Thus the system of hydrodynamical equations for the description of laminar motions in multicomponent reacting continuous media, considered in the present chapter, is assumed to be also valid for the treatment of instantaneous motions in turbulent gaseous mixtures.

2.1. INITIAL BALANCE EQUATIONS AND CONSERVATION LAWS FOR REGULAR MOTIONS IN GASEOUS MIXTURES

Let us consider a reacting gaseous mixture in the upper atmosphere of a planet consisting of N components. The Greek alphabet characters α,β and γ $(\alpha,\beta,\gamma=1,2,...,N)$ will be assigned to the hydrodynamic parameters as indexes of different mixture components. The mixture will macroscopically be viewed as a continuum with complicated properties characterized by the system of *state variables*. This system involves the mean mass density $\rho(\boldsymbol{r},t)$, the temperature $T(\boldsymbol{r},t)$, the thermodynamic pressure $p(\boldsymbol{r},t)$ and the number densities $n_\alpha(\boldsymbol{r},t)$ $(\alpha = 1,2,...,N)$ of chemically active species in the mixture. The state variables are functions of time t and spatial coordinates x, y, z in an arbitrary coordinate system fixed with respect to the planet.

As is known (*Landau and Lifshits, 1988*), the hydrodynamic model of a reacting mixture is based on a bound set of equations including: non-stationary differential equations of the mechanics of continua featuring mass, energy, and momentum conservation laws; relevant equations of state for pressure and internal energy, called *thermal and caloric* equations, respectively; and defining (*rheological*) relations for various thermodynamic fluxes, including dissipation and heat fluxes, viscous stress tensors, etc). Besides, it is necessary to know expressions for every possible thermodynamic function, such as internal energy, enthalpy, different heat capacities of the components, etc., as well as formulas for various molecular exchange coefficients and, for chemically non-equilibrium media, also chemical reaction rate coefficients. To solve the set of partial differential equations, the initial and boundary conditions describing the geometry of the thermodynamic system (for a material object having clearly defined boundaries) and the mass, momentum, and energy exchanges between the system and the environment must be formulated. Such boundary conditions are to be taken *ad hoc* for each specific hydrodynamic problem.

A General View of the Balance Equation. Let us introduce parameters describing the state of a continuous medium and its dynamics, in particular, an instantaneous state of a turbulized mixture, as *defining parameters*. The equation of substantial balance for any defining parameter of a mixture $A(\boldsymbol{r},t)$ can be written in the following general way (*Dyarmati, 1970*):

$$\rho \frac{dA}{dt} = -\partial_j J_{(A)j} + \sigma_{(A)} \tag{2.1.1}$$

Here $\rho(\boldsymbol{r},t)$ is the total mass density of the mixture (see (2.1.9)); A is the specific (local) value of any scalar quantity $\Lambda = \int_w \rho A dw$ (for example, mass, energy, entropy, etc.), which is to be balanced; w is the arbitrary volume of the system; $J_{(A)j}(\boldsymbol{r},t)$ are the vector components of the substantial molecular flux density of parameter A (that is, the total value of all kinds of transport of property A across a unit surface, including the convective component $\rho A V_j$, as well as fluxes defining the surface impact on an integral quantity Λ); $\sigma_{(A)}(\boldsymbol{r},t)$ is the volume density of the source of feature A, in other words, the rate of generation ($\sigma_{(A)} > 0$) or destruction ($\sigma_{(A)} < 0$) of quantity A in a unit volume per unit time.

The substantial (*Lagrangian*) derivative with respect to time (the *operator equation*) is written in the form

$$\frac{d(..)}{dt} = \partial_t(..) + V_j \partial_j(..), \qquad \partial_j(..) \equiv \frac{\partial(..)}{\partial x_j}, \quad \partial_t \equiv \frac{\partial(..)}{\partial t}, \tag{2.1.2}$$

where $V_j(\boldsymbol{r},t)$ are the hydrodynamic velocity components of the mixture (see (2.1.9)). To reduce the writing of formulas with respect to repeated indexes *i, j* and *k* (*i, j, k* =1,2,3), relating to the Cartesian coordinates of the point $\boldsymbol{r}$, summation is carried out from now on. Parameter A is a scalar, being the Cartesian component of a vector or tensor; in the case of turbulent flow, it corresponds to the instantaneous value of a field quantity of an arbitrary tensor rank). The flux $J_{(A)j}$ represents a tensor whose rank is one order higher than that of the property *A,* while the generation rate $\sigma_{(A)}$ is a tensor of the same order.

Let us notice, that, generally speaking, quantities $J_{(A)j}$ and $\sigma_{(A)}$ can not be defined unequivocally, as substitution of the flux $J_{(A)j}$ into $(J_{(A)j}+B_j)$ and the source $\sigma_{(A)}$ into $(\sigma_{(A)}+\partial_j B_j)$ (where B_j is an arbitrary vector) will not change the balance equation (2.1.1). In each particular case the vector B_j should be chosen with regard to the physical sense of quantity A and, with other things being equal, one should be guided by reason of convenience.

The following *operator relation* will be widely used later on

$$\rho\frac{dA}{dt}=\partial_t(\rho A)+\partial_j(\rho A V_j), \tag{2.1.3}$$

which supports the linkage between substantial and local changes of feature A.

When $A\equiv 1$ relation (2.1.3) is reduced to the total continuity equation for mass local balance (the *continuity equation*)

$$\partial_t\rho+\partial_j(\rho V_j)=0, \tag{2.1.4}$$

whose consequence is just (2.1.3), which is easy to check using (2.1.2) and (2.1.4) in order to derive (2.1.3). Equation (2.1.4) reflects total mass conservation in an arbitrary volume *w* within the system during its motion, i.e. the fact that the total mass can vary only in case of matter outflow through the boundaries of *w*.

Many important relations between averaged features of a moving turbulized multicomponent continuum will be used in the following chapters to derive various balance equations for instantaneous (not averaged after Reynolds) values of thermohydrodynamical parameters of a mixture A written in the substantial form (see 2.1.1), i.e. indicating explicitly the divergence $(\partial_j J_{(A)j})$ and the source $(\sigma_{(A)})$ parts. Among others, this relations will serve to obtain evolutionary transfer equations for second correlation moments of different fluctuating-in-a-flow variables of state. We will address them here incorporating some details necessary for further purposes.

2.1.1. Differential Equations for Material Balance

Let us successively consider the cases of various defining parameters A in (2.2.1), describing the state of the *N*-component reacting gas mixture.

Specific Volume Balance. In (2.1.1) we shall accept $A \equiv v(\boldsymbol{r},t)$, where $v \equiv 1/\rho$ is the specific volume of the system, and define the substantial flux and source of the quantity $v(\boldsymbol{r},t)$ with

$$J_{(v)j} \equiv -V_j, \qquad \sigma_{(v)} \equiv 0. \tag{2.1.5}$$

Then from (2.1.1) we obtain

$$\rho\frac{dv}{dt} = \partial_j V_j, \tag{2.1.6}$$

that is, the balance equation of the specific volume of a total material continuum.

Continuity Equations for Concentrations of Chemical Mixture Components. The diffusion equation for the α-th component of a medium (when components and chemical reactions involved can be considered as continuously distributed in space) in substantial form looks as follows (*De Groot and Mazur, 1962*)

$$\rho\frac{dZ_\alpha}{dt} = -\partial_j J_{\alpha j} + \sum_{s=1}^{r} \nu_{\alpha s}\xi_s, \qquad (\alpha = 1,2,\ldots,N), \tag{2.1.7}$$

where $Z_\alpha(\boldsymbol{r},t) \equiv n_\alpha/\rho$ is the specific (that is, per unit mass of the total continuum) number density of the α-component; $n_\alpha(\boldsymbol{r},t)$ is the number density (per unit volume) of α-molecules; $J_{\alpha j}(\boldsymbol{r},t)$ is the (molecular) diffusion flux of α-substance in a reference system traveling with mean mass velocity of the mixture V_j (i.e. $J_{\alpha j} \equiv n_\alpha(V_{\alpha j} - V_j)$); $V_{\alpha j}(\boldsymbol{r},t)$ is the hydrodynamic velocity of α-particles; $\sigma_\alpha = \sum_{s=1}^{r} \nu_{\alpha s}\xi_s$ is the source term expressing the total increase (decrease) of the number of α-particles per unit volume per unit time at the expense of chemical reactions (for details see section 3.2.3); $\xi_s(\boldsymbol{r},t)$ is the s-th chemical reaction rate ($s = 1,2,\ldots,r$), which should be determined from additional physicochemical laws (it is believed that $\xi_s > 0$ if a reaction proceeds); and $\nu_{\alpha s}$ are the stoichiometric coefficients of α-substance in the s-th reaction. Note that the coefficients $\nu_{\alpha s}$ determine mass shares of respective components participating in reactions; they are negative for components entering into reactions, i.e. "*reagents*" in the s-th reaction, and positive for the products of a reaction proceeding in the right direction.

The equations (2.1.7) are linearly dependent, if the following relations are valid:

$$\sum_{\alpha=1}^{N} M_\alpha Z_\alpha = 1, \quad (^1) \qquad \sum_{\alpha=1}^{N} M_\alpha J_{\alpha j} = 0, \quad (j = 1,2,3), \quad (^2)$$

$$\sum_{\alpha=1}^{N} M_\alpha \sigma_\alpha = \sum_{s=1}^{r} \xi_s \left[\sum_{\alpha=1}^{N} M_\alpha \nu_{\alpha s}\right] = 0. \quad (^3) \tag{2.1.8}$$

Here M_α is the molecular mass of a particle of sort α and the identities (2.1.8 $^{(1,2)}$) are a corollary of the definition of the complete mass density ρ and the hydrodynamic (mean-mass) velocity V_j for the multicomponent continuum:

$$\rho \equiv \sum_{\alpha=1}^{N} M_\alpha n_\alpha, \qquad V_j \equiv \rho^{-1} \sum_{\alpha=1}^{N} M_\alpha n_\alpha V_{\alpha j}, \tag{2.1.9}$$

while the relation (2.1.8 $^{(3)}$) represents a consequence of the mass conservation law in chemical reactions:

$$\sum_{\alpha=1}^{N} M_\alpha \nu_{\alpha s} = 0, \qquad (s = 1,2,\ldots,r).$$

Having in mind this connection, one of the differential equations (2.1.7) can always be eliminated from the consideration by replacing it with the algebraic integral (2.1.8 $^{(1)}$).

Comparing (2.1.7) and (2.1.1) for $A \equiv Z_\alpha$, we obtain the following expressions for the flux and source values:

$$J_{(Z_\alpha)j} \equiv J_{\alpha j}, \qquad \sigma_{(Z_\alpha)} \equiv \sigma_\alpha. \tag{2.1.10}$$

Conservation Equations for Chemical Elements in a Gaseous Mixture. With the phenomenological description of a reacting multicomponent mixture, it is expedient to use the different material balance equations in various particular cases. Of importance in aeronomical studies are the so-called *diffusion equations for chemical elements* (involved in the system under study) which were obtained in a form convenient for geophysical applications (*Kolesnichenko and Tirskiy 1979*). Designating the number of γ-atoms in an "α-molecule" by η_α^γ, we have

$$n^\gamma = \sum_{\alpha=1}^{N} \eta_\alpha^\gamma n_\alpha, \tag{2.1.11}$$

which is the total number density of γ-elements per unit volume in the multicomponent continuum.

For the sake of compact writing of the relations used in what follows, we suppose that all chemical elements available in the system (in unbound or bound form) are involved in N mixture components and that they are attributed with numbers from N_m, where N_m is the number of different molecular (polyatomic) components in the mixture. This does not imply loss of generality regarding the mathematical description because it is always possible to supplement the actual multicomponent mixture with new components, or *chemical elements*, which are not initially involved in the mixture as individual components but compose the molecules of the system under consideration. With this in mind, we suppose that for these "*element-components*" of the N-component medium the respective thermohydrodynamical parameters (such as n_γ and $J_{\gamma j}$) are equal to zero. Then we have

$$\eta^{\gamma}_{\alpha} = \delta^{\gamma}_{\alpha} = \begin{cases} 1, & \alpha = \gamma \\ 0, & \alpha \neq \gamma \end{cases} \qquad (\alpha, \gamma = N_m + 1, N_m + 2, ..., N).$$

Multiplying (2.1.7) by the coefficients η^{γ}_{α} and summing the product with respect to α $(\alpha = 1,2,..., N)$, we obtain conservation laws for the individual chemical elements of the system as

$$\rho \frac{dZ^{\gamma}}{dt} = -\partial_j J^{\gamma}_j \qquad (\gamma = N_m + 1, N_m + 2, ..., N), \tag{2.1.12}$$

where the following designations are introduced:

$$Z^{\gamma} \equiv \frac{n^{\gamma}}{\rho} = \sum_{\alpha=1}^{N} \eta^{\gamma}_{\alpha} Z_{\alpha} = Z_{\gamma} + \sum_{\alpha=1}^{N_m} \eta^{\gamma}_{\alpha} Z_{\alpha}, \tag{2.1.13}$$

$$J^{\gamma}_j = \sum_{\alpha=1}^{N} \eta^{\gamma}_{\alpha} J_{\alpha j} = J_{\gamma j} + \sum_{\alpha=1}^{N_m} \eta^{\gamma}_{\alpha} J_{\alpha j}, \qquad (\gamma = N_m + 1, N_m + 2, ..., N). \tag{2.1.14}$$

Here $Z^{\gamma}(\boldsymbol{r},t)$, $J^{\gamma}_j(\boldsymbol{r},t)$ are the specific number density and the total substantial diffusion flux of a chemical element γ in the multicomponent continuum, respectively. The flux $J^{\gamma}_j(\boldsymbol{r},t)$ comprises transport of element γ by all diffusion fluxes of the molecular components in the medium containing this element. Note that it is important to avoid confusing the quantities $Z^{\gamma}(\boldsymbol{r},t)$, $J^{\gamma}_j(\boldsymbol{r},t)$ with the density $Z_{\gamma}(\boldsymbol{r},t)$ and the flux $J_{\gamma j}(\boldsymbol{r},t)$ of the "*element-component*" of sort γ. When writing (2.1.12), the very important notion was taken into account that the source term $\sigma_{(Z^{\gamma})}$ for chemical elements disappears, because the elements remain in the system where chemical reactions occur (nuclear reactions are not considered here), i.e.

$$\sigma_{(Z^{\gamma})} \equiv \sum_{\alpha=1}^{N} \eta^{\gamma}_{\alpha} \sigma_{\alpha} = 0, \quad (\gamma = N_m + 1, N_m + 2, ..., N). \tag{2.1.15}$$

The conservation equations for elements (2.1.12) without volume sources are convenient to use instead of any of the $(N - N_m)$ diffusion equations for multi-element components of the system (2.1.7).

It is necessary to note, that the following relations are valid here also (compare to (2.1.8))

$$\sum_{\gamma=N_m+1}^{N} M_{\gamma} Z^{\gamma} = 1, \quad (^1) \qquad \sum_{\gamma=N_m+1}^{N} M_{\gamma} J^{\gamma}_j = 0, \quad (^2) \tag{2.1.16}$$

which are corollaries of (2.1.8) and for this reason equations (2.1.12) being linearly dependent. Indeed, we have

$$1=\sum_{\alpha=1}^{N} M_\alpha Z_\alpha=\sum_{\alpha=1}^{N_m} M_\alpha Z_\alpha+\sum_{\alpha=N_m+1}^{N} M_\alpha Z_\alpha=\sum_{\alpha=1}^{N_m} M_\alpha Z_\alpha+\sum_{\gamma=N_m+1}^{N} M_\gamma\Big[Z^\gamma-$$

$$-\sum_{\alpha=1}^{N_m}\eta_\alpha^\gamma Z_\alpha\Big]=\sum_{\gamma=N_m+1}^{N} M_\gamma Z^\gamma+\sum_{\alpha=1}^{N_m}\Big[M_\alpha-\sum_{\gamma=N_m+1}^{N}\eta_\alpha^\gamma M_\gamma\Big]=\sum_{\gamma=N_m+1}^{N} M_\gamma Z^\gamma .$$

And also:

$$0=\sum_{\alpha=1}^{N} M_\alpha J_{\alpha j}=\sum_{\alpha=1}^{N_m} J_{\alpha j}\left[\sum_{\gamma=N_m+1}^{N}\eta_\alpha^\gamma M_\gamma\right]+\sum_{\gamma=N_m+1}^{N} M_\gamma J_{\gamma j}=$$

$$=\sum_{\gamma=N_m+1}^{N} M_\gamma\left[J_{\gamma j}+\sum_{\alpha=1}^{N_m}\eta_\alpha^\gamma J_{\alpha J}\right]=\sum_{\gamma=N_m+1}^{N} M_\gamma J_j^\gamma .$$

The obvious relation

$$M_\alpha=\sum_{\gamma=N_m+1}^{N}\eta_\alpha^\gamma M_\gamma,\qquad (\alpha=1,2,\ldots,N_m) \tag{2.1.17}$$

was used here for the mass of α-molecules, M_α, expressed in terms of the mass of the chemical elements, M_γ.

Thus, if $A\equiv Z^\gamma$ is substituted *a priori* in (2.1.1), it would be necessary to put

$$J_{(Z^\gamma)j}\equiv J_j^\gamma,\qquad \sigma_{(Z^\gamma)}\equiv 0. \tag{2.1.18}$$

2.1.2. The Conservation Equation for the Momentum of the Total Continuum

In geophysical and aeronomical applications, problems are used to be focused on relative motions in gaseous atmospheric media, which are studied in a coordinate system fixed on the rotating planetary surface. This results in auxiliary terms appearing in the respective equations of motion that take into account the *Coriolis* acceleration effect as well as the centripetal acceleration, although the latter relating to the planet's rotation usually is small in comparison with the acceleration due to gravity. In this case, the substantial form of the complete conservation equation of momentum for multicomponent gaseous mixtures takes the form

$$\rho\frac{dV_i}{dt}=-\partial_i p+\partial_j\pi_{ij}+2\rho\varepsilon_{kij}V_k\Omega_j+\rho\sum_{\alpha=1}^{N} Z_\alpha F_{\alpha i}, \tag{2.1.19}$$

where $p(\boldsymbol{r},t)$ is the thermodynamical pressure of the mixture; $\pi_{ij}(\boldsymbol{r},t)$ is the viscous stress tensor related to molecular momentum transport of all mixture components (here written in the so-called *diffusion approach* (*Sedov, 1984; Nigmatulin, 1987*); see also section 2.1.3); $2\rho\varepsilon_{kij}V_k\Omega_j$ are the Coriolis force vector components; Ω_j are the angular rotation rate vector components of the coordinate system, in reference to an absolute coordinate system in which equation (2.1.19) has been written (it is supposed everywhere below that the rotation rate of a planet $\Omega_j \equiv const$); $F_{\alpha j}(\boldsymbol{r},t)$ is the total external force acting on a single α-particle; and ε_{kij} is the *Levy-Chivita* alternating tensor:

$$\varepsilon_{ilk} = \begin{cases} 1, & i,l,k = 1,2,3; \quad 3,1,2; \quad 2,3,1; \\ 0, & i = l; \quad i = k; \quad l = k; \\ -1, & i,l,k = 2,1,3; \quad 3,2,1; \quad 1,3,2. \end{cases} \tag{2.1.20}$$

Now we explicitly extract the gravitational force from $F_{\alpha j}$, which yields

$$F_{\alpha j} \equiv M_\alpha g_j + F^*_{\alpha j}, \qquad g_j = -g\delta_{j3}, \tag{2.1.21}$$

where the z-axis is directed upwards, normal to a gravitational equipotential surface (the coordinate $x_3 \equiv z = 0$ laying at some initial equipotential surface, e.g., sea level in the case of the Earth); $g = |\boldsymbol{g}|$ is the acceleration due to gravity reduced with regard to the centrifugal force ($\boldsymbol{g} = \nabla\left[\tfrac{1}{2}|\boldsymbol{\Omega}\times\boldsymbol{R}|^2 - G\dfrac{M_{pl}}{|\boldsymbol{R}|}\right]$; G is the gravitational constant, M_{pl} is the mass of a planet; $\boldsymbol{R}$ is the radius-vector passing from planet's center to the considered spatial point; $F^*_{\alpha i}$ is the total non-gravitational force (for instance, the *Lorentz* electromagnetic force $F^*_{\alpha i} = e_\alpha(E_i + c^{-1}\varepsilon_{kij}V_{\alpha k}B_j)$); and δ_{ij} is the *Kronecker* tensor.

Thus, in the case of a substantial flux and a three-dimensional source of quantity V_i, comparing (2.1.19) with (2.1.1) at $A \equiv V_i$ gives the following expressions:

$$J_{(V_i)j} \equiv -\pi_{ij}, \qquad \sigma_{(V_i)} \equiv -\partial_i p + 2\rho\varepsilon_{kij}V_k\Omega_j + \rho\sum_{\alpha=1}^{N} Z_\alpha F_{\alpha i}. \tag{2.1.22}$$

As was already noticed, since the choice of flux $J_{(V_i)j}$ and source $\sigma_{(V_i)}$ of the momentum is ambiguous, a different approach can be used. In particular, for geophysical applications it is sometimes convenient to represent the total pressure of a mixture as two addends

$$p = p^d + p_0, \tag{2.1.23}$$

where p^d is the so-called *impact pressure*, and p_0 is the part complying with the (hydrostatic) equation

$$\partial_j p_0 = \rho_0 g_j = -\rho_0 g\delta_{3j}, \tag{2.1.24}$$

where ρ_0 is the stationary mass density characteristic of the atmosphere (for example, at sea level), and $\Delta\rho = \rho - \rho_0$. Then

$$J^*_{(V_i)j} \equiv p^d \delta_{ij} - \pi_{ij}, \qquad \sigma^*_{(V_i)} \equiv -\delta_{i3} g \Delta\rho + \rho \sum_{\alpha=1}^{N} Z_\alpha F^*_{\alpha i} + 2\rho \varepsilon_{kji} V_k \Omega_j . \tag{2.1.25}$$

2.1.3. Diverse Energy Equations for Multicomponent Media

When developing phenomenological models on various levels of complexity, adapted for simulated multicomponent turbulent media, diverse energy balance equations for reacting mixtures are to be used. In this chapter, we will derive these equations for the case of conservative external fields.

Potential Energy Balance in Multicomponent Mixtures. Let us define the specific potential energy of a gaseous mixture by

$$\Psi(\boldsymbol{r},t) = \sum_{\alpha=1}^{N} \psi_\alpha Z_\alpha , \tag{2.1.26}$$

where $\psi_\alpha(\boldsymbol{r})$ is the scalar potential (per single α-particle), matching the requirement

$$F_{\alpha j} = -\partial_j \psi_\alpha , \quad \partial_t \psi_\alpha = 0. \tag{2.1.27}$$

The respective balance equation can be obtained proceeding from the operator relation (2.1.3). Setting for this purpose $A \equiv \Psi$, we have

$$\rho \frac{d\Psi}{dt} = \partial_t \left[\rho \sum_{\alpha=1}^{N} \psi_\alpha Z_\alpha \right] + \partial_j \left[\rho \sum_{\alpha=1}^{N} \psi_\alpha Z_\alpha V_j \right] = \sum_{\alpha=1}^{N} \psi_\alpha \left[\partial_t (\rho Z_\alpha) + \partial_j (\rho Z_\alpha V_j) \right] +$$

$$+ \sum_{\alpha=1}^{N} (\rho Z_\alpha \partial_t \psi_\alpha + \rho Z_\alpha V_j \partial_j \psi_\alpha) = \sum_{\alpha=1}^{N} \psi_\alpha \rho \frac{dZ_\alpha}{dt} - \rho V_j \sum_{\alpha=1}^{N} Z_\alpha F_{\alpha j} = \sum_{\alpha=1}^{N} \psi_\alpha \left[-\partial_j J_{\alpha j} + \right.$$

$$\left. + \sigma_\alpha \right] - \rho V_j \sum_{\alpha=1}^{N} Z_\alpha F_{\alpha j} = -\partial_j \left[\sum_{\alpha=1}^{N} \psi_\alpha J_{\alpha j} \right] + \sum_{\alpha=1}^{N} J_{\alpha j} \partial_j \psi_\alpha - \rho V_j \sum_{\alpha=1}^{N} Z_\alpha F_{\alpha j} + \sum_{\alpha=1}^{N} \psi_\alpha \sigma_\alpha .$$

Supposing next that the density of the internal source of potential energy vanishes

$$\sigma_{(\Psi)}^{(i)} \equiv \sum_{\alpha=1}^{N} \psi_\alpha \sigma_\alpha = 0, \tag{2.1.28}$$

which is valid only provided that the potential energy is conserved in chemical reactions (*Dyarmati, 1970*):

$$\sum_{\alpha=1}^{N} \psi_\alpha \nu_{\alpha s} = 0, \qquad (s = 1,2,\dots,r), \tag{2.1.29}$$

we finally obtain

$$\rho \frac{d\Psi}{dt} = -\partial_j \left[\sum_{\alpha=1}^{N} \psi_\alpha J_{\alpha j} \right] + \sum_{\alpha=1}^{N} J_{\alpha j} \partial_j \psi_\alpha - \rho V_j \sum_{\alpha=1}^{N} Z_\alpha F_{\alpha j}. \tag{2.1.30}$$

Now comparing (2.1.30) to (2.1.1), for $A \equiv \Psi$, we come to the following expressions for the substantial flux $J_{(\Psi)j}$ and the volume source $\sigma_{(\Psi)}$ of the potential energy of the mixture:

$$J_{(\Psi)j} \equiv \sum_{\alpha=1}^{N} \psi_\alpha J_{\alpha j}, \tag{2.1.31}$$

$$\sigma_{(\Psi)} \equiv -\sum_{\alpha=1}^{N} J_{\alpha j} F_{\alpha j} - \rho V_j \sum_{\alpha=1}^{N} Z_\alpha F_{\alpha j}. \tag{2.1.32}$$

Total Energy Conservation of a Mixture. The conservation law for the total energy of a multicomponent mixture in the substantial form looks like

$$\rho \frac{dE}{dt} = -\partial_j J_{(\mathrm{E})j}, \tag{2.1.33}$$

where

$$E(\boldsymbol{r},t) = \tfrac{1}{2} V_j V_j + \Psi + \varepsilon \tag{2.1.34}$$

is the specific total energy of the multicomponent continuum; $\varepsilon(\boldsymbol{r},t)$ is specific internal energy of the medium (in fact, this quantity is determined by the relation (2.1.34));

$$J_{(E)j} \equiv q_j + (p\delta_{ij} - \pi_{ij}) V_i + \sum_{\alpha=1}^{N} \psi_\alpha J_{\alpha j} \tag{2.1.35}$$

are the vector components of the substantial total energy flux in the moving mixture; and $q_j(\boldsymbol{r},t)$ is the molecular heat flux. Note that the relation (2.1.35) should also be considered as an accurate definition of the heat flux which, as is known, is commonly in use in both experimental physics and thermophysics.

We have to admit that, generally speaking, the value of the internal energy $\varepsilon(\boldsymbol{r},t)$ introduced by the relation (2.1.34) and written here in the so-called *diffusion approximation* (*Sedov, 1984*) is not fully correct. The reason is that not only the terms related to thermal motion of molecules and to short-range intermolecular force interactions (in agreement with the conventional understanding of internal energy), but also the terms depending on the macroscopic *kinetic diffusion energy of the components* in a center-of-mass system are to be included. Indeed, the additive in the form $\frac{1}{2}\sum_{\alpha=1}^{N} M_\alpha Z_\alpha (V_\alpha - V)^2$, which is of second order of smallness for a multicomponent mixture, should be introduced explicitly in the case of a multiphase medium (such as fine grained gas-dust, etc.), as it was argued in (*Kolesnichenko 1978*).

Balance Equation for the Mechanical Energy of a Mixture. In order to derive the balance equation for the system's internal energy $\varepsilon(\boldsymbol{r},t)$ we will take advantage of (2.1.33). But first let us write the balance equation for the specific kinetic energy of progressive motion of the center of mass. For this purpose, scalarly multiplying the equation of motion (2.1.19) by $V_i(\boldsymbol{r},t)$ and performing standard transformations, we obtain the *work-energy theorem* for a multicomponent mixture in the following (substantial) form:

$$\rho\frac{d(V_iV_i/2)}{dt} = \rho V_j \sum_{\alpha=1}^{N} Z_\alpha F_{\alpha j} - \partial_j\left[\left(p\delta_{ij} - \pi_{ij}\right)V_i\right] + p\partial_j V_i - \pi_{ij}\partial_j V_i. \tag{2.1.36}$$

Now, combining (2.1.30) and (2.1.36), we obtain the following equation for substantial balance of the specific mechanical energy $\varepsilon_m \equiv \frac{1}{2}V_iV_i + \Psi$:

$$\rho\frac{d\varepsilon_m}{dt} = -\partial_j J_{(\varepsilon_m)j} + \sigma_{(\varepsilon_m)}, \tag{2.1.37}$$

where

$$J_{(\varepsilon_m)j} \equiv (p\delta_{ij} - \pi_{ij})V_i + \sum_{\alpha=1}^{N} \psi_\alpha J_{\alpha j} \tag{2.1.38}$$

is the substantial energy flux density ε_m;

$$\sigma_{(\varepsilon_m)} \equiv p\partial_j V_j - \pi_{ij}\partial_j V_i - \sum_{\alpha=1}^{N} J_{\alpha j} F^*_{\alpha j} \tag{2.1.39}$$

is the source density of mechanical energy where the last addend $\sigma^e_{(\varepsilon_m)} \equiv -\sum_{\alpha=1}^{N} J_{\alpha j} F^*_{\alpha j}$ conforms to the density of an external source of mechanical energy.

Let us notice, that the physical interpretation of the individual terms in various energy equations is used to be the same for laminar (instantaneous) and averaged turbulent motions in a multicomponent continuum. Therefore, in order to avoid unnecessary repetition, we put such a discussion aside until Chapter 3.

Balance Equation for the Internal Energy of a Multicomponent Continuum. The balance equation of the internal energy of a gaseous mixture can be obtained by subtracting (2.1.37) from (2.1.33):

$$\rho \frac{d\varepsilon}{dt} = -\partial_j q_j - p\partial_j V_j + \pi_{ij}\partial_j V_i + \sum_{\alpha=1}^{N} J_{\alpha j} F_{\alpha j}^{*}. \tag{2.1.40}$$

In geophysical and aeronomical applications, it is often convenient to employ the equation for the internal energy of gaseous mixtures written in terms of the total specific enthalpy $h(\boldsymbol{r},t)$ of a substance defined by (*Prigogine and Defay, 1954*):

$$h = \sum_{\alpha=1}^{N} h_\alpha Z_\alpha = \sum_{\alpha=1}^{N} (\varepsilon_\alpha + p_\alpha / n_\alpha) Z_\alpha = \varepsilon + p/\rho. \tag{2.1.41}$$

Then, in view of the transformation

$$\rho \frac{d\varepsilon}{dt} + p\partial_j V_j = \rho \frac{dh}{dt} - \frac{dp}{dt}, \tag{2.1.42}$$

which follows from the definition (2.1.41) and the continuity equation (2.1.4), we have

$$\rho \frac{dh}{dt} = \frac{dp}{dt} - \partial_j q_j + \pi_{ij}\partial_j V_i + \sum_{\alpha=1}^{N} J_{\alpha j} F_{\alpha j}^{*}. \tag{2.1.43}$$

For the main goal of this book, it is sufficient to approximate the partial (per single particle) enthalpy $h_\alpha(\boldsymbol{r},t)$ of the α-component using

$$h_\alpha \cong C_{p\alpha} T + h_\alpha^0, \tag{2.1.44}$$

which allows us to obtain

$$h = \sum_{\alpha=1}^{N} Z_\alpha h_\alpha \cong C_p T + \sum_{\alpha=1}^{N} Z_\alpha h_\alpha^0, \tag{2.1.45}$$

$$C_p = \sum_{\alpha=1}^{N} C_{p\alpha} Z_\alpha, \tag{2.1.46}$$

where $C_{p\alpha}$ is the partial heat capacity (at constant pressure) of the α-component; h_α^0 is its enthalpy at zero temperature (the so-called *heat production*); and $C_p(\boldsymbol{r},t)$ is the specific heat capacity of the mixture at constant pressure. Hereafter the quantities $C_{p\alpha}$ and h_α^0, approximating the real changes of the heat capacity $C_{p\alpha}(T)$ and the partial heat production $h_\alpha^0(T)$ in a limited temperature range, are assumed to be constant.

Now let us write down the equation (2.1.43) in terms of the variables $T(\boldsymbol{r},t)$ and $p(\boldsymbol{r},t)$. Using the relations (2.1.7), (2.1.41), (2.1.44), and (2.1.46), we obtain

$$\rho\frac{dh}{dt}=\rho\frac{d}{dt}\left[\sum_{\alpha=1}^{N}Z_\alpha h_\alpha\right]=\rho\sum_{\alpha=1}^{N}Z_\alpha\frac{dh_\alpha}{dt}+\rho\sum_{\alpha=1}^{N}h_\alpha\frac{dZ_\alpha}{dt}=$$

$$=\rho\left[\sum_{\alpha=1}^{N}C_{p\alpha}Z_\alpha\right]\frac{dT}{dt}+\sum_{\alpha=1}^{N}h_\alpha[-\partial_j J_{\alpha j}+\sigma_\alpha]=\rho C_p\frac{dT}{dt}-\partial_j\left[\sum_{\alpha=1}^{N}h_\alpha J_{\alpha j}\right]+$$

$$+\partial_j T\sum_{\alpha=1}^{N}C_{p\alpha}J_{\alpha j}+\sum_{s=1}^{r}q_s\xi_s, \qquad (2.1.47)$$

where by means of the relation

$$q_s(T)=\sum_{\alpha=1}^{N}h_\alpha \nu_{\alpha s}=q_s^0+\sum_{\alpha=1}^{N}C_{p\alpha}\nu_{s\alpha}, \qquad (s=1,2,\ldots,r) \qquad (2.1.48)$$

the so-called *reaction heat* at constant T and p is introduced. It is equal to the difference between the sum of the products of partial reaction product enthalpies and the respective stoichiometric coefficients, and the similar sum for reactants (reagents). The quantity $q_s^0=\sum_{\alpha=1}^{N}h_\alpha^0\nu_{\alpha s}$ is interpreted as the heat of the s-th chemical reaction at zero temperature. Then, from (2.1.43) and (2.1.47) the equation for the internal energy of a mixture follows

$$\rho C_p\frac{dT}{dt}=-\partial_j J_{qj}+\frac{dp}{dt}+\pi_{ij}\partial_j V_i-\sum_{s=1}^{r}q_s\xi_s+\sum_{\alpha=1}^{N}J_{\alpha j}F_{\alpha j}^*-\partial_j T\sum_{\alpha=1}^{N}C_{p\alpha}J_{\alpha j}, \qquad (2.1.49)$$

where

$$J_{qj}=q_j-\sum_{\alpha=1}^{N}h_\alpha J_{\alpha j} \qquad (2.1.50)$$

is the *reduced heat flux*.

The last summation in the right-hand part of (2.1.49) represents the effect of the so-called *diffusing heat capacities*. In the majority of all practically important cases, the specific isobaric heat capacities of the individual components in a gaseous mixture $C_{p\alpha}^*\equiv C_{p\alpha}/M_\alpha$ are quite similar (so, $C_{p\alpha}^*\approx idem$, $C_{p\alpha}\approx idem\cdot M_\alpha$); therefore, by virtue of (2.1.8 [(2)]), this term in (2.1.49) often can be dropped out. Evidently, the equation for the internal energy of a reacting gaseous mixture written as (2.1.49) permits to single out explicitly the contribution of the individual chemical reaction heats to the overall energetics of the medium. In particular, the height distribution of the kinetic temperature of an atmospheric gas can be obtained thereby.

Comparing (2.1.43) with (2.1.1), for $A \equiv h$, yields

$$J_{(h)j} \equiv q_j, \tag{2.1.51}$$

which is the substantial enthalpy flux of a mixture, as well as

$$\sigma_{(h)} \equiv \frac{dp}{dt} + \pi_{ij}\partial_j V_i + \sum_{\alpha=1}^{N} J_{\alpha j} F_{\alpha j}^{*} \tag{2.1.52}$$

which is the source density of the specific enthalpy of a mixture for regular flows.

2.1.4. The Equation of State for a Mixture of Ideal Gases

We will use the *baroclinic* equation of state for a mixture of perfect gases as the thermal equation of state for an aggregate multicomponent gas continuum (an equation for pressure):

$$p = \sum_{\alpha=1}^{N} p_\alpha = \mathrm{k}T\sum_{\alpha=1}^{N} n_\alpha = R^{*}\rho T = R^{*}T/v, \tag{2.1.53}$$

where

$$R^{*} = \mathrm{k}\sum_{\alpha=1}^{N} Z_\alpha. \tag{2.1.54}$$

is the so-called "gas constant" for a mixture; k is Boltzmann's constant; and $p_\alpha(\boldsymbol{r},t)$ is the partial pressure of the α-component, $p_\alpha = \mathrm{k}Tn_\alpha$. The first equality in (2.1.53) representing *the Dalton law* is a corollary of the definition of p_α, irrespective of the fact whether the individual gases constituting the mixture are perfect or not. This quantity is the product of the total thermodynamic pressure of a medium p by the molar ratio $x_\alpha(\boldsymbol{r},t)$ of the α-component

$$p_\alpha = x_\alpha p, \tag{2.1.55}$$

$$x_\alpha \equiv n_\alpha / n = n_\alpha / \sum_{\beta=1}^{N} n_\beta, \qquad \sum_{\alpha=1}^{N} x_\alpha = 1, \tag{2.1.56}$$

where $n(\boldsymbol{r},t)$ is the total number density of the medium.

2.1.5. The Complete System of Hydrodynamic Equations for Gas Mixtures

Let us write down the complete set of multicomponent hydrodynamic equations in a form suitable to address the problems of the structure, thermal regime, and dynamics of the middle and upper atmosphere of a planet (*Marov and Kolesnichenko, 1987*):

$$\rho\frac{d}{dt}\left(\frac{1}{\rho}\right)=\partial_j V_j, \tag{2.1.57}$$

$$\rho\frac{dZ_\alpha}{dt}=-\partial_j J_{\alpha j}+\sum_{s=1}^{r}\nu_{\alpha s}\xi_s, \qquad (\alpha=1,2,\ldots,N-1), \tag{2.1.58}$$

$$\rho\frac{dV_i}{dt}=-\partial_i p+\partial_j\pi_{ij}+2\rho\varepsilon_{kij}V_k\Omega_j+\rho g_i+\rho\sum_{\alpha=1}^{N}Z_\alpha F^*_{\alpha i}, \tag{2.1.59}$$

$$\rho C_p\frac{dT}{dt}=-\partial_j\left[q_{Rj}+q_j-\sum_{\alpha=1}^{N}h_\alpha J_{\alpha j}\right]+\frac{dp}{dt}+\pi_{ij}\partial_j V_i-\sum_{s=1}^{r}q_s\xi_s+$$

$$+\sum_{\alpha=1}^{N}J_{\alpha j}F^*_{\alpha j}-\partial_j T\sum_{\alpha=1}^{N}C_{p\alpha}J_{\alpha j}+Q_d, \tag{2.1.60}$$

$$p=\mathrm{k}T\sum_{\alpha=1}^{N}n_\alpha=R^*\rho T. \tag{2.1.61}$$

Here

$$q_{Rj}=\int_0^\infty (q_{R\nu})_j\,d\nu=\int_0^\infty \omega_j I_\nu\,d\omega\,d\nu \tag{2.1.62}$$

is the radiative energy vector (the *Pointing vector*); I_ν is the spectral radiation intensity, that is, the amount of radiative energy per unit frequency at wavelength ν and per unit solid angle, passing per unit time through a unit area at point $\boldsymbol{r}$, the area being normal to the direction ω of energy propagation; Q_d are the possible additional local heating sources of the atmospheric gas treated in more detail by Marov and Kolesnichenko (*1987*); see also Chapter 6.

The set of hydrodynamic equations (2.1.57)–(2.1.61) should be supplemented by the initial and boundary conditions as well as several relevant expressions including: the source terms in diffusion equations regarding chemical components; the radiative heat flux vector q_{Rj}; and the defining relations for coupling the viscous stress tensor π_{ij}, the heat flux q_j, and the diffusion fluxes $J_{\alpha j}$ with the gradients of the thermohydrodynamical parameters of the medium, such as the mass density, velocity, temperature, and concentrations of the chemical components. The defining relations for laminar flow will be obtained in section 2.3.3 using non-equilibrium thermodynamical methods, while the contribution of chemical reactions to the balance equations (2.1.58) will be studied in detail in Chapter 3.

We will now focus on some specific features of the set (2.1.57)–(2.1.61) and make the following remarks:

- It is often possible to introduce two simplifications into the diffusion equations for individual chemical mixture components (2.1.58). If dynamical and diffusion transport in the examined height range are not of great importance, the so-called *photochemical approximation* can be obtained:

$$\frac{\partial n_\alpha}{\partial t} = \sigma_\alpha = \sum_{s=1}^{r} \nu_{\alpha s} \xi_s, \quad (\alpha = 1, 2, \ldots, N). \tag{2.1.58*}$$

Besides, if the direct and inverse aeronomical reaction rates are sufficiently high (in other words, the time for reactions to proceed is fairly short), such that the quantity $\partial n_\alpha / \partial t$ is smaller than other terms, then (2.1.58*) takes the form $\sigma_\alpha = 0$, representing the equation of photochemical equilibrium (*Marov and Kolesnichenko, 1987*).

- The equation of motion (2.1.59) describing the motion of the medium as a whole, does not allow to reveal the motions of the individual mixture components (electrons, ions, and neutrals). The complex multicomponent (molecular and ionic) composition of the medium must be evaluated elaborately, however, only when dealing with atmospheric chemistry problems, whereas the system of equations describing one- or two-fluid hydrodynamics is quite pertinent to describe the dynamical problems. The equations of motion for the individual neutral mixture components in the diffusion approximation (*Sedov, 1984*) are considered in subsection 2.3.6.

- When modeling such a medium as the upper atmosphere of a planet, an important and complex problem regards the description of the incident heat flux into studied region of the medium. Balance is ensured by the direct absorption of solar electromagnetic and corpuscular radiation by atmospheric components, followed by the subsequent transformation owing to aeronomical reactions and *Joule heating*, by various dynamical processes including energy dissipation of wave motions at various spatial scales, and different heat sink mechanisms. These processes result in the redistribution of heat from sources non-uniformly distributed in the atmosphere. Basically, in each particular case it is expedient to decipher accurately such energy sources (see *Gordiets et al., 1979, 1982; Marov and Kolesnichenko, 1987*), which have been introduced in (2.1.60) in a general form (the term Q_d).

- The lack of local thermodynamic equilibrium in the rarefied gas envelope of a planet prevents us from using only one (aggregate) energy equation (2.1.60), written in terms of the *kinetic temperature* T of the mixture (the temperature of translational degrees of freedom). As to the latter, the lack of thermal equilibrium manifests itself in the different kinetic temperatures of ions, electrons, and neutral particles ($T_e >> T_i$, $T_e >> T_n$), which is observed, for instance, in a considerable height range in the Earth's thermosphere above 150 km, especially around noon (*Brace et al., 1969; Bauer, 1973; Marov et al., 1997*). As to the internal degrees of freedom, this lack exhibits itself in the difference, under some conditions, of the oscillatory temperature of a component from its kinetic temperature, as well as in the permanent presence of chemically active atmospheric components and particles with excited energy levels (*Rishbeth and Garriott, 1969*; *Marov et al., 1996; 1997*). The *multi-temperature* structure of the thermosphere can be modeled as three-fluid hydrodynamics of a poorly ionized gas mixture (*Marov and Kolesnichenko, 1987*). Because there is only a small difference in the kinetic temperatures among neutral species, one is allowed to deal with a single energy equation for neutral component containing a major part of the thermal energy of the atmospheric gas.

- The hydrodynamic equations (2.1.57)–(2.1.61) can be considered valid for turbulent mixture flow as well, assuming that in this case they describe the true (instantaneous) state of the medium because the smallest characteristic scale of turbulent fluctuations is usually much larger than the free path of molecules (see subsection. 3.1.1 for further details).

Finally, we will briefly address basic criteria for hydrodynamic equations for mixtures in terms of their applicability to describe the middle and upper atmosphere. As is known, the continuous hydrodynamic equations are appropriate in the sense that they yield physically justified results only when the local macroscopic properties of the gas do not noticeably vary at the scale of the free path of molecules l (*Ferziger and Kaper, 1972*). This is just the case when the particle velocity distribution is close to the Maxwellian distribution $f_\alpha^{(0)}$. To secure only small deviations from the Maxwellian distribution, the relaxation time of the equilibrium distribution $f_\alpha^{(0)}$ conditioned by the time lapse τ between particle collisions, should be small enough compared to the hydrodynamic time t_0, during which characteristic changes of the basic macroscopic parameters occur. In other words, the following inequalities should be true simultaneously:

$$t_0 \equiv \left|\partial \ln \Theta / \partial t\right|^{-1} >> \tau, \; (^1) \qquad L_0 \equiv \left|\nabla \ln \Theta\right|^{-1} >> l \; (^2). \tag{2.1.63}$$

Here Θ designates any thermohydrodynamic parameter. Since the particle content decreases with height, specifically in the thermosphere, the mean free path time τ grows as $\tau \approx (n\sigma v_t)^{-1}$, where v_t is the thermal velocity. This time τ is essentially of the same order of magnitude as the relaxation time for translational degrees of freedom. Nevertheless, throughout the whole thermosphere τ remains insignificant and the restriction $\tau \le 10^{-3}$ s is valid. Anyway, according to the data available, τ is much shorter than the duration t of processes responsible for variations of basic structural parameters describing the upper atmosphere, $t \ge 1$ s (*Rishbeth and Garriott, 1969*). Thus, the first criterion for the hydrodynamic approach to be valid (1.1.63 (^{1})) is satisfied.

The second inequality (1.1.63(^{2})) can be re-written as $l / L_0 \equiv \boldsymbol{Kn} << 1$ and reduced to the requirement for the *Knudsen* number $\boldsymbol{Kn}$ to be small. Although the free path of molecules l grows with height varying from 10 cm to 5 km between 100 and 400 km in the Earth's thermosphere, the scales of thermospheric non-homogeneities use to be even larger. For example, at the level of the lower thermosphere with a temperature gradient of 1 K/km and the temperature measuring about 1000 K (*Hedin, 1991*), the characteristic scale is $L_0 \sim 10^8$ cm and $\boldsymbol{Kn} << 1$. Because the hydrodynamic description is applicable if $\boldsymbol{Kn} \leq 0.2$ (*Grad, 1949, 1963*) this means that this approach can be used everywhere in the thermosphere up to exobase from which atoms and/or molecules may escape into outer space. The system of the Boltzmann equations should be employed, however, for modeling physicochemical processes in the exosphere located in the Earth's atmosphere above 400–500 km (*Marov and Kolesnichenko, 1987; Marov et al., 1997*).

2.2. THE SECOND LAW OF THERMODYNAMICS: THE RATE OF ENTROPY ORIGIN IN GASEOUS MIXTURES

Molecular diffusion transport encompasses diffusion itself, thermal conduction, thermodiffusion, and viscosity. These phenomena are featured by the respective terms of the equations of mass, momentum, and heat conservation (2.1.57)–(2.1.60). Every equation involves the flow divergence of some quantity related, although implicitly, to the gradients of thermohydrodynamic parameters (the so-called *thermodynamic forces*).

There are two ways to derive linear links (*defining relations*) between these flows and the associated thermodynamic forces based on macroscopic (phenomenological) and kinetic approaches. The kinetic approach, based on the system of the generalized Boltzmann equations for multicomponent gaseous mixtures, was thoroughly developed only for gases of moderate density when the interaction potential between particles is known (see, e.g., *Chapman and Couling, 1960*; *Ferziger and Kaper, 1972*; *Marov and Kolesnichenko, 1987*). The phenomenological approach, based on the laws of mechanics for a continuous media and non-equilibrium thermodynamics in application to the macroscopic volume of a mixture, is not bound to the postulation of a particular microscopic model describing interactions between particles and is suitable for a wide class of media. Within the phenomenological approach, *kinetic coefficients* (coefficients for gradients of thermohydrodynamic parameters in defining relations) are not explicitly deciphered, but their physical meaning can often be clarified (for example, for rarefied gases) within the molecular-kinetic theory approach.

An attempt to deduce phenomenologically the defining relations for non-ideal multicomponent continuous media, including the *Stephan-Maxwell relations* for multicomponent diffusion, was undertaken (*Kolesnichenko and Tirskiy, 1976*). The defining relations obtained in this study turned out to be identical in structure to similar relations derived using the kinetic approach (*Hirschfelder et al., 1961*). However, the latter accepts rather infelicitous definitions of the multicomponent diffusion factors (such as quantities asymmetrical in indexes) and thermal diffusion factors, because both definitions are not in accordance with the *Onsager-Kasimir reciprocal relations* in non-equilibrium thermodynamics (*de Groot and Mazur, 1962*; *Dyarmati, 1970*). The principle of "reciprocity" established empirically (though it can be also deduced based on the methods of statistical mechanics) is fundamental *per se* and can be referred to as *the fourth law of thermodynamics* (the third law on inaccessibility of absolute zero temperature is beyond the scope of this book). For this reason, the concordance of molecular exchange coefficients with the Onsager-Kasimir principle of reciprocity should be regarded as mandatory. This concordance is of extreme importance when employing theoretical results of the kinetic theory of monatomic rarefied gases, such as design formulas for transport coefficients, in modeling real polyatomic gaseous mixtures, in which intercomponent transitions between states of different internal degrees of freedom occur or chemical reactions proceed. In these cases it is important to adopt such a definition of the kinetic coefficients that they would be in agreement with the reciprocal relations (*Van de Ree, 1967*).

The procedure to arrive at the concordance in the framework of the kinetic approach mentioned above, was most successfully implemented to by Ferziger and Kaper, who defined, for example, the multicomponent diffusion factors as being symmetric (*Ferziger and Kaper, 1972*). Instead, in this book the phenomenological approach is developed to obtain the defining relations for thermodynamic flows and all major algebraic formulas linking kinetic transport coefficients. In particular, this includes the Stephan-Maxwell relations for multicomponent diffusion and a correlated expression for the total heat flow. The obtained defining relations and transport coefficients, are completely identical to the respective results of the kinetic theory. They also argue for the universal character of the thermodynamic approach because it is applicable to the evaluation of both monatomic gases and more complicated continuous media, such as polyatomic chemically active gas mixtures or liquid solutions (electrolytes, suspensions, etc.), for which the respective kinetic treatment is not available.

2.2.1. The Onsager Principle

Before applying the formalism of non-equilibrium thermodynamics for continuous media to describe the processes of heat and mass transfer in laminar (as well as in turbulent, see Chapter 5) flows in a multicomponent mixture, we will briefly discuss the essence of some basic postulates. They underlie the theory and can be used practically in the thermodynamic treatment of any irreversible process like that pertinent to a turbulized continuum.

As is known (*de Groot and Mazur, 1962*), the phenomenological relations of irreversible processes (the *Onsager relations*) are used in linear non-equilibrium thermodynamics as defining (*rheological*) relations, which supplement the system of hydrodynamic conservation equations

$$J_k = \sum_{k=1}^{Q} L_{kl} X_l, \quad (k = 1,2,\ldots,Q), \qquad (2.2.1)$$

where Q is the number of independent physical processes, and L_{kl} is the matrix of *phenomenological* (kinetic) coefficients linking the fluxes J_k and the thermodynamic forces X_l. The values J_k correspond to the rates of change of *extensive* quantities, such as mass and energy, for which there are conservation laws, or transferred quantities, such as heat, which are related to fluxes in the conservation laws. In turn, the values X_l are proportional to the gradients of *intensive* parameters diverging a thermodynamic system from an equilibrium.

The fluxes and thermodynamic forces in (2.2.1), generally, are tensors of any rank. As was mentioned earlier, in the framework of the phenomenological theory the explicit view of the kinetic coefficients in (2.2.1) is not deciphered, though their physical meaning can be revealed using molecular-kinetic theory (*Ferziger and Kaper, 1972*). The number of non-zero kinetic coefficients in (2.2.1) is restricted by the *Curie principle*. According to this principle, by virtue of symmetry properties of the considered material medium, the flux components (in our particular case vector components along the coordinate axes) will depend only on some of the thermodynamic forces. So, for an isotropic system whose properties are identical in all directions under equilibrium conditions, processes with different tensor dimensions do not interact with each other.

Besides, the *Onsager-Kasimir* symmetry relations (*the principle of reciprocity*) are accepted as an independent postulate in the axiomatic approach:

$$L_{kl}(\boldsymbol{B},\boldsymbol{\Omega}) = \varepsilon_k \varepsilon_l L_{lk}(-\boldsymbol{B},-\boldsymbol{\Omega}). \tag{2.2.2}$$

They permit to minimize the number of phenomenological coefficients in the linear relations (2.2.1). Here $\boldsymbol{B}$ is the magnetic induction, $\boldsymbol{\Omega}$ is the angular rotation rate of the system, whereas $\varepsilon_k = 1$ is taken for even (energy, concentrations) and $\varepsilon_k = -1$ is taken for odd (momentum density) macroscopic parameters which are the respective particle velocity functions. For an isotropic, non-rotating system without an external magnetic field, the symmetry relations (1.2.2) take on a simpler form (*de Groot and Mazur, 1962*):

$$L_{kl} = L_{lk} \tag{2.2.3}$$

where L_{kl} are scalar quantities. The symmetry relations (2.2.2) can be regarded as an empirical axiom, irrespective of the proof based on statistical mechanics (*Miller, 1974*). According to (*Mason, 1974*), experimental corroboration of the reciprocity principle is as convincing as corroboration of the first, second, and third principles of thermodynamics. This allows one to address postulate (2.2.2) as a paradigm and to use it to analyze a broad class of phenomena.

To define fluxes and their conjugated thermodynamic forces, a specific representation of the *rate of production* (source density) of the entropy $\sigma_{(S)}$ within the system with irreversible processes under consideration is usually employed as the bilinear form

$$T\sigma_{(S)} = \sum_{k=1}^{Q} J_k X_k \geq 0, \tag{2.2.4}$$

Then forces X_k, conjugated with the previously defined fluxes J_k, can be unambiguously found as coefficients at the respective fluxes in this expression.

To enable deciphering of formula (2.2.4) for density of the entropy source within the phenomenological theory, the equation of evolution of specific entropy S of a continuous system must be obtained in an explicit form. It looks as follows

$$\rho\frac{dS}{dt}=\partial_j J_{(S)j}+\sigma_{(S)},\qquad (\sigma_{(S)}\geq 0),\tag{2.2.5}$$

Here $J_{(S)j}$ are the components of the vector of substantial flux density of entropy, while divergence $\partial_j J_{(S)j}$ describes reversible heat exchange between considered system and the environment. Obviously, the inequality $\sigma_{(S)}\geq 0$ meets the second law of thermodynamics stating that the entropy of a closed system cannot decrease.

The explicit form of the equation (2.2.5) can be obtained, with due regard for the balance equations for specific volume (2.1.6), specific numerical densities of chemical components (2.1.7), and specific internal energy of a mixture (2.1.40), from the following *Gibbs identity* for these quantities written along the trajectory of motion of center of mass of an elementary physical volume:

$$T\frac{dS}{dt}=\frac{d\varepsilon}{dt}+p\frac{d}{dt}\left(\frac{1}{\rho}\right)-\sum_{\alpha=1}^{N}\mu_\alpha\frac{dZ_\alpha}{dt}.\tag{2.2.6}$$

Here μ_α is the chemical potential of α-component calculated per one particle of matter,

$$\mu_\alpha=h_\alpha-TS_\alpha=\varepsilon_\alpha+pv_\alpha-TS_\alpha;\tag{2.2.7}$$

ε_α, v_α, S_α are partial internal energy, partial volume, and partial entropy of α-component, respectively (*Prigogine, and Defay, 1954*). Note, that below the expression for specific free Gibbs' energy (*specific thermodynamic potential*) will be used which follows from (2.2.7):

$$G=\sum_{\alpha=1}^{N}Z_\alpha\mu_\alpha=\varepsilon+pv-TS=h-TS,\tag{2.2.8}$$

where

$$\varepsilon=\sum_{\alpha=1}^{N}Z_\alpha\varepsilon_\alpha\ (^1),\quad v=\sum_{\alpha=1}^{N}Z_\alpha v_\alpha\ (^2),\quad S=\sum_{\alpha=1}^{N}Z_\alpha S_\alpha\ (^3).\tag{2.2.9}$$

For the mixture of perfect gases obeying to the equation of state (2.1.53), one has $v=(kT/p)\sum_{\beta=1}^{N}Z_\beta$, whence

$$v_\beta=kT/p=1/n.\tag{2.2.10}$$

In section 2.2.2. we shall focus on the deduction and analysis of the evolutionary equation of transfer of specific entropy (2.2.5) in the explicit form for a laminar flow of a multicomponent gaseous mixture. Similar consideration for turbulent continuum will be given in Chapter 5. Before those discussions, however, it is worthwhile to make some remarks concerning applicability of the Gibbs identity to nonequilibrium processes in a continuous thermodynamic system.

According to the principle of *quasi-local equilibrium* (the basic postulate of nonequilibrium thermodynamics), the whole system can be splitted into fairly small macroscopic areas, each being considered as an equilibrium (more precisely, quasi-equilibrium) thermodynamic system. If specific density of internal energy $\varepsilon(\boldsymbol{r},t)$, specific volume $v(\boldsymbol{r},t)$, and specific concentrations $Z_\alpha(\boldsymbol{r},t)$ $(\alpha = 1,2,\ldots,N)$ of various chemical components are selected as *variables of state* of a mixture, then the thermodynamic state of an elementary physical volume in the vicinity of a point $\boldsymbol{r}$ (with coordinates x_j $(j = 1,2,3)$) in an instant time t is comprehensively described by the key functional dependence, that is, by specific entropy $S = S(\varepsilon(\boldsymbol{r},t),\ v(\boldsymbol{r},t),\ Z_1(\boldsymbol{r},t),\ldots,Z_N(\boldsymbol{r},t))$, for which the Gibbs identity (2.2.6) is valid.

On the basis of this identity, the conjugated quantities

$$\frac{1}{T} = \left\{\frac{\partial S}{\partial \varepsilon}\right\}_{V,\{Z_\alpha\}}, \quad \frac{p}{T} = \left\{\frac{\partial S}{\partial v}\right\}_{\varepsilon,\{Z_\alpha\}}, \quad -\frac{\mu_\alpha}{T} = \left\{\frac{\partial S}{\partial Z_\alpha}\right\}_{\varepsilon,V,\{Z_\beta \neq Z_\alpha\}} \tag{2.2.11}$$

can be attributed with the meaning of local temperature $T(\boldsymbol{r},t)$, local pressure $p(\boldsymbol{r},t)$, and local chemical potential $\mu_\alpha(\boldsymbol{r},t)$ of α-component ($\alpha = 1,2,\ldots,N$). In other words, these quantities are actually determined by the relations (2.2.11). Within the phenomenological approach, the area of applicability of the postulate about local thermodynamic equilibrium is established from experimental evidence. As a rule, this postulate is valid, if dissipative processes in a system are important and eliminate an occurrence of large gradients of state variables.

It is necessary also to emphasize, that the hypothesis of local equilibrium of a medium is equivalent to the supposition on an acceptance, in addition to the Gibbs' relation, all the rest of thermostatic relations for infinitesimally small areas of a nonequilibrium system. It can be assumed, for example, that specific thermodynamic potential (2.2.8) exists and the known relation (*de Groot and Mazur, 1962*)

$$dG = -SdT - vdp + \sum_{\beta=1}^{N} \mu_\beta dZ_\beta, \tag{2.2.12}$$

is valid. Then, from (2.2.12) with the help of (2.2.6) it is possible to obtain the *Gibbs-Duhem* relation

$$\rho \sum_{\alpha=1}^{N} Z_\alpha d\mu_\alpha = -\rho\, SdT + dp. \tag{2.2.13}$$

It is of fundamental importance for thermodynamics of multicomponent systems and will be widely used below.

Concluding this subsection, let us now summarize the basic statements which are adopted as independent postulates in the applications of nonequilibrium thermodynamics to any irreversible process:

- the principle of quasi-local thermodynamic equilibrium is fair;
- the inequality $\sigma_{(S)} \geq 0$ expressing the second law of thermodynamics, is valid for entropy production related to irreversible processes in a system;
- when restricted to linear area, the phenomenological relations (2.2.1) for fluxes and thermodynamic forces, involved in (2.2.4) for entropy production, are valid;
- relations of symmetry (2.2.2) for kinetic coefficients involved in the linear laws (2.2.1) hold true.

In addition to the above items, the correct axiomatic statement of nonequilibrium thermodynamics demands that the Onsager-Kasimir reciprocity principle should be supplemented by the description of the set of independent thermodynamic fluxes (*Meixner, 1969*).

2.2.2. The Evolutionary Entropy Transfer Equation for Multicomponent Gaseous Mixtures

To take advantage of the Onsager linear relations (2.2.1), it is first necessary to find the particular form of the entropy balance equation (2.2.5) for the model of multicomponent thermodynamic system under consideration. For this purpose, we exclude the relevant derivatives with respect to time in the right part of the Gibbs identity (2.2.6), using the hydrodynamic equations of mixture (2.1.6), (2.1.7), and (2.1.40) for quantities $\varepsilon,\ v, Z_\alpha$ $(\alpha = 1,2,..., N)$. We then obtain

$$\rho \frac{dS}{dt} = -\frac{\partial_j q_j}{T} + \frac{1}{T}\left(\pi_{ij}\partial_j V_i + \sum_{\alpha=1}^{N} J_{\alpha j} F_{\alpha j} - \sum_{\alpha=1}^{N} \mu_\alpha \sigma_\alpha \right) + \frac{1}{T}\sum_{\alpha=1}^{N} \mu_\alpha \partial_j J_{\alpha j} \,. \qquad (2.2.14)$$

Such form of the relation (2.2.14) does not match the equation of the substantial balance such as (2.1.1). However, with the help of transformations

$$\partial_j q_j / T = \partial_j (q_j / T) - q_j \partial_j (1/T) = \partial_j (q_j / T) + (q_j / T^2)\partial_j T, \qquad (2.2.15)$$

$$(\mu_\alpha / T)\partial_j J_{\alpha j} = \partial_j (\mu_\alpha J_{\alpha j} / T) - J_{\alpha j}\partial_j (\mu_\alpha / T), \qquad (\alpha = 1,2,..., N) \qquad (2.2.16)$$

relation (2.2.14) is easy to reduce to the form of the substantial balance equation

$$\rho\frac{dS}{dt}=-\partial_j\left\{\frac{1}{T}\left(q_j-\sum_{\beta=1}^{N}\mu_\beta J_{\beta j}\right)\right\}+\frac{1}{T}\left\{-J_{qj}\frac{\partial_j T}{T}+\pi_{ij}\partial_j V_i-\right.$$

$$\left.-\sum_{\beta=1}^{N}\left[T\partial_j\left(\frac{\mu_\beta}{T}\right)-F_{\beta j}+h_\beta\frac{\partial_j T}{T}\right]J_{\beta j}+\sum_{s=1}^{r}A_s\xi_s\right\}, \tag{2.2.17}$$

where

$$A_s=-\sum_{\alpha=1}^{N}\mu_\alpha\nu_{\alpha s},\qquad (s=1,2,...,r) \tag{2.2.18}$$

is the so-called *chemical affinity of s-th* reaction (*de Groot and Mazur, 1962*) , and J_{qj} is the effective heat flux defined by formula (2.1.50). Comparing (2.2.15) with the general equation of entropy balance (2.2.5), we obtain the following expression for the substantial density of the entropy flux

$$J_{(S)j}\equiv\frac{1}{T}\left[q_j-\sum_{\alpha=1}^{N}\mu_\alpha J_{\alpha j}\right]=\frac{1}{T}\left[J_{qj}-\sum_{\alpha=1}^{N}S_\alpha J_{\alpha j}\right], \tag{2.2.19}$$

while the entropy production multiplied by T is determined as

$$0\le T\sigma_{(S)}\equiv-J_{qj}\frac{\partial_j T}{T}+\pi_{ij}\partial_j V_i+\sum_{\alpha=1}^{N}\left[-T\partial_j\left(\frac{\mu_\alpha}{T}\right)-h_\alpha\frac{\partial_j T}{T}+F_{\alpha j}\right]J_{\alpha j}+\sum_{s=1}^{r}A_s\xi_s, \tag{2.2.20}$$

and this quantity *(dissipation of energy)*, according to the second law of thermodynamics, is positively defined for irreversible processes.

Let us introduce designations for thermodynamic forces:

$$X_{0j}\equiv\frac{\partial_j T}{T},\quad X^*_{\alpha j}\equiv-F_{\alpha j}+T\partial_j\left(\frac{\mu_\alpha}{T}\right)+h_\alpha\frac{\partial_j T}{T},\quad (\alpha=1,2,...,N),$$

$$e_{ij}\equiv\tfrac{1}{2}(\partial_j V_i+\partial_i V_j),\qquad A_s\equiv-\sum_{\beta=1}^{N}\mu_\beta\nu_{\beta s}\quad (s=1,2,...,r). \tag{2.2.21}$$

The dissipation of energy $T\sigma_{(S)}$, being local measure of non-equilibrium of a system, looks like the total of paired products of thermodynamic forces (2.2.21) by fluxes

$$J_{qj},\quad J_{\alpha j}\quad (\alpha=1,2,...,N),\quad \pi_{ij},\quad \xi_s\ (s=1,2,...,r), \tag{2.2.22}$$

corresponding to four sources of nonequilibrium processes of diverse physical nature. Therefore, if in the following one postulates a linear dependence (2.2.1) between the fluxes (2.2.22) and thermodynamic forces (2.2.21), it is possible to take advantage of the reciprocity relations (2.2.2) in order to reduce the number of unknown phenomenological coefficients and to eventually obtain the necessary defining relations.

The exposed formalism of the nonequilibrium thermodynamics will be applied most comprehensively to a turbulized multicomponent continuum in Chapter 5. Nonetheless, here we shall address the processes of molecular diffusion and heat conduction as the phenomena described by polar vectors, i.e. by vectors $\boldsymbol{b}$, which, when reflected at the point $\boldsymbol{r}=0$, behave as $\boldsymbol{b}^{ref}(\boldsymbol{r},t)=-\boldsymbol{b}(-\boldsymbol{r},t)$. Such isolated examination of processes of heat and mass transfer is especially justified in isotropic systems, in which phenomena, described by thermodynamic forces and fluxes of various tensor dimensions, do not affect each other (*the Curie principle*).

2.3. DEFINING RELATIONS FOR DIFFUSION AND HEAT FLUXES IN CONTINUOUS MULTICOMPONENT MIXTURES

Now we will derive the defining relations for molecular fluxes of diffusion and heat using the methods of thermodynamics of irreversible processes. The *Stefan-Maxwell relations* for multicomponent diffusion will be also obtained, together with the respective expression for full heat flux, which are pertinent for describing the processes of heat and mass transfer in a multicomponent gaseous medium.

2.3.1. Linear Constituent Relations for Molecular Fluxes of Diffusion and Heat

According to (2.2.20), the following expression for dissipation of energy $T\sigma^*_{(S)}$ due to diffusion and heat conduction can be written for an arbitrary N-component heat-conducting mixture

$$0\le T\sigma^*_{(S)}=-J_{qj}\frac{\partial_j T}{T}+\sum_{\alpha=1}^{N}\left\{-T\partial_j\left(\frac{\mu_\alpha}{T}\right)-h_\alpha\frac{\partial_j T}{T}+F_{\alpha j}\right\}J_{\alpha j}=$$

$$=J_{qj}X_{0j}-\sum_{\alpha=1}^{N}J_{\alpha j}X^*_{\alpha j}, \tag{2.3.1}$$

with the following designations:

$$X_{0j}\equiv-\frac{\partial_j T}{T},\quad X^*_{\alpha j}\equiv-F_{\alpha j}+T\partial_j\left(\frac{\mu_\alpha}{T}\right)+h_\alpha\frac{\partial_j T}{T}.$$

To deduce the defining relations for molecular diffusion and heat fluxes, $J_{\alpha j}$ and J_{qj}, respectively, and to compare the results obtained with those emerging from the gas kinetic theory, let us express the chemical potential gradient μ_α of the α-component using the gradients of thermohydrodynamic parameters. Next, let us view the mixture under consideration as an ideal mixture of perfect gases. Then, the chemical potential of any of the N components looks like (*Prigogine and Defay, 1954*):

$$\mu_\alpha = \mu_\alpha^0(T,p) + \mathrm{k}T\ln(x_\alpha), \quad (\alpha = 1,2,\ldots,N), \tag{2.3.2}$$

where $\mu_\alpha^0(T,p)$ is the chemical potential of the pure α-component at temperature T and pressure p (see (3.2.13)), and $x_\alpha = n_\alpha / n$ is the molar ratio of the α-component. Therefore it is possible to write

$$\partial_j\left(\frac{\mu_\alpha}{T}\right) = \frac{\partial\left[\mu_\alpha^0(T,p)/T\right]}{\partial T}\partial_j T + \frac{\partial\left[\mu_\alpha^0(T,p)\right]}{\partial p}\frac{\partial_j p}{T} + \mathrm{k}\frac{\partial_j x_\alpha}{x_\alpha}. \tag{2.3.3}$$

As the thermodynamic relations

$$\frac{\partial\left[\mu_\alpha^0(T,p)/T\right]}{\partial T} = -\frac{h_\alpha}{T^2}, \qquad \frac{\partial\left[\mu_\alpha^0(T,p)\right]}{\partial p} = v_\alpha = \frac{1}{n}, \tag{2.3.4}$$

are applied, then (2.3.3) can be rewritten as

$$T\partial_j\left(\frac{\mu_\alpha}{T}\right) = -\frac{h_\alpha}{T}\partial_j T + \frac{\partial_j p}{n} + \mathrm{k}T\frac{\partial_j x_\alpha}{x_\alpha}. \tag{2.3.5}$$

Hence, for the thermodynamic force $X^*_{\alpha j}$ $(\alpha = 1,2,.., N)$ we obtain the following expression:

$$X^*_{\alpha j} = -F_{\alpha j} + \frac{\partial_j p}{n} + \mathrm{k}T\frac{\partial_j x_\alpha}{x_\alpha}. \tag{2.3.6}$$

As was said earlier, when a thermodynamic flux is close to thermodynamic equilibrium, such a flux can be represented as a symmetric function of thermodynamic forces (this is the basic postulate of thermodynamics of irreversible processes):

$$J_{qj} = L_{00}^{jk} X_{0k} - \sum_{\beta=1}^{N} L_{0\beta}^{jk} X^*_{\beta k}, \qquad (^1)$$

$$J_{\alpha j} = L_{\alpha 0}^{jk} X_{0k} - \sum_{\beta=1}^{N} L_{\alpha\beta}^{jk} X^*_{\beta k}, \quad (\alpha = 1,2,\ldots,N). \qquad (^2) \tag{2.3.7}$$

Here the indexes $j,k = 1,2,3$ relate to the coordinate system while the kinetic coefficients $L_{\alpha\beta}^{jk}$ $(\alpha,\beta = 0,1,\ldots,N)$ represent tensors which depend on state variables and parameters, describing the geometrical symmetry of the medium.

Next, we will examine isotropic media in reference to the complete group of orthogonal coordinate transformations. According to the general tensor function theory (*Sedov, 1984*), the symmetry properties of isotropic media are characterized fairly by the metric tensor g^{jk} and all tensors are tensor functions of the metric tensor only:

$$L_{\alpha\beta}^{jk} = L_{\alpha\beta} g^{jk}, \tag{2.3.8}$$

where $L_{\alpha\beta}$ are scalar quantities. In the orthogonal coordinate system $g^{jk} = \delta^{jk}$, and the defining relations (2.3.7) will take the form

$$J_{qj} = L_{00} X_{0j} - \sum_{\beta=1}^{N} L_{0\beta} X_{\beta j}^{*}, \qquad (^1)$$

(2.3.9)

$$J_{\alpha j} = L_{\alpha 0} X_{0j} - \sum_{\beta=1}^{N} L_{\alpha\beta} X_{\beta j}^{*}, \quad (\alpha = 1,2,\ldots,N). \qquad (^2)$$

As an immediate consequence of postulating the linear relations (2.3.9), the Onsager-Kasimir condition of symmetry yields

$$L_{\alpha\beta} = L_{\beta\alpha} \quad (\alpha, \beta = 0, 1, 2,\ldots, N), \tag{2.3.10}$$

which reduces the number of independent phenomenological coefficients $L_{\alpha\beta}$ from $(N+1)^2$ to $\frac{1}{2}(N+1)(N+2)$.

Furthermore, additional ($N+1$) connections imposed on the Onsager coefficients follow from the fact that the thermodynamic forces X_{0j} and $X_{\beta j}^{*}$ (β=1,2..., N) are linearly independent, as well as from the relation (2.1.8). Indeed, substituting relation (2.3.9 $^{(2)}$) into identity (2.1.8 $^{(2)}$), we obtain

$$0 = \sum_{\alpha=1}^{N} M_{\alpha} J_{\alpha j} = \left[\sum_{\alpha=1}^{N} M_{\alpha} L_{\alpha 0}\right] X_{0j} - \sum_{\beta=1}^{N}\left[\sum_{\alpha=1}^{N} M_{\alpha} L_{\alpha\beta}\right] X_{\beta j}^{*},$$

whence, by virtue of the indicated linear independence , it follows that

$$\sum_{\alpha=1}^{N} M_{\alpha} L_{\alpha 0} = 0, \qquad (^1)$$

(2.3.11)

$$\sum_{\alpha=1}^{N} M_{\alpha} L_{\alpha\beta} = 0, \quad (\beta = 1,2,\ldots, N). \qquad (^2)$$

Thus, only $\frac{1}{2} N(N+1)$ of the $(N+1)^2$ independent coefficients $L_{\alpha\beta}$ $(\alpha, \beta = 0,1,2,\ldots, N)$ remain.

Among other properties of the positive semi-definite matrix of the phenomenological coefficients $L_{\alpha\beta}$ we will specify the following *Sylvester conditions*: $L_{\alpha\beta} > 0$ (non-negativity of all matrix elements $L_{\alpha\beta}$ in the main diagonal), $(L_{\alpha\beta})^2 \le L_{\alpha\alpha} L_{\beta\beta}$ (each minor of the non-negative definite matrix $L_{\alpha\beta}$, containing elements of its main diagonal as the own main diagonal, also should be non-negative), etc. These conditions are a corollary of the non-negativity of the dissipative function $T\sigma^*_{(S)}$ written with regard to (2.3.9) as the uniform quadratic form of forces. Within the limits of these specified restrictions, the introduced Onsager's kinetic coefficients vary in a very wide range according to a coupling degree between the processes of heat and mass transfer.

Using the phenomenological approach, in order to define the molecular exchange coefficients similar to the respective coefficients obtained from the kinetic theory, we will introduce a new set of linearly dependent vectors $d_{\beta j}$ instead of the linearly independent vectors $X^*_{\beta j}$ $(\beta = 1,2,..., N)$. This may be done, for instance, as follows. Let us assume

$$d_{\beta j} = \frac{n_\beta}{p}(X^*_{\beta j} + M_\beta R^*_j), \quad (\beta = 1,2,..., N), \tag{2.3.12}$$

and define the vector R^*_j, shared by all of the components, from the condition

$$\sum_{\beta=1}^{N} d_{\beta j} = 0. \tag{2.3.13}$$

Then, taking into account (2.3.6), we can write

$$R^*_j = -\sum_{\beta=1}^{N} Z_\beta X^*_{\beta j} = -\frac{\partial_j p}{p} + \sum_{\beta=1}^{N} Z_\beta F_{\beta j}, \tag{2.3.14}$$

whence, finally, for the quantities $d_{\beta j}$ $(\beta = 1,2,..., N)$ we find

$$d_{\beta j} = \partial_j x_\beta + (x_\beta - M_\beta Z_\beta)\frac{\partial_j p}{p} + \left[-F_{\beta j} + M_\beta \sum_{\alpha=1}^{N} Z_\alpha F_{\alpha j} \right] \frac{\partial_j n}{p}. \tag{2.3.15}$$

Note that the expression (2.3.15) perfectly coincides with the definition of *diffusive thermodynamic force vectors* emerging from the kinetic theory of rarefied gases (*Hirschfelder et al., 1954*). Now substituting the formula $X^*_{\beta j} = (p/n_\beta) d_{\beta j} - M_\beta R^*_j$ in (2.3.9) and taking into account the relations (2.3.11), we finally obtain

$$J_{qj} = L_{00} X_{0j} - p \sum_{\beta=1}^{N} L_{0\beta} \frac{d_{\beta j}}{n_\beta}, \tag{2.3.16}$$

$$J_{\alpha j} = L_{\alpha 0} X_{00} - p \sum_{\beta=1}^{N} L_{\alpha\beta} \frac{d_{\beta j}}{n_\beta}, \qquad (\alpha = 1,2,..., N). \qquad (2.3.17)$$

Let us compare these rheological relations with the equivalent relations given by the kinetic theory. Within the frame of the basic hypothesis that the gradients of thermohydrodynamic quantities and external forces give rise to small departures of the distribution functions from the equilibrium Maxwellian distribution, the formal kinetic theory describing multicomponent gaseous mixtures of monatomic gases with moderate density as in the first order *Chapman-Enskog* method (*Chapman and Couling, 1952*) leads to the following expressions for the total heat flux q_j and diffusion rate fluxes $w_{\alpha j} \equiv V_{\alpha j} - V_j$ $(\equiv J_{\alpha j}/n_\alpha)$ (see also *Marov and Kolesnichenko, 1987*):

$$q_j = -\lambda_0(\xi)\partial_j T - p\sum_{\beta=1}^{N} D_{T\beta} d_{\beta j} + \sum_{\beta=1}^{N} h_\beta n_\beta w_{\beta j}, \qquad (2.3.18)$$

$$w_{\alpha j} = -D_{T\alpha} \frac{\partial_j T}{T} - p\sum_{\beta=1}^{N} D_{\alpha\beta}(\xi) d_{\beta j}, \qquad (\alpha = 1,2,..., N), \qquad (2.3.19)$$

where $D_{\alpha\beta}(\xi)$ and $D_{T\alpha}(\xi)$ are the symmetric multicomponent diffusion coefficients $(D_{\alpha\beta} = D_{\beta\alpha})$ and thermal diffusions coefficients, respectively, and $\lambda_0(\xi)$ is the partial thermal conductivity of the multicomponent gaseous mixture in the absence of all diffusion forces. The order of approximation $\xi = 1,2,...$, for which the transport coefficients are defined here, matches to the number of terms in the expansion of the perturbed distribution functions coefficients of the components in the *Sonin-Laguerre* polynomial series. Because the coefficients $D_{\alpha\beta}(\xi)$ and $D_{T\alpha}(\xi)$ are not linearly independent,

$$\sum_{\alpha=1}^{N} M_\alpha Z_\alpha D_{T\alpha}(\xi) = 0, \qquad (^1)$$

$$\sum_{\alpha=1}^{N} M_\alpha Z_\alpha D_{\alpha\beta} = 0, \quad (\beta = 1,2,..., N), \qquad (^2)$$

(2..3.20)

there are ½ *N*(*N-1*) independent diffusion factors and (*N-1*) independent thermal diffusion factors for an *N*-component mixture.

By analogy with the kinetic theory of gases (formulas (2.3.18)–(2.3.20)), within the phenomenological approach we introduce (in terms of the Onsager coefficients) the symmetric multicomponent diffusion factors $D_{\alpha\beta}$ $(\alpha, \beta = 1,2,..., N)$, the thermal diffusion factors $D_{T\alpha}$ $(\alpha = 1,2,..., N)$, and the partial thermal conductivity λ_0 (which will be used below to express true heat conduction within a mixture; see (2.3.73)) with the help of the following formulas:

$$\lambda_0 \equiv \frac{1}{T} L_{00}, \qquad D_{T\alpha} \equiv \frac{1}{n_\alpha} L_{0\alpha} \quad (\alpha = 1,2,\ldots, N),$$

$$D_{\alpha\beta} \equiv \frac{p}{n_\alpha n_\beta} L_{\alpha\beta}, \quad (D_{\alpha\beta} = D_{\beta\alpha}) \quad (\alpha, \beta = 1,2,\ldots, N). \tag{2.3.21}$$

Then, the defining relations (2.3.16) and (2.3.17) will take the following standard form for the fluxes of diffusion and heat:

$$J_{qj} = -\lambda_0 \partial_j T - p \sum_{\beta=1}^{N} D_{T\beta} d_{\beta j}, \tag{2.3.22}$$

$$J_{\alpha j} = -n_\alpha D_{T\beta} \partial_j (\ln T) - n_\alpha \sum_{\beta=1}^{N} D_{\alpha\beta} d_{\beta j}, \quad (\alpha = 1,2,\ldots, N). \tag{2.3.23}$$

Here the molecular transfer coefficients $D_{T\beta}$ and $D_{\alpha\beta}$ introduced earlier fit to the relations

$$\sum_{\beta=1}^{N} M_\beta Z_\beta D_{T\beta} = 0, \qquad (^1)$$

$$\sum_{\beta=1}^{N} M_\beta Z_\beta D_{\alpha\beta} = 0, \quad (\alpha = 1,2,\ldots, N) \qquad (^2) \tag{2.3.24}$$

by virtue of (2.3.10) and (2.3.11).

As to the coefficient λ_0, the following remark must be made. As can be seen from (2.3.22) λ_0 can not be determined from direct experimental data, as the temperature gradient in a gas mixture invokes thermodiffusion and, therefore, gives rise to concentration gradients. Therefore, the quantities $d_{\beta j}$ are not equal to zero even in stationary processes, and hence, the heat flux owing to the temperature gradient, is also accompanied by a heat flux due to the concentration gradient.

Let us note that if the gaseous medium is anisotropic, for instance at the expense of its interaction with the strong electromagnetic field of a planet, the theory developed above becomes more complicated. In that case, in addition to the metric tensor g^{jk}, other tensors describing the anisotropy of the medium should be introduced as arguments of the functions $L^{jk}_{\alpha\beta}$ (*Sedov, 1984*).

2.3.2. The Stefan-Maxwell Relations for Multicomponent Diffusion

The gas-kinetic defining relations (2.3.18) and (2.3.19) are valid for any approximation number of the Chapman-Enskog method and, in general, the molecular exchange coefficients $D_{\alpha\beta}(\xi)$, $D_{T\alpha}(\xi)$, and $\lambda_0(\xi)$ are expressed in terms of the ratios of the order

$N\xi$ determinants with composite elements (*Hirschfelder et al., 1954*). The necessity of taking into account the higher order approximations leads to an enormous increase of computations when evaluating such coefficients. This is, for example, the case for flows of ionized gaseous mixtures in the planetary ionosphere (ξ=3 and more (*Devoto, 1966*)), where both temperature and elementary chemical composition vary within the flow.

An expression for the true thermal conductivity λ appears to be even more cumbersome. To evaluate the latter, it is necessary to eliminate the vectors $d_{\alpha j}$ from (2.3.18) using the diffusion relations (2.3.19). Their formal solution with respect to $d_{\alpha j}$ gives

$$d_{\beta j} = -\sum_{\alpha=1}^{N} \mathrm{E}_{\beta\alpha} w_{\alpha j} - \sum_{\alpha=1}^{N} D_{T\alpha}\, \mathrm{E}_{\beta\alpha} \partial_j (\ln T)\,, \tag{2.3.25}$$

where $\mathrm{E}_{\beta\alpha}$ are the elements of the inversed $D_{\beta\alpha}$ matrix (*Kolesnichenko and Tirskiy, 1976*). Substituting (2.3.25) into (2.3.18), we find

$$q_j = -\lambda \partial_j T + \sum_{\alpha=1}^{N} n_\alpha h_\alpha w_{\alpha j} + p \sum_{\alpha=1}^{N}\sum_{\beta=1}^{N} D_{T\beta}\, \mathrm{E}_{\beta\alpha}\, D_{T\alpha}\,, \tag{2.3.26}$$

where the true thermal conductivity is given by the expression

$$\lambda = \lambda_0 - \mathrm{k}\, n \sum_{\alpha=1}^{N}\sum_{\beta=1}^{N} D_{T\beta}\, \mathrm{E}_{\beta\alpha}\, D_{T\alpha}\,. \tag{2.3.27}$$

If transport properties are mainly determined by the neutral components of the gaseous mixture (like in aeronomical problems) the quantity λ_0 is only slightly different from the true thermal conductivity λ and the second term in (2.3.27) can often be neglected. However, if charged components occur additionally (as in the ionosphere), the contribution of the second term in (2.3.27) becomes rather important and can reach about 30% (*Ferziger and Kaper, 1972; Marov and Kolesnichenko, 1987*). It is extremely inconvenient to use the formula (2.3.27) in the case of partially ionized gas mixtures, as the necessary calculations require double inversion of the higher order matrices: the first inversion relates to finding the coefficients λ_0, $D_{\beta\alpha}$ and $D_{T\alpha}$ from the relevant systems of algebraic equations from the kinetic theory, while the second one relates to resolving the system (2.3.19) with respect to the vectors $d_{\alpha j}$. Besides, the system of the diffusion equations emerging after substituting the diffusion rates $w_{\alpha j}$ from (2.3.19) into the balance equations (2.1.7), is proven not to be resolved with respect to higher derivatives.

As is known, the numerical evaluation of such systems encounters certain difficulties. Therefore, it is more convenient to use the defining relations (2.3.23) in the form resolved with respect to thermodynamic diffusion forces $d_{\alpha j}$ in terms of the fluxes $w_{\alpha j}$ (*Marov and Kolesnichenko, 1987*). The inverse transformation of this kind can be written as the so-called *generalized Stefan-Maxwell relations*

$$d_{\beta j} = \sum_{\substack{\alpha=1 \\ \alpha\neq\beta}}^{N} R_{\beta\alpha}(w_{\alpha j} - w_{\beta j}) + \frac{\partial_j T}{T} \sum_{\substack{\alpha=1 \\ \alpha\neq\beta}}^{N} R_{\beta\alpha}(D_{T\alpha} - D_{T\beta}), \quad (\beta = 1,2,\ldots,N), \tag{2.3.28}$$

which involve the diffusion constants of binary gaseous mixtures $\mathcal{D}_{\alpha\beta}$ or the diffusion resistance coefficients $R_{\alpha\beta}$, intimately related with $\mathcal{D}_{\alpha\beta}$ (see formulas (2.3.59) and (2.3.62)) instead of the multicomponent diffusion coefficients $D_{\alpha\beta}$. Availability of the precise Stefan-Maxwell relations, valid for higher approximations of the molecular exchange coefficients, would also allow one to obtain the simplest formulas for the true thermal conductivity λ and the thermal diffusion factor $D_{T\alpha}$ in any approximation of the Chapman-Enskog method (*Kolesnichenko, 1979*).

Historically, the relations (2.3.28) have been obtained phenomenologically by Stefan (*Stefan, 1871*) and Maxwell (*Maxwell, 1890*) assuming that the force, exerted on an α-particle by a β-particle, is proportional to the difference of their diffusion rates, and that the total resistance force on an α-particle in the mixture is equal to the sum of the independent resistance forces from all other particle species. They additionally neglected thermodiffusion ($D_{T\alpha} = 0$) and also supposed that the resistance coefficient matrix $R_{\alpha\beta}$ is symmetric for multicomponent mixtures.

The generalized Stefan-Maxwell relations, taking into account thermodiffusion and the effects of external mass forces, were obtained using the kinetic theory for monatomic gases in the first approximation of the Chapman-Enskog method for the multicomponent diffusion coefficients $\lfloor D_{\alpha\beta} \rfloor_1$ and the second approximation for the thermal diffusion coefficients $[D_{T\alpha}]_2$ in the following form (*Hirschfelder et al., 1954*):

$$d_{\beta j} = \sum_{\substack{\alpha=1 \\ \alpha\neq\beta}}^{N} \frac{x_\alpha x_\beta}{[\mathcal{D}_{\alpha\beta}]_1}(w_{\alpha j} - w_{\beta j}) + \sum_{\substack{\alpha=1 \\ \alpha\neq\beta}}^{N} \frac{x_\alpha x_\beta}{[\mathcal{D}_{\alpha\beta}]_1}\left([D_{T\alpha}]_2 - [D_{T\beta}]_2\right) \quad (\beta = 1,2,\ldots,N) \tag{2.3.29}$$

Here the sampling function involving the only *Sonin-Laguerre* polynomial was utilized in the variation representation of the integral equations defining the first Chapman-Enskog iteration. In this approximation, substituting the diffusion forces from (2.3.29) into (2.3.18) yields the following expression for the thermal conductivities

$$\lambda = [\lambda_0]_2 - \frac{\mathrm{k}n}{2} \sum_{\alpha=1}^{N} \sum_{\beta=1}^{N} \frac{x_\alpha x_\beta}{[\mathcal{D}_{\alpha\beta}]_1} \left\{[D_{T\alpha}]_2 - [D_{T\beta}]_2\right\}. \tag{2.3.30}$$

The relations (2.3.29) and the true thermal conductivity were also derived based on the complete second approximation of the Chapman-Enskog theory, although for this case the symmetry of the resistance coefficients $R_{\alpha\beta}$ has not been established (*Muckenfuss and Curtiss, 1958*). Moreover, it was supposed (*Truesdell, 1962*) that in the second approximation the matrix $R_{\alpha\beta}$ is asymmetric; in other words, the Stefan-Maxwell relations (2.3.29) are not thermodynamically universal in their character but represent a mathematical phenomenon inherent only to the first approximation of the Chapman-

Enskog theory. Later on, an attempt to obtain the relations (2.3.28) from the gas kinetic theory in an arbitrary approximation was undertaken (*Muckenfuss, 1973*). However, it was wrongly concluded that the correction factors to the binary diffusion coefficients (taking into account higher approximations when expanding perturbed the distribution functions of the individual components by *Sonin-Laguerre* polynomials) depend only on the number of approximations N of the Chapman-Enskog theory, with N the number of components in the system, but not on the interacting components themselves. Besides, the explicit form of this correction was not obtained.

The generalized Stefan-Maxwell relations and formulas for inserting corrections to the binary diffusion constants in an arbitrary approximation of the molecular transport coefficients for partially ionized mixtures were deduced in the following studies. For the first time the extreme cases of zero magnetic field (*Kolesnichenko, 1979*) and a strong magnetic field bringing about anisotropy in the transport coefficients were examined (*Kolesnichenko, 1982; Kolesnichenko and Marov, 1982*). The resistance coefficient symmetries have been shown, in those studies, to be in perfect agreement with the respective results of irreversible thermodynamic processes.

2.3.3. The Generalized Stefan-Maxwell Relations Based on the Methods of Thermodynamics of Irreversible Processes

Now we will proceed to the phenomenological deduction of the Stefan-Maxwell relations for regular motions in a gas mixture. For the purpose, let us resolve the equations (2.3.16) and (2.3.17) with respect to the generalized thermodynamic forces X_{0j} and $X_{\beta j} \equiv -(p/n_\beta)d_{\beta j}$ $(\beta = 1,2,...,N)$ in terms of the fluxes J_{qj} and $J_{\alpha j}$ $(1,2,...,N)$. With this in mind, omitting the last equation of system (2.3.17) and using (2.3.11), the relations (2.3.16) and (2.3.17) can be represented as

$$J_{qj} - L_{00}X_{0j} = \sum_{\beta=1}^{N-1} L_{0\beta}\left[X_{\beta j} - \frac{M_\beta}{M_N}X_{Nj}\right], \tag{2.3.31}$$

$$J_{\alpha j} - L_{\alpha 0}X_{0j} = \sum_{\beta=Б}^{N-1} L_{\alpha\beta}\left[X_{\beta j} - \frac{M_\beta}{M_N}X_{Nj}\right], \quad (\alpha = 1,2,...,N-1), \tag{2.3.32}$$

with the following designations:

$$X_{0j} \equiv \frac{\partial_j T}{T}, \quad X_{\beta j} \equiv -\frac{p}{n_\beta}d_{\beta j} \quad (\beta = 1,2,...,N-1).$$

Now resolving the system (2.3.32) with respect to the differences $X_{\beta j} - \dfrac{M_\beta}{M_N}X_{Nj}$ $(\beta = 1,2,...,N)$, we find

$$X_{\beta j}-\frac{M_\beta}{M_N}X_{Nj}=\sum_{\alpha=1}^{N-1}\mathcal{M}_{\beta\alpha}(J_{\alpha j}-L_{\alpha 0}X_{0j}),\quad (\beta=1,2,...,N-1). \tag{2.3.33}$$

Here $\mathcal{M}_{\beta\alpha}$ are elements of the inverse matrix matching the relations

$$\sum_{\alpha=1}^{N-1}\mathcal{M}_{\beta\alpha}L_{\alpha\gamma}=\delta_{\beta\gamma}=\begin{cases}1, & \beta=\gamma\\ 0, & \beta\neq\gamma\end{cases}. \tag{2.3.34}$$

Symmetry of the coefficients $\mathcal{M}_{\beta\alpha}$ follows from symmetry of the phenomenological coefficients $L_{\alpha\beta}$:

$$\mathcal{M}_{\alpha\beta}=\mathcal{M}_{\beta\alpha}\quad (\alpha,\beta=1,2,...,N-1). \tag{2.3.35}$$

From the relation (2.3.31), with due regard to (2.3.33), we obtain

$$J_{qj}=\left[L_{00}-\sum_{\alpha=1}^{N-1}\sum_{\beta=1}^{N-1}L_{0\beta}\mathcal{M}_{\beta\alpha}L_{\alpha 0}\right]X_{0j}+\sum_{\alpha=1}^{N-1}\left[\sum_{\beta=1}^{N-1}L_{0\beta}\mathcal{M}_{\beta\alpha}\right]J_{\alpha j}. \tag{2.3.36}$$

Now, multiplying each of the equations (2.3.33) by Z_β (note that $\sum_{\beta=1}^{N}M_\beta Z_\beta=1$; $\sum_{\beta=1}^{N}Z_\beta X_{\beta j}=0$) and adding them up from 1 to $(N-1)$, we find the desired relations for the diffusion force vectors in the following form:

$$X_{Nj}=\left[M_N\sum_{\alpha=1}^{N-1}\sum_{\gamma=1}^{N-1}Z_\gamma\mathcal{M}_{\gamma\alpha}L_{\alpha 0}\right]X_{0j}-M_N\sum_{\alpha=1}^{N-1}\left[\sum_{\gamma=1}^{N-1}Z_\gamma\mathcal{M}_{\gamma\alpha}\right]J_{\alpha j}, \tag{2.3.37}$$

$$X_{\beta j}=\sum_{\alpha=1}^{N-1}L_{\alpha 0}\left[M_\beta\sum_{\gamma=1}^{N-1}Z_\gamma\mathcal{M}_{\gamma\alpha}-\mathcal{M}_{\beta\alpha}\right]X_{0j}-\sum_{\alpha=1}^{N-1}\left[\mathcal{M}_{\beta\alpha}-M_\beta\sum_{\gamma=1}^{N-1}Z_\gamma\mathcal{M}_{\gamma\alpha}\right]J_{\alpha j}. \tag{2.3.38}$$

$$(\beta=1,2,...,N)$$

The thermodynamic force X_{0j} can be retrieved from the expression (2.3.36). The equations (2.3.37) and (2.3.38) represent the inversed relations (2.3.32). To write these equations as the generalized Stefan-Maxwell relations, let us add the identity $\sum_{\alpha=1}^{N}M_\alpha J_{\alpha j}=0$, multiplied by the constants a_0, a_N and a_β $(\beta=1,2,...,N-1)$, and (2.3.36), (2.3.37), and (2.3.38), respectively, and define the free parameters a_0 and a_β from the symmetry condition for the coefficients A. For this purpose, it is necessary to put

$$a_0 = -\sum_{\alpha=1}^{N-1}\sum_{\gamma=1}^{N-1} Z_\gamma \mathcal{M}_{\gamma\alpha} L_{\alpha 0}, \tag{2.3.39}$$

$$a_\beta = \left(M_\beta / M_N\right) a_N - \sum_{\gamma=1}^{N-1} Z_\gamma \mathcal{M}_{\gamma\beta}, \quad (\beta = 1,2,\ldots, N-1) \tag{2.3.40}$$

Then we obtain

$$J_{qj} = A_{00} X_{0j} + \sum_{\alpha=1}^{N} A_{0\alpha} J_{\alpha j}, \tag{2.3.41}$$

$$-X_{\beta j} = A_{\beta 0} X_{0j} + \sum_{\alpha=1}^{N} A_{\beta\alpha} J_{\alpha j}, \qquad (\beta = 1,2,\ldots, N), \tag{2.3.42}$$

where the coefficients A are

$$A_{00} = L_{00} - \sum_{\beta=1}^{N-1}\sum_{\gamma=1}^{N-1} L_{0\beta} \mathcal{M}_{\beta\gamma} L_{\gamma 0}, \tag{2.3.43}$$

$$A_{0N} = A_{N0} = a_0 M_N = -M_N \sum_{\alpha=1}^{N-1}\sum_{\gamma=1}^{N-1} Z_\gamma \mathcal{M}_{\gamma\alpha} L_{\alpha 0}, \tag{2.3.44}$$

$$A_{0\alpha} = A_{\alpha 0} = a_0 M_\alpha + \sum_{\beta=1}^{N-1} L_{0\alpha} \mathcal{M}_{\beta\alpha} = \sum_{\beta=1}^{N-1} L_{\beta 0} \left[\mathcal{M}_{\beta\alpha} - M_\alpha \sum_{\gamma=1}^{N-1} Z_\gamma \mathcal{M}_{\gamma\beta} \right], \quad (\alpha = 1,2,\ldots N-1) \tag{2.3.45}$$

$$A_{\beta\alpha} = A_{\alpha\beta} = -(M_\alpha M_\beta / M_N) a_N + \sum_{\gamma=1}^{N-1} Z_\gamma \left(M_\alpha \mathcal{M}_{\gamma\beta} + M_\beta \mathcal{M}_{\gamma\alpha} \right) - \mathcal{M}_{\beta\alpha}, \quad (\alpha, \beta = 1,2,\ldots N-1) \tag{2.3.46}$$

$$A_{N\alpha} = A_{\alpha N} = -a_N M_\alpha + M_N \sum_{\gamma=1}^{N-1} Z_\gamma \mathcal{M}_{\gamma\alpha}, \tag{2.3.47}$$

$$A_{NN} = -a_N M_N \,. \tag{2.3.48}$$

Herewith the identity

$$\sum_{\alpha=1}^{N} Z_\alpha A_{\alpha 0} = 0 \tag{2.3.49}$$

is valid.

Thus, the coefficients $A_{\alpha\beta}$ $(\alpha,\beta=0,1,...,N)$ are defined to the precision of the constant a_N. Generally, the latter may be chosen arbitrarily. Here we define the constant assuming that the fluxes $J_{\alpha j}$ are arbitrary. Then rewriting the relations (2.3.41) and (2.3.42) as

$$X_{0j}=\frac{1}{A_{00}}J_{qj}-\sum_{\alpha=1}^{N}\frac{A_{0\alpha}}{A_{00}}J_{\alpha j}, \tag{2.3.50}$$

$$X_{\beta j}=-\frac{A_{\beta 0}}{A_{00}}J_{qj}+\sum_{\alpha=1}^{N}\left[\frac{A_{\beta 0}A_{0\alpha}}{A_{00}}-A_{\beta\alpha}\right]J_{\alpha j}, \qquad (\beta=1,2,...N) \tag{2.3.51}$$

and using the identity $\sum_{\beta=1}^{N}Z_\beta X_{\beta j}=0$ and the arbitrariness of the vectors $J_{\alpha j}$, we obtain, in view of (2.3.49) , the following important relations:

$$\sum_{\beta=1}^{N}Z_\beta A_{\beta\alpha}=0, \qquad (\alpha=1,2,...,N). \tag{2.3.52}$$

Finally, substituting the coefficients (2.3.46) and (2.3.47) into the relations (2.3.52), we find the expression

$$a_N=M_N\sum_{\beta=1}^{N-1}\sum_{\gamma=1}^{N-1}Z_\beta Z_\gamma \mathcal{M}_{\beta\gamma}. \tag{2.3.53}$$

for the parameter a_N. Thus, the constants in (2.3.43) to (2.3.48) are completely defined through the elements of the symmetrical matrix with the phenomenological coefficients $L_{\alpha\beta}$ and the elements of its inverse symmetrical matrix $\mathcal{M}_{\alpha\beta}$.

Let us now bring the expressions (2.3.42) into the form of the generalized Stefan-Maxwell relations for multicomponent diffusion. For this purpose, we subtract the equalities (2.3.52) multiplied by $J_{\beta j}/Z_\beta$ from (2.3.42):

$$-X_{\beta j}=A_{\beta 0}X_{0j}+\sum_{\alpha=1}^{N}A_{\beta\alpha}n_\alpha\left(w_{\alpha j}-w_{\beta j}\right), \qquad (\beta=1,2,...,N), \tag{2.3.54}$$

or in a more usual notation:

$$d_{\beta j}=\sum_{\alpha=1}^{N}\frac{n_\beta n_\alpha}{p}A_{\beta\alpha}\left(w_{\alpha j}-w_{\beta j}\right)-\frac{A_{\beta 0}n_\beta}{p}\frac{\partial_j T}{T}, \qquad (\beta=1,2,...,N). \tag{2.3.55}$$

We will then show that the following expression

$$A_{\beta 0}=-\sum_{\alpha=1}^{N}n_\alpha A_{\beta\alpha}\left[\frac{L_{0\alpha}}{n_\alpha}-\frac{L_{0\beta}}{n_\beta}\right], \qquad (\beta=1,2,...,N). \tag{2.3.56}$$

holds. This can be done with using (2.3.45), (2.3.46) and the identities (2.3.49) and (2.3.52), which results in

$$\sum_{\alpha=1}^{N} n_\alpha A_{\beta\alpha}\left[\frac{L_{0\alpha}}{n_\alpha}-\frac{L_{0\beta}}{n_\beta}\right]=\sum_{\alpha=1}^{N} A_{\beta\alpha}L_{0\alpha}-\frac{L_{0\beta}}{n_\beta}\sum_{\alpha=1}^{N} n_\alpha A_{\beta\alpha}=A_{\beta N}L_{0N}+\sum_{\alpha=1}^{N-1} A_{\beta\alpha}L_{0\alpha}=$$

$$=\sum_{\alpha=1}^{N-1} L_{0\alpha}\left[A_{\beta\alpha}-(M_\alpha/M_N)A_{\beta N}\right]=\sum_{\alpha=1}^{N-1} L_{0\alpha}\left[-\mathcal{M}_{\beta\alpha}+M_\beta\sum_{\gamma=1}^{N-1} Z_\gamma \mathcal{M}_{\gamma\alpha}\right]=-A_{0\beta}. \tag{2.3.57}$$

Substituting (2.3.56) into (2.3.55) and using the designations (2.3.21) for the thermal diffusion factor, we finally find

$$d_{\beta j}=\sum_{\substack{\alpha=1\\ \alpha\neq\beta}}^{N}\frac{n_\beta n_\alpha}{p}A_{\beta\alpha}\left(w_{\alpha j}-w_{\beta j}\right)+\frac{\partial_j T}{T}\sum_{\substack{\alpha=1\\ \alpha\neq\beta}}^{N}\frac{n_\beta n_\alpha}{p}A_{\beta\alpha}\left(D_{T\alpha}-D_{T\beta}\right), \tag{2.3.58}$$

$$(\beta=1,2,...N)$$

which completely coincides with the Stefan-Maxwell relations (2.3.28) if one assumes that

$$\frac{n_\alpha n_\beta}{p}A_{\beta\alpha}\equiv R_{\beta\alpha},\quad (\alpha,\beta=1,2,...,N)\ . \tag{2.3.59}$$

Thus, if the Onsager-Kasimir reciprocity principle (2.3.10) is valid for the phenomenological coefficients $L_{\alpha\beta}$, the resistance coefficient matrix in the thermodynamically derived Stefan-Maxwell relations is symmetrical:

$$R_{\alpha\beta}=R_{\beta\alpha}\quad (\alpha,\beta=1,2,...,N). \tag{2.3.60}$$

Comparing (2.3.58) with the similar relations obtained using the methods of the kinetic theory for rarefied gases (*Hirschfelder et al., 1954; Kolesnichenko, 1979*)

$$d_{\beta j}=\sum_{\substack{\alpha=1\\ \alpha\neq\beta}}^{N}\frac{x_\beta x_\alpha}{\mathcal{D}_{\alpha\beta}}\left(w_{\alpha j}-w_{\beta j}\right)+\frac{\partial_j T}{T}\sum_{\substack{\alpha=1\\ \alpha\neq\beta}}^{N}\frac{x_\beta x_\alpha}{\mathcal{D}_{\alpha\beta}}\left(D_{T\alpha}-D_{T\beta}\right),\quad (\beta=1,2,...,N), \tag{2.3.61}$$

where $\mathcal{D}_{\alpha\beta}=\left[\mathcal{D}_{\alpha\beta}\right]_1 f^N_{\alpha\beta}(\xi)$ and $f^N_{\alpha\beta}(\xi)$ are correction factors taking into account higher order approximations in the binary diffusion coefficients of multicomponent monotonic gas mixtures), we may conclude that (2.3.58) are more general in their form than (2.3.61).

Unfortunately, thermodynamics of irreversible processes does not provide a method for defining the phenomenological coefficients $A_{\alpha\beta}$ $(\alpha,\beta=1,2,...,N)$ and $A_{0\beta}$ $(\beta=1,2,...,N)$. In the general case of an arbitrary multicomponent gas-fluid medium, these coefficients should be experimentally assessed; for rarefied gases, they can be found by considering particle collision dynamics in detail with the help of the kinetic theory (*Hirschfelder et al., 1954; Ferziger and Kaper, 1972*). To take advantage of this approach, we identify (2.3.55) and (2.3.61); then, for an ideal gaseous mixture this yields:

$$\mathcal{D}_{\alpha\beta} = \frac{\mathrm{k}T}{n} A_{\alpha\beta} \quad (\alpha, \beta = 1,2,\ldots, N), \tag{2.3.62}$$

where $\mathcal{D}_{\alpha\beta}$ are the binary diffusion coefficients whose values can be adopted from the gas kinetic theory in any approximation of the Chapman-Enskog method (or taken empirically).

It is convenient to define the so-called *thermal diffusion ratios* $K_{T\beta}$ in terms of the phenomenological coefficients $A_{0\beta}$ using the formula

$$K_{T\beta} = \frac{n_\beta}{p} A_{\beta 0} \qquad (\beta = 1,2,\ldots, N). \tag{2.3.63}$$

In this case, incorporating the relations (2.3.21), (2.3.57) and (2.3.62), it is possible to obtain the following expression for the introduced transport coefficients $K_{T\beta}$:

$$K_{T\beta} = \sum_{\alpha=1}^{N} \frac{x_\beta x_\alpha}{\mathcal{D}_{\alpha\beta}} \left(D_{T\beta} - D_{T\alpha} \right), \quad (\beta = 1,2,\ldots, N). \tag{2.3.64}$$

This expression connects $K_{T\beta}$ and the thermal diffusion coefficient $D_{T\beta}$ for a multicomponent mixture, introduced earlier through formula (2.3.21). With this, by virtue of (2.3.49), the following equality fulfills:

$$\sum_{\alpha=1}^{N} K_{T\alpha} = 0. \tag{2.3.65}$$

Besides, it is easy to verify that the relations

$$\sum_{\beta=1}^{N} L_{\alpha\beta} A_{\beta 0} = L_{0\alpha}, \quad (\alpha = 1,2,\ldots, N), \tag{2.3.66}$$

are valid from which, in view of the definitions (2.3.21) and (2.3.63), the system of equations for finding the thermal diffusion ratios $K_{T\beta}$ in terms of the multicomponent thermal diffusion coefficients $D_{\alpha\beta}$ and $D_{T\alpha}$, respectively, follows:

$$\sum_{\beta=1}^{N} D_{\alpha\beta} K_{T\beta} = D_{T\alpha}, \qquad (\alpha = 1,2,\ldots, N). \tag{2.3.67}$$

Indeed, using the relations (2.3.11), (2.3.34), (2.3.44) and (2.3.45), it is possible to obtain successively

$$\sum_{\beta=1}^{N} L_{\alpha\beta}A_{\beta 0} = L_{\alpha N}A_{N0} + \sum_{\beta=1}^{N-1} L_{\alpha\beta}A_{\beta 0} = -a_0 \sum_{\beta=1}^{N-1} M_{\beta}L_{\beta\alpha} +$$

$$+\sum_{\beta=1}^{N-1} L_{\alpha\beta}\left[a_0 M_{\beta} + \sum_{\gamma=1}^{N-1} L_{0\gamma}\mathcal{M}_{\gamma\beta}\right] = \sum_{\beta=1}^{N-1}\sum_{\gamma=1}^{N-1} L_{\alpha\beta}\mathcal{M}_{\gamma\beta}L_{0\gamma} = L_{0\beta}. \quad (2.3.68)$$

Thus, the relations (2.3.64), (2.3.65), and (2.3.67) for thermal diffusion ratios that appear in the kinetic theory for rarefied gaseous mixtures are also universal (phenomenological) in their character, since the deduction of these relations is possible within the developed thermodynamic approach.

The introduced thermal diffusion ratios $K_{T\beta}$ must be commented further. By virtue of the relations (2.3.24$^{(2)}$), the determinant of the coefficients of the system (2.3.67) becomes zero. The N- component vector with respect to the concentrations Z_{α} is the unique solution for the system of the homogeneous equations corresponding to (2.3.67). However, according to (2.3.24$^{(1)}$), this vector is orthogonal to the N-component vector of thermal diffusion coefficients, i.e. $\sum_{\alpha=1}^{N} M_{\alpha}Z_{\alpha}D_{T\alpha} = 0$. Whence it follows, that the equations (2.3.67) have a solution in the form of some vector having components $K_{T\alpha}$ being defined with an accuracy to a constant multiplied by the concentrations Z_{α}, the condition (2.3.65) ensuring uniqueness of the coefficients $K_{T\beta}$.

Using the thermal diffusion ratios, the expression (2.3.23) for the diffusion flux $J_{\alpha j}$ can be written as the following generalized *Fick law*:

$$J_{\alpha j} = -n_{\alpha}\sum_{\beta=1}^{N} D_{\alpha\beta}\left[d_{\beta j} + K_{T\beta}\frac{\partial_j T}{T}\right], \qquad (\alpha = 1,2,\ldots,N), \qquad (2.3.23^*)$$

where

$$d_{\beta j} = \partial_j x_{\beta} + (x_{\beta} - M_{\beta}Z_{\beta})\frac{\partial_j p}{p} + \frac{n_{\beta}}{p}\left[-F_{\beta j} + M_{\beta}\sum_{\alpha=1}^{N} Z_{\alpha}F_{\alpha j}\right]$$

are the thermodynamic diffusion force vectors.

Finally, using (2.3.62) and (2.3.63) the Stefan-Maxwell generalized relations (2.3.55) can be rewritten as follows:

$$\sum_{\substack{\alpha=1\\ \alpha\neq\beta}}^{N} \frac{n_{\beta}J_{\alpha j} - n_{\alpha}J_{\beta j}}{n^2\mathcal{D}_{\alpha\beta}} = d_{\beta j} + K_{T\beta}\frac{\partial_j T}{T}, \qquad (\beta = 1,2,\ldots,N). \qquad (2.3.69)$$

With regard to these relations one should aware that the resistance coefficient matrix $R_{\alpha\beta}$, defined by the expression (2.3.59), is degenerate by virtue of the identity (2.3.52). For this reason, not all of the N different diffusion fluxes $J_{\alpha j}$ can be retrieved from (2.3.69). The *algebraic integral* (2.1.8) is a necessary additional condition.

2.3.4. The Total Heat Flux in an Ideal Multicomponent Media

As was found above, the total heat flux q_j of a multicomponent mixture due to heat conduction and diffusion is (see (2.1.50) and (2.3.22))

$$q_j = J_{qj} + \sum_{\alpha=1}^{N} h_\alpha J_{\alpha j} = -\lambda_0 \partial_j T - p\sum_{\alpha=1}^{N} D_{T\alpha} d_{\alpha j} + \sum_{\alpha=1}^{N} h_\alpha J_{\alpha j}. \tag{2.3.70}$$

On the other hand, from (2.3.41), in view of (2.3.63), it is possible to obtain another expression for the reduced heat flux J_{qj}:

$$J_{qj} = A_{00} X_{0j} + \sum_{\alpha=1}^{N} A_{0\alpha} J_{\alpha j} = -\lambda \partial_j T + p\sum_{\alpha=1}^{N} K_{T\alpha} w_{\alpha j}, \tag{2.3.71}$$

which is correlated with the Stefan-Maxwell relations (2.3.69). Here in terms of

$$\lambda = \frac{1}{T} A_{00} \tag{2.3.72}$$

the *true thermal conductivity* is defined which is connected with the earlier introduced coefficient λ_0 through the relation

$$\lambda = \lambda_0 - \mathrm{k}T\sum_{\alpha=1}^{N} K_{T\alpha} D_{T\alpha} = \lambda_0 - \mathrm{k}T\sum_{\alpha=1}^{N}\sum_{\beta=1}^{N} K_{T\alpha} D_{\alpha\beta} K_{T\beta}. \tag{2.3.73}$$

Indeed, by virtue of (2.3.34), (2.3.43), and (2.3.68), we have

$$A_{00} \equiv L_{00} - \sum_{\alpha=1}^{N-1}\sum_{\delta=1}^{N-1} L_{0\delta} M_{\delta\alpha} L_{0\alpha} = L_{00} - \sum_{\alpha=1}^{N-1}\sum_{\delta=1}^{N-1} L_{0\delta} M_{\delta\alpha} \Big[L_{0N} A_{N0} +$$

$$+ \sum_{\beta=1}^{N-1} L_{\alpha\beta} A_{\beta 0} \Big] = L_{00} - \sum_{\beta=1}^{N-1} L_{0\beta} A_{\beta 0} - A_{N0} L_{0N} = L_{00} - \sum_{\beta=1}^{N} L_{0\beta} A_{\beta 0}, \tag{2.3.74}$$

whence the expression (2.3.73) follows if the designations (2.3.21), (2.3.63) and (2.3.72) are used.

Thus, the total heat flux in a multicomponent medium can be written as

$$q_j = -\lambda \partial_j T + p\sum_{\alpha=1}^{N} K_{T\alpha} w_{\alpha j} + \sum_{\alpha=1}^{N} h_\alpha n_\alpha w_{\alpha j}, \qquad (2.3.75)$$

in full agreement with the gas kinetic theory (*Ferziger and Kaper, 1972*).

Unlike λ_0, the coefficient λ can be measured directly in a stationary system because all diffusion rates $w_{\alpha j}$ are equal to zero under steady-state conditions (when the gas as a whole is at rest). As is seen from (2.3.73), the difference between coefficients λ and λ_0 is of order $D_{T\alpha}^2$. Note that for atmospheric gas this difference can be neglected because of the smallness of thermal diffusion factors (*Marov and Kolesnichenko, 1987*). The addend $p\sum_{\alpha=1}^{N} K_{T\alpha} w_{\alpha j}$ in the expression (2.3.71) for the heat flux vector corresponds to the diffusion heat flux.

2.3.5. Formulas for the Multicomponent Diffusion Coefficients

Now let us obtain the algebraic equations enabling the calculation of the multicomponent diffusion coefficients $D_{\alpha\beta}$ in terms of the binary diffusion coefficients $\mathcal{D}_{\alpha\beta}$. It is easy to verify the validity of the relation

$$\sum_{\substack{\beta=1 \\ \beta\neq\alpha}}^{N} A_{\beta\alpha}(n_\beta L_{\gamma\alpha} - n_\alpha L_{\gamma\beta}) = n_\gamma\left(\delta_{\alpha\gamma} - M_\alpha Z_\alpha\right) \quad (\alpha,\gamma = 1,2,...,N). \qquad (2.3.76)$$

Indeed, by virtue of (2.3.11), (2.3.46), and (2.3.52), we have

$$\sum_{\substack{\beta=1 \\ \beta\neq\alpha}}^{N} A_{\beta\alpha}\left(n_\beta L_{\gamma\alpha} - n_\alpha L_{\gamma\beta}\right) = \sum_{\beta=1}^{N} A_{\beta\alpha}\left(n_\beta L_{\gamma\alpha} - n_\alpha L_{\gamma\beta}\right) = -n_\alpha \sum_{\beta=1}^{N} A_{\beta\alpha} L_{\gamma\beta} =$$

$$= -n_\alpha\left[\sum_{\beta=1}^{N-1} A_{\beta\alpha} L_{\gamma\beta} + A_{\beta N} L_{\gamma N}\right] = -n_\alpha\left[M_\alpha \sum_{\delta=1}^{N-1} Z_\delta\left(\sum_{\beta=1}^{N-1} \mathcal{M}_{\delta\beta} L_{\beta\gamma}\right) - \sum_{\beta=1}^{N-1} \mathcal{M}_{\beta\alpha} L_{\gamma\beta}\right] =$$

$$= -n_\alpha\left[M_\alpha \sum_{\delta=1}^{N-1} Z_\delta \delta_{\delta\gamma} - \delta_{\alpha\gamma}\right] = n_\gamma\left(\delta_{\alpha\gamma} - M_\alpha Z_\alpha\right)$$

When using the designations (2.3.21) and (2.3.62) for the quantities $A_{\beta\alpha}\ (= \mathrm{k}T/n\mathcal{D}_{\alpha\beta})$ and $L_{\gamma\alpha}\ (= n_\gamma n_\alpha D_{\gamma\alpha}/p)$ the relations (2.3.76) can be rewritten as

$$\sum_{\substack{\beta=1 \\ \beta\neq\alpha}}^{N} \frac{x_\alpha x_\beta}{\mathcal{D}_{\alpha\beta}}\left(D_{\gamma\alpha} - D_{\gamma\beta}\right) = \delta_{\gamma\alpha} - M_\alpha Z_\alpha, \quad (\gamma,\alpha = 1,2,...,N), \qquad (2.3.77)$$

which are appropriate to define the multicomponent diffusion coefficients $D_{\gamma\beta}$ $(\gamma,\beta=1,2,...,N)$ of a mixture in terms of the binary diffusion coefficients $\mathcal{D}_{\gamma\beta}$ $(\gamma,\beta=1,2,...,N)$. The equations (2.3.77) are linearly dependent (since the summation of relation (2.3.76) with respect to α results in identity), therefore they should be supplemented with one more equation, namely (2.3.24).

Let us shape the equations (2.3.77) and (2.3.24) into a form convenient to practically calculation the diffusion coefficients $D_{\gamma\beta}$. For this purpose we first rewrite (2.3.77) as follows:

$$\sum_{\substack{\beta=1\\ \beta\neq\alpha}}^{N} D_{\gamma\beta}\left[\frac{x_\alpha x_\beta}{\mathcal{D}_{\alpha\beta}}-\frac{x_\alpha x_\beta}{\mathcal{D}_{\alpha\beta}}\frac{D_{\gamma\alpha}}{D_{\gamma\beta}}\right]=M_\alpha Z_\alpha-\delta_{\gamma\alpha}\,. \tag{2.3.78}$$

Then, observing that from relations (2.3.24) follows

$$M_\alpha Z_\alpha D_{\gamma\alpha}=-\sum_{\substack{\delta=1\\ \delta\neq\alpha}}^{N} M_\delta Z_\delta D_{\delta\gamma}\qquad(\alpha,\gamma=1,2,...,N)\,, \tag{2.3.79}$$

we can reduce (2.3.78) to

$$\sum_{\substack{\beta=1\\ \beta\neq\alpha}}^{N}\left[\frac{x_\alpha x_\beta}{\mathcal{D}_{\alpha\beta}}+\sum_{\substack{\delta=1\\ \delta\neq\alpha}}^{N}\frac{M_\beta}{M_\alpha}\frac{x_\alpha x_\delta}{\mathcal{D}_{\alpha\delta}}\right]D_{\gamma\beta}=M_\alpha Z_\alpha-\delta_{\gamma\alpha}\quad(\alpha,\gamma=1,2,...,N) \tag{2.3.80}$$

In the particular case of a three-component mixture we have the following typical expressions for the multicomponent diffusion coefficients:

$$D_{11}=\frac{n^2}{\rho^2 n_1^2}\left(\frac{n_1 n_3 M_3^2\mathcal{D}_{23}\mathcal{D}_{31}+n_1 n_2 M_2^2\mathcal{D}_{12}\mathcal{D}_{23}+(\rho_2+\rho_3)^2\mathcal{D}_{12}\mathcal{D}_{31}}{n_1\mathcal{D}_{23}+n_2\mathcal{D}_{31}+n_3\mathcal{D}_{12}}\right),$$

$$D_{12}=\frac{n^2}{\rho^2}\left(\frac{n_3 M_3^2\mathcal{D}_{23}\mathcal{D}_{31}-M_2(\rho_1+\rho_2)\mathcal{D}_{12}\mathcal{D}_{23}-M_1(\rho_2+\rho_3)\mathcal{D}_{31}\mathcal{D}_{11}}{n_1\mathcal{D}_{32}+n_2\mathcal{D}_{31}+n_3\mathcal{D}_{12}}\right), \tag{2.3.81}$$

where $\rho_\alpha=M_\alpha n_\alpha$ is the mass density of α-particles. The expressions for the remaining coefficients $D_{\alpha\beta}$ can be obtained from (2.3.81), properly rearranging the indices.

Let us notice that formulas like (2.3.80) and (2.3.81) were obtained for the first time in the framework of the gas kinetic theory for monatomic gases approximation of the Chapman-Enskog method (*Curtiss, 1968*). Instead, phenomenological deducing these formulas was done here, which allowed us to establish the universal character of these correlations.

2.3.6. Multicomponent Diffusion in the Upper Atmosphere

The generalized Stefan-Maxwell relations (2.3.69) serve as the basic ingredients for modeling numerically the processes of heat and mass transfer in the planetary upper atmosphere (see Part II). Let us rewrite them in a somewhat modified form pertinent for the evaluation of some problems of aeronomy, specifically those in the heterosphere of a planet.

The total external macroscopic force acting on a gas particle of sort α in a relative coordinate system fixed to the planet's surface and rotating relative to an inertial system with constant angular velocity Ω_j, is given by the expression

$$F_{\alpha j} = M_\alpha g_j - 2M_\alpha \varepsilon_{kji} V_{\alpha k} \Omega_i, \tag{2.3.82}$$

where

$$g_j = \boldsymbol{\Omega}^2 R_j + \left(GM_{pl} / |\boldsymbol{R}|^3\right) R_j$$

is the reduced acceleration due to gravity and R_j is the central radius vector. Substituting (2.3.82) into (2.3.15) and taking into account the relation $p_\alpha = x_\alpha p$ for the partial pressure, the expression for the thermodynamic diffusion force vector $d_{\alpha j}$ can be written as

$$d_{\alpha j} = \frac{1}{p}\left[\partial_j p_\alpha - M_\alpha Z_\alpha \partial_j p + 2M_\alpha n_\alpha \varepsilon_{kji} w_{\alpha k} \Omega_j\right]. \tag{2.3.83}$$

With the use of (2.3.83) and (2.1.59) it is possible to shape the Stefan-Maxwell generalized relations (2.3.69) as equations of motion for individual mixture components in the relative coordinate system (written in the *diffusion approximation*):

$$\rho_\alpha \left[\frac{dV_j}{dt} + 2\varepsilon_{kji} V_k \Omega_i\right] = -\partial_j p_\alpha + \frac{\rho_\alpha}{\rho} \partial_i \pi_{ij} + \sum_{\substack{\beta=1 \\ \beta\neq\alpha}}^{N} \frac{x_\alpha x_\beta}{\mathcal{D}_{\alpha\beta}} \left(w_{\beta j} - w_{\alpha j}\right) -$$

$$- pK_{T\alpha} \partial_j (\ln T) + \rho_\alpha g_j - 2\rho_\alpha \varepsilon_{kji} w_{\alpha k} \Omega_i. \tag{2.3.84}$$

In the right part of (2.3.84) there are terms featuring the effect of various forces applied to moving α-particles . In addition to the above mentioned external forces per unit volume $n_\alpha F_{\alpha j}$, they also include the partial pressure gradient, the viscous friction force, friction force due to interacting one-component media, traveling with different velocities, and the thermal force causing thermal diffusion. It should be noted that, because of only slightly differing molecular weights of the diffusing gases in the upper atmosphere, only taking into account the influence of thermodiffusion of the light minor species (such as hydrogen, deuterium, and helium) on the distribution of the neutral mixture components is sufficient (*Marov and Kolesnichenko, 1987*).

Formulas from the Kinetic Theory for Calculating the Frictional Force. Introducing the effective collision frequency for α - and β-particles

$$\tau_{\alpha\beta}^{-1} \equiv \mathrm{k}\, x_\beta / M_{\alpha\beta} \mathcal{D}_{\alpha\beta}, \qquad \mathcal{D}_{\alpha\beta} = \left[\mathcal{D}_{\alpha\beta}\right]_1, \tag{2.3.85}$$

where

$$\left[\mathcal{D}_{\alpha\beta}\right]_1 = \frac{3}{16n}\left(\frac{2\pi \mathrm{k} T}{M_{\alpha\beta}}\right)^{1/2} \frac{1}{\pi\sigma_{\alpha\beta}^2 \Omega_{\alpha\beta}^{(1,1)*}} \tag{2.3.86}$$

are binary diffusion coefficients taken in the first approximation of the Chapman-Enskog method, let us represent the collisional term in (2.3.84) as

$$P_{\alpha\beta j} \equiv p\frac{x_\alpha x_\beta}{\mathcal{D}_{\alpha\beta}}\left(w_{\beta j} - w_{\alpha j}\right) = n_\alpha M_{\alpha\beta} \tau_{\alpha\beta}^{-1}\left(w_{\beta j} - w_{\alpha j}\right). \tag{2.3.87}$$

Here $M_{\alpha\beta} \equiv M_\alpha M_\beta / (M_\alpha + M_\beta)$ is the reduced mass of a pair of α- and β-particles; $\sigma_{\alpha\beta} = \left(\sigma_\alpha + \sigma_\beta\right)/2$ is the spacing between the centers of the two molecules with diameters σ_α и σ_β at the moment of the collision;

$$\Omega_{\alpha\beta}^{(l,r)*} = \frac{4(l+1)}{(r+1)!\left[2l+1-(-1)^l\right]\pi\sigma_{\alpha\beta}^2} \int_0^\infty \exp\left[-\gamma_{\alpha\beta}^2\right]\gamma_{\alpha\beta}^{3+2r} Q_{\alpha\beta}^l \, d\gamma_{\alpha\beta} \tag{2.3.88}$$

is the reduced collisional integral of all pairs of particles (*Hirschfelder et al., 1954*), expressing the measure of departure from the model considering gas molecules as solid spheres (for such a model $\Omega_{\alpha\beta}^* = 1$); $\gamma_{\alpha\beta}^2 = M_{\alpha\beta}\, g_{\alpha\beta}^2 / 2\mathrm{k}T$; $\gamma_{\alpha\beta}$ is the relative initial velocity of the colliding molecules; $Q_{\alpha\beta}^{(l)}$ is the generalized collisional cross-section accompanied by momentum transfer between α- and β-particles having a relative thermal velocity $\left|\boldsymbol{g}_{\alpha\beta}\right|$

$$Q_{\alpha\beta}^{(l)} = Q_{\beta\alpha}^{(l)} = 2\pi\int_0^\infty \left(1 - \cos^l \chi_{\alpha\beta}\right)\boldsymbol{b}\, d\boldsymbol{b}\ ; \tag{2.3.89}$$

and $\chi_{\alpha\beta} = \chi_{\beta\alpha}\left(\boldsymbol{b}, \left|\boldsymbol{g}_{\alpha\beta}\right|\right)$ is the scattering angle (the deflection angle for molecules in the coordinate system fixed at the gravitational center). Collisional dynamics is ultimately involved in the transport coefficients through the collisional integrals (2.3.88). To evaluate the reduced integrals Ω^*, it is necessary to know χ as function of the initial relative velocity $\boldsymbol{g}_{\alpha\beta}$ and the aiming distance $\boldsymbol{b}$ (*Hirschfelder et al., 1954*). The dependence of χ on $\boldsymbol{g}_{\alpha\beta}$ and $\boldsymbol{b}$ is determined by the molecular interaction potential $\varphi(r)$. Thus, by setting the interaction potential, it is possible to calculate the integrals Ω^*. The most satisfactory (and convenient) calculations of Ω^* are those which have been carried out on the basis of either the *Lennard-Jones* (6-12) potential:

$$\varphi_\alpha(r) = 4\in_\alpha\left[\left(\frac{\sigma_\alpha}{r}\right)^{12} - \left(\frac{\sigma_\alpha}{r}\right)^6\right], \quad (2.3.90)$$

or the modified *Buckingham (*6-exp) potential:

$$\varphi_\alpha(r) = \begin{cases} \dfrac{\in_\alpha}{1-(6/\alpha_\alpha)}\left\{\dfrac{6}{\alpha_\alpha}\exp\left[\alpha_\alpha\left(1-\dfrac{r}{\sigma_\alpha}\right)\right] - \left(\dfrac{\sigma_\alpha}{r}\right)^6\right\}, & r > r_{\max} \\ \infty, & r < r_{\max}\,. \end{cases} \quad (2.3.91)$$

Here $\in_\alpha, \sigma_\alpha, \alpha_\alpha$ are the parameters of the interaction potentials for homogeneous α-molecules (power constants). As is shown in the gas kinetic theory, the Ω^*– integrals for these potentials only depend on the characteristic temperature $T^* = kT/\in$, which expresses the *law of corresponding states* for transport phenomena (*Hirschfelder et al., 1961*). The study in the field was addressed specifically to calculating the Ω^*-integrals as functions of the characteristic temperature *T** for particular particle interaction potentials (see, e.g., *Chapman and Couling*, 1952; *Hirschfelder et al., 1954; Devoto, 1966*). Tables for the Ω^*-integrals have been composed for the Lennard-Jones (6-12) potential and the modified Buckingham (6-exp) potential, as well as for a number of other models (*Hirschfelder et al., 1954*). The results of later calculations of the Ω^*-integrals for these potentials (*Mason, 1954; Munn et al., 1965*), were comprehensively summarized in (*Ferziger and Kaper, 1972*).

Using (2.3.86) and (2.3.88), in the first approximation of the Chapman-Enskog theory we obtain for the quantity $\tau^{-1}_{\alpha\beta}$

$$\tau^{-1}_{\alpha\beta} = n_\beta \frac{4}{3}\sqrt{\frac{8kT}{\pi M_{\alpha\beta}}}\pi\sigma^2_{\alpha\beta}\Omega^{(1,1)*}_{\alpha\beta} = n_\beta < g_{\alpha\beta} >< Q_{\alpha\beta} >. \quad (2.3.92)$$

Here $< g_{\alpha\beta} >= (8kT/\pi M_{\alpha\beta})^{1/2}$ is the mean relative particle velocity from the Maxwellian velocity distribution, and $< Q_{\alpha\beta} >$ is the mean effective collisional cross-section involving momentum transfer between α- and β-particles. From the relations (2.3.88) and (2.3.89) it follows that

$$< Q_{\alpha\beta} >= \frac{4}{3}\pi\sigma^2_{\alpha\beta}\Omega^{(1,1)*}_{\alpha\beta} = \frac{4}{3}\left(\frac{M_{\alpha\beta}}{2kT}\right)^3 \int_0^\infty Q_{\alpha\beta}(g)\exp\left[-\left(\frac{M_{\alpha\beta}}{2kT}\right)g^2\right]g^5\,dg, \quad (2.3.93)$$

where $Q_{\alpha\beta}(g)$ is the collisional cross-section with momentum transfer between α- and β-particles having a relative thermal velocity *g*. Note that for the main components of the Earth's upper atmosphere the mean collisional cross-sections $<Q_{\alpha\beta}>$ can be found in (*Banks, 1966*; *Banks and Holzer, 1969*).

The Complete Equation of Motion for a Neutral Atmospheric Component. While the planetary upper atmosphere is a partially ionized multicomponent gas mixture, the use of the Stefan-Maxwell relations enables one to obtain the equation of motion only for the neutral atmospheric components. In the case of a hydrodynamic system velocity V_j close to the velocity of neutral gas V_{nj} without mutually diffused components ($w_{\alpha_j}=0$), summing up the relations (2.3.84) with respect to index α related only to neutral mixture components gives rise to the following complete equation of motion for the neutral upper atmosphere :

$$\rho_n\left[\frac{dV_{nj}}{dt}+2\varepsilon_{kji}V_{nk}\Omega_i\right]=-\partial_j p_n+\frac{\rho_n}{\rho}\partial_i\pi_{ij}+\sum_\alpha{}^{(n)}\sum_\beta{}^{(i)}n_\alpha M_{\alpha\beta}\tau^{-1}_{\alpha\beta}\left(V_{\beta j}-V_{nj}\right)-$$

$$-p\sum_\alpha{}^{(n)}K_{T\alpha}\frac{\partial_j T}{T}+\rho_n g_j. \qquad (2.3.94)$$

Here $\rho_n=\sum_\alpha{}^{(n)}\rho_\alpha$ is the mass density of neutral particles in the system, $p_n=\sum_\alpha{}^{(n)}p_\alpha$ is the neutral gas pressure, and the upper indexes (n) and (i) denote summation over the neutral or charged mixture components, respectively. The third term in the right part of the equation (2.3.94) is responsible for the ionic deceleration; this term appears when the neutral atmosphere interacts with the ionosphere.

Finally, we will address another useful form of the generalized Stefan-Maxwell relations. If one employs the equation of state $p_\alpha=\mathrm{k}Tn_\alpha$ for the α-th component and the equations of motion of the total continuum (2.1.59) where $\Omega_j=0$ is being assumed, the relations (2.3.69) can be easily transformed as follows:

$$\partial_j n_\alpha=-\frac{n_\alpha}{H_\alpha}k_{zj}-\left(1+\frac{K_{T\alpha}}{x_\alpha}\right)\frac{n_\alpha}{T}\partial_j T+\sum_{\substack{\beta=1\\ \beta\neq\alpha}}^N\frac{n_\alpha n_\beta(V_{\beta j}-V_{\alpha j})}{n\mathcal{D}_{\alpha\beta}}+\frac{n_\alpha M_\alpha}{kT}\Psi_j, \qquad (2.3.95)$$

where $g_j=-gk_{zj}$, k_{zj} is the unit vector in the vertical direction, $H_\alpha=\mathrm{k}T/M_\alpha g$ is the local scale height for α-particles, and $\rho\Psi_j\equiv-\rho dV_j/dt+\partial_i\pi_{ij}$. As a rule, the condition $\Psi_j k_{zj}\approx 0$ is valid in the upper atmosphere of a planet (*Rishbeth and Garriott, 1969; Akasofu and Chapman, 1972*). This means that the equation of hydrostatic equilibrium $\partial p/\partial z=-\rho g$ is valid in the vertical direction. In that case the diffusion relations (2.3.95) take a rather simple form in the direction of the vector k_{zj}:

$$\frac{\partial n_\alpha}{\partial z}=-\frac{n_\alpha}{H_\alpha}-\left(1+\frac{K_{T\alpha}}{x_\alpha}\right)\frac{n_\alpha}{T}\frac{\partial T}{\partial z}+\sum_{\substack{\beta=1\\ \beta\neq\alpha}}^N\frac{n_\alpha n_\beta(V_{\beta z}-V_{\alpha z})}{n\mathcal{D}_{\alpha\beta}}, \qquad (\alpha=1,2,\ldots,N). \qquad (2.3.96)$$

This form was widely used in many early studies of the height distribution of the chemical components in the Earth's thermosphere (see, e.g., *Akasofu and Chapman, 1972; Bauer, 1973*).

Summary

- *The hydrodynamic model of regular motion in an ideal multicomponent mixture was developed using methods of mechanics of continuum media taking into account chemical reactions, heat and mass transfer, and external conservative forces. Different balance equations, intimately related with the conservation laws of matter, energy, and momentum, were thoroughly analyzed with the goal to find constraints for theoretical examining and modeling the various dynamic and physicochemical processes involved. The main focus is given to the phenomenological approach to deduce the defining relations for the thermodynamic diffusion and heat fluxes, as well as to derive convenient algebraic formulas linking together various molecular transport coefficients, based on the methods of non-equilibrium thermodynamics.*

- *When modeling heat and mass transfer processes, it is convenient to have the defining relations in the form of the Stefan-Maxwell relations involving the diffusion coefficients for binary gaseous mixtures, rather than the multicomponent diffusion coefficients for which the kinetic theory of rarefied gases gives too cumbersome formulas. These relations and the respective expression for the full heat flux in a multicomponent mixture were obtained based on the methods of thermodynamics of irreversible processes using the Onsager-Kasimir reciprocity principle. The phenomenological deduction of the generalized Stefan-Maxwell relations justifies the validity of their utilization jointly with the semi-empirical expressions for the binary diffusion and thermal diffusion coefficients.*

- *Important phenomenological formulas were found as a consequence of deducing the generalized Stefan-Maxwell relations for multicomponent diffusion. They include: the algebraic equations for calculating the multicomponent diffusion coefficients in terms of the binary diffusion coefficients; the formulas linking the thermal diffusion ratios with the thermal and multicomponent diffusion coefficients of a mixture; the formulas linking the true and partial heat conduction. These formulas are fully identical in their structure to the expressions obtained within the first approximation of the Chapman-Enskog method in the kinetic theory of multicomponent monoatomic gas mixtures and are in accordance with the results of this theory. At the same time, in contrast to the gas-kinetic approach, which is completely adapted only for gases of moderate density when the interaction potential between gas particles is known, the phenomenological approach is more advantageous because it is not confined to postulate a particular microscopic model of the medium. Therefore, the results obtained are universal in their character and are pertinent for the description of a broad class of media, such as polyatomic gaseous mixtures, dense gases, fluid solutions, etc.*

CHAPTER 3

TURBULENT MOTION OF MULTICOMPONENT MIXTURES WITH VARIABLE THERMOPHYSICAL PROPERTIES

In the last decades considerable attention was focused on the study of developed turbulent flows of compressible gases (see, e.g., *Townsend, 1956; Hinze, 1959; Van Miegham, 1973; Ievlev, 1975, 1990; Lushchik et al., 1978, 1986; Kompaniets, 1979; Bruyatsky, 1986; Turbulence: Principles and Applications, 1980; Turbulent shear flows - I, 1982; Lesieur, 1990; Chorin, 1994; Zakharov et al., 1992; Frisch, 1995*). Historically, it was stimulated by the numerous problems put forward by rocketry and space engineering (*Avduevsky, 1962; Avduevsky and Medvedev, 1968; Donaldson, 1972; Zhelazni et al., 1973; Mellor and Herring, 1973; Maslov and Petrovskaya, 1975; Petukhov et al., 1984*), chemical technology (*Williams, 1971; Kumantsev et al., 1974; Polak et al., 1975; Kompaniets et al., 1977*), and natural phenomena and environment protection (*Gibson and Launder, 1976; Israel, 1979; Lumley and Panofski, 1966; Marchuk, 1982; Meteorology and Atomic Energy, 1959, 1971; Earth Study From Space, 1998*). Great progress in space exploration of the gas envelopes of Earth and other planets in the solar system imparted momentum to the development of turbulence models taking into account the multicomponent character and compressibility of the medium and numerous chemical reactions (*Marov and Kolesnichenko, 1971, 1981, 1983,1987; Kolesnichenko and Marov, 1980, 1984, 1994, 1998*).

We will begin this chapter with deducing the averaged differential balance equations for substance, momentum, and energy. These equations are addressed as a benchmark for the model intended for the description of developed turbulent flows in multicomponent mixtures of chemically active gases. The physical sense of individual terms in these equations is thoroughly analyzed. Special attention is paid to deducing closing rheological relations for the turbulent fluxes of diffusion, heat and Reynolds' turbulent stress tensor, using a routine technique based on the notion of mixing length. Advances in the development and application of first closure order, semi-empirical turbulence models for homogeneous compressible fluids, which are referred to as the so-called *gradient models* (see, for example, *Townsend, 1956; Bruyatskiy, 1986; Van Miegham, 1973*), made possible to extend some of these to free stratified flows in a reacting multicomponent mixture with a transversal shear velocity (*Marov and Kolesnichenko, 1987*).

In general, the closure problem for reacting media is complicated because it is necessary to average strongly non-linear substance producing source terms as a result of chemical reactions bearing an exponential character, and to model the large number of additional second order single-point moments for the temperature and the concentration of chemical components in fluctuating mixture flows. In other words, it is mandatory to take into account the mutual influence of turbulence and the kinetics of chemical reactions. A procedure for averaging the chemical reaction rates of any order is offered, and an approach to semi-empirical modeling these additional correlations is outlined.

3.1. MEAN MOTION OF A TURBULENT MULTICOMPONENT MIXTURE WITH VARIABLE DENSITY

Solution of the equations of multicomponent hydrodynamics (2.1.57)–(2.1.61) describing the true (instantaneous and unsteady) state of the gaseous mixture for given initial and boundary conditions encounters great difficulties. This is because their numerical evaluation entails the finite-difference approximation of enormous spatial-temporal flow fields by the final number of grid nodes.

Currently the only economically justified approach is to find solutions of the hydrodynamic equations for large spatial-temporal scales of motion which determine the averaged structural parameters of the *stochastic* (turbulized) medium, and thus to phenomenologically model the small scale motions. In this case, stochasticity means that an ensemble of probable field flow realizations occurs, which is defined as a *statistically averaged* (mathematically expected) entity for all fluctuating thermohydrodynamic characteristics. Averaging any flow parameter can also be carried out over a set of realizations occurring at various instants at a given spatial point, or over a set of values at various spatial points in some volume Δw at a fixed time. Basically, the flow parameters in the averaged hydrodynamic equations are defined as time-averaged while they are represented in these equations as derivatives with respect to time.

So, as was mentioned in Chapter 1, in order to eliminate the obvious inconsistency within these equations, the time intervals Δt, with respect to which the averaging is performed, should be longer than duration of individual fluctuations though, if the average motion is time dependent, should be shorter than the time scale at which the mean magnitudes change noticeably. Accordingly, the volume Δw should satisfy conditions similar to those for the time period Δt. In particular, for atmospheric dynamics it is agreed to distinguish mean zonal motions with horizontal extensions with scales of 10^4 km, and departures from these mean motions called *pulsations, fluctuations*, and *vortices*. These fluctuations may have various spatial scales ranging from several meters up to thousands of kilometers. Thus, in aeronomy the word *"turbulent"* is frequently understood simply as a departure from the mean magnitude irrespective of its scale (*Brasseur and Solomon, 1987*).

One may see that the partition of real stochastic motion into slowly varying mean and rapidly fluctuating turbulent (irregular, random, fluctuating about mean values) motions completely depends on the choice of the spatial-temporal region for which the mean values are defined. The size of this region fixes the scale of mean motion. All large vortices contribute to the mean motion defined by the mean values of the parameters ρ, V_j, T, Z_α $(\alpha = 1,2,\ldots,N)$. All smaller vortexes, filtered out with averaging, contribute to the turbulent motion defined by respective fluctuations of the same structural parameters. To obtain *representative* mean values and respective fluctuations of physical quantities, the spatial-temporal averaging region must comprise a very large number of vortices whose sizes are smaller than the averaging region, only a minor part being allowed to be larger than this region.

3.1.1. Choice of the Averaging Operator

In turbulence theories various ways to average a physical quantity $A(\boldsymbol{r},t)$ are utilized. For instance, the time averaging

$$\overline{A}=(1/\Delta t)\int_{t-\Delta t/2}^{t+\Delta t/2}A(t)dt, \tag{3.1.1}$$

when the averaging interval Δt is assumed to be sufficiently long compared to the characteristic fluctuation time of the field and substantially short compared to the time scale at which the averaged field changes. Other above mentioned approaches are spatial averaging by means of integration over a spatial volume, or stochastic averaging over an ensemble of probable realizations. The latter approach is the most fundamental. It is based on the concept of an ensemble, that is, an infinite aggregation of identical hydrodynamic systems distinguished from each other by the state of the velocity field and/or other parameters of motion at a given instant. According to the well known *ergodicity* hypothesis (*Monin and Yaglom, 1971*), the time-average is identical to the ensemble-average in the case of a stationary stochastic process. Not analyzing advantages and deficiencies of the various averaging methods in detail, let us only notice that "the practice of designing phenomenological models for the study of turbulent motions shows that the techniques for introducing the averaged properties of motion are, generally speaking, incidental for the composition of the complete system of averaged hydrodynamic equations if one pursues satisfaction of Reynolds' postulates when averaging whatever" (*Sedov, 1980*):

$$\overline{A+B}=\overline{A}+\overline{B},\quad \overline{aA}=a\overline{A},\quad \overline{\overline{A}B}=\overline{A}\,\overline{B}. \qquad (^1) \tag{3.1.2}$$

Here $A(\boldsymbol{r},t)$ and $B(\boldsymbol{r},t)$ are some fluctuation characteristics of a turbulent field, $\overline{A}(\boldsymbol{r},t)$ and $\overline{B}(\boldsymbol{r},t)$ being their mean values, and a is a constant. Next we assume that any averaging operator employed in (3.1.2 (^{1})) commutes with the differentiation and integration operators both with respect to space and time:

$$\overline{\partial A(\boldsymbol{r},t)/\partial t}=\partial\overline{A(\boldsymbol{r},t)}/\partial t,\qquad \overline{\int A(\boldsymbol{r},t)dt}=\int\overline{A(\boldsymbol{r},t)}dt, \qquad (^2)$$

$$\overline{\partial A(\boldsymbol{r},t)/\partial x_j}=\partial\overline{A(\boldsymbol{r},t)}/\partial x_j,\qquad \overline{\int A(\boldsymbol{r},t)dx_j}=\int\overline{A(\boldsymbol{r},t)}dx_j. \qquad (^3) \tag{3.1.2}$$

Note that some of the relations (3.1.2) are only fulfilled approximately, though the less the changes of the mean values $\overline{A}(\boldsymbol{r},t)$ in time (and/or space) within the considered region of integration are, the more accurate the relations (3.1.2) are.

By now, the classical turbulence theories for homogeneous incompressible fluids are rather complete (see, e.g., *Townsend, 1956; Monin and Yaglom, 1971; Turbulence: Principles and Applications, 1980; Ievlev, 1990*). In these theories, the averaging is usually performed in a similar way for all hydrothermodynamical parameters without any exceptions and, as a rule, without weighting factors. In case of ensemble averaging probable realizations,

$$\overline{A} = \lim_{N\to\infty} \frac{1}{N} \sum_{\text{p}=1}^{N} A^{(\text{p})} \tag{3.1.3}$$

the summation is performed over the set of realizations, while the respective average field $\overline{A}$ is defined as the expected value of A for an ensemble of identical systems. With time (3.1.2) or ensemble (3.1.3) averaging the instantaneous value of the parameter A is represented as the sum of averaged $\overline{A}$ and fluctuation A' components:

$$A = \overline{A} + A', \quad (\overline{A'} = 0)\,. \tag{3.1.4}$$

At the same time, averaging like this, which is identical for all state variables, results in unwieldy hydrodynamic equations for the mean motion in case of a multicomponent continuum with varying density ρ because of the necessity to keep correlators like $\overline{\rho V_j'}$, $\overline{\rho' V_i' V_j'}$, $\overline{\rho Z_\alpha'}$ etc. within the equations. Moreover, this makes it hard to physically interpret each individual term in the averaged equations. Therefore, when developing turbulence models for chemically active gaseous media, we will use further both the "ordinary" mean value of a fluctuating quantity $A(\boldsymbol{r},t)$ and the so-called *weighted-mean* value of this quantity referred to as the *Favre mean* (*Favre, 1969*), which is defined, for example, by the relation

$$< A >= \overline{\rho A}/\overline{\rho} = \left[\lim_{N\to\infty} \frac{1}{N} \sum_{p=1}^{N} \rho^{(p)} A^{(p)} \right] / \left[\lim_{N\to\infty} \frac{1}{N} \sum_{p=1}^{N} \rho^{(p)} \right]. \tag{3.1.5}$$

Note that in aeronomical applications, the time average or zonal, meridional, and vertical mean values can be used as the "ordinary" mean (*Brasseur and Solomon, 1987*); in this case

$$A =< A > + A'', \quad (\overline{A''} \neq 0)\ ; \tag{3.1.6}$$

A'' is the respective turbulent pulsation. Thus, to denote the mean quantities, two symbols will be used below: an over-bars designate ensemble averaging (in time and/or space), while angle brackets designate weighted-mean averaging. Double accents label fluctuations with regard to quantities averaged according to Favre.

If $\rho \cong \overline{\rho} - const$ as, for example, in the case of a fluid with Boussinesque properties (*Boussinesque, 1977*) both averaging procedures coincide. At the same time, in the case of a compressible multicomponent gaseous continuum, using (3.1.5) for a number of fluctuating thermohydrodynamical parameters substantially simplifies both structure and analysis of the averaged hydrodynamic equations (*Van Mieghem, 1973; Marov and Kolesnichenko, 1987*). Besides, averaging like this is convenient because the respective mean values emerge from experimental studies of turbulent flows using conventional methods (see, e.g., *Kompaniets et al., 1979* for further discussion). It is worth noticing the possibility to use weighted-mean flow parameters while modeling turbulent motion of a homogeneous fluid with variable density as was indicated by (*Van Driest, 1952*) and later on studied thoroughly for multicomponent chemically active continuous media based on non-equilibrium thermodynamics (*Kolesnichenko, 1980*).

Weighted-mean values. Some properties of the weighted-mean characteristics of physical quantities widely used below can easily be derived from the definition (3.1.5) and Reynolds' relations (3.1.2) (*Kolesnichenko and Marov, 1979*):

$$\overline{\rho A''}=0,\quad \overline{A''}=-\frac{1}{\bar\rho}\overline{\rho' A''},\quad \overline{\rho AB}=\bar\rho<A><B>+\overline{\rho A''B''},$$

$$\overline{\partial_j<A>}=\partial_j<A>,\quad (AB)''=<A>B''+<B>A''+A''B''-\frac{1}{\bar\rho}\overline{\rho A''B''},$$

$$\overline{\rho A\,\partial_j B}=\bar\rho<A>\partial_j<B>+\overline{\rho A\,\partial_j B''},$$

$$\overline{\rho\frac{dA}{dt}}=\bar\rho\frac{D<A>}{Dt}+\partial_j(\overline{\rho\ A''V_j''}),\tag{3.1.7}$$

where

$$\frac{D<A>}{Dt}=<V_j>\partial_j<A>+\partial_t<A>$$

is the substantial derivative for average motion.

The averaged continuity equation. The average density $\bar\rho$ and the weighted-mean hydrodynamic velocity $<V_j>=\overline{\rho V_j}/\bar\rho$ comply with the continuity equation for mean motion

$$\partial_t\bar\rho+\partial_j\left(\bar\rho<V_j>\right)=0.\tag{3.1.8}$$

This equation can be obtained by applying the averaging operator (3.1.1) to the continuity equation (2.1.4) for the true values of the density and the hydrodynamic velocity, provided that the averaging is extended over the interval Δt which is large with respect to the interval of fast turbulent changes in the fluctuation A' but is small compared to the characteristic fluctuation time of the averaged hydrodynamic quantity $\bar A(\boldsymbol r,t)$. As the turbulent mass flux is $\overline{\rho V_j''}=0\quad(\overline{V_j''}\neq 0)$, then there is no mass transport due to turbulence on the average if averaging is performed according to Favre. At the same time, because of known difficulties when modeling the correlations $\overline{\rho' V_j'}$, arising while averaging (2.1.4) without weighting, it is desirable to preserve the standard form of the continuity equation by formally replacing the true values of the density and the velocity by their averaged values. This is a strong argument in favor of using the weighted-mean averaging $<V_j>$ for the hydrodynamic flow velocity (*Van Mieghem, 1977*).

Note that the *Reynolds averaging* of the hydrodynamic mixture equations can also be carried out using another method; in what follows we will employ the stochastic averaging operator (3.1.3) when developing models for averaged turbulent multicomponent continua, if no other approach is specified.

The averaged operator equation. Averaging the operator relation (2.1.3) when using the properties (3.1.7) of Favre averaging) and the averaged continuity equation (3.1.8) leads to the following identity

$$\overline{\rho\frac{dA}{dt}}=\partial_t\left(\overline{\rho A}\right)+\partial_j\left(\overline{\rho A V_j}\right)=\partial_t\left(\bar{\rho}<A>\right)+\partial_j\left(\bar{\rho}<A><V_j>\right)+\partial_j\left(\overline{\rho A'' V_j''}\right)=$$

$$=\bar{\rho}\,\partial_t<A>+\bar{\rho}<V_j>\partial_j<A>+\overline{\partial_j\left(\rho A'' V_j''\right)}. \qquad (3.1.9)$$

Taking advantage of the formula

$$J^T_{(A)j}\equiv\overline{\rho A'' V_j''}=\bar{\rho}<A'' V_j''>, \qquad (3.1.10)$$

let us define the turbulent flux density of characteristic $A(\boldsymbol{r},t)$ as the single-point second moment (pair correlation, correlator) representing the transport of property A by turbulent velocity fluctuations. We also introduce the designation

$$\frac{D(..)}{Dt}\equiv\partial_t(..)+<V_j>\partial_j(..) \qquad (3.1.11)$$

for the substantial derivative with respect to time for the averaged motion. Then the identity (3.1.9) may be rewritten as

$$\overline{\rho\frac{dA}{dt}}=\bar{\rho}\frac{D<A>}{Dt}+\partial_j J^T_{(A)j}. \qquad (3.1.12)$$

Besides, by virtue of (3.1.8) the operator relation

$$\bar{\rho}\frac{DA}{Dt}=\partial_t\left(\bar{\rho}A\right)+\partial_j\left(\bar{\rho}A<V_j>\right) \qquad (3.1.13)$$

is valid which links both the substantial and local changes in $A(\boldsymbol{r},t)$ in the averaged flow. It should be emphasized that the parameter $A(\boldsymbol{r},t)$ in (3.1.13) on the one hand can imply the instantaneous value of some specific field characteristic of the flow (a scalar, vector or tensor component, etc.), and on the other hand can represent both its averaged $<A>$ and fluctuation component (A' or A'').

3.1.2. Averaged Conservation Laws for a Turbulized Mixture

Let us consider a turbulized multicomponent gaseous mixture as a continual medium whose elementary (instantaneous, true) states of motion can be described by the system of hydrodynamic equations (2.1.57)–(2.1.61) with a random sample of initial and boundary conditions. This is possible for spatial-temporal between scales of molecular motions and the smallest turbulence scales represented by the linear dimensions and life time of the smallest vortices. The latter, as a rule, are at least three orders of magnitude larger than the scales of molecular motions, i.e. the mean distance between molecules, and especially the molecular sizes (*Van Mieghem, 1977*). The only exclusion occurs when dealing with extremely rarefied gases, which is beyond the scope of this study. This is why we further on will use the results of Chapter 2 in order to obtain the balance equations and conservation laws averaged according to Reynolds.

The General View of the Averaged Balance Equation. Using the identity (3.1.9) when averaging over an ensemble of identical systems described by the equation (2.1.1), we obtain the substantial balance equation of a structural parameter $A(\boldsymbol{r},t)$ for an averaged continuum in a differential form:

$$\bar{\rho}\frac{D<A>}{Dt}=\partial_j J^{\Sigma}_{(A)j}+\overline{\sigma_{(A)}}\,. \tag{3.1.14}$$

Here

$$J^{\Sigma}_{(A)j}\equiv\overline{J_{(A)j}}+J^{T}_{(A)j} \tag{3.1.15}$$

is the substantial total flux density including the averaged regular and turbulent fluxes of characteristic $A(\boldsymbol{r},t)$, and $\overline{\sigma_{(A)}}$ is the averaged volume density of the source of parameter A. Let us note that defining the fluxes (3.1.15) relates to the main problem of the phenomenological turbulence theory, i.e. the closure problem that will be addressed in the Chapters 4 and 5.

If one transforms the left-hand part of the equation (3.1.14) using relation (3.1.10), then the local form of the differential balance equation for the field quantity $A(\boldsymbol{r},t)$ averaged according to Favre can be obtained:

$$\partial_t\left(\bar{\rho}<A>\right)+\partial_j\left[\bar{\rho}<A><V_j>+J^{\Sigma}_{(A)j}\right]=\overline{\sigma_{(A)}}\,. \tag{3.1.16}$$

Here

$$J^{\Sigma 0}_{(A)j}\equiv\bar{\rho}<A><V_j>+J^{\Sigma}_{(A)j}=\bar{\rho}<A><V_j>+\overline{J_{(A)j}}+J^{T}_{(A)j} \tag{3.1.17}$$

is the local total flux density of the characteristic $<A>$ in the averaged turbulized continuum involving the convective term $\bar{\rho}<A><V_j>$. The flux density $J^{\Sigma 0}_{(A)j}$ is the amount of quantity $<A>$ passing across a unit surface area per unit time for a surface fixed in the external coordinate system and in such a way that the local flux density

$J^{\Sigma 0}_{(A)j}$ is normal to the surface.

When dealing with deducing macro equations for turbulent motion of a multicomponent mixture, we will average over an ensemble of identical systems of hydrodynamic conservation equations which are valid at micro scale and related to the specific volume (2.1.6), the specific concentrations (2.1.7), the momentum (2.1.19), and the energy (2.1.43). The same procedure will be applied to the equation of state for pressure of the mixture (2.1.53). In doing so, we shall introduce in a natural fashion the physical parameters of the averaged turbulized continuum $<Z_\alpha>$, $<V_j>$, $<h>$, etc. as weighted-mean values of the respective micro characteristics. However, the pressure p, the density ρ and all molecular thermodynamic fluxes q_j, $J_{\alpha j}$, π_{ij}, ξ_s will be averaged without using weighting factors.

The Equation for the Specific Volume of a Turbulized Medium. Let us assume that $A \equiv v$ in (3.1.14) and use the expressions (2.1.5) for the quantities $J_{(v)j}$ and $\sigma_{(v)}$. Then we obtain

$$\bar{\rho}\frac{D<v>}{Dt} = -\partial_j J^{\Sigma}_{(v)j}, \tag{3.1.18}$$

where $<v> = 1/\bar{\rho}$ is the average specific volume of a medium, and

$$J^{\Sigma}_{(v)j} \equiv \overline{J_{(v)j}} + J^{T}_{(v)j} \tag{3.1.19}$$

is the substantial total flux density of the specific volume in a turbulized continuum. With this, the averaged regular and turbulent fluxes of the quantity v are determined by the relations

$$\overline{J_{(v)j}} \equiv -\overline{\rho v V_j} = -\overline{V_j} = -<V_j> - \overline{V''_j}, \tag{3.1.20}$$

$$J^{T}_{(v)j} \equiv \bar{\rho}<v''V''_j> = \overline{\rho v V''_j} = \overline{V''_j} = -\frac{1}{\bar{\rho}}\overline{\rho' V''_j}, \tag{3.1.21}$$

respectively. Then, for the total flux of the specific volume $J^{\Sigma}_{(v)j}$ we have

$$J^{\Sigma}_{(v)j} = -\overline{V_j} + \overline{V''_j} = -<V_j> - \overline{V''_j} + \overline{V''_j} = -<V_j>. \tag{3.1.19*}$$

Hence the balance equation for the averaged specific volume in its substantial form becomes

$$\bar{\rho}\frac{D<v>}{Dt} = \partial_j <V_j>. \tag{3.1.22}$$

Further, we will also use widely the relation

$$v'' = -\rho' / \bar{\rho}\rho, \qquad \overline{\rho v''} = -\overline{\rho'} / \bar{\rho}\ , \tag{3.1.23}$$

which links fluctuations of the density ρ' and the specific volume v''. This relation immediately follows from the definition of the quantity v'':

$$v'' = v - \langle v \rangle = 1/\rho - 1/\bar{\rho} = (\rho - \bar{\rho})/\rho\bar{\rho} = -\rho'/\rho\bar{\rho}.$$

Diffusion Equations for Chemical Components in a Turbulized Mixture. Applying the averaging operator (3.1.3) to the diffusion equation (2.1.7) with regard to the identity (3.1.9) leads to the following balance equation for the averaged quantity $\langle Z_\alpha \rangle$:

$$\bar{\rho}\frac{D\langle Z_\alpha \rangle}{Dt} = -\partial_j J^{\Sigma}_{\alpha j} + \overline{\sigma_\alpha}, \qquad (\alpha = 1,2,...,N), \tag{3.1.24}$$

where

$$J^{\Sigma}_{\alpha j} \equiv \overline{J_{\alpha j}} + J^{T}_{\alpha j} \tag{3.1.25}$$

is the total diffusion flux of the α-component in the averaged turbulized medium; and

$$J^{T}_{\alpha j} \equiv \overline{\rho Z''_\alpha V''_j} = \bar{\rho}\langle Z''_\alpha V''_j \rangle \tag{3.1.26}$$

is the turbulent diffusion flux for the α-substance.

The influence of chemical reactions inherent to the system on the spatial-temporal distribution of the specific concentration $\langle Z_\alpha \rangle$ of α-particles is taken into account using the averaged source σ_α:

$$\overline{\sigma_\alpha} = \sum_{s=1}^{r} \nu_{\alpha s}\overline{\xi_s}. \tag{3.1.27}$$

The method to average the production source term $\overline{\sigma_\alpha}$ due to chemical reactions will be analyzed specially in subsection 3.2.3.

It is worthwhile to notice that by using the weighted-mean averaging properties (3.1.7) it is easy to obtain another (conventional) notation for the turbulent diffusion flux: $J^{T}_{\alpha j} = \overline{n'_\alpha V'_j} - (\overline{n_\alpha}/\bar{\rho})\,\overline{\rho' V'_j}$. Obviously, awkwardness of this expression in comparison with (3.2.13) once again testifies the efficiency of using weighted averaging for multicomponent media with variable density.

By applying the averaging operator (3.1.3) to (2.1.8) and (2.1.9) we obtain the equivalents of these relations for the averaged motion:

$$\sum_{\beta=1}^{N} M_\beta \langle Z_\beta \rangle = 1 \quad (^1), \qquad \sum_{\beta=1}^{N} M_\beta \overline{J_{\beta j}} = 0 \quad (^2), \qquad \sum_{\beta=1}^{N} M_\beta \overline{\sigma_\beta} = 0 \quad (^3) \tag{3.1.28}$$

In addition, the identity $\sum_{\beta=1}^{N} M_\beta J^T_{\beta j} = \sum_{\beta=1}^{N} M_\beta \overline{\rho Z_\beta V''_j} = \overline{\rho\left(\sum_{\beta=1}^{N} M_\beta Z_\beta\right) V''_j} = \overline{\rho V''_j} = 0$ is valid for the turbulent diffusion fluxes $J^T_{\beta j}$, i.e.

$$\sum_{\beta=1}^{N} M_\beta J^T_{\beta j} = 0 \quad (^1) \qquad \text{and hence,} \qquad \sum_{\beta=1}^{N} M_\beta J^\Sigma_{\beta j} = 0. \quad (^2) \tag{3.1.29}$$

Thus the averaged diffusion equations (3.1.24), like their non-averaged analogs (2.1.7), are linearly dependent and one of them should be replaced by the algebraic integral (3.1.29 (^{2})).

Averaged Conservation Equations for the Chemical Elements. Proceeding from (3.1.14) and assuming that $A \equiv Z^\gamma$ $(\gamma = N_m + 1, \ldots, N)$ we will now find the differential form of the conservation laws for the individual chemical elements in averaged turbulized multicomponent continua. Using the formulas (2.1.18) for the quantities J^γ_j and $\sigma^\gamma_{(Z)}$ yields

$$\bar{\rho} \frac{D < Z^\gamma >}{Dt} = -\partial_j J^{\gamma\Sigma}_j, \qquad (\gamma = N_m + 1, \ldots, N), \tag{3.1.30}$$

where

$$< Z^\gamma > = \sum_{\alpha=1}^{N} \eta^\gamma_\alpha < Z_\alpha >, \quad (^1) \qquad J^{\gamma\Sigma}_j \equiv \bar{J}^\gamma_j + J^{\gamma T}_j, \; (\gamma = N_m + 1, \ldots, N). \quad (^2) \tag{3.1.31}$$

With this, the averaged regular diffusion flux density of the γ-element complies with the relation

$$\bar{J}^\gamma_j = \sum_{\alpha=1}^{N} \eta^\gamma_\alpha \bar{J}_{\alpha j} = \bar{J}_{\gamma j} + \sum_{\alpha=1}^{N_m} \eta^\gamma_\alpha \bar{J}_{\alpha j}, \quad (\gamma = N_m + 1, \ldots, N), \tag{3.1.32}$$

which is the average of (2.1.14). A similar relation is observed for the turbulent diffusion flux density of the γ-element:

$$J^{\gamma T}_j \equiv \overline{\rho Z^\gamma V''_j} = \sum_{\alpha=1}^{N} \eta^\gamma_\alpha \overline{\rho Z_\alpha V''_j} = \sum_{\alpha=1}^{N} \eta^\gamma_\alpha J^T_{\alpha j} = J^T_{\gamma j} + \sum_{\alpha=1}^{N_m} \eta^\gamma_\alpha J^T_{\alpha j}. \tag{3.1.33}$$

Furthermore, like in the case of molecular diffusion, the identities

$$\sum_{\gamma=N_m+1}^{N} M_\gamma < Z^\gamma > = 1, \; (^1) \qquad \sum_{\gamma=N_m+1}^{N} M_\gamma \bar{J}^\gamma_j = 0, \; (^2) \qquad \sum_{\gamma=N_m+1}^{N} M_\gamma J^{\gamma T}_j = 0 \; (^3) \tag{3.1.34}$$

are valid. At the same time, the first two of these identities represent the averaged equalities (2.1.16), while the third one results from the definition (3.1.33) of the turbulent diffusion flux $J_j^{\gamma T}$ of the γ-element and from the relation (2.1.17), which specifies the molecular mass through the masses of the individual chemical elements.

The Equation of Motion for Turbulized Mixtures. Let us assume that $A \equiv V_j$ in (3.1.14) and use the expressions (2.1.22) for the flux $J_{(V_i)j}$ and the momentum source $\sigma_{(V_i)}$ in which fluctuations of the quantities Ω_j and $F_{\alpha j}$ in the turbulized flow are neglected. Then we obtain the equation of motion

$$\bar{\rho}\frac{D<V_i>}{Dt} = -\partial_i \bar{p} + \partial_j \pi^{\Sigma}_{ij} + 2\bar{\rho}\,\varepsilon_{kij} <V_k> \Omega_j + \bar{\rho}\sum_{\alpha=1}^{N} <Z_\alpha> F_{\alpha j} \tag{3.1.35}$$

for averaged turbulized multicomponent continua. Here

$$\pi^{\Sigma}_{ij} \equiv -J^{\Sigma}_{(V_i)j} = -\overline{J_{(V_i)j}} - J^{T}_{(V_i)j} = \overline{\pi_{ij}} + R_{ij} \tag{3.1.36}$$

is the complete (aggregate) stress tensor for turbulent flows, and

$$R_{ij} \equiv -J^{T}_{(V_i)j} = -\overline{\rho V_i'' V_j''} = -\bar{\rho} <V_i'' V_j''> \tag{3.1.37}$$

is the tensor representing complementary (turbulent) stresses which are referred to as Reynolds stresses. This second rank symmetric tensor features the mean momentum transport at the expense of turbulent velocity fluctuations. If turbulent motions dominate the flow, the averaged viscous stresses $\overline{\pi_{ij}}$ can, as a rule, be neglected against the Reynolds stresses R_{ij} (provided the viscous sublayer adjoining a solid wall is not considered). Obviously, because the components of the turbulent stress tensor R_{ij} are unknown quantities, developing turbulence models is bound to specifying definitions for these quantities.

When using (2.1.25) for the flux $J_{(V_i)j}$ and the source $\sigma_{(V_i)}$ one obtains another notation for the averaged equation of motion:

$$\bar{\rho}\frac{D<V_i>}{Dt} = -\partial_j p^d + \partial_j \pi^{\Sigma}_{ij} - \delta_{i3} g \overline{\Delta\rho} + 2\bar{\rho}\varepsilon_{kij} <V_k> \Omega_j + \bar{\rho}\sum_{\alpha=1}^{N} <Z_\alpha> F^{*}_{\alpha i}\,. \tag{3.1.38}$$

Using the dynamic pressure p^d, this form is especially convenient for geophysical applications. Furthermore, when dealing with flows in a free stratified atmosphere where buoyancy forces (the third addend in the right hand part of (3.1.38)) are important, all terms of the equation (3.1.38) are usually of the order of $g\overline{\Delta\rho}$ or smaller. Then, since the total pressure gradient is expressed by the sum the dynamic and hydrostatic pressure gradients, as well as because of (2.1.24), we have the following approximate equality:

$$\partial_j \overline{p} = \partial_j \overline{p}^d + \partial_j p_0 \approx -\delta_{j3} g \overline{\Delta \rho} - \delta_{j3} \rho_0 g = -\delta_{j3} \rho_0 g (1 + \overline{\Delta \rho} / \rho_0).$$

Consequently, when the estimation $\overline{\Delta \rho} / \rho_0 << 1$ is correct, the total pressure gradient may be expressed approximately as

$$\partial_j \overline{p} \approx -\delta_{j3} \rho_0 g \,. \tag{3.1.39}$$

This particular expression will be used in Chapter 4.

Basically, in averaged turbulized flows, unlike in its laminar analog, there is a wide diversity of possible mechanisms to exchange (*transition rates*) between various kinds of particle motion energies contributing to the total, unchangeable energy of the continuum. Therefore, for the sake of most comprehensively interpretating the individual energy balance addends, it is important to consider the full system of energy equations describing the averaged field of fluctuating thermohydrodynamic parameters in a mixture including the kinetic energy balance equation for turbulent fluctuations.

The Balance Equation for the Averaged Potential Energy of a Mixture. Averaging the equation (2.1.30) according to Reynolds taking into account the identity (3.1.9) results in the following balance equation for the averaged specific potential energy of a turbulized mixture:

$$\overline{\rho} \frac{D < \Psi >}{Dt} = -\partial_j J^{\Sigma}_{(\Psi)j} + \overline{\sigma_{(\Psi)}}, \tag{3.1.40}$$

where

$$J^{\Sigma}_{(\Psi)j} \equiv \overline{J}_{(\Psi)j} + J^{T}_{(\Psi)j} = \sum_{\alpha=1}^{N} \psi_\alpha J^{\Sigma}_{\alpha j} \tag{3.1.41}$$

is the total substantial potential energy flux in a turbulized multicomponent continuum;

$$\overline{J}_{(\Psi)j} = \sum_{\alpha=1}^{N} \psi_\alpha \overline{J}_{\alpha j} \tag{3.1.42}$$

is the averaged regular potential energy flux in a mixture; and

$$J^{T}_{(\Psi)j} = \overline{\rho} < \Psi V''_j > = \sum_{\alpha=1}^{N} \psi_\alpha \overline{\rho Z_\alpha V''_j} = \sum_{\alpha=1}^{N} \psi_\alpha J^{T}_{\alpha j} \tag{3.1.43}$$

is the turbulent potential energy flow in a mixture.

The averaged potential energy source for a multicomponent mixture is defined by the relation

$$\overline{\sigma}_{(\Psi)} = -\sum_{\alpha=1}^{N} J^{\Sigma}_{\alpha j} F^{*}_{\alpha j} - \overline{\rho} < V_j > \sum_{\alpha=1}^{N} < Z_\alpha > F_{\alpha j} \,. \tag{3.1.44}$$

Note that the quantity $\sum_{\alpha=1}^{N} J^{\Sigma}_{\alpha j} F^{*}_{\alpha j}$ represents the total transformation rate of the mean motion potential energy into other energy forms per unit volume; this follows from comparing the equations (3.1.44) and (3.1.57). In turn, the quantity $\bar{\rho} < V_j > \sum_{\alpha=1}^{N} < Z_\alpha > F_{\alpha j}$ relates to the transformation rate of the mean potential energy into mean motion kinetic energy (see (3.1.45)), this process being reversible (*adiabatic*).

The Balance Equation for the Mean Motion Kinetic Energy of a Turbulized Medium. Multiplying scalarly the equation of motion (3.1.35) by $< V_j >$, taking into account the asymmetrical properties of the Levi-Civita tensor (2.1.22), and using quite simple transformations we will obtain the following substantial form of the equation for the averaged motion of a mixture (the work-energy theorem) similar to (2.1.36):

$$\bar{\rho} \frac{D(< \mathbf{V} >^2 / 2)}{Dt} = - \partial_j \left[(\bar{p} \delta_{ij} - \pi^{\Sigma}_{ij}) < V_i > \right] + \bar{p} \partial_j < V_j > -$$
$$- \pi^{\Sigma}_{ij} \partial_j < V_i > + \bar{\rho} < V_j > \sum_{\alpha=1}^{N} < Z_\alpha > F_{\alpha j}, \qquad (3.1.45)$$

where $< \mathbf{V} >^2 / 2$ is the specific kinetic energy of the mean motion, and the first, second, and third terms in the right hand part of the equations multiplied by Dt represent the work densities of external surface forces, internal surface forces, and external mass forces, respectively. Although (3.1.45) is an energy equation by its nature, it does not represent the energy conservation law for turbulized continua. Instead, this equation describes the reversible transformation law for the mean motion kinetic energy into the work of external mass and surface forces and into the work of internal forces, with no account for non-reversible transition of mechanical energy into thermal or other forms of energy.

Let us explain the physical meaning of the individual terms in (3.1.45). The quantity $\partial_j (\bar{p} < V_j >)$ relates to the outflow of mechanical energy from a unit volume during a unit time; the divergence $\partial_j (\pi^{\Sigma}_{ij} < V_i >)$ represents the rate at which the complete surface stress π^{Σ}_{ij} does work within a unit volume of the mean traveling system; the quantity $\bar{p} \partial_j < V_j >$ (> 0 or < 0) is connected to the rate at which the averaged internal energy (heat) inverse adiabatically transforms into mechanical energy (see (3.1.57)) and represents the work done by the traveling mixture flow against the average pressure $\bar{p}$ in a unit volume during a unit time; the sign of the quantity $\bar{p} \partial_j < V_j >$ depends on whether the mixture flow expands $(\partial_j < V_j > > 0)$ or compresses $(\partial_j < V_j > < 0)$. Finally, the quantity $\pi^{\Sigma}_{ij} \partial_j < V_i >$ represents the full non-reversible transformation rate of the mean motion kinetic energy into other forms of energy in a unit volume (see (3.1.57) and (3.1.68)), energy dissipation of the mean motion occurring under the influence of both molecular viscosity at rate $\overline{\pi}_{ij} \partial_j < V_i >$ and turbulence at rate

$R_{ij}\partial_j <V_i> (<0)$.

Now, combining the equations (3.1.40) and (3.1.45), we can obtain the balance equation for the mean motion mechanical energy of turbulized continua:

$$\bar{\rho}\frac{D(<V>^2/2+<\Psi>)}{Dt}=-\partial_j\left[(\bar{p}\delta_{ij}-\pi^{\Sigma}_{ij})<V_i>+\sum_{\alpha=1}^{N}\psi_\alpha J^{\Sigma}_{\alpha j}\right]+$$

$$+\bar{p}\partial_j<V_j>-\pi^{\Sigma}_{ij}\partial_j<V_i>-\sum_{\alpha=1}^{N}J^{\Sigma}_{\alpha j}F_{\alpha j}. \qquad (3.1.46)$$

The Balance Equation for the Averaged Internal Energy of Turbulized Mixtures. This equation we will derive from (3.1.14) assuming that $A \equiv h$ and also using (2.1.51) for the heat flux $(J_{(h)j} \equiv q_j)$ and (2.1.52) for the energy source $(\sigma_{(h)} \equiv dp/dt + \pi_{ij}\partial_j V_i + \sum_{\alpha=1}^{N} J_{\alpha j}F^{*}_{\alpha j})$. As a result, we have

$$\bar{\rho}\frac{D<h>}{Dt}=-\partial_j q^{\Sigma}_j+\bar{\sigma}_{(h)}, \qquad (3.1.47)$$

where

$$q^{\Sigma}_j \equiv \bar{q}_j + q^T_j \qquad (3.1.48\ (^a))$$

is the total heat flux in an averaged multicomponent turbulized continuum;

$$q^T_j \equiv \overline{\rho\, h'' V''_j} \cong <C_p> \overline{\rho T'' V''_j} + \sum_{\alpha=1}^{N} <h_\alpha> J^T_{\alpha j} \qquad (3.1.48\ (^b))$$

is the turbulent heat flux (explicit and latent heat being expressed by the first and the second terms, respectively). This flux arises due to the correlation between fluctuations of the specific enthalpy h'' of a mixture and the hydrodynamic flow velocity V''_j. The approximate equality (3.1.47 ([b])) is accurate up to the terms involving triple correlations. This equality is easy to obtain using the properties (3.1.7) of Favre's averaging and the formula

$$h''=\sum_{\alpha=1}^{N}\left[<Z_\alpha>h''_\alpha+<h_\alpha>Z''_\alpha+(Z''_\alpha h''_\alpha)''\right]=<C_p>T''+\sum_{\alpha=1}^{N}<h_\alpha>Z''_\alpha+(C''_p T'')'' \qquad (3.1.49)$$

for fluctuations of the specific enthalpy. Here

$$h''_\alpha = C_{p\alpha}T'' \qquad (^1) \qquad (3.1.50)$$

is fluctuation of the partial enthalpy of the α-component $(C_{p\alpha}=const)$, whereas

$$< C_p > = \sum_{\beta=1}^{N} C_{p\beta} < Z_\beta >, \quad (^2) \qquad \text{and} \qquad C''_p = \sum_{\beta=1}^{N} C_{p\beta} Z''_\beta, \quad (^3) \qquad (3.1.50)$$

are the averaged and fluctuation components of the specific heat capacity of the mixture at constant pressure. The averaged specific enthalpy of a mixture in (3.1.47) is determined by the following precise relation:

$$< h > = < C_p >< T > \left[1 + < C''_p T'' > / < C_p >< T >\right] + \sum_{\alpha=1}^{N} h^0_\alpha < Z_\alpha >, \quad (^4) \qquad (3.1.50)$$

which is obtained by averaging the expression (2.1.41).

It is convenient to represent the substantial derivative of the total pressure of a mixture in the expression for $\overline{\sigma}_{(h)}$ as

$$\frac{dp}{dt} = \frac{Dp}{Dt} + V''_j \partial_j p = \frac{Dp}{Dt} + V''_j \partial_j \overline{p} + V''_j \partial_j p' =$$

$$= \frac{Dp}{Dt} + \partial_j (p' V''_j) - p' \partial_j V''_j + V''_j \partial_j \overline{p}.$$

Then it is possible to write

$$\overline{\frac{dp}{dt}} = \frac{D\overline{p}}{Dt} + \partial_j (\overline{p' V''_j}) - \overline{p' \partial_j V''_j} + J^T_{(v)j} \partial_j \overline{p}. \qquad (3.1.51)$$

Besides, we have

$$\overline{\pi_{ij} \partial_j V_i} = \overline{\pi}_{ij} \partial_j < V_i > + \overline{\pi_{ij} \partial_j V''_i} = \overline{\pi}_{ij} \partial_j < V_i > + \overline{\rho} < \varepsilon_e >, \qquad (3.1.52(^1))$$

where

$$\overline{\rho} < \varepsilon_e > \equiv \overline{\pi_{ij} \partial_j V''_i} \qquad (3.1.52(^2))$$

is the mean dissipation rate of turbulent kinetic energy into heat due to molecular viscosity in a unit volume during a unit time. This specific rate is the key characteristic in the turbulence theory.

Substituting (3.1.51) and (3.1.52 (^{1})) into (3.1.47), we finally obtain

$$\overline{\rho} \frac{D < h >}{Dt} + \partial_j \left(\overline{q}_j + q^{T*}_j\right) = \frac{D\overline{p}}{Dt} + \overline{\pi}_{ij} \partial_j < V_i > -$$

$$- \overline{p' \partial_j V''_j} + J^T_{(v)j} \partial_j \overline{p} + \sum_{\alpha=1}^{N} \overline{J}_{\alpha j} F^*_{\alpha j} + \overline{\rho} < \varepsilon_e >, \qquad (3.1.47^*)$$

where the quantity

$$q_j^{T*} \equiv q_j^T - \overline{p'V_j''} \tag{3.1.54}$$

determines the so-called *ordinary* turbulent heat flux, this quantity being intimately related to the turbulent entropy flux (see subsection 3.3.4 and Chapter 5 for more details).

It is also possible to write down the equation (3.1.47) in terms of the averaged internal energy $<\varepsilon>$ of a turbulized continuum. Indeed, the quantity $<\varepsilon>$ is defined by the relation

$$<\varepsilon>=<h>-\frac{\overline{p}}{\overline{\rho}}=<C_V><T>\left[1+\frac{<C_V''T''>}{<C_V><T>}\right]+\sum_{\alpha=1}^{N} h_\alpha^0 <Z_\alpha>, \tag{3.1.55 (a)}$$

which is the precise result of averaging the expression (2.1.41). Here

$$<C_V>=\sum_{\alpha=1}^{N} C_{V\alpha} <Z_\alpha>, \quad \left(<C_p>-<C_V>=<R^*>=\mathrm{k}\sum_{\alpha=1}^{N}<Z_\alpha>\right) \tag{3.1.55 (b)}$$

is the averaged specific heat capacity of a mixture at constant volume (a variable thermophysical quantity). Note that the second equality in (3.1.55 (a)) is valid only when

$$\varepsilon_\alpha = C_{V\alpha}T + h_\alpha^0, \quad (C_{V\alpha} = const). \tag{3.1.55 (c)}$$

Using the transformation (compare with (2.1.42))

$$\overline{\rho}\frac{D<\varepsilon>}{Dt}+\overline{p}\partial_j<V_j>=\overline{\rho}\frac{D<h>}{Dt}-\frac{D\overline{p}}{Dt}, \tag{3.1.56}$$

resulting from (3.1.55 (a)) and averaging the continuity equation (3.1.5), we finally obtain

$$\overline{\rho}\frac{D<\varepsilon>}{Dt}+\partial_j\left(\overline{q}_j+q_j^{T*}\right)=-\overline{p}\partial_j<V_j>+\overline{\pi}_{ij}\partial_j<V_i>-\overline{p'\partial_j V_j''}+$$
$$+J_{(v)j}^T\partial_j\overline{p}+\sum_{\alpha=1}^{N}\overline{J}_{\alpha j}F_{\alpha j}^*+\overline{\rho}<\varepsilon_e>. \tag{3.1.57}$$

In this equation the quantity $\overline{p'\partial_j V_j''}$ relates to the transformation rate of kinetic energy of turbulent vortices into internal energy (see equation (3.1.68)) and represents the work on vortexes done by the ambient medium in a unit volume during a unit time, as a consequence of the occurring pressure fluctuations p' and the expansion or compression of vortices $(\partial_j V_j > 0$ or $\partial_j V_j < 0)$. Comparing the equations (3.1.57) and (3.1.40) shows that the quantity $\sum_{\alpha=1}^{N} \bar{J}_{\alpha j} F^{*}_{\alpha j}$ is connected to the transition rate between the averaged internal and the averaged potential energies as a result of work done by non-gravitational external forces. Similarly, comparing (3.1.57) and (3.1.45) indicates that the quantities $\bar{p}\,\partial_j < V_j >$ and $\bar{\pi}_{ij}\,\partial_j < V_i >$ are connected to the transition rate between the internal and mean motion kinetic energies.

The correlation $\bar{\rho} < \varepsilon_e > = \overline{\pi_{ij}\partial_j V_i''} \approx \overline{\pi'_{ij}\partial_j V_i'}$ in a developed turbulent field (see section 4.2.3)) represents the mean value of the work done in a unit volume during a unit time by viscous stress fluctuations on turbulent vortices with a non-zero velocity shear $(\partial_j V_j'' \neq 0)$. This work is always positive because $< \varepsilon_e >$ represents the dissipation rate of turbulent kinetic energy into heat under the influence of molecular viscosity.

Finally, let us consider the transformation rate of $J^{T}_{(v)j}\partial_j \bar{p}$. In the presence of buoyancy forces this quantity is convenient to extrapolate taking into account (3.1.39) and using the following expression:

$$J^{T}_{(v)j}\partial_j \bar{p} \approx g(\rho_0 / \bar{\rho})\overline{\rho' V_3''} \,. \qquad (3.1.58)$$

Generally, for turbulized mixture flows the following two cases can be implemented (see, e.g., *Van Miegham, 1973*):

- For large vortices (like in the planetary atmosphere) the quantity $g\overline{\rho' V_3''} < 0$. This is caused by the fact that fluctuations ρ' of thermal origin at the considered scales determine the sign of the vortex displacement in the vertical direction due to the buoyancy. Indeed, light and heavy vortexes use to coincide with warm and cold ones, respectively, and, for example, for light warm vortexes ($\rho' < 0$) ascending ($V_3'' > 0$) in the gravitational field we obtain $g\overline{\rho' V_3''} <$ 0. Thus, large vortices transform thermal (internal) energy of the flow into turbulent kinetic energy.

- For small-scale turbulence the quantity $g\overline{\rho' V_3''} > 0$. As was mentioned in Chapter 1, the Brent-Vaisala frequency $N^2 = (-g/\bar{\rho})\partial_3\bar{\rho}$ serves as a local stratification characteristic. In this case $N^2 > 0$ and the buoyancy force serves as a restoring factor, i.e. turbulence expends its energy for work against buoyancy forces. This means that the quantity $g\overline{\rho' V_3''}$ represents the transformation rate of turbulent energy into the averaged internal energy in a unit volume of the medium. In other words, small-scale vortices convert turbulence energy into heat (see equation (3.1.68)).

The Total Energy Conservation Law for Turbulized Mixtures. Let us now obtain substantial form of the conservation law for the averaged total energy of a turbulized continuum. This equation will allow us to deduce the evolutionary transfer equation for turbulent energy, which is fundamental in the turbulence theory.

Applying the averaging operator (3.1.3) to (1.1.33) and using the relations (2.1.34) and (2.1.35) for the quantities $E(\boldsymbol{r},t)$ and $J_{(E)j}$ we obtain the following substantial form of the conservation law for the averaged total energy of a multicomponent mixture:

$$\bar{\rho}\frac{D<E>}{Dt}=\partial_j J^{\Sigma}_{(E)j}, \quad (J^{\Sigma}_{(E)j}\equiv \bar{J}_{(E)j}+J^{T}_{(E)j}), \tag{3.1.59}$$

where

$$<E>=<V_jV_j/2>+<\Psi>+<\varepsilon> \tag{3.1.60}$$

is the averaged total specific energy of the mixture;

$$J^{T}_{(E)j}\equiv \bar{\rho}<E''V''_j>=\overline{\rho(V_jV_j/2+\Psi+\varepsilon)V''_j} \tag{3.1.61}$$

is the turbulent total energy flux in the mixture; and

$$\bar{J}_{(E)j}=\overline{q_j+(p\,\delta_{ij}-\pi_{ij})V_i+\sum_{\alpha=1}^{N}\psi_\alpha J_{\alpha j}}=\bar{q}_j+(\bar{p}\delta_{ij}-\bar{\pi}_{ij})\partial_j<V_i>+$$
$$+\overline{pV''_j}-\overline{\pi_{ij}V''_j}+\sum_{\alpha=1}\psi_\alpha\bar{J}_{\alpha j} \tag{3.1.62}$$

is the averaged regular total energy flux in the multicomponent system.

Then it is convenient to modify the kinetic energy of instantaneous motion as follows:

$$\rho V_jV_j/2=\rho(<V_j>+V''_j)(<V_j>+V''_j)/2=$$
$$=\rho<V_j><V_j>/2+\rho<V_j>V''_j+\rho\, V''_jV''_j/2.$$

Statistically averaging this equality and taking into account (3.1.4) we obtain

$$\overline{\rho V_jV_j/2}=\bar{\rho}<V_j><V_j>/2+\overline{\rho V''_jV''_j/2}, \tag{3.1.63}$$

or

$$<V_jV_j>/2=<V_j><V_j>/2+<e>, \tag{3.1.64}$$

where one more quantity,

$$<e>\equiv\overline{\rho e}/\bar{\rho}=\overline{\rho V''_jV''_j}/2\bar{\rho} \tag{3.1.65}$$

is introduced. The latter is of key importance because it represents the averaged value of the kinetic energy of turbulent fluctuations (or simply *turbulent energy*), $e \equiv V_j'' V_j''/2$ being the specific fluctuation kinetic flow energy.

As a result, if the transformation (3.1.63) is employed, the expressions (3.1.60) and (3.1.61) become

$$<E>=<V_j><V_j>/2+<\Psi>+<\varepsilon>+<e>, \tag{3.1.66}$$

$$J^T_{(E)j}=\overline{\rho}<eV_j''>-<V_i>R_{ij}+J^T_{(\Psi)j}+J^T_{(\varepsilon)j}, \tag{3.1.67}$$

where the correlation

$$J^T_{(\varepsilon)j}\equiv\overline{\rho}<\varepsilon V_j''>=\overline{\rho(h-p/\rho)V_j''}=q_j^T-\overline{pV_j''}=q_j^{T*}+\overline{p'V_j''}-\overline{pV_j''} \tag{3.1.68}$$

defines the turbulent flux of specific internal energy in the mixture.

Finally, combining the formulas (3.1.62), (3.1.61*) and (3.1.66) for the full turbulent flux $J^{\Sigma}_{(E)j}$ of the total energy, we find:

$$J^{\Sigma}_{(E)j}=\overline{J}_{(E)j}+J^T_{(E)j}=\overline{q}_j+q_j^{T*}+\overline{\rho(e+p'/\rho)V_j''}-\overline{\pi_{ij}V_j''}+$$
$$+(\overline{p}\delta_{ij}-\pi^{\Sigma}_{ij})<V_i>+\sum_{\alpha=1}^{N}\psi_\alpha J^{\Sigma}_{\alpha j}. \tag{3.1.67*}$$

Here $q_j^{\Sigma}\equiv\overline{q}_j+q_j^T$ is the total heat flux due to averaged molecular and turbulent transport; $\overline{p}<V_j>$ is the mechanical energy flux; $\pi^{\Sigma}_{ij}<V_i>$ is the total energy flux due to the work done by viscous and turbulent stresses; $\left(\overline{\rho eV_j''-\pi_{ij}V_i''}\right)$ is the turbulent vortex energy flux as a consequence of turbulent diffusion; and $\sum_{\alpha=1}^{N}\psi_\alpha J^{\Sigma}_{\alpha j}$ is the total potential energy flux due to averaged molecular and turbulent diffusion.

It is important to note that the term $\overline{p'V_j''}$ in (3.1.67) must not be associated with the energy flux as it drops out of the complete energy equation when the formula (3.1.54) for the *ordinary* turbulent heat flux is taken into account. This term was introduced just for convenience and is used as such hereafter (see section 3.3 and Chapter 5).

The Transfer Equation for Turbulent Energy of Multicomponent Compressible Mixtures. The equation for turbulent energy transfer for a multicomponent compressible mixture (or some of its modifications) is the fundamental equation which underlies many modern semi-empirical turbulence models. It can be deduced using different techniques one of which is given in Chapter 4. Here we will derive it from the balance equations (3.1.46), (3.1.57), and (3.1.59).

Subtracting (3.1.46) and (3.1.57) from the equation (3.1.59) where the quantities $<E>$ and $J^{\Sigma}_{(E)j}$ are determined by (3.1.60*) and (3.1.67), respectively, we obtain the substantial balance equation for the averaged kinetic energy of turbulent fluctuations in a mixture:

$$\bar{\rho}\frac{D<e>}{Dt} = -\partial_j\left(\overline{\rho(e+p'/\rho)V''_j - \pi_{ij}V''_i}\right) + R_{ij}\,\partial_j <V_i> +$$

$$+\overline{p'\partial_j V''_j} - J^T_{(v)j}\partial_j \bar{p} + \sum_{\alpha=1}^{N} J^T_{\alpha j}F^{*}_{\alpha j} - \bar{\rho}<\varepsilon_e>. \tag{3.1.68}$$

The term in the left hand part characterizes the temporal variation (as well as the convective transport by the averaged motion) of the averaged kinetic energy of turbulent fluctuations $<e>$; the first term in the right hand part expresses the kinetic turbulence energy transport at the expense of turbulent "diffusion"; the third term represents the work done by pressure forces in a fluctuating motion; the fourth and the fifth terms express the turbulence energy generation rate due to buoyancy forces and to forces of non-gravitational origin; at last, the sixth term represents the dissipation rate of turbulent kinetic energy into thermal internal energy owing to the molecular viscosity. Because the quantity $R_{ij}\,\partial_j <V_i>$ in the right hand part of the equations (3.1.45) and (3.1.68) has the opposite sign, it can be interpreted as the transition rate of the mean motion kinetic energy into the kinetic energy of turbulent fluctuations. It should once again be emphasized that this energy transition is the only kinematic process depending just on the choice of the averaging procedure for turbulent motion. It is known that in the case of small-scale turbulence one has $R_{ij}\partial_j <V_i> > 0$; thus small-scale turbulence always converts the mean motion kinetic energy into kinetic energy of turbulent fluctuations. This is the so-called dissipative effect of small-scale turbulence. Large-scale turbulence, however, is able to transform the kinetic turbulence energy into mean motion energy (*Van Mieghem, 1977*).

The Equation for the Averaged Internal Energy of a Mixture Expressed Through the Temperature and Pressure. We will first obtain the expressions necessary for the averaged ($<h>$) and fluctuation (h'') enthalpy components of a mixture. The complete specific enthalpy for the true motion of a turbulized continuum is determined by the relation (2.1.41) which can be written here as $h=\sum_{\alpha=1}^{N} Z_\alpha h_\alpha \approx C_p T + \sum_{\alpha=1}^{N} Z_\alpha h^0_\alpha$, the second form of the notation being valid when the simple expression $h_\alpha = C_{p\alpha}T + h^0_\alpha$ is used for the partial enthalpy of the α-component only. Note that in such a case the quantities $C_{p\alpha}$ and h^0_α, being constant in a flow, approximate the partial heat capacity per particle and the formation heat of α-substance, respectively. Using one more property of the weighted-mean averaging, which is

$$(AB)'' = <B>A'' + (AB'')'' = <A>B'' + <B>A'' + (A''B'')'', \tag{3.1.69}$$

as well as the relations (3.1.50([1])) and (3.1.50([2])) for the averaged and fluctuation components of the quantities h_α and C_p we obtain from (2.1.45) the required expression for the complete enthalpy fluctuations in a mixture:

$$h'' = \sum_{\alpha=1}^{N} (Z_\alpha h_\alpha)'' = \sum_{\alpha=1}^{N} \left[< Z_\alpha > h''_\alpha + < h_\alpha > Z''_\alpha + (Z''_\alpha h''_\alpha)'' \right] =$$

$$= \sum_{\alpha=1}^{N} \left[C_{p\alpha} < Z_\alpha > T'' + Z''_\alpha < h_\alpha > + Z''_\alpha h''_\alpha - < Z''_\alpha h''_\alpha > \right] =$$

$$= < C_p > T'' + \sum_{\alpha=1}^{N} Z''_\alpha h_\alpha - < C''_p T'' > . \quad (3.1.70)$$

From (3.1.70) the expression for temperature fluctuations follows:

$$< C_p > T'' = h'' - \sum_{\alpha=1}^{N} Z''_\alpha h_\alpha + < C''_p T'' > = h'' - \sum_{\alpha=1}^{N} Z''_\alpha h_\alpha + \sum_{\alpha=1}^{N} < Z''_\alpha h''_\alpha > , \quad (3.1.71)$$

which is precisely considered to be based on the assumption made above on the character of the temperature dependence of the quantity h_α. This formula, being repeatedly utilized, allows to substitute temperature fluctuations by enthalpy and mixture composition fluctuations.

Multiplying (3.1.71) by $\rho A''$, where $A(\boldsymbol{r},t)$ is whatever parameter fluctuating in a flow and averaging the result over an ensemble of probable realizations, the following general expression can be obtained:

$$< C_p > < A'' T'' > = < A'' h'' > - \sum_{\alpha=1}^{N} < h_\alpha > < A'' Z''_\alpha > - < A'' C''_p T'' > \approx$$
$$\approx < A'' h'' > - \sum_{\alpha=1}^{N} < h_\alpha > < A'' Z''_\alpha > . \quad (3.1.72)$$

The second approximate equality in (3.1.72) is accurate up to the 3-rd order correlation and paired products of the 2-nd order correlation (composition and enthalpy fluctuations). When specifying $A(\boldsymbol{r},t)$, some useful relations emerging from (3.1.72) follow:

$$q_j^T \approx < C_p > \overline{\rho T'' V''_j} + \sum_{\alpha=1}^{N} < h_\alpha > J_{\alpha j}^T , \quad (A \equiv V_j),$$

$$< C_p > \overline{T''} \approx \overline{h''} - \sum_{\alpha=1}^{N} < h_\alpha > \overline{Z''_\alpha} , \quad (A \equiv v),$$

$$<C_p><Z''_\alpha T''> \approx <Z''_\alpha h''> - \sum_{\beta=1}^{N} <h_\beta><Z''_\beta Z''_\alpha>, \qquad (A \equiv Z_\beta),$$

$$<C_p>^2<T''^2> \approx <h''^2> - 2\sum_{\beta=1}^{N} <h_\beta><h'' Z''_\beta> + \sum_{\alpha=1}^{N}\sum_{\beta=1}^{N} <h_\alpha><h_\beta><Z''_\alpha Z''_\beta>, \quad (A \equiv T).$$

(3.1.73)

Let us obtain another representation for the averaged complete enthalpy of the mixture different from (3.1.50([3])). Applying the averaging operator to (2.1.45) one finds

$$<h> = \sum_{\alpha=1}^{N} \left[<Z_\alpha><h_\alpha> + <Z''_\alpha h''_\alpha> \right] = <C_p><T> + \sum_{\alpha=1}^{N} <Z_\alpha> h_\alpha^0 + <C''_p T''>, \quad (3.1.74)$$

or, taking into account (3.1.73 ([4])),

$$<h>\left[1 - <C''_p h''>/<C_p><h>\right] = <C_p><T>\left[1 - <C''^2_p>/<C_p>^2\right] +$$

$$+ \sum_{\beta=1}^{N} <Z_\beta> h_\beta^0 \left[1 - <C''_p Z''_\beta>/<C_p><Z_\beta>\right]. \quad (3.1.75)$$

This expression may be simplified naturally enough for simple multicomponent turbulence models where the paired correlations of temperature and composition are omitted. In particular, this is applicable to the averaged equation of state for pressure when fluctuations of the heat capacity C_p in the turbulized mixture flow can be neglected and hence $C''_p = 0$. A similar expression for $<h>$ can also be obtained more generally when correlations such as $<A'' C''_p>$ for the parameters $A \equiv h,\ C_p,\ Z_\beta$ are small compared to the first order terms $<A><C_p>$. This yields

$$<h> \approx <C_p><T> + \sum_{\alpha=1}^{N} <Z_\alpha> h_\alpha^0 = \sum_{\alpha=1}^{N} <h_\alpha><Z_\alpha>. \quad (3.1.75^*)$$

We thus have obtained all necessary relations allowing us to express the energy equation (3.1.47*) in terms of the variables $<T>$ and $\overline{p}$. In the case when the approximation (3.1.75*) is applicable and the diffusion equation (3.1.24) is taken into account, we can write the following expression for the substantial derivative of $<h>$ as (compare with (2.1.47))

$$\overline{\rho}\frac{D<h>}{Dt} = \overline{\rho}\sum_{\alpha=1}^{N} <h_\alpha>\frac{D<Z_\alpha>}{Dt} + \overline{\rho}\sum_{\alpha=1}^{N} <Z_\alpha>\frac{D<h_\alpha>}{Dt} = \sum_{\alpha=1}^{N} <h_\alpha>\left[-\partial_j J^\Sigma_{\alpha j} + \overline{\sigma}_\alpha\right] +$$

$$+ \overline{\rho}\sum_{\alpha=1}^{N} <Z_\alpha C_{p\alpha}>\frac{D<T>}{Dt} = \overline{\rho}<C_p>\frac{D<T>}{Dt} - \partial_j\left\{\sum_{\alpha=1}^{N} <h_\alpha> J^\Sigma_{\alpha j}\right\} +$$

$$+\partial_j < T > \sum_{\alpha=1}^{N} C_{p\alpha} J_{\alpha j}^{\Sigma} + \sum_{s=1}^{r} < q_s > \overline{\xi}_s, \tag{3.1.76}$$

where

$$< q_s > = \sum_{\alpha=1}^{N} \nu_{\alpha s} < h_{\alpha} >, \qquad (s = 1,2,\ldots,r) \tag{3.1.77}$$

is the averaged heat of the s-th reaction (2.1.48). Then the equation for the internal energy of the averaged turbulized continuum (3.1.47*), being expressed through temperature and pressure, becomes

$$\overline{\rho} < C_p > \frac{D < T >}{Dt} = -\partial_j \left\{ \overline{q}_j + q_j^{T*} - \sum_{\alpha=1}^{N} < h_{\alpha} > J_{\alpha j}^{\Sigma} \right\} + \frac{D\overline{p}}{Dt} + \overline{\pi}_{ij} \partial_j < V_i > - \overline{p' \partial_j V_j''} -$$

$$- \partial_j < T > \sum_{\alpha=1}^{N} C_{p\alpha} J_{\alpha j}^{\Sigma} - \sum_{s=1}^{r} < q_s > \overline{\xi}_s + J_{(v)j}^{T} \partial_j \overline{p} + \sum_{\alpha=1}^{N} \overline{J}_{\alpha j} F_{\alpha j}^{*} + \overline{\rho} < \varepsilon_e >. \tag{3.1.78}$$

This equation can be used to evaluate the turbulence models based on simple gradient closure procedures.

More generally, in view of (3.1.75) for the averaged enthalpy $< h >$, it is also possible to obtain the following expression for the substantial derivative $\overline{\rho} D < h > / Dt$:

$$\overline{\rho} \frac{D < h >}{Dt} = \overline{\rho} < C_p > \left[1 - \frac{< C_p''^2 >}{< C_p >^2} \right] \frac{D < T >}{Dt} + \frac{\overline{\rho}}{< C_p >} \frac{D < C_p'' h'' >}{Dt} -$$

$$- \sum_{\alpha=1}^{N} \left[\overline{\rho} H_{\alpha}^{*} \frac{D < Z_{\alpha} >}{Dt} \right] - \frac{\overline{\rho}}{< C_p >} \sum_{\alpha=1}^{N} \left[< h_{\alpha} > \frac{D < C_p'' Z_{\alpha}'' >}{Dt} \right], \tag{3.1.79}$$

where $H_{\alpha}^{*} \equiv < h_{\alpha} > - C_{p\alpha} < C_p'' T'' > / < C_p >$. Evidently, an attempt to express the averaged internal energy equation in terms of the temperature $< T >$ only, can not be successful in the case of complicated multicomponent turbulence models, when the evolutionary transfer equations for the paired correlations of enthalpy (temperature) and composition fluctuations are involved. The latter are related, in particular, to the correlators $< C_p'' Z_{\alpha}'' >$ and $< C_p'' h'' >$. This circumstance is generally regarded as a powerful argument against the use of temperature as the basic state variable for a turbulized continuum. Instead, it is more appropriate to use the mixture enthalpy $< h >$ as the basic gas state variable defining the thermo-hydrodynamics for averaged turbulized multicomponent media.

3.2. TURBULENT FLOWS OF REACTING GASEOUS MIXTURES

The derived in section 3.1 partial differential equations of hydrodynamics based on the general conservation laws for continuous media are applicable for evaluating the averaged turbulent motions of gas-phase reacting mixtures. The closure problem of these equations poses some additional difficulties. The *first* difficulty is the necessity to take into account the compressibility of chemically active continua and therefore, significant changes in the mass density. In particular, in meteorology, convective compressible flows are considered the Boussinesque approximation exclusively. In this approach, the density change is taken into account only in the terms depending on acceleration due to gravity. However, such an approach is fully inapplicable, for example, to describe turbulent deflagration burning because multiple density variations may occur in the flow. The *second* difficulty (which will be addressed in more detail in Chapter 4) relates to the necessity to model the large number of auxiliary paired correlations of temperature and concentration fluctuations. These fluctuations appear when averaging the production source terms of substance in the equations describing mixture composition variations. In case of compressible reacting flows the evolutionary transfer equations for such correlations become dramatically complicated.

3.2.1. The Average Thermal Equation of State for Ideal Gas Mixtures

The set of averaged hydrodynamic equations for mixture should be supplemented by an averaged equation of state for pressure. For the compressible baroclinic media, the equation of state is the one for ideal gas mixtures (2.1.53). Then, applying the statistical averaging operator (3.1.3) to (2.1.53), we obtain the following exact expression:

$$\bar{p} = \sum_{\alpha=1}^{N} \bar{p}_\alpha = \mathrm{k} < T > \sum_{\alpha=1}^{N} \bar{n}_\alpha + \mathrm{k} \sum_{\alpha=1}^{N} < T'' n_\alpha > = \bar{\rho}\, \mathrm{k} < T > \sum_{\alpha=1}^{N} < Z_\alpha > + \bar{\rho}\, \mathrm{k} \sum_{\alpha=1}^{N} < T'' Z_\alpha > =$$

$$= \bar{\rho}\, \mathrm{k} < T > \sum_{\alpha=1}^{N} < Z_\alpha > \left[1 + < T'' Z''_\alpha > / < T > < Z_\alpha >\right], \tag{3.2.1}$$

which includes a large number of paired correlations of temperature T'' and concentrations Z''_α fluctuating in the flow. We shall focus on two simple cases: first, when the correlation terms $< T'' Z''_\alpha >$ are small enough compared to the first order terms, i.e. $< T'' Z''_\alpha > \ll < T > < Z_\alpha >$; and second, when the molecular masses of the mixture components are comparable, i.e. $M_\alpha = idem$. Then we have $1 = \sum_{\beta=1}^{N} M_\beta Z_\beta \approx idem \sum_{\beta=1}^{N} Z_\beta$ and $\sum_{\beta=1}^{N} < T'' Z_\beta > \approx (1/idem) < T'' > = 0$, whence it follows that the averaged equation of state for pressure takes the form

$$\overline{p}=\overline{\rho}\,\mathrm{k}<T>\sum_{\alpha=1}^{N}<Z_\alpha>=\overline{\rho}<R^*><T>. \qquad (3.2.2)$$

Here

$$<R^*>=\mathrm{k}\sum_{\alpha=1}^{N}<Z_\alpha>=\mathrm{k}\overline{n}/\overline{\rho} \qquad (3.2.3)$$

is the Favre averaged "gas constant", and $\overline{n}=\sum_{\alpha=1}^{N}\overline{n}_\alpha$ is the averaged total number of particles per unit volume. We will use the thermal equation of state (3.2.2) in relatively simple multicomponent turbulence models based on the gradient closure hypothesis.

3.2.2. The Averaged Hydrodynamic Equations for a Mixture

The differential balance equations (3.1.22), (3.1.24), (3.1.30), (3.1.35), and (3.1.57) together with the equation of state (3.2.1) make up the system of exact hydrodynamic equations for a mixture pertinent for mean motion scales. These equations were derived without any simplifications, such as *a priori* omitting some terms. For convenience in what follows, we here summarize the set of these equations:

$$\overline{\rho}\frac{D<v>}{Dt}=\partial_j<V_j>, \qquad (<v>=1/\overline{\rho}), \qquad (3.2.4)$$

$$\overline{\rho}\frac{D<Z_\alpha>}{Dt}=-\partial_j\left(\overline{J}_{\alpha j}+J^T_{\alpha j}\right)+\sum_{s=1}^{r}\nu_{\alpha s}\overline{\xi}_s, \qquad (<Z_\alpha>=\overline{n}/\overline{\rho}) \qquad (3.2.5)$$

$$\overline{\rho}\frac{D<V_i>}{Dt}=-\partial_i\overline{p}+\partial_j\left(\overline{\pi}_{ij}+R_{ij}\right)+2\overline{\rho}\varepsilon_{kji}<V_k>\Omega_j+\overline{\rho}\sum_{\alpha=1}^{N}<Z_\alpha>F_{\alpha i}, \qquad (3.2.6)$$

$$\overline{\rho}\frac{D<\varepsilon>}{Dt}+\partial_j\left(\overline{q}_j+q_j^{T*}\right)=-\overline{p}\,\partial_j<V_j>+\overline{\pi}_{ij}\partial_j<V_i>-\overline{p'\partial_jV_j''}+$$
$$+J^T_{(v)j}\,\partial_j\overline{p}+\sum_{\alpha=1}^{N}\overline{J}_{\alpha j}F^*_{\alpha j}+\overline{\rho}<\varepsilon_e>, \qquad (3.2.7)$$

$$\overline{p}=\overline{\rho}\,\mathrm{k}<T>\sum_{\alpha=1}^{N}<Z_\alpha>\left[1+<Z''_\alpha T''>/<Z_\alpha><T>\right]. \qquad (3.2.8)$$

However, the system turns out to be non-closed because it contains new unknown quantities along with the mean values of the thermohydrodynamic parameters $\overline{\rho},\overline{p},<T>,<V_j>,<Z_\alpha>$ and their derivatives. These unknowns appear owing to the non-linearity of the basic hydrodynamic equations for multicomponent mixtures describing instantaneous motions. They characterize the auxiliary (correlation) terms stipulated by the presence of turbulent fluctuations. It follows from the system (3.2.4)–(3.2.8) that the averaged motion of a reacting mixture is featured by the various averaged molecular thermodynamic fluxes ($\overline{q}_j, \overline{\pi}_{ij}, \overline{J}_{\alpha j}$, and $\overline{\xi}_s$), as well as by the mixed single-point second order moments.

With respect to the molecular fluxes, it should be emphasized that their regular analogs can not easily be obtained through Favre averaging. For example, directly averaging the expression (2.1.62) for the viscous stress tensor considerably complicates its structure. This is why in developing a phenomenological multicomponent turbulence model it is more reasonable to deduce the defining relations for these quantities from methods of non-equilibrium thermodynamics, rather than the relevant analogs for instantaneous values. This approach is treated in detail in section 5.2.

As to the mixed single-point second order moments (*paired correlations, correlators*), they represent the transport of hydrodynamic characteristics through a medium by turbulent pulsation. First of all, these are the turbulent specific volume flux $J^T_{(v)j}$ (3.1.21); the turbulent α-substance fluxes $J^T_{\alpha j}$ $(\alpha = 1,2,\ldots,N)$ (3.1.26) (*eddy diffusion fluxes*); the turbulent specific enthalpy flux q^T_j (3.1.48) (*turbulent heat flux*); the turbulent Reynolds stress tensor R_{ij} (3.1.37); and a large number of paired correlations such as $< Z''_\alpha T''>$ and $< Z''_\alpha Z''_\beta>$ $(\alpha, \beta = 1,2,\ldots,N)$. The latter appear in the diffusion equations for concentrations (3.2.5) in the case of chemically active gaseous mixtures due to averaging of the source terms (see (3.2.41)). They are also explicitly included in the averaged equation of state for pressure (3.2.8). All these quantities are needed in order to close the system of hydrodynamic equations for multicomponent media when Reynolds averaging is utilized. The correlation terms comprising pressure fluctuations ($\overline{p'\partial_j V''_j}$ and $\overline{\partial_j (p' V''_j)}$), as well as the viscous dissipation rate of turbulent kinetic energy $< \varepsilon_e >$ should also be specified.

In the turbulence models of incompressible homogeneous liquids (generally, with passive admixture not effecting the dynamic turbulence conditions), the simplest closure procedures based on the Boussinesque gradient hypothesis (*Boussinesque, 1977*) are widely accepted. Such an approach linearly relates the turbulent fluxes $q^T_j, R_{ij}, J^T_{\alpha j}$ to the gradients of the averaged quantities $< h >, < V_j >, < Z_\alpha >$, respectively, in terms of some local proportionality constants, i.e. the turbulent exchange coefficients (turbulent transport). These rheological gradient relations, earlier obtained in their most general forms (*Kolesnichenko and Marov, 1984*), are discussed in more detail below. Their utilization enables us to reduce the averaged equations of motion for turbulized fluid flows to the same form as for laminar flows. Respectively, this makes it possible to solve jointly the problems of both viscous laminar and turbulent flows.

Basically, one must be aware that the closure problem of the averaged system of hydrodynamic equations can not be solved with using the gradient hypothesis only. Some supplementary assumptions on the turbulent exchange coefficients should be adopted and the methods of their calculation clarified. Moreover, such an approach is fully unjustified when turbulent energy transport due to convection and diffusion significantly effects the flow features at an individual point. In other words, when this process is rooted in the turbulence set up it is impossible to introduce local turbulent exchange coefficients (*Ievlev, 1990).*

At the same time, following the Keller-Friedmann method developed for homogeneous compressible turbulized fluids (*Keller and Friedmann, 1924*; see also section 4.1) it is possible to obtain the differential equations describing the spatial-temporal evolution of the second correlation moments q_j^T, R_{ij}, $J_{\alpha j}^T$, etc., through the third order moments. However, in order to find these moments, supplementary equations with the fourth order correlations are required. This means that extending this process results in an infinite number of equations with an infinite number of unknown high order correlations .

Therefore, it is clear that, when developing a multicomponent turbulence model, it is necessary to establish a closure hypothesis. This would allow one to terminate the hierarchy of such equations and unknowns by introducing algebraic relations between the moments of order $(m+1)$ and m. One may invoke the Millionshchikov hypothesis (*Millionshchikov, 1941*) as such an approach, which relates the fourth moments (for example, for fluctuations of the velocity V') to the second moments assuming the fluctuations of V' to be distributed *Gaussian.* However, this hypothesis has no sufficient physical substantiation and also does not take into account the so-called *feasibility requirements* imposed by the lowest orders moments. These requirements suggest that if the first m moments are known, then the $(m+1)$-th moment should fit some particular inequalities. An elementary example is the *Schwarz inequality* $\left|\overline{A}\right| \le \sqrt{\overline{A'^2}}$. Nevertheless, there is an infinite number of turbulent flows for which some restricted closure procedures is appropriate and the problem is to find those adequate to the physical nature of the phenomenon under consideration.

Recently enough, second order shear turbulence models for homogeneous fluids were advanced. Some authors believe (see, e.g., *Turbulence: Principles and Applications, 1980; Turbulent shear flows - I, 1982*) that in these models an optimum level of complexity has been achieved. Here all the second order correlation moments involved in the averaged hydrodynamic equations are described by the system of evolutionary transport equations similar to the equation (3.2.6) for the mean velocity field. Closing these equations is also based on the gradient hypothesis, i.e. on using some local algebraic relations to model the third and higher orders correlation moments. For instance, the transfer equations for the components of the Reynolds stress tensor in which the quantities R_{ij} are dependent variables in the system of partial differential equations, belong to such models. It should be noted, however, that numerical realization of these differential equations used to consume too much computing time and do not ensure better results against those obtained with simpler gradient closure procedures. This problem is addressed in more detail in Chapter 4 when discussing complicated multicomponent turbulence models.

3.2.3. The Average Chemical Reaction Rate in Turbulent Flows

If in a gaseous mixture, containing minor species, chemical reactions with high rates occur without considerable heat consumption or release, the temperature and concentration of the reacting admixtures can often be considered as passive (and conservative) turbulent flow components, i.e. as quantities whose influence on the dynamic characteristics of the flow is negligible. In that case, the source term of mass production in chemical reactions, $\sigma_\alpha = \sum_{s=1}^{r} \nu_{\alpha s} \xi_s$, in the diffusion equations (3.2.5) and the source of chemical heating, $\sum_{s=1}^{r} q_s \xi_s$, in the energy equation (3.2.7) can be omitted. However, if the temperature and concentration of species in a reacting gas can not be treated as passive admixtures then chemical reactions and heat release sources should be added to the averaged hydrodynamic equations given in subsection 3.2.2. An example is modeling burning accompanied by strong interrelation between chemical kinetic and gas dynamic processes.

A question of vital importance in the problem of interaction between chemical reactions and turbulence is finding an explicit representation of the averaged chemical substance sources (sinks), $\overline{\sigma_\alpha}$, and the averaged heat release (heat consumption) magnitude, $\sum_{s=1}^{r} < q_s > \overline{\xi}_s$, in the hydrodynamic equations for the scales of mean motion. Another problem is how to define additional terms featuring influence of chemical reactions on the structure of the evolutionary transfer equations in the closure models at the level of second order moments.

The procedure to average the instantaneous values of the source intensity σ_α in equation (3.2.5) is a non-trivial task which requires special treatment. Generally speaking, the averaged chemical reaction rate in turbulent flows is not determined by the *Arrhenius* kinetics for averaged parameters but rather significantly depends on their fluctuations (*Kuznetsov, 1969; Kompaniwts et al., 1979; Ievlev, 1990*). The averaging difficulties are stipulated by the fact that analytical expressions linking instantaneous reaction rates to fluctuating flow characteristics (such as density, temperature, and concentrations of components) are always substantially non-linear. Obviously, the degree of complexity is dependent on the aggregate chemical reaction order, as well as on the non-linear temperature dependence of the specific reaction rates.

Let us derive an expression for the mean rate, $\overline{\xi}_s$, of homogeneous chemical reaction while turbulent burning is occurring. First of all, consider some basic ideas from the formal theory of chemical reaction kinetics. The resulting s-th reaction rate,

$$\beta_{1s}[1] + \beta_{2s}[2] + \ldots \underset{\mathcal{K}_{rs}}{\overset{\mathcal{K}_{fs}}{\Leftrightarrow}} \eta_{1s}[1] + \eta_{2s}[2] + \ldots \qquad (s = 1, 2, \ldots, r) \tag{3.2.9}$$

for ideal thermodynamic mixtures (see section 2.3.1) is determined by the direct reaction rate parameters ω_{fs} and the equilibrium constant $\mathcal{K}_s$ by the formula (*de Groot and Mazur, 1962*):

$$\xi_s = \mathcal{K}_{fs} \prod_{\alpha=1}^{N} n_\alpha^{\beta_{\alpha s}} \left\{ 1 - \frac{1}{\mathcal{K}_s} \prod_{\alpha=1}^{N} n_\alpha^{\nu_{\alpha s}} \right\} = \omega_{fs} \left[1 - \exp(-A_s / \mathrm{k} T) \right]. \tag{3.2.10}$$

In (3.2.9) and (3.2.10) $\beta_{\alpha s}$, $\eta_{\alpha s}$ are the stoichiometric coefficients of the α-component related to the s-th chemical reaction, $[\alpha]$ are the chemical reactant figures, $\mathcal{K}_{fs}$, $\mathcal{K}_{rs}$ are the direct and inverse reaction rate constants, respectively, $\mathcal{K}_s = \mathcal{K}_{fs}/\mathcal{K}_{rs}$, $\nu_{\alpha s} = \eta_{\alpha s} - \beta_{\alpha s}$, $\omega_f = \mathcal{K}_{fs} \prod_{\alpha=1}^{N} n_\alpha^{\beta_{\alpha s}}$ and $A_s = -\sum_{\alpha=1}^{N} \nu_{\alpha s}\mu_\alpha$ represent the rate and chemical affinity of the s-th reaction, respectively, and the symbol $\Leftrightarrow$ means that the reaction may proceed in both directions. The total direct reaction order ω_{fs} is determined as

$$\Delta\beta_s = \sum_{\alpha=1}^{N} \beta_{\alpha s} .$$

As is known, in the Arrhenius kinetics approach the rate constant $\mathcal{K}_{fs}$ is approximated by the expression (*Williams, 1971*)

$$\mathcal{K}_{fs} = \mathcal{K}_{fs}^{0}\, T^{a_{fs}} \exp\left(-\frac{E_{fs}}{\mathrm{k}T}\right), \tag{3.2.11}$$

in which the frequency constant *(pre-exponent)*, the temperature exponent of the frequency factor of the chemical reaction, and the *activation energy* are designated as $\mathcal{K}_{fs}^{0}$, a_{fs}, and E_{fs}, respectively.

Next we will examine multicomponent ideal systems and take the chemical potential of the α-component, μ_α (2.3.2), in the following general form (*Prigogine and Defay, 1954*):

$$\mu_\alpha = \mu_\alpha^0(T,p) + \mathrm{k}T\ln(x_\alpha) \qquad (\alpha = 1,2,\ldots,N), \tag{3.2.12}$$

where

$$\mu_\alpha^0(T,p) = \mathrm{k}T\ln p + h_\alpha^0 - TC_{p\alpha}^0 \ln T - T\int_0^T \frac{dT}{T}\int_0^T C_{p\alpha}^{\bullet}(T)dT - \mathrm{k}T\gamma_\alpha \tag{3.2.13}$$

is the chemical potential of a pure α-substance at a given temperature T and pressure p, γ_α is the chemical constant of α-particles, $C_{p\alpha}^0$ and $C_{p\alpha}^{\bullet}(T)$ are the progressive and oscillatory components of the heat capacity $C_{p\alpha} \equiv (\partial h_\alpha/\partial T)_{p,\{n_\beta\}}$, respectively; and h_α^0 is the partial enthalpy of the α-component extrapolated to zero temperature. Note that, for the sake of generality, we temporarily neglect the condition of the partial heat capacities $C_{p\alpha}$ being constant, adopted elsewhere in the book. Then the equilibrium constant, $\mathcal{K}_s$, and the chemical affinity, A_s, of the s-th reaction can be written as follows:

$$A_s = -\sum_{\alpha=1}^{N} \nu_{\alpha s}\mu_\alpha = \mathrm{k}T\ln\left[\mathcal{K}_s(T) \Big/ \prod_{\alpha=1}^{N} n_\alpha^{\nu_{\alpha s}}\right], \qquad (s = 1,2,\ldots,r) \tag{3.2.14}$$

$$\mathcal{K}_s(T) = n^{\Delta\nu} \exp\left[-\sum_{\alpha=1}^{N} \nu_{\alpha s}\mu_\alpha^0 / \mathrm{k}T\right] = \Delta\nu_s \ln(\mathrm{k}T) - q_s^0/\mathrm{k}T +$$

$$+\frac{1}{\mathrm{k}}\sum_{\alpha=1}^{N}\nu_{\alpha s}C_{p\alpha}^0 \ln T + \frac{1}{\mathrm{k}}\int_0^T \frac{dT}{T}\int_0^T \sum_{\alpha=1}^{N}\nu_{\alpha s}C_{p\alpha}^\bullet(T)\,dT + \sum_{\alpha=1}^{N}\nu_{\alpha s}\gamma_\alpha, \tag{3.2.15}$$

where $\Delta\nu_s = \sum_{\alpha=1}^{N}\nu_{\alpha s}$ is the algebraic sum of stoichiometric coefficients of the s-th reaction, and $q_s^0 = \sum_{\alpha=1}^{N}\nu_{\alpha s}h_\alpha^0$ is the s-th reaction heat extrapolated to zero temperature.

The expression in brackets at the right hand side of (3.2.14) is of exceptional importance. It follows from purely thermodynamic reasons (see formula (2.2.20)) that both ξ_s and A_s vanish under equilibrium conditions. Thus, the following equation *(the mass action law)* results from (3.2.14):

$$\left[\mathcal{K}_s(T) / \prod_{\alpha=1}^{N} n_\alpha^{\nu_{\alpha s}}\right] = 1, \qquad (s = 1,2,\ldots,r), \tag{3.2.15*}$$

enabling to assess the composition of reaction products under chemical equilibrium.

As was already mentioned, averaging the resulting instantaneous chemical reaction rate, ξ_s, (3.2.10) with subsequent rejecting terms involving the third and higher order moments represents, in general, a difficult problem. The complexity is ensured by the strong non-linearity of expression (3.2.10), the degree of sophistication depending on the total order of the reaction $\Delta\beta_s$ and the non-linear temperature dependence of the specific reaction rate $\mathcal{K}_{fs}$ (formula (3.2.11)). Besides, the necessity arises to model successively a large number of correlation terms involving turbulent temperature and composition fluctuations. Basically, these terms should be derived from the respective evolutionary transfer equations containing, in turn, the higher order correlations. Closing such equations is possible when additional assumptions on the relations of higher order correlation terms with the "known" lower order correlations are introduced (see Chapter 4).

We shall further suppose that (3.2.10)–(3.2.15) represent the instantaneous state of chemically active flows in turbulized gas mixture. Then, according to the Reynolds assumption, the physical quantities T, p and n_α can be regarded as the sum of the averaged values $<T>, <p>$ and $<n_\alpha>$ and their turbulent fluctuations $\delta T, \delta p, \delta n_\alpha$. Here for the sake of notation uniformity, the symbols $<A>$ and δA designate the averaged and fluctuation components of any physical parameter A irrespective of the averaging method applied.

As $\xi_s = \xi_s(T, n_\alpha)$, and being restricted to only second order approximation, one can accept

$$\xi_s(T, n_\alpha) \approx \xi_s(<T>, <n_\alpha>) + \delta\xi_s + \tfrac{1}{2}\delta^2\xi_s, \tag{3.2.16}$$

where

$$\delta\xi_s = \left(\frac{\partial \xi_s}{\partial T}\right)_0 \delta T + \sum_{\alpha=1}^{N}\left(\frac{\partial \xi_s}{\partial n_\alpha}\right)_0 \delta n_\alpha, \qquad (3.2.17)$$

$$\delta^2\xi_s = \left(\frac{\partial^2 \xi_s}{\partial T^2}\right)_0 (\delta T)^2 + \sum_{\beta=1}^{N}\left(\frac{\partial^2 \xi_s}{\partial T \partial n_\beta}\right)_0 \delta T \delta n_\beta + \sum_{\alpha=1}^{N}\sum_{\beta=1}^{N}\left(\frac{\partial^2 \xi_s}{\partial n_\alpha \partial n_\beta}\right)_0 \delta n_\alpha \delta n_\beta. \qquad (3.2.18)$$

Here the index "0" marks the values of the derivatives calculated for $T = <T>,\ n_\alpha = <n_\alpha>$.

In order to obtain a general form of the averaged chemical reaction rate, we first address a typical, directly proceeding, second order chemical reaction, $[1]+[2]\xrightarrow{\mathcal{K}_f}[3]$ to illustrate the complexity of the problem. The averaged rate of component $[1]$ disappearance can be written as:

$$-\overline{\sigma}_1 = \overline{\mathcal{K}_f n_1 n_2} = \overline{\rho^2 \mathcal{K}_f Z_1 Z_2} = <\rho\, \mathcal{K}_f> \left[\overline{\rho} <Z_1><Z_2> + \overline{\rho\, Z_1'' Z_2''}\right] +$$
$$+ <Z_1> \overline{\rho\, (\rho\, \mathcal{K}_f)'' Z_2''} + <Z_2> \overline{\rho\, (\rho\, \mathcal{K}_f)'' Z_1''} + \overline{\rho\, (\rho\, \mathcal{K}_f)'' Z_1'' Z_2''}. \qquad (*)$$

The first term at the right hand side of this expression contains the product of averaged quantities. It can be obtained by substituting the respective mean flow parameters into the expression for the source σ_1. The second term describes the distributional inhomogeneity of reactant concentrations at the effective value of the reaction rate constant complying with the adopted averaging method. The remaining terms are conditioned by the influence of concentration and reaction rate constant fluctuations. Because of the temperature dependence of the chemical reaction rate constant (see formula (3.2.11)), rate constant fluctuations in case of non-isothermal processes depend on temperature fluctuations in an extremely complicated fashion, which in turn considerably complicates formula (*). But even for isothermal processes in strongly diluted systems when the density and the reaction rate coefficient may be regarded as constant with a sufficient degree of accuracy, expression (*) still contains an additional term leading to a change in the chemical reaction rate compared with its value calculated from the average values.

Now, let us proceed to the general case of turbulent flows with chemical reactions. By varying the independent variables T and n_β, it is easy to obtain the first variation of the resulting rate, $\xi_s = \xi_s(T, n_\alpha)$, from expression (3.2.10) and to write it as:

$$\delta\xi_s = \left[1 - \exp(-A_s/\mathrm{k}T)\right]\delta\omega_{fs} + \omega_{fs}\exp(-A_s/\mathrm{k}T)\,\delta(A_s/\mathrm{k}T), \qquad (3.2.19)$$

where

$$\delta\omega_{fs} = \omega_{fs}\left(\frac{E_{fs} + \mathrm{k}Ta_{fs}}{\mathrm{k}T^2}\delta T + \sum_{\alpha=1}^{N}\frac{\beta_{\alpha s}}{n_\alpha}\right). \qquad (3.2.20)$$

We then shall take advantage of the known thermodynamic relations (*de Groot and Mazur, 1962*) in order to obtain the variation $\delta(A_s/\mathrm{k}T)$:

$$\left(\frac{\partial(A_s/T)}{\partial T}\right)_p=\sum_{\alpha=1}^{N}\nu_{\alpha s}\left(\frac{\partial(\mu_\alpha/T)}{\partial T}\right)_p=\frac{q_s(T,p)}{T^2}, \tag{3.2.21}$$

$$\left(\frac{\partial A_s}{\partial p}\right)_T=-\sum_{\alpha=1}^{N}\nu_{\alpha s}\left(\frac{\partial\mu_\alpha}{\partial p}\right)_T=V_s(T,p), \tag{3.2.22}$$

$$\left(\frac{\partial A_s}{\partial n_\alpha}\right)_{T,p}=-\sum_{\beta=1}^{N}\nu_{\beta s}\left(\frac{\partial\mu_\beta}{\partial n_\alpha}\right)_{T,p}=-\sum_{\beta=1}^{N}\nu_{\beta s}\mu_{\beta\alpha}=\sum_{\beta=1}^{N}\mu_{\beta\alpha}n_\beta\left(\frac{\nu_{\beta s}}{n_\beta}-\frac{\nu_{\alpha s}}{n_\alpha}\right), \tag{3.2.23}$$

in which the following designations are introduced:

$$q_s(T,p)=\sum_{\alpha=1}^{N}\nu_\alpha h_\alpha=\sum_{\alpha=1}^{N}\nu_{\alpha s}\int_0^T C_{p\alpha}(T)dT+q_s^0, \tag{3.2.24}$$

$$V_s(T,p)=\sum_{\alpha=1}^{N}\nu_{\alpha s}v_\alpha, \tag{3.2.25}$$

$$\mu_{\beta\alpha}=\left(\frac{\partial\mu_\beta}{\partial n_\alpha}\right)_{T,p,\{n_\beta\}(\beta\neq\alpha)}. \tag{3.2.26}$$

Here v_α is the partial molar volume of the α-component, q_s is the *heat of reaction s* at constant T and p, which is equal to the difference between the sum of the partial enthalpies of the reaction products multiplied by the respective stoichiometric coefficients and the analogous sum for the original reactants (see formula (2.1.48)); $\mu_{\beta\alpha}$ is the partial derivative of the chemical potential μ_β with respect to the numerical density n_α at constant temperature T, pressure p, and all other n_β $(\sum_{\alpha=1}^{N}n_\alpha\mu_{\beta\alpha}=0)$; and $V_s(T,p)$ is the volume change in due course of the s-th reaction at constant temperature and pressure.

For a mixture of ideal gases we have $v_\beta=1/n$, $\mu_{\beta\alpha}=\mathrm{k}T(\delta_{\beta\alpha}/n_\alpha-1/n)$ (see formula (2.3.4)). So,

$$\delta\left(\frac{A_s}{\mathrm{k}T}\right)=\frac{q_s(T,p)}{\mathrm{k}T^2}\delta T-\frac{V_s(T,p)}{\mathrm{k}T}\delta p+\sum_{\alpha=1}^{N}\left(\frac{\Delta\nu_s}{n}-\frac{\nu_{\alpha s}}{n_\alpha}\right)\delta n_\alpha=$$

$$=\frac{q_s(T,p)-\Delta\nu_s\mathrm{k}T}{\mathrm{k}T^2}\delta T-\sum_{\alpha=1}^{N}\frac{\nu_{\alpha s}}{n_\alpha}\delta n_\alpha. \tag{3.2.27}$$

According to (3.2.20) and (3.2.27) the first variation $\delta\xi_s$ of the chemical reaction rate ξ_s can be written as:

$$\delta\xi_s = \Lambda_{s0}\delta T + \sum_{\beta=1}^{N} \Lambda_{s\beta}\delta n_\beta , \tag{3.2.28}$$

where

$$\Lambda_{s0} = \omega_{fs}\left\{\frac{E_{fs} + a_{fs}\,\mathrm{k}T}{\mathrm{k}T^2}\left(1 - \frac{1}{\mathcal{K}_s}\prod_{\alpha=1}^{N} n_\alpha^{\nu_{\alpha s}}\right) + \frac{q_s - \Delta\nu_s\,\mathrm{k}T}{\mathrm{k}T^2\mathcal{K}_s}\prod_{\alpha=1}^{N} n_\alpha^{\nu_{\alpha s}}\right\}, \tag{3.2.29}$$

$$\Lambda_{s\gamma} = \omega_{fs}\left(\frac{\beta_{\gamma s}}{n_\gamma} - \frac{\eta_{\gamma s}}{n_\gamma \mathcal{K}_s}\prod_{\alpha=1}^{N} n_\alpha^{\nu_{\alpha s}}\right). \tag{3.2.30}$$

To obtain the second variation, $\delta^2\xi_s$, of the rate $\xi_s = \omega_{fs} - \omega_{rs}$, where $\omega_{rs} = \omega_{fs}\exp(-A_s/\mathrm{k}T)$, first it is convenient to represent it as:

$$\begin{aligned}\delta^2\xi_s &= \delta(\delta\omega_{fs} - \delta\omega_{rs}) = \delta\left[\omega_{fs}\delta(\ln\omega_{fs}) - \omega_{rs}\delta(\ln\omega_{rs})\right] = \\ &= \omega_{fs}\left[(\delta\ln\omega_{fs})^2 + \delta^2(\ln\omega_{fs})\right] - \omega_{rs}\left[(\delta\ln\omega_{rs})^2 + \delta^2(\ln\omega_{rs})\right].\end{aligned} \tag{3.2.31}$$

We also use formula (3.2.20) together with the relations

$$\delta(\ln\omega_{rs}) = \frac{E_{fs} + a_{fs}\,\mathrm{k}T - q_s + \Delta\nu_s\,\mathrm{k}T}{\mathrm{k}T^2}\delta T + \sum_{\alpha=1}^{N}\frac{\eta_{\alpha s}}{n_\alpha}\delta n_\alpha , \tag{3.2.32}$$

$$\delta^2(\ln\omega_{fs}) = -\frac{2E_{fs} + a_{fs}\,\mathrm{k}T}{\mathrm{k}T^3}(\delta T)^2 - \sum_{\alpha=1}^{N}\frac{\beta_{\alpha s}}{n_\alpha^2}(\delta n_\alpha)^2 , \tag{3.2.33}$$

$$\delta^2(A_s/\mathrm{k}T) = \frac{\Delta\nu_s\,\mathrm{k}T - q_s - q_s^0}{\mathrm{k}T^3}(\delta T)^2 + \sum_{\alpha=1}^{N}\frac{\nu_{\alpha s}}{n_\alpha^2}(\delta n_\alpha)^2 , \tag{3.2.34}$$

$$\delta^2(\ln\omega_{rs}) = \delta^2(\ln\omega_{fs}) - \delta^2(A_s/\mathrm{k}T), \tag{3.2.35}$$

which are easily deduced taking into account (3.2.20) and (3.2.27).

Then, finally, we obtain (*Kolesnichenko and Marov, 1984*):

$$\tfrac{1}{2}\delta^2\xi_s = B_{s0}(\delta T)^2 + \sum_{\alpha=1}^{N} B_{s\alpha}(\delta T\delta n_\alpha) + \sum_{\alpha=1}^{N}\sum_{\beta=1}^{N} B_{s\alpha\beta}(\delta n_\alpha \delta n_\beta), \tag{3.2.36}$$

where

$$B_{s\alpha} \equiv \omega_{fs} \frac{E^*_{fs}}{\mathrm{k} T^2 n_\alpha} \left\{ \beta_{\alpha s} - \left(1 - \frac{q^*_s}{E^*_{fs}}\right) \frac{\eta_{\alpha s}}{\mathcal{K}_s} \prod_{\beta=1}^{N} n_\beta^{\nu_{\beta s}} \right\}, \qquad (3.2.37)$$

$$B_{s\alpha\beta} \equiv \frac{\omega_{fs}}{2 n_\alpha n_\beta} \left\{ \beta_{\alpha s} (\beta_{\alpha s} - \delta_{\alpha\beta}) - \frac{\eta_{\alpha s}}{\mathcal{K}_s} (\eta_{\alpha s} - \delta_{\alpha\beta}) \prod_{\gamma=1}^{N} n_\gamma^{\nu_{\gamma s}} \right\}, \qquad (3.2.38)$$

$$B_{s0} \equiv \omega_{fs} \left\{ \left(1 - \frac{1}{\mathcal{K}_s} \prod_{\alpha=1}^{N} n_\alpha^{\nu_{\alpha s}} \right) \frac{E^{*2}_{fs} - \mathrm{k} T (2 E^*_{fs} - a_{fs} \mathrm{k} T)}{2 \mathrm{k}^2 T^4} + \right.$$
$$\left. + \frac{q^*_s (q^*_s - 2 E^*_{fs} + \mathrm{k} T) + \mathrm{k} T q^*_s}{2 \mathrm{k}^2 T^4 \mathcal{K}_s} \prod_{\beta=1}^{N} n_\beta^{\nu_{\beta s}} \right\}. \qquad (3.2.39)$$

Here the following designations were used:

$$E^*_{fs} \equiv E_{fs} + a_{fs} \mathrm{k}, \qquad q^*_s \equiv q_s(T, p) - \Delta \nu_s \mathrm{k} T. \qquad (3.2.40)$$

The thermodynamic variables p, T, n_α in (3.2.29), (3.2.30), (3.2.37), (3.2.38), and (3.2.39) are averaged in a similar way but the averaging symbols are omitted to simplify the notation.

Now, averaging the approximate equality (3.2.16) over an ensemble of probable realizations and using the relations (3.2.28) and (3.2.36), in a turbulent flow we obtain the following expression for the effective value of chemical reaction rate (3.2.9):

$$< \xi_s (T, n_\alpha) > = \xi_s (< T >, < n_\alpha >) + B_{s0} < (\delta T)^2 > +$$
$$+ \sum_{\alpha=1}^{N} B_{s\alpha} < \delta T \, \delta n_\alpha > + \sum_{\alpha=1}^{N} \sum_{\beta=1}^{N} B_{s\alpha\beta} < \delta n_\alpha \delta n_\beta > . \qquad (3.2.41)$$

It is evident that in order to improve the accuracy of determining the source terms $\overline{\sigma_\alpha}$ in the averaged diffusion equations (3.2.5), when it is necessary to take into account how the intensity of the turbulent fluctuation field affects the chemical reaction pattern, the evolutionary transfer equations for the paired correlations of temperature and composition fluctuations should be incorporated into (3.2.41).

In this regard the following must be noticed. Usually, by virtue of the exponential form of expression (3.2.10) and the poor convergence of the respective exponential series in equation (3.2.16), more detailed information concerning the structure of turbulent flow fields in multicomponent mixtures is needed. In other words, the knowledge of only paired correlations for fluctuating temperatures and compositions (and possibly some higher order moments) is insufficient to evaluate satisfactorily the average value of the source term $\overline{\sigma_\alpha}$. More complete information could be obtained by introducing, for instance, a single-point function of a joint probability density $P(\rho, V_j, T, Z_\alpha \ (\alpha = 1, 2, ..., N); \boldsymbol{r}, t)$ for the velocity V_j and the basic thermodynamic parameters of reacting mixture flows, followed by solving the respective evolutionary equation for this function (see, e.g., *Borgi, 1984; O'Brian, 1983; Shrinivisan et al., 1977*). Currently, this modern approach is being advanced. However, there is no complete theory enabling to obtain directly the probability density functions based on the definition of the evolutionary integro-differential equation. In addition, the number of solved test problems of turbulent heat and mass transfer is insufficient. This prevents us to estimate adequately the efficiency of the above approach. Meanwhile, approximating the averaged chemical reactions rates $\overline{\xi_s}$ at the level of the second order moments, which we have focused on, allows one to asses the mutual influence of chemical kinetics and dynamics in turbulent flows. Apparently, although such an approach is limited enough, it is more advantageous than completely ignoring this complicated problem.

Finally, let us emphasize that analyzing how turbulent flows influence chemical reaction rates (and *vice versa*) requires inter-comparison of the following characteristic times:

- the characteristic time of hydrodynamic phenomena, t_{hydr};
- the turbulence time scale, t_{turb} (for instance, the dissipative time scale $< e > / \varepsilon_e$);
- the time scale of chemical reactions, t_{chem}.

The relationships between the times t_{hydr}, t_{turb}, and t_{chem} may vary in a wide range; usually $t_{turb} << t_{hydr}$. For example, for flows in a boundary layer $t_{turb} / t_{hydr} \sim 10^{-2}$ s (see, e.g., *Meteorology and Atomic Energy, 1971*). When dealing with aeronomical problems and, in particular, treating the relations between turbulence and chemical kinetics, the following possible cases are of principal interest and importance (*Ievlev, 1990; Marov and Kolesnichenko, 1987*):

- $t_{chem} > t_{hydr}$ (and thus $t_{chem} >> t_{turb}$): the chemical reactions are in non-equilibrium with respect to the averaged motion and *frozen* with respect to the pulsations; in this case chemical transformation processes do not affect the gas-dynamic properties of the flow, but the mean gas composition should be determined from the averaged continuity equations for the individual chemical components in which the source terms are calculated with allowance for temperature and composition fluctuations;
- $t_{chem} < t_{hydr}$: the chemical reactions are independent of the averaged motion and pulsations; in this case the mutual influence of both chemical kinetics and heat and mass transfer becomes significant in turbulent flows;
- $t_{chem} << t_{hydr}$: the averaged flow can be treated as a thermochemically equilibrium flow, though chemical reactions may be in non-equilibrium with regard to pulsations; in this case the mixture composition is determined by the mass action law (3.2.15*) with regard to the influence of temperature and composition fluctuations, the chemical transformation processes effecting the turbulent exchange coefficients.

3.3. DEFINING RELATIONS FOR TURBULENT FLOWS IN MULTICOMPONENT MEDIA

When developing the turbulence model for multicomponent reacting gas mixtures it is necessary to have rheological relations for turbulent fluxes of diffusion ($J^T_{\alpha j} \equiv \bar{\rho} < Z''_\alpha V''_j >$ $(\alpha = 1,2,..., N)$), heat ($q^T_j \equiv \bar{\rho} < h'' V''_j >$), specific volume ($J^T_{(v)j} \equiv \bar{\rho} < v'' V''_j > = -\overline{\rho' V''_j} / \bar{\rho}$), and the Reynolds stress tensor ($R_{ij} \equiv -\bar{\rho} < V''_i V''_j >$) in order to close the averaged hydrodynamic equations (3.2.4), (3.2.5), (3.2.6), and (3.2.7). Several approaches to model such correlation quantities which can be distinguished by their degree of complexity were discussed in section 3.2; more detail will be given in Chapters 4 and 5. The linear rheological relations linking turbulent substance, momentum, and energy fluxes to the respective gradients of the averaged thermohydrodynamic parameters of the turbulized mixture can be obtained in particular, by a conventional technique based on the concept of *mixing length* (intermixing). Such a technique, applied in this section, allows us to derive the relations which extend the classical results obtained for homogeneous fluids with constant density beyond those for turbulent flows of multicomponent compressible mixtures.

3.3.1. The Gradient Hypothesis

Let us suppose that the transport of scalar properties by turbulent fluctuations occurs as a diffusion process and that some mixing path scale exists, which represents the distance that an elementary volume of gas in a turbulent flow should pass before it irreversibly merges with the surrounding medium. Then, designating the Lagrangian turbulent fluctuation of a field quantity A corresponding to the Eulerian fluctuation A'' as A''_L, and the effective mixing length of property A which passes a turbulent vortex in the flow before it is destroyed due to interaction with other perturbations as $\xi_{(A)j}$, we have (*Van Mieghem, 1977*)

$$A''_L = A'' + \xi_{(A)j} \partial_j < A > . \tag{3.3.1}$$

Turbulent Diffusion Fluxes. Assuming that the composition of a turbulized medium is conservative (i.e. assuming that vortices traveling over the distance ξ_j preserve the same specific density of the α-component in the *Lagrangian* volume they originally possessed) we obtain

$$(Z_\alpha)''_L = 0, \quad Z''_\alpha = -\xi_k \partial_k < Z_\alpha > . \tag{3.3.2}$$

This yields the following rheological relation for the turbulent substance flow $J^T_{\alpha j}$ $(\alpha = 1,2,..., N)$:

$$J^T_{\alpha j} \equiv \bar{\rho} < Z''_\alpha V''_j > = -\bar{\rho} < \xi_k V''_j > \partial_k < Z_\alpha > = -\bar{\rho} D^T_{kj} \partial_k < Z_\alpha >, \tag{3.3.3}$$

where the expression $D^T_{kj} \equiv <\xi_k V''_j>$ defines the asymmetrical eddy diffusion coefficient tensor which in the anisotropic case accounts for the distinction between the intensities of turbulent velocity and composition fluctuations for different coordinate axes. This is equivalent to the statement that turbulent fluxes of α-substance is proportional to the gradient of the mean concentration $<Z_\alpha>$ and is inversely directed in reference to the gradient.

It should be noticed that in general it is necessary to consider the proper effective mixing length $\xi_{\alpha k}$ for the transport of every mixture component. Besides, the conservativeness hypothesis (*Lagrange invariance*) for concentrations Z_α in a chemically active flow is used to be not justified because chemical kinetics and turbulent heat and mass transfer processes has substantial influence on the finite chemical reaction rates (*Ievlev, 1975; Marov and Kolesnichenko, 1987*). However, because this problem is yet poorly investigated, we will admit $\xi_{\alpha k} \equiv \xi_k$ and $(Z_\alpha)''_L \equiv 0$. A thermodynamic approach to the closure problem of the hydrodynamic equations for the mean mixture motion (developed in Chapter 5) allows us to describe substance transport processes through several turbulent exchange coefficients (ultimately related to the intrinsic length scales and mixing rates of the components) and to obtain the most general rheological relations for turbulent diffusion fluxes, in particular in the form of the generalized *Stefan-Maxwell* relations. Such an approach will be developed in Chapter 5.

Turbulent Substance Transport in a Stratified Medium. Let us consider turbulent mass transport in a resting gaseous medium (when $<V_j>=0$, but $V''_j \neq 0$) situated in a gravitational field (a still atmosphere is just an example). As is known from classical hydrodynamics, a mixture can be in mechanical equilibrium when no diffusion and thermal equilibrium conditions occur. It is known (*Landau and Lifshits, 1988*) that in such a case the pressure, the density, the concentrations of individual components, and the temperature depend only on the height x_3, i.e. the fluid is *stratified.* Then, in view of the equation of state (3.2.2) and the equation of hydrostatics,

$$\partial_3 \bar{p} = -\bar{\rho} g \tag{3.3.4}$$

(the x_3-axis being directed upward), expression (3.3.3) for the turbulent diffusion flux in the vertical direction can be transformed into the following form:

$$J^T_{\alpha 3} = -\bar{\rho}\, D^T \partial_3 <Z_\alpha> = -D^T \left\{ \partial_3 \bar{n}_\alpha + \bar{n}_\alpha \left(\frac{1}{<H>} + \frac{\partial_3 <T>}{<T>} - \frac{\partial_3 M^*}{M^*} \right) \right\}. \tag{3.3.3*}$$

Here $D^T(x_3) \equiv \overline{\rho \xi_3 V''_3} / \bar{\rho}\,(>0)$ is the eddy diffusion coefficient; $<H> = \mathrm{k} <T> / M^* g = \bar{p} / \bar{\rho} g$ is the mean local scale height (see (2.3.95)); ξ_3 is the mixing length of the composition in the vertical direction. Evidently, the last term in (3.3.3*) (missing, by the way, in the well known work of (*Lettau, 1951*)) plays an important role in the diffusion process in planetary upper atmospheres where the average molecular mass $M^* = \bar{\rho} / \bar{n}$ of the turbulized gaseous mixture strongly varies with height.

Indeed, using the expression $\overline{\rho}\partial_3 < Z_\alpha > = \overline{\rho}\partial_3(\overline{n}_\alpha / \overline{\rho}) = \partial_3 \overline{n}_\alpha - \overline{n}_\alpha \partial_3(\ln \overline{\rho})$ and the equation of state for mixtures (2.2.2), it is possible to write

$$\partial_3(\ln \overline{\rho}) = \partial_3(\ln \overline{p}) - \partial_3(\ln < T >) + \partial_3(\ln M^*) = -\overline{\rho} g / \overline{\rho} < R^* >< T > -$$

$$-\partial_3(\ln < T >) + \partial_3(\ln M^*) = -1/ < H > - \partial_3 < T > / < T > + \partial_3 M^* / M^* .$$

Then, for the logarithmic derivative with respect to the density $\overline{\rho}$ in the vertical direction we have

$$\overline{\rho}\partial_3 < Z_\alpha > = \partial_3 \overline{n}_\alpha + \overline{n}_\alpha \left(\frac{1}{< H >} + \frac{\partial_3 < T >}{< T >} - \frac{\partial_3 M^*}{M^*} \right). \quad (3.3.5)$$

Substitution of (3.3.5) into the relation $J^T_{\alpha 3} = -\overline{\rho} D^T \partial_3 < Z_\alpha >$ results in (3.3.3 *).

The eddy diffusion coefficients D^T_{kj}, in contrast to the molecular diffusion coefficients, not only describe the fluid properties but also the properties of the specific turbulent patterns in the mixture flow. Therefore, they directly depend both on the intensity of the turbulent field and on the averaging scale (or method) of the fluctuating velocity and composition. It is the way of introducing the mean turbulent motion properties which provides the basis for measurement techniques. The latter are critical because they allow to compare the theoretical results on the turbulent exchange coefficients with experimental data. The coefficients $D^T_{kj}(\boldsymbol{r})$ are functions of the coordinates and usually exceed their laminar analogs by many orders of magnitude. This remark also holds for all other turbulent exchange coefficients in what follows.

Turbulent Heat Flux. Now, we will obtain the rheological relation for the turbulent heat flux, $q^T_j \equiv \overline{\rho h'' V''_j}$, in a multicomponent mixture defined by formula (3.1.48). Using the relation (*Prigogine and Defay, 1954*)

$$G \equiv \sum_{\alpha=1}^{N} Z_\alpha \mu_\alpha = h - TS \quad (3.3.6)$$

for the *Gibbs free energy* together with property (3.1.69) of Favre's averaging the following precise expression for the turbulent fluctuation G'' emerges:

$$G'' \equiv \sum_{\alpha=1}^{N} \left[Z''_\alpha < \mu_\alpha > + (Z_\alpha \mu''_\alpha)'' \right] = h'' - < T > S'' - (ST'')'' ,$$

whence it follows that

$$h'' - < T > S'' - \sum_{\alpha=1}^{N} Z''_\alpha < \mu_\alpha > = \left(ST'' + \sum_{\alpha=1}^{N} Z_\alpha \mu''_\alpha \right)'' . \quad (3.3.7)$$

Assuming that the same thermodynamic relations are valid for the averaged thermodynamic quantities and for those conventionally defined for a laminar mixture flow, one may argue that the pulsating components of the pressure, the temperature, and the chemical potential satisfy the Gibbs-Dugham identity (2.2.13)

$$\sum_{\alpha=1}^{N} Z_\alpha \mu''_\alpha = -ST'' + p'/\rho \ . \tag{3.3.8}$$

Now, multiplying identity (3.3.7) by $\rho V''_j$ and statistically averaging over an ensemble of identical systems, we have

$$q_j^T - <T> J_{(S)j}^T - \sum_{\alpha=1}^{N} <\mu_\alpha> J_{\alpha j}^T = \overline{\rho(p'/\rho)''V''_j} = \overline{\rho(p'/\rho)V''_j} = \overline{p'V''_j} \ ,$$

whence, using (3.1.54), we eventually obtain

$$J_{(S)j}^T = \frac{1}{<T>}\left(q_j^T - \overline{p'V''_j} - \sum_{\alpha=1}^{N} <\mu_\alpha> J_{\alpha j}^T\right) = \frac{1}{<T>}\left(q_j^{T*} - \sum_{\alpha=1}^{N} <\mu_\alpha> J_{\alpha j}^T\right) =$$

$$= \frac{1}{<T>} J_{qj}^{T*} + \sum_{\alpha=1}^{N} <S_\alpha> J_{\alpha j}^T, \qquad (q_j^{T*} \equiv q_j^T - \overline{p'V''_j}\ ) \tag{3.3.9}$$

for the turbulent entropy flux in a mixture defined by the relation $J_{(S)j}^T \equiv \overline{\rho} <S''V''_j>$.

Note that this formula has the same structure as that for the entropy flux in a laminar flow of a multicomponent gas (compare (3.3.9) with (2.2.17)). The turbulent entropy flux includes the reduced heat fluxes, J_{qj}^{T*} (J_{qj}^T) (see (2.1.50)):

$$J_{qj}^{T*} \equiv q_j^{T*} - \sum_{\alpha=1}^{N} <h_\alpha> J_{\alpha j}^T = q_j^T - \overline{p'V''_j} - \sum_{\alpha=1}^{N} <h_\alpha> J_{\alpha j}^T = J_{qj}^T - \overline{p'V''_j}\ , \tag{3.3.10 (a)}$$

$$J_{qj}^T \equiv q_j^T - \sum_{\alpha=1}^{N} <h_\alpha> J_{\alpha j}^T \tag{3.3.10 (b)}$$

and the transport of the partial entropies, $<S_\alpha>$, by turbulent diffusion fluxes, $J_{\alpha j}^T$. The relation

$$<\mu_\alpha> = <h_\alpha> - <T><S_\alpha>, \quad (\alpha = 1,2,...,N), \tag{3.3.11}$$

for the averaged chemical potential of the α-component was used when deducing the expression for $J_{(S)j}^T$ in (3.3.9).

Rewriting now (3.3.9) as

$$J_{qj}^{T*} = <T> J_{(S)j}^{T} - <T> \sum_{\alpha=1}^{N} <S_\alpha> J_{\alpha j}^{T} = <T> \overline{\rho V_j''(S'' - \sum_{\alpha=1}^{N} <S_\alpha> Z_\alpha'')} \qquad (3.3.12)$$

and substituting S'' and Z''_α by the expressions $S'' = S''_L - \xi_{(S)k}\partial_k <S>$ and $Z''_\alpha = Z''_{\alpha L} - \xi_k \partial_k <Z_\alpha>$, and linking together the *Eulerian* and *Lagrangian* turbulent fluctuations of the quantities S and Z_α, respectively, we obtain

$$J_{qj}^{T*} = <T> \overline{\rho V_j''\left(S''_L - \sum_{\alpha=1}^{N} <S_\alpha> (Z_\alpha)''_L\right)} - <T> \overline{\rho V_j'' \xi_{(S)k}}\, \partial_k <S> +$$

$$+ <T> \overline{\rho V_j'' \sum_{\alpha=1}^{N} <S_\alpha> \xi_k}\, \partial_k <Z_\alpha> = <T> \overline{\rho V_j''\left(S''_L - \sum_{\alpha=1}^{N} <S_\alpha> (Z_\alpha)''_L\right)} -$$

$$- <T> \bar{\rho}\left(\chi_{jk}^{T}\partial_k <S> - D_{jk}^{T}\sum_{\alpha=1}^{N} <S_\alpha> \partial_k <Z_\alpha>\right) =$$

$$= <T> \overline{\rho V_j''\left(S''_L - \sum_{\alpha=1}^{N} <S_\alpha> (Z_\alpha)''_L\right)} - \bar{\rho} <C_p> D_{jk}^{T}\left(<T> - \frac{\partial_k \bar{p}}{\bar{\rho} <C_p>}\right), \qquad (3.3.13)$$

or, in view of (3.3.10 ([b])),

$$q_j^T = \overline{p' V_j''} + \sum_{\alpha=1}^{N} <h_\alpha> J_{\alpha j}^{T} - \bar{\rho} <C_p> D_{jk}^{T}\left(\partial_k <T> - \frac{\partial_k \bar{p}}{\bar{\rho} <C_p>}\right) +$$

$$+ <T> \overline{\rho V_j''\left(S''_L - \sum_{\alpha=1}^{N} <S_\alpha> (Z_\alpha)''_L\right)}. \qquad (3.3.14)$$

The expression $\chi_{jk}^{T} \equiv <\xi_{(S)k} V_j''>$ determines the turbulent thermal conductivity tensor.

Generally, the entropy and concentrations of the chemically active mixture components are not the Lagrangian invariants of turbulent transport (i.e. $S''_L \neq 0$, $Z''_{\alpha L} \neq 0$), because vortex motions can be accompanied by various thermal effects and/or by changes in chemical composition. Thermal effects could be caused, for example, by local heat release at the expense of chemical reactions or small-scale turbulent heating due to viscous dissipation. One may suppose, however, that parameters S and Z_α are conservative properties of a medium. This means that the turbulent motion of a Lagrangian vortex particle, between some level $\boldsymbol{r}(x_j,t)$ where the particle has separated from the flow, and the next level $\boldsymbol{r}(x_j,t) + \xi_j$, occurs isoentropically, and also no changes in the spatial distribution of the gas components take place. Then the Lagrangian fluctuations become $S''_L = 0$ and $(Z_\alpha)''_L = 0$ (adiabaticity of the process) and expression (3.3.14) for the turbulent heat flux in a mixture, q_j^T, appears to be much simpler:

$$q_j^T = \overline{p'V_j''} + \sum_{\alpha=1}^{N} < h_\alpha > J_{\alpha j}^T - \lambda_{jk}^T \left(\partial_k < T > - \frac{\partial_k \overline{p}}{\overline{\rho} < C_p >} \right), \tag{3.3.15}$$

where the turbulent heat conductivity coefficient tensor is introduced by the formula $\lambda_{jk}^T \equiv \overline{\rho} < C_p > D_{jk}^T$.

As indicated by (3.3.15), there are two mechanisms to transfer thermal energy across the turbulized multicomponent gas effected by the averaged temperature gradient (the potential temperature in a stratified medium) and by the turbulent diffusion fluxes, $J_{\alpha j}^T$. In the latter case, each α-particle transports an average amount of $< h_\alpha >$ thermal energy. It should be underlined that the first term in (3.3.15) does not play the role of an energy flux because the quantity $\overline{p'V_j''}$ disappears from the complete energy equation (3.1.78) if (3.1.54) for the flux q_j^{T*}, i.e. an "ordinary" turbulent heat flux, is taken into account This term is retained in (3.3.15) for the convenience (see Chapter 5 for more details).

It is also worthwhile to comment the deduction of (3.3.13). The last equality of this expression was obtained assuming that the so-called turbulent *Lewis numbers* $\boldsymbol{Le}_{jk}^T \equiv \chi_{jk}^T / D_{jk}^T = 1$. This is the usual approach in the turbulence theory (see, e.g., *Lapin, 1989*), which is equivalent to the assumption that the mixing lengths for substance, ξ_j, and for entropy, $\xi_{(S)j}$, are identical (see (3.3.3) and (3.3.14)). However, in general these scales should be distinguished because turbulent vortices can participate more actively in heat transport than in substance transport, and vice versa. Besides, the following relation was used:

$$\partial_j < S > - \sum_{\alpha=1}^{N} < S_\alpha > \partial_j < Z_\alpha > = \frac{< C_p >}{< T >} \left(\partial_j < T > - \frac{\partial_j \overline{p}}{\overline{\rho} < C_p >} \right), \tag{3.3.16}$$

which is easily derived from the averaged Gibbs relation (2.2.6). When substituting the identity (see (3.1.75*))

$$\partial_j < h > \approx \partial_j \left(\sum_{\alpha=1}^{N} < Z_\alpha > < h_\alpha > \right) = \left(\sum_{\alpha=1}^{N} C_{p\alpha} Z_\alpha \right) \partial_j < T > + \sum_{\alpha=1}^{N} < h_\alpha > \partial_j < Z_\alpha > =$$

$$= < C_p > \partial_j < T > + \sum_{\alpha=1}^{N} < h_\alpha > \partial_j < Z_\alpha >$$

into the averaged Gibbs relation

$$< T > \partial_j < S > + \sum_{\alpha=1}^{N} < \mu_\alpha > \partial_j < Z_\alpha > = \partial_j < \varepsilon > + \overline{p} \partial_j < v > = \partial_j < h > - < v > \partial_j \overline{p},$$

we can rewrite the latter as

$$<T>\partial_j<S>+\sum_{\alpha=1}^{N}(<\mu_\alpha>-<h_\alpha>)\partial_j<Z_\alpha>=<C_p>\partial_j<T>-<v>\partial_j\bar{p},$$

or, in view of (3.3.11), as

$$<T>\left[\partial_j<S>-\sum_{\alpha=1}^{N}<S_\alpha>\partial_j<Z_\alpha>\right]=<C_p>\partial_j<T>-<v>\partial_j\bar{p}.$$

From this relation (3.3.16) follows.

Vertical Turbulent Heat Flux. When applying (3.3.15) to a stratified atmosphere and incorporating the hydrostatic equation (3.3.4) it is possible to rewrite this defining relation for vertical component of the turbulent heat flux as

$$q_3^T=\overline{p'V_3''}+\sum_{\alpha=1}^{N}<h_\alpha>J_{\alpha 3}^T-\lambda^T\left(\partial_3<T>+\frac{g}{<C_p>}\right). \qquad (3.3.15^*)$$

Here $\lambda^T(x_3)$ is the turbulent thermal conductivity in the vertical direction (this parameter depends on the averaging method and the intensity of the turbulent temperature and velocity fields); $\gamma_a=g/<C_p>$ is the dry adiabatic temperature gradient ($\gamma_a=0.98$ degrees /100 m in the terrestrial troposphere); and $\Theta\equiv<T>+\gamma_a x_3$ is analogous to the potential temperature.

Relation (3.3.15) can also be used in a more convenient form. Indeed, in case of isotropic turbulence the turbulent exchange coefficient tensors are reduced to the diagonal form $D_{jk}^T=D^T\delta_{jk}$, $\chi_{jk}^T=\chi^T\delta_{jk}$. As a result, and using (3.3.15) and (3.3.17), one can write

$$q_j^T=\overline{p'V_j''}-\bar{\rho}D^T\partial_j<h>+\left(\bar{\rho}D^T<C_p>-\lambda^T\right)\partial_j<T>+\lambda^T\frac{\partial_j\bar{p}}{\bar{\rho}<C_p>}=$$

$$=\overline{p'V_j''}-\bar{\rho}\chi^T\left(\partial_j<h>-\frac{\partial_j\bar{p}}{\bar{\rho}}\right), \qquad (3.3.17)$$

where $\chi^T\equiv\lambda^T/\bar{\rho}<C_p>$ is the turbulent thermal diffusion coefficient. Here it was also supposed that the Lewis number $\boldsymbol{Le}^T\equiv\chi^T/D^T=1$. Although in its appearance expression (3.3.17) coincides with the expression for the heat flux in a homogeneous medium, it is completely different because for a multicomponent mixture the averaged enthalpy, $<h>$, is defined by (3.1.75*). In other words, the quantity $\partial_j<h>$ depends not only on $\partial_j<T>$, as for homogeneous media, but also on $\partial_j<Z_\alpha>$.

Reynolds' Tensor. The Reynolds turbulent stress tensor, defined by the relation (3.1.37),

$$R_{ij} \equiv -\overline{\rho V_i'' V_j''} = \begin{pmatrix} -\overline{\rho V_1''^2} & -\overline{\rho V''_1 V_2''} & -\overline{\rho V_1'' V_3''} \\ -\overline{\rho V_2'' V_1''} & -\overline{\rho V_2''^2} & -\overline{\rho V_2'' V_3''} \\ -\overline{\rho V_3'' V_1''} & -\overline{\rho V_3'' V_2''} & -\overline{\rho V_3''^2} \end{pmatrix}, \tag{3.3.18}$$

represents a second rank, symmetric tensor. It describes turbulent stresses caused by the interaction of traveling turbulent gas parcels (eddies). Like molecular stresses, the turbulent ones, in fact result from momentum transport but in this case at the expense of turbulent velocity fluctuations. Using Lagrangian fluctuations of the velocity vector components it is possible to show that this tensor is connected to the weighted-mean flow velocity gradients by the following linear relation:

$$\begin{aligned} R_{ij} &= -\tfrac{2}{3}\overline{\rho} < e > \delta_{ij} + 2\overline{\rho}\nu^T_{ijsl} \overset{o}{e}_{sl} = \\ &= -\tfrac{2}{3}\overline{\rho} < e > \delta_{ij} + \overline{\rho}\nu^T_{ijsl}\left(\partial_l < V_s > + \partial_s < V_l > - \tfrac{2}{3}\delta_{sl}\,\partial_k < V_k >\right), \end{aligned} \tag{3.3.19}$$

where $< e > \equiv \overline{\rho V_j'' V_j''}/2\overline{\rho}$ is the turbulent flow energy defined by (3.1.65); $\overset{o}{e}_{ij} \equiv e_{ij} - \tfrac{1}{3}\left(e_{kl}\delta_{kl}\right)\delta_{ij} = e_{ij} - \tfrac{1}{3}\delta_{ij}\,\partial_k < V_k >$ is the symmetrical zero trace part of the strain tensor, $e_{ij} \equiv \tfrac{1}{2}\left[\partial_j < V_i > + \partial_i < V_j >\right]$; ν^T_{ijsl} is the turbulent (kinematic) viscosity tensor. The fourth rank tensor ν^T_{ijsl}, symmetric with respect to the indices i, j and s, l, should satisfy the condition $\nu^T_{jisl}\overset{o}{e}_{sl} = 0$, which follows from (3.3.19) and in general can also depend on invariants in the strain tensor (this is why relation (3.3.19) is *quasi-linear*). It is expedient to use the formula $\nu^T_{ijsl} = \tfrac{1}{2}\left(\nu^T_{is}\delta_{jl} + \nu^T_{js}\delta_{il}\right)$ based on assuming tensor degeneration in such a way that it can be expressed through the second rank, symmetric tensor $\nu^T_{ij} = < e >^{1/2} l_{ij}$ (*Monin, 1950*). Relation (3.3.19) can also be obtained using the methods from non-equilibrium thermodynamics.

In the simplest case (such as a flat shear flow along the x_3-axis), the horizontal component of the Reynolds stress tensor (3.3.19) becomes

$$R_{13} = -\overline{\rho V_1'' V_3''} = \overline{\rho}\nu^T \partial_3 < V_1 >, \tag{3.3.19*}$$

where $\nu^T = <\xi_3 V_3''>$ is the vertical coefficient of the turbulent viscosity defined as the ratio of the apparent internal stress and the respective averaged deformation rate. When writing (3.3.19*) it is supposed that $V_1'' = -\xi_3\,\partial_3 < V_1 >$, i.e. eddies shifting in vertical direction over a distance ξ_3 keep the same momentum at the level $(x_3 + \xi_3)$ they had at the initial level x_3 (the *Prandtl hypothesis*).

Let us emphasize once more that the linear equations, with respect to the gradient equations, (3.3.3), (3.3.15), and (3.3.19) for the turbulent fluxes of diffusion, heat, and the turbulent stress tensor, respectively, are not always valid. Their justification would require, strictly speaking, the turbulent field of fluctuating velocities and other thermohydrodynamic parameters to be constrained by the equilibrium between generation and dissipation of turbulent energy at any point in space (*Ievlev, 1975*). Otherwise, if convective and diffusion terms in the balance equation for the energy of turbulent fluctuations (see (3.1.68)) are essential, i.e. the flow parameters at every point depend on the characteristics of the turbulent flow as a whole, the local formulas (3.3.3), (3.3.15), and (3.3.19) become inaccurate (see Chapter 4 for more details).

3.3.2. Modeling the Turbulent Transport Coefficients

Evidently, the rheological relations (3.3.3), (3.3.15), (3.3.19) for turbulent fluxes do not solve the problem of modeling multicomponent turbulence but determine the character (structure) of such a model only. Formulas of type $D^T_{kj} \equiv <\xi_k V''_j>$ do not allow one to assess D^T_{kj} experimentally because the mixing length ξ_k is an uncertain quantity which is impossible to measure. Therefore the closure problem of the averaged hydrodynamic equations for mixtures (3.2.4)–(3.2.8) reduces to the problem of finding approximation formulas for the turbulent transport coefficients. Such an approach is known as *the semi-empirical theory of turbulence*. The expressions for the eddy diffusion coefficient, D^T_{kj}, the turbulent thermal conductivity, λ^T_{kj}, and the turbulent viscosity, ν^T_{ijsl}, can be obtained by various methods differing in complexity. Here we will address one of the simple approaches to semi-empirically determining the turbulent exchange coefficients. More complicated methods based on the evolutionary transfer equations, that are reduced to algebraic relations involving single-point pair correlators of velocity, temperature, and composition fluctuating in a flow, will be discussed in Chapters 7 and 8.

Prandtl's Model. Let us consider an averaged flow where gravitational forces determine the primary flow direction. A standard practice is to adopt the following hypothesis (which complies with the principle of local equilibrium of the turbulent flow): the turbulent transport coefficients at any point only depend on the local parameters of the flow at the same point. Accordingly, the turbulent viscosity coefficients can be written, for example, as $\nu^T = \nu^T(\bar{\rho}, \nu, L, f^i, \partial_3 <V_1>)$. Here $L(\boldsymbol{r})$ is the external turbulence scale at a given point in the flow, and $f^i(\boldsymbol{r})$ are the local characteristics of the volume force fields (in particular, the inertial forces related to acceleration of the fluid in the longitudinal direction, being $f \propto \partial_1 <V_1>$).

This dependence takes a different form if one takes into account the effect of local flow properties on both the first and second velocity derivatives. The scale $L(\boldsymbol{r})$ characterizes the geometrical structure of the turbulent field or the characteristic size of large vortexes involved in turbulent transport, carrying the major portion of the kinetic flow energy. In this case it is referred to as the integral turbulence scale, L. Sometimes $L(\boldsymbol{r})$ may be treated as the mean mixing length, $\Lambda(x_3)=\sqrt{\overline{\xi^2}}$, as was originally done by Prandtl (*Prandtl, 1925, 1942*). Then it coincides with the order of magnitude of the velocity correlation radius.

Some auxiliary ideas must be invoked in order to determine the external turbulence scale, $L(\boldsymbol{r})$. Because of such an uncertainty one has the opportunity to account for the integrated flow properties and its pre-history, especially in the local formulas for the turbulent exchange coefficients. In particular, in the case of free shear layers, the parameter Λ may be regarded as a constant proportional to its thickness within the entire layer. However, the factor of proportionality depends on the type of the free flow. For instance, it was found for a stream flowing around an infinite flat wall that the mean mixing length, Λ, is proportional to the distance to the wall: $\Lambda(x_3) = \kappa x_3$, where κ is the *Karman constant* which can be taken equal to ~ 0.4.

In regions with strong turbulence far from a wall where turbulence only slightly depends on the molecular viscosity, v, and $f^i = 0$, the relation for the turbulent viscosity coefficient, $\nu^T = \nu^T(\overline{\rho}, L\partial_3 < V_1 >)$, is possible to establish using the dimensionality approach. Hence, apart from a constant factor, the *Prandtl formula* follows:

$$\nu^T = L^2 |\partial_3 < V_1 >| \qquad (\text{or } \nu^T = \Lambda^2 |\partial_3 < V_1 >|, \quad \Lambda = \kappa x_3). \tag{3.3.20}$$

The constant factor is being assigned for every particular type of motion based experimental evidence. In some cases it is convenient to drop it and to redefine the turbulence scale, L, accordingly.

Near a solid wall, where the effects of molecular viscosity, v, are substantial, it turns out that $\nu^T = \nu\varphi\left(\frac{\Lambda^2 \partial_3 < V_1 >}{\nu}\right)$. Further improving this expression can be carried out with the help of both theoretical (more precisely, semi-empirical) reasons and purely experimental methods (*Lapin, 1989; Monin and Yaglom, 1971*). If the longitudinal inertial forces are incorporated as arguments, the relations for v^T mentioned above change and may be written as

$$\nu^T = L^2 (\partial_1 < V_1 > / \partial_3 < V_1 >) |\partial_3 < V_1 >|; \qquad \nu^T = \nu\varphi\left(\frac{\Lambda^2 \partial_3 < V_1 >}{\nu}, \frac{\partial_1 < V_1 >}{\partial_3 < V_1 >}\right).$$

The turbulent thermal conductivity coefficient, λ^T, and eddy diffusion coefficient, D^T, can be determined using the simplest and frequently used assumption that the turbulent *Prandtl* and *Schmidt* numbers,

$$\boldsymbol{Pr}^T = \nu^T / \chi^T = \overline{\rho} < C_p > \nu^T / \lambda^T, \qquad \boldsymbol{Sc}^T \equiv \nu^T / D^T, \tag{3.3.21}$$

are approximately constant. This is due to the fact that, in contrast to the turbulent transport coefficients themselves, these relations vary only slightly both within a turbulized flow and among flows. As a rule, it is assumed that $\chi^T = D^T$ (the Lewis number $\boldsymbol{Le}^T = \chi^T / D^T = 1$), such that the Prandtl and Schmidt numbers coincide: $\boldsymbol{Pr}^T = \boldsymbol{Sc}^T = const$ with $\boldsymbol{Pr}^T = 0.86 \div 0.90$ for flows near a wall and $\boldsymbol{Pr}^T = 0.5$ for flat jets and mixing layers.

Using (3.3.20), (3.3.3 *), and (3.3.15 *), we can now write down the rheological relations for the Reynolds stress tensor and the turbulent diffusion and heat fluxes as

$$R_{13} = \bar{\rho} L^2 |\partial_3 < V_1 >| \partial_3 < V_1 >, \qquad (3.3.19^{**})$$

$$J^T_{\alpha 3} = -\frac{1}{\boldsymbol{Pr}^T} \bar{\rho} L^2 |\partial_3 < V_1 >| \partial_3 < Z_\alpha >, \quad (\alpha = 1,2,\ldots,N), \qquad (3.3.3^{**})$$

$$q^T_3 = -\frac{1}{\boldsymbol{Pr}^T} \bar{\rho} < C_p > L^2 |\partial_3 < V_1 >| \cdot \left(\partial_3 < T > + \frac{g}{< C_p >} \right) + \sum_{\alpha=1}^{N} < h_\alpha > J^T_{\alpha 3} =$$

(3.3.15**)

$$= -\frac{1}{\boldsymbol{Pr}^T} \bar{\rho} L^2 |\partial_3 < V_1 >| (\partial_3 < h > + g).$$

These relations describe the vertical transfer of momentum, substance, and thermal energy during the turbulent mixing of the multicomponent mixture.

Among the assumptions adopted when deriving these formulas, the *Lagrange invariance* hypothesis for transported substance is of the main importance. As it was mentioned above, this hypothesis is not generally valid for a chemically active gas mixture stratified in a gravitational field. Therefore, the relations (3.3.19**), (3.3.3**), and (3.3.15**) should be corrected for the effect of non-uniformity of entropy (temperature) and composition distribution on the efficiency of turbulent mixing. Such a correction to the turbulent transport coefficients for multicomponent mixture can be found, generally speaking, by using the so-called *K-theory* for multicomponent turbulence (see subsection 4.3.9.). In a homogeneous stratified medium (for example, in the well mixed lower atmosphere of a planet) this effect arises only because of the presence of vertical temperature gradients within some individual regions. This gives rise to the appearance of additional buoyancy forces (the *Archimedean forces*) which facilitate or prevent generation of turbulence energy.

To account for this effect, Prandtl proposed a dimensionless criterion, namely, the Richardson gradient number, $\boldsymbol{Ri} = (g/<T>)(\partial_3 <T> + g/<C_p>)/(\partial_3 <V_1>)^2$ (see (4.2.32)). As follows from the similarity theory, it is natural to assume that all dimensionless characteristics of the turbulent flow are particular functions of the number $\boldsymbol{Ri}$. To take into account the buoyancy forces in (3.3.20), (3.3.3**), and (3.3.15**), it is possible to utilize the following corrections to the scale L:

- in case of stable stratification ($\boldsymbol{Ri} > 0$) preventing the development of turbulence: $L = L^*(1 - \beta_1 \boldsymbol{Ri})$, $5 < \beta_1 < 10$ (usually $\beta_1 \cong 7$) (*Monin and Yaglom, 1971*);

- in case of unstable stratification ($\boldsymbol{Ri} < 0$) increasing turbulence energy at the expense of instability energy: $L = L^*(1 - \beta_2 \boldsymbol{Ri})^{-1/4}$ ($\beta_2 \cong 14$) (*Lumley and Panofsky, 1966*); the formula $L = L^*(1 - c\boldsymbol{Ri})^{0,25}$ where c is an empirical coefficient (*Bradshaw, 1969*) is also appropriate for this case;

- in the extreme case ($\boldsymbol{Ri} = 0$), when the adiabatic height distribution of temperature ($\partial_3 < T > = -g / < C_p > \equiv \gamma_a$) not exerting influence on turbulence development: $L = L^*$, the latter being the mixing length in the absence of buoyancy forces.

At the same time, the Archimedean forces change the Prandtl-Schmidt number (*Munk and Anderson, 1948*):

$$\boldsymbol{Pr}^T = \boldsymbol{Pr}^{T*}(1+3.33\boldsymbol{Ri})^{1.5}/(1+10\boldsymbol{Ri})^{0.5} .$$

One should be aware that, as it was earlier mentioned, the rheological relations (3.3.19**), (3.3.3**), and (3.3.15**) are not always valid. In particular, in the turbulent flow behind a grid can be areas where the velocity of the averaged flow is constant and its gradient is $\partial_3 < V_1 >= 0$, while the correlator $< V_1''V_3''>$ is non-zero because turbulence is generated directly behind the grid and is further transported downstream by the averaged flow. However, the mixing length hypothesis (3.3.20) demands zero magnitudes for ν^T and, according to Prandtl's model (*Prandtl, 1942*), there should be no turbulence. This circumstance reveals the basic deficiency of this model: the mixing length hypothesis assumes local equilibrium in a turbulent field. Fortunately, it occurs that the mutual shifting of points in space where $< V_1''V_3''> \neq 0$ and $\partial_3 < V_1 >= 0$, is often small, and this is why using formula (3.3.19*) does not yield considerable errors when numerically modeling a flow.

The constraints of the mixing length hypothesis made necessary the development of turbulence models allowing to take into account the turbulence "diffusion" by solving the evolutionary transfer equations for the second order moments. A fundamental role in developing such turbulence theories was played by Kolmogorov (*Kolmogorov, 1942*). The hypothesis he proposed related the turbulent viscosity, ν^T, and the turbulent energy dissipation rate, $\varepsilon_e \equiv \overline{\pi'_{ij}\partial_j V'_i}$, (see (3.1.52 (^{2})) to the averaged kinetic energy of turbulent fluctuations, $<e> \equiv < V_s''V_s''>/2$, in the following way: $\nu^T = c_\mu L\sqrt{<e>}$, $\varepsilon_e =<e>^{3/2}/L$. Here by virtue of uncertainty in L, the constant in the second relation can be assumed to be equal to unity. Then the empirical constant, c_μ, is close to 0.09, as follows from experimental data (see (4.2.25) and (4.2.29)). There is a well known group of semi-empirical *Prandtl-Kolmogorov-Lounder* turbulence models for homogeneous media (see, e.g., *Turbulence: Principles and Applications, 1980*) where these relations are utilized to determine the turbulent transport coefficients in free shear flows (flat or rotationally symmetric) by solving numerically the system of interdependent differential equations for the turbulent kinetic energy, $<e>$, and the dissipation rate, ε_e, or for the external turbulence scale, L.

3.3.3. The Definition of Correlations Involving Density Fluctuations

In contrast to homogeneous turbulized gaseous continua where the effects of compressibility use to be negligible (*Monin and Yaglom, 1971*), the total mass density, ρ, in the general case of multicomponent reacting gaseous mixtures considerably varies from one point to another as a result, for example, of local heat release from chemical reactions. The correlations of type $\overline{\rho' A''}/\overline{\rho}$ $(\equiv -\overline{\rho} <v''A''>$ $A = V_j,\ h,\ Z_\alpha)$ appear in the balance equations for the averaged internal mixture energy (3.1.57) and the turbulent kinetic energy (3.1.68), as well as in the evolutionary transport equations for turbulent fluxes of heat, q_j^T, substance, $J_{\alpha j}^T$, and momentum, R_{ij}. Their definition requires additional balance equations for correlation functions of type $J_{(v)j}^T$, $\overline{\rho v'' Z''_\alpha}$, and $\overline{\rho v'' h''}$, to be used with numerous new correlation terms which are however, poorly modeled, as it will be discussed in detail in section 4.3.

At the same time, for a large number of turbulent modes the relative density fluctuations in multicomponent mixtures caused by pressure fluctuations are negligible against those caused by temperature and component concentration changes, as it is in particular, the case for the middle planetary atmosphere. Therefore there is a simpler possibility to define correlations $\overline{\rho v'' A''}$ in terms of known turbulent diffusion and heat fluxes based on some algebraic relation deduced from the baroclinic equation of state for pressure.

To derive this relation, let us first obtain an accurate expression for turbulent fluctuations of density ρ' in gaseous mixture whose equation of state is taken analogue to the equation of state for ideal gases (2.1.53),

$$p = \mathrm{k}\, T n = \mathrm{k}\, \rho T \sum_{\alpha=1}^{N} Z_\alpha = R^* \rho T\,,$$

where

$$R^* = \mathrm{k}\, n/\rho = \mathrm{k} \sum_{\alpha=1}^{N} n_\alpha / \rho = \mathrm{k} \sum_{\alpha=1}^{N} Z_\alpha \tag{3.3.22}$$

is the "gas constant" for the mixture.

Substituting in (2.1.53) the true (instantaneous) values of the quantities R^* and T denoted as the sum of the averaged and fluctuation values $(R^* = <R^*> + (R^*)'')$, $T = <T> + T'')$ yields

$$\begin{aligned} p &= <R^*> \rho <T> + (R^*)'' \rho <T> + <R^*> \rho T'' + (R^*)'' \rho\, T'' = \\ &= <R^*> \rho <T> + \mathrm{k}\rho <T> \sum_{\alpha=1}^{N} Z''_\alpha + <R^*> \rho\, T'' + \mathrm{k}\rho \sum_{\alpha=1}^{N} (Z''_\alpha T''). \end{aligned} \tag{3.3.23}$$

Here the designations

$$<R^*> = \mathrm{k} \sum_{\alpha=1}^{N} <Z_\alpha> = \mathrm{k}\overline{n}/\overline{\rho}, \quad (R^*)'' = \mathrm{k} \sum_{\alpha=1}^{N} Z''_\alpha \tag{3.3.24}$$

were introduced.

Applying the statistical averaging operator (3.1.3) to (3.3.23) we obtain the averaged equation of state for pressure (3.3.1) which is convenient to write in a somewhat different form:

$$\overline{p} = < R^* > \overline{\rho} < T > + \mathrm{k}\,\overline{\rho} \sum_{\alpha=1}^{N} < Z''_\alpha T'' > . \tag{3.3.25}$$

It follows from (3.3.25) that

$$< R^* >< T > = \overline{p} / \overline{\rho} - \mathrm{k} \sum_{\alpha=1}^{N} < Z''_\alpha T'' > . \tag{3.3.26}$$

Using now (3.3.26) to eliminate the product $< R^* >< T >$ from (3.3.23) we obtain

$$p = \overline{p}\rho / \overline{\rho} + < R^* > \rho\, T'' + \mathrm{k}\,\rho < T > \sum_{\alpha=1}^{N} Z''_\alpha + \mathrm{k}\,\rho \sum_{\alpha=1}^{N} (Z''_\alpha T'') -$$

$$- \mathrm{k}\,\rho \sum_{\alpha=1}^{N} < Z''_\alpha T'' > = \overline{p} + \overline{p}\rho' / \overline{\rho} + < R^* > \rho\, T'' + \mathrm{k}\,\rho < T > \sum_{\alpha=1}^{N} Z''_\alpha + \mathrm{k}\,\rho \sum_{\alpha=1}^{N} (Z''_\alpha T'')'' ,$$

whence the precise relation for turbulent fluctuations of the mass density, ρ', for multicomponent gas mixtures due to fluctuations of pressure, temperature, and concentrations of the individual components follows:

$$\frac{\rho'}{\overline{\rho}} = \frac{p'}{\overline{p}} - \frac{< R^* > \rho T''}{\overline{p}} - \frac{\mathrm{k} < T > \rho \sum_{\alpha=1}^{N} Z''_\alpha}{\overline{p}} - \frac{\mathrm{k}\,\rho \sum_{\alpha=1}^{N} (Z''_\alpha T'')''}{\overline{p}} . \tag{3.3.27}$$

It is important to underline that the averaged pressure, $\overline{p}$, in (3.3.27) is defined by the accurate equation of state (3.3.25).

Note that for flows with small *Mach* numbers, $\boldsymbol{Ma}$, it is possible to neglect in (3.3.27) the relative turbulent pressure fluctuations in comparison with the relative density and/or temperature fluctuations. This condition is known as the *Morkovin hypothesis* for which the validity is confirmed down to $\boldsymbol{Ma} = 5$ for flows without chemical reactions (*Morkovin, 1961*). For example, the forced convection in the Earth's atmosphere clearly manifests itself only in those jet flows where the wind velocity gradients reach rather high values. In this case the ratios between the fluctuations ρ', p', and T'' and their respective mean values $\overline{\rho}$, $\overline{p}$, and $< T >$ have the following orders of magnitude (*Van Mieghem, 1977*):

$$|\rho'| / \overline{\rho} \approx |T''| / < T > \approx 10^{-4}, \quad |p'| / \overline{p} \approx 10^{-5}. \tag{3.3.28}$$

At the same time, the scale at which turbulent pressure fluctuations occur for free atmospheric convection is, as a rule, much larger than for the forced convection scale, and therefore the terms containing pressure fluctuations, p', in (3.3.27) generally can not be omitted.

Considering multicomponent chemically active mixtures, the temperature fluctuation, T'', in (3.3.27) is convenient to express in terms of enthalpy fluctuations using for this purpose the accurate relation (3.1.71) which we rewrite as

$$<C_p> T'' = h'' - \sum_{\alpha=1}^{N} <h_\alpha> Z''_\alpha - \sum_{\alpha=1}^{N} C_{p\alpha}(Z''_\alpha T'')''. \tag{3.3.29}$$

This is because for multicomponent mixtures the evolutionary transfer equations for the correlations containing temperature fluctuations are much more complicated than the similar equations containing total enthalpy fluctuations. For example, the equation for the mean square of enthalpy fluctuations, $<h''^2>$, in contrast to the similar equation for the variance $<T''^2>$, does not include a large number of additional pair correlations of temperature and composition fluctuations, whose presence in the equation for $<T''^2>$ is ultimately related to the presence of chemical thermal energy sources in the regular equation for temperature (2.1.49) (*Kolesnichenko and Marov, 1980*).

Thus, when substituting relation (3.2.29) in (3.3.27), we have

$$\frac{\rho'}{\bar{\rho}} = -\rho v'' = \frac{p'}{\bar{p}} - \frac{\rho <R^*>}{\bar{p} <C_p>}\left(h'' - \sum_{\alpha=1}^{N} <h_\alpha> Z''_\alpha\right) - \frac{\rho \mathrm{k} <T>}{\bar{p}} \sum_{\alpha=1}^{N} Z''_\alpha +$$

$$+ \frac{\rho <R^*>}{\bar{p} <C_p>} \sum_{\alpha=1}^{N} C^*_{p\alpha}(Z''_\alpha T'')'' = \frac{p'}{\bar{p}} - \frac{\rho <R^*>}{\bar{p} <C_p>}\left(h'' - \sum_{\alpha=1}^{N} h^0_\alpha Z''_\alpha\right) +$$

$$+ \frac{\rho <R^*>}{\bar{p} <C_p>} \sum_{\alpha=1}^{N} C^*_{p\alpha}(Z''_\alpha T)''. \tag{3.3.30}$$

Here the following designations were introduced:

$$C^*_{p\alpha} \equiv C_{p\alpha} - \frac{\bar{\rho} <C_p>}{\bar{n}} = \frac{1}{\bar{n}} \sum_{\beta=1}^{N} \left(C_{p\alpha} - C_{p\beta}\right) \bar{n}_\beta \approx 0, \tag{3.3.31}$$

the approximate equality being valid when the specific heat capacities of the individual components do not strongly differ from each other.

Now, let us obtain an exact formula for the unknown correlation $\overline{\rho v'' A''}$. Multiplying (3.3.30) by A'' and statistically averaging over an ensemble of probable realizations, we obtain

$$\overline{\rho v''A''} = -\frac{\overline{p'A''}}{\overline{p}} + \frac{<R^*>}{\overline{p}<C_p>}\left[\overline{\rho}<h''A''> - \sum_{\alpha=1}^{N}<h_\alpha>\overline{\rho}<Z''_\alpha A''>\right] -$$

$$-\frac{k<T>}{\overline{p}}\sum_{\alpha=1}^{N}\overline{\rho}<Z''_\alpha A''> + \frac{<R^*>}{\overline{p}<C_p>}\sum_{\alpha=1}^{N}C^*_{p\alpha}\overline{\rho}<Z''_\alpha T''A''>. \tag{3.3.32}$$

To eliminate the temperature fluctuations, T'', from (3.3.30), we recur successively the relation (3.3.29) twice. As a result, for the correlations $(Z''_\alpha T'')$ we have

$$(Z''_\alpha T'') = \frac{1}{<C_p>}\left((h''Z''_\alpha) - \sum_{\beta=1}^{N}<h_\beta>(Z''_\alpha Z''_\beta) - Z''_\alpha\sum_{\beta=1}^{N}C_{p\beta}(Z''_\beta T'')''\right) =$$

$$= \frac{1}{<C_p>}\left\{(h''Z''_\alpha) - \sum_{\beta=1}^{N}<h_\beta>(Z''_\alpha Z''_\beta) - \frac{Z''_\alpha}{<C_p>}\left((C''_p h'') - \sum_{\beta=1}^{N}<h_\beta>(C''_p Z''_\beta) -\right.\right.$$

$$\left.\left.-(C''_p\sum_{\beta=1}^{N}C_{p\beta}(Z''_\beta T'')'')''\right)\right\} = \frac{1}{<C_p>}\left\{(h''Z''_\alpha) - \sum_{\beta=1}^{N}<h_\beta>(Z''_\alpha Z''_\beta)...\right. . \tag{3.3.33}$$

Then, in view of this relation it is possible for the third order correlator in (3.3.32) to write

$$<Z''_\alpha T''A''> = <C_p>^{-2}<A''Z''_\alpha>\left[<C''_p h''> - \sum_{\beta=1}^{N}<h_\beta><C''_p Z''_\beta>\right] +$$

$$+<C_p>^{-1}<Z''_\alpha h''A''> - <C_p>^{-1}\sum_{\beta=1}^{N}<h_\beta><A''Z''_\alpha Z''_\beta> -$$

$$-<C_p>^{-2}<A''h''C''_p Z''> + <C_p>^{-2}<A''Z''_\alpha[C''_p(C''_p T'')'']''>, \tag{3.3.34}$$

where

$$<A''Z''_\alpha[C''_p(C''_p T'')'']''> = <A''Z''_\alpha C''^2_p T''> - <C''_p T''><A''Z''_\alpha C''_p> - <A''Z''_\alpha><C''^2_p T''>. \tag{3.3.35}$$

The closure problem for multicomponent turbulence is solved here at the level of second order moments. Thus, limiting ourselves by accounting only for pair correlations in (3.3.34), we obtain the following relation for the unknown quantity $\overline{\rho}<v''A''>$, which is of key value for the developed approach:

$$\overline{\rho}<v''A''> = \overline{A''} = -\frac{\overline{p'A''}}{\overline{p}} + \frac{<R^*>}{\overline{p}<C_p>}\left(\overline{\rho h''A''} - \sum_{\alpha=1}^{N}<h_\alpha>\overline{\rho Z''_\alpha A''}\right) +$$

$$+\frac{\mathrm{k}}{\bar{p}<C_p>}\sum_{\alpha=1}^{N} H_\alpha \overline{\rho Z''_\alpha A''}. \quad (3.3.36)$$

Here the following designation was used:

$$H_\alpha \equiv <C_p><T>+\frac{C^*_{p\alpha}}{<C_p>^2}\left(-<C''_p h''>+\sum_{\beta=1}^{N}<h_\beta><C''_p Z''_\beta>\right). \quad (3.3.37)$$

If $C''_p \cong 0$, formula (3.3.36) takes a simpler form:

$$<v''A''>=\frac{\overline{p'A''}}{\bar{p}\bar{\rho}}+\frac{<R^*>}{\bar{p}<C_p>}\left(<h''A''>-\sum_{\alpha=1}^{N}<h_\alpha><Z''_\alpha A''>\right)+\frac{\mathrm{k}<T>}{\bar{p}}\sum_{\alpha=1}^{N}<Z''_\alpha A''>. \quad (3.3.38)$$

Relation (3.3.38) will be used in multicomponent turbulence models based on the simple *(gradient)* closure approach when the pair correlations for the temperature and composition fluctuations are neglected (see (3.1.79)). If $C^*_{p\alpha} \approx 0$ (and hence $C''_p \propto (n/\rho)'' \neq 0$) the formula

$$<v''A''>=\frac{\overline{p'A''}}{\bar{p}\bar{\rho}}+\frac{<R^*>}{\bar{p}<C_p>}\left(<h''A''>-\sum_{\alpha=1}^{N}h^0_\alpha<Z''_\alpha A''>\right) \quad (3.3.38^*)$$

is valid. We will use this in the more complicated turbulence models based on the differential transfer equations for the second moments of the fluctuating thermohydrodynamic quantities.

Identifying now parameter A with the hydrodynamic flow velocity, V_j, in (3.3.38), we obtain the expression for the turbulent specific volume flux, $J^T_{(v)j}$:

$$J^T_{(v)j}=-\frac{\overline{p'V''_j}}{\bar{p}}+\frac{<R^*>}{\bar{p}<C_p>}\left(q^T_j-\sum_{\alpha=1}^{N}<h_\alpha>J^T_{\alpha j}\right)+\frac{\mathrm{k}}{\bar{p}<C_p>}\sum_{\alpha=1}^{N}H_\alpha J^T_{\alpha j} \approx$$

$$\approx\frac{<R^*>}{\bar{p}<C_p>}J^T_{qj}+\frac{\mathrm{k}<T>}{\bar{p}}\sum_{\alpha=1}^{N}J^T_{\alpha j}. \quad (3.3.39)$$

This is regarded as the most important characteristic of a turbulent field when semi-empirical multicomponent turbulence model using weighted Favre averaging is utilized. The second, approximate relation is valid provided $C^*_{p\alpha} \approx 0$. The turbulent specific volume flux, $J^T_{(v)j}$, appears both in the earlier derived averaged hydrodynamic equations and in many evolutionary transfer equations for single-point, second moments, in particular, in the conservation equation for turbulent energy (3.2.7).

Similarly, if we set consecutively $A \equiv h,\ v,\ Z_\alpha$ in (3.3.36), we obtain the following relations:

$$\overline{h''} = -\frac{\overline{p'h''}}{\overline{p}} + \frac{<R^*>}{\overline{p}<C_p>}\left(\overline{\rho\, h''^2} - \sum_{\alpha=1}^{N} <h_\alpha> \overline{\rho\, Z''_\alpha h''}\right) + \frac{\mathrm{k}}{\overline{p}<C_p>}\sum_{\alpha=1}^{N} H_\alpha \overline{\rho\, Z''_\alpha h''}, \qquad (^1)$$

$$\overline{v''} = -\frac{\overline{p'v''}}{\overline{p}} + \frac{<R^*>}{\overline{p}<C_p>}\left(\overline{h''} - \sum_{\alpha=1}^{N} <h_\alpha> \overline{Z''_\alpha}\right) + \mathrm{k}\sum_{\alpha=1}^{N} H_\alpha \overline{Z''_\alpha}, \qquad (^2)$$

$$\overline{Z''_\beta} = -\frac{\overline{p'Z''_\beta}}{\overline{p}} + \frac{<R^*>}{\overline{p}<C_p>}\left(\overline{\rho\, h''Z''_\beta} - \sum_{\alpha=1}^{N} <h_\alpha> \overline{\rho\, Z''_\alpha Z''_\beta}\right) + \sum_{\alpha=1}^{N} H_\alpha \overline{\rho\, Z''_\alpha Z''_\beta} / \overline{p}<C_p>. \qquad (^3)$$

(3.3.40)

These are further employed in complicated reacting turbulence models for eliminating correlation terms such as $<v''A''>$ from the respective transfer equations for the second correlation moments (see Chapter 4).

In summary, let us enumerate the basic (reference) differential equations and the final relations characterizing a rather simple model of turbulized multicomponent continua. The model corresponds to the case when all second correlations such as $<A''B''>$ for the fluctuating thermohydrodynamic parameters A and B, being different from the hydrodynamic flow velocity, V_j, in the used formulas, are small compared to the first order terms $<A><B>$, and can be omitted. First of all, these are the hydrodynamic equations for mean motion (3.2.4)–(3.2.7); the averaged equations of state for pressure in the form (3.2.2); relation (3.2.41) for chemical reaction rates taken as $\overline{\xi_s} = \xi_s(<T>, \overline{n}_\alpha)$ (this means that the quantities $\overline{\xi}_s$ are calculated from the averaged values of temperature and composition only); relation (3.3.39), defining the turbulent specific volume flux, $J^T_{(v)j}$; and the rheological relations (3.3.3), (3.3.15), and (3.3.19) for the turbulent fluxes of diffusion, $J^T_{\alpha j}$, heat, q^T_j , and Reynolds stress tensor, R_{ij}.

This conditioned system of differential equations and the final relations should be supplemented by a set of chemical components, their gas-dynamic, thermophysical, and chemical properties being clarified by the universal laws of kinetics and thermodynamics including the equations of state and the expressions for the various thermodynamic functions preserving their usual form within the approach under consideration, by the formulas for the molecular and turbulent transport coefficients, and by the appropriate initial and boundary conditions. This system forms a "simplified" continual model for reacting multicomponent turbulence. Various geophysical and aeronomical problems based on such a modeling approach are drawn as an example in Chapters 6 and 7.

Summary

- *A system of closed relations including the differential equations for mean motion, the equations for chemical kinetics and state under conditions of turbulent mixing, as well as the defining relations for different turbulent fluxes of substance, momentum, and energy was obtained. This system takes into account the multicomponent nature and compressibility of the gaseous mixture, diffusive heat and mass transport, chemical reactions and gravitational effects. It is suitable for evaluating a wide class of motions and physicochemical processes in multicomponent reacting media.*

- *When averaging the initial equations of multicomponent hydrodynamics given in Chapter 2, the weighted-mean Favre averaging was employed systematically, along with the conventional averaging procedure without weighting (for example, over an ensemble of probable realizations). The former permits to simplify substantially the notation and analysis of the averaged hydrodynamic equations for gaseous mixtures with variable density. Differential equations for the scale of mean motion were derived. These are: the balance equation for the specific volume; the balance equations for concentrations of the molecular components in the mixture; the conservation equations for the chemical elements; the equation of motion for turbulized media; the balance equation for the potential energy; the balance equation for the kinetic energy of the mean motion; the balance equation for the internal energy of the turbulized medium, this equation being denoted, among others, in terms of temperature and pressure; the transfer equation for the turbulent energy of compressible mixtures; and the conservation equation for the total energy of turbulent multicomponent continua. The physical sense of the individual terms in these equations, including the energy balance components, has been analyzed.*

- *An averaging procedure for the chemical reaction rates in a reacting medium was proposed. It allows us to obtain the relevant kinetic laws under conditions of turbulent motion and offers an approach to solve the problem of modeling such media. The complexity of the problem lies in the necessity to average strongly nonlinear (exponential) mass production source terms from chemical reactions and subsequently determining the large number of additional correlations connected with temperature and composition fluctuations.*

- *The rheological relations for the turbulent fluxes of diffusion, heat, and the Reynolds stress tensor were obtained by using the conventional method based on the concept of mixing length. These relations extend the results found for homogeneous incompressible fluids to the multicomponent case. The correlation relations were derived, which involve density fluctuations and allow us to close the system of averaged hydrodynamic equations.*

CHAPTER 4

EVOLUTIONARY TRANSFER MODELS FOR THE SECOND CORRELATION MOMENTS

We will now proceed to the deduction of differential transfer equations to reveal how the single-point second moments, $< A''B'' >$, of turbulent fluctuations of thermohydrodynamic parameters in a chemically active multicomponent medium having variable density and thermal properties evolve. Such equations for a homogeneous fluid within the Boussinesque approximation (*Boussinesque, 1877*) underlie the *method of invariant modeling* in many turbulence theories at different levels of sophistication.

It is recognized that, although the equations are semi-empirical in their nature and include approximate expressions with empirical coefficients for the description of higher order correlation functions, they ensure the development of rather flexible models. They allow us to take into account how convection and diffusion, as well as the mechanisms of generation, redistribution, and dissipation of turbulent field energy affect the spatial-temporal distribution of the averaged thermohydrodynamic parameters in a medium. This is why these equations are widely used in numerical simulations of those fluids where the pre-history of the flow substantially influences the turbulence properties at a certain point. Besides, they are pertinent to finding the turbulent exchange coefficients within the free flows with a lateral shear (velocity gradient), specifically applicable to modeling some natural media (*Ievlev, 1975, 1990; Turbulence: Principles and Applications, 1980*; *Marov and Kolesnichenko, 1987*).

At the same time, there are some constraints to this approach when modeling turbulent flows. As a matter of fact, the very existence of this particular form of the approximating relations for higher order correlations in the transfer equations for the second moments is possible only provided some "equilibrium" turbulence spectrum under the given conditions occurs. In addition, the modeling relations should respond to the same tensor symmetry properties as the modeled terms and have the same dimensionality. Moreover, it is often assumed that empirical constants preserve unchanged and there is no need to match them for every particular flow, whereas the form of the approximating relations and the constant values may strongly vary depending on the flow regimes. Nonetheless, the closure procedures using the evolutionary transfer equations for the second moments are addressed as potentially more promising than the first order procedures order examined in section 3.3.

Let us also notice that the problems raised by the complex geometry (3D-flows, rip flows, etc.) can not be calculated virtually using semi-empirical theories. This is because the semi-empirical theories are insufficiently universal. In particular, when investigating turbulent transport, they do not take into account the role of large eddies which are commensurable with the flow cross-section and characterized by parameters substantially depending on both the problem's geometry and the conditions in the flow. For evaluating such problems direct numerical simulations of large-scale turbulence (numerical experiments) are usually applied (*Belotserkovskii, 1984, 1985, 1997; Ievlev, 1990*).

Occurring concentration gradients represents one of the major properties of chemically reacting flows which were usually neglected by the classical "constant" density turbulence models. Gradients of density, temperature, and concentrations resulting from local heat release of chemical reactions can significantly affect the hydrodynamic velocity field of the fluid due to turbulent heat and mass transfer. In such a way feedback of chemical kinetics with hydrodynamics is implemented. In case of turbulized mixtures, the fluctuations (pulsations) of mass density, temperature, and concentrations of individual components occur in addition to the velocity fluctuations.

Basically, the approach developed in this chapter comprises a coherent exposition of the ideas used in the phenomenological turbulence theories for homogeneous fluids, as applied to compressible multicomponent mixtures. The turbulent flow of reacting mixture with variable density can be modeled at different levels of complexity (see *Turbulence: Principles and Applications, 1980; Turbulent shear flows - I, 1982*). Here we focus on the closure problem for the system of the averaged multicomponent hydrodynamics equations at the level of second order coupling moments, invoking only the evolutionary transfer equations for single-point pair (mixed) correlators. Advances in the development of the second order turbulence models pertinent to homogeneous fluids with constant density (see, e.g., *Donaldson, 1972; Deardorff, 1973; Andre et al., 1976; Turbulence: Principles and Applications, 1980*), imply that some of them could be efficiently extended to flows in compressible multicomponent media. However, one should keep in mind that the quality of any employed model is ultimately determined by the comparison with the available experimental data.

4.1. THE GENERAL FORM OF THE TRANSFER EQUATION FOR PAIR CORRELATIONS IN A COMPRESSIBLE FLOW

The system of the averaged multicomponent hydrodynamics equations (3.2.4)–(3.2.8) comprising single-point pair correlations turns out to be non-closed after being averaged according to Reynolds. To close such a system, it is necessary to invoke a large number of additional evolutionary (prognostic) transfer equations for the second moments. In these equations the higher moments can be approximated by the gradient relations written in analogy to those utilized in turbulence models for flows of non-reacting gases with constant density. Therefore we face the problem to derive additional relations for the turbulent flows of substance, momentum, energy ($J^T_{\nu j}$, $J^T_{\alpha j}$, q^T_j, R_{ij}) and some other second order moments.

Obviously, using the statistical averaging (3.1.3) results in loss of some information about the turbulent flow. The missing information involves the equations for the fluctuating components of density, velocity, temperature (enthalpy), composition, etc. In the *Keller-Friedmann* method (*Keller and Friedmann, 1924*) these equations were utilized as the basic ones when deriving the differential equations for the correlators of random (stochastic), arbitrary order quantities. Here we offer the deduction of the general form of the transfer evolutionary equation for the single-point second moments of the fluctuating thermohydrodynamic parameters in a incompressible mixture flow obtained with the weighted averaging (3.1.5).

4.1.1. Differential Transfer Equations for Fluctuations

The statistical approach to describing turbulent transport processes in homogeneous compressible fluids is based on the analysis of a chain of engaging prognostic equations for the correlation moments for coupling the incremental orders. Let us briefly address a general approach to composing these equations based on the example of an incompressible fluid with constant density.

A multipoint and alternative velocity fluctuation correlator of k-th order is determined by statistically averaging the product of k fluctuation velocities,

$$\mathcal{B}_{A_1 \ldots A_k}(\boldsymbol{r}_1, t_1; \ldots; \boldsymbol{r}_k, t_k) = \overline{V'_{A_1}(\boldsymbol{r}_1, t_1) \ldots V'_{A_k}(\boldsymbol{r}_k, t_k)},$$

where $\boldsymbol{r}_k$ is the radius vector of a point in space at time t_k, and A_k is an index running through x, y, z. The complete statistical description of a random velocity field is equivalent to defining of all velocity fluctuation correlators of arbitrary order, which is virtually unfeasible. Practically, only the lowest orders correlators are utilized. For example, the pair correlator $\mathcal{B}_{ij}(\boldsymbol{r}, \boldsymbol{r}_1, t) = \overline{V'_i(\boldsymbol{r}, t) V'_j(\boldsymbol{r}_1, t)}$ can be used, which is the multipoint alternative second order moment of the velocity fluctuations. When studying shear turbulence in incompressible fluids, it is often convenient to use only the single-point pair correlators of the velocity fluctuations, $\mathcal{B}_{ij}(\boldsymbol{r}, t) = \overline{V'_i(\boldsymbol{r}, t) V'_j(\boldsymbol{r}, t)}$, which is the Reynolds tensor.

By virtue of commutability of the averaging and differentiation operations (the Reynolds postulate, see (3.1.2)), the derivative of the correlator $\mathcal{B}_{ij}(\boldsymbol{r}, t)$ with respect to time can be represented as

$$\partial_t \mathcal{B}_{ij}(\boldsymbol{r}, t) = \overline{V'_i(\boldsymbol{r}, t)\, \partial_t V'_j(\boldsymbol{r}, t)} + \overline{V'_j(\boldsymbol{r}, t)\, \partial_t V'_i(\boldsymbol{r}, t)}.$$

For deriving the evolutionary equations for the second moments, $\mathcal{B}_{ij}(\boldsymbol{r}, t)$, it is necessary to exclude the derivatives with respect to time in the right hand part of the last equality using the relevant hydrodynamic equations for the velocity fluctuations. Then the third order correlation functions for the velocity fluctuations will appear in the obtained equations. Similarly, it is possible to deduce even more complicated evolutionary equations, for example, for the third order correlators in which the fourth order correlation functions will appear, etc. Breaking this chain at any step leads to the an open system. This circumstance represents the main problem of the *Keller-Friedmann method.*

New difficulties arise when this approach is extended to a multicomponent chemically reacting compressible medium. One should be aware that even for a homogeneous fluid the system of equations for linking moments is so complicated that in the quoted method the equations were not written explicitly; instead, only the basic idea of their deduction was suggested including comments on the number and type of the equations. Basically, for a turbulized flow of multicomponent reactive mixtures, actually deducing the chain of equations to evaluate the dynamics of the incremental correlation moments through weighted averaging, is of key importance both for solving the closure problem in the models with diverse complexities and for deeper understanding the processes of kinetics and turbulent transport in such media (*Marov and Kolesnichenko, 1987*).

The Transfer Equation for Density Fluctuations. First, let us obtain the balance equation for the mean density fluctuation, ρ'. The density and the velocity of a fluid have the averaged and fluctuation components

$$\rho(\boldsymbol{r},t)=\overline{\rho}(\boldsymbol{r},t)+\rho'(\boldsymbol{r},t), \qquad V_j(\boldsymbol{r},t)=<V_j(\boldsymbol{r},t)>+V_j''(\boldsymbol{r},t). \tag{4.1.1}$$

Substituting (4.1.1) into the continuity equation (2.1.4), we have

$$\partial_t\overline{\rho}+\partial_j(\overline{\rho}<V_j>)+\partial_t\rho'+\partial_j(\rho'<V_j>)+\partial_j(\rho V_j'')=0.$$

The total of the first two addends in this relation vanishes by virtue of the continuity equation for the mean motion (3.1.8). Therefore, when using the operator relation (3.1.8) for the substantial derivative with respect to time in the averaged motion, one can write

$$\frac{D\rho'}{Dt}=-\partial_j(\rho V_j'')-\rho'\partial_j<V_j>. \tag{4.1.2}$$

Let us write (4.1.2) also in its substantial balance form. Using (3.1.8) for this purpose, we obtain

$$\overline{\rho}\frac{D}{Dt}(\rho'/\overline{\rho})=-\partial_j(\rho V_j''). \tag{4.1.3}$$

The Differential Transfer Equation for the Quantity $\rho A''/\overline{\rho}$. Let $A(\boldsymbol{r},t)$ be the specific value of any scalar quantity (in particular, this may be tensor components) whose substantial balance looks like (2.1.1). Then, by virtue of (3.1.14), it may be written as

$$-\partial_j J_{(A)j}+\sigma_{(A)}=\rho\frac{dA}{dt}\equiv(\overline{\rho}+\rho')\frac{D<A>}{Dt}+\rho V_j''\partial_j<A>+\rho\frac{dA''}{dt}=$$

$$=\rho'\frac{D<A>}{Dt}-\partial_j J^{\Sigma}_{(A)j}+\overline{\sigma_{(A)}}+\rho V_j''\partial_j<A>+\rho\frac{dA''}{dt}.$$

From here it follows that

$$\rho\frac{dA''}{dt}=-\rho' V_j''\partial_j<A>-\rho'\frac{D<A>}{Dt}-\partial_j J'_{(A)j}+\sigma'_{(A)}, \tag{4.1.4}$$

or, in view of the expression for the quantity $\partial_j(\rho V_j'')$ emerging from (4.1.2) and of the averaged continuity equation (3.1.8), we have

$$\rho\frac{DA''}{Dt}=-\partial_j(\rho V_j''A'')+A''\partial_j(\rho V_j'')-\rho V_j''\partial_j<A>-\rho'\frac{D<A>}{Dt}-$$

$$-\partial_j J'_{(A)j}+\partial_j J^T_{(A)j}+\sigma'_{(A)}=-\partial_j[\rho V_j''A''+J'_{(A)j}-J^T_{(A)j}]-\rho V_j''\partial_j<A>-$$

$$-\overline{\rho}A''\frac{D}{Dt}(\rho/\overline{\rho})-\rho'\frac{D<A>}{Dt}+\sigma'_{(A)}. \tag{4.1.5}$$

Thus, for the substantial balance of the quantity $\rho A''/\overline{\rho}$ we obtain the following evolutionary transfer equation:

$$\overline{\rho}\frac{D}{Dt}(\rho A''/\overline{\rho})+\partial_j\left[\rho V_j''A''+J'_{(A)j}-J^T_{(A)j}\right]=$$

$$=-\rho V_j''\partial_j<A>-\rho'\left(\overline{\sigma_{(A)}}-\partial_j J^{\Sigma}_{(A)j}\right)+\sigma'_{(A)}. \tag{4.1.6}$$

Considering the different instances of the quantity $A(\boldsymbol{r},t)$, it is possible from (4.1.6) to obtain the transfer equations for the fluctuation component of the hydrodynamic velocity, V_j'', the specific enthalpy, h'', the concentration of various mixture components, Z_α'', etc. Such equations can be utilized to derive the important evolutionary transfer equations for correlators (of any order) of fluctuating parameters at various spatial-temporal points (see, e.g., *Monin and Yaglom, 1971; Grafov et al., 1990*)).

4.1.2. Evolutionary Transfer Equation: The General Form

To derive the evolutionary transfer equation for single-point pair correlations of fluctuating mixture parameters, let us multiply (4.1.4) by the fluctuation B'' of some thermo-hydrodynamic parameter $B(\boldsymbol{r},t)$ for which a balance equation such as (2.1.1) is available as well, and average the result according to Reynolds. Then we have

$$\overline{\rho B''\frac{dA''}{dt}}=-J^T_{(B)j}\partial_j<A>-\frac{\overline{\rho'B''}}{\overline{\rho}}\left(\partial_j J^T_{(A)j}+\overline{\rho}\frac{D<A>}{Dt}\right)+\overline{B\left(\sigma'_{(A)}-\partial_j J'_{(A)j}\right)},$$

or

$$\overline{\rho B''\frac{dA''}{dt}}=-J^T_{(B)j}\partial_j<A>+\overline{B''\left(\sigma_{(A)}-\partial_j J_{(A)j}\right)}. \tag{4.1.7}$$

Here the averaged balance equation in its general form (3.1.14) is taken into account. It is possible to write a similar expression by interchanging the position of the parameters

A and B in (4.1.7). This yields

$$\overline{\rho A''\frac{dB''}{dt}} = -J^T_{(A)j}\,\partial_j <B> + \overline{A''\left(\sigma_{(B)} - \partial_j J_{(B)j}\right)}. \tag{4.1.8}$$

Adding now (4.1.7) to (4.1.8) and taking into account identity (3.1.12), we obtain

$$\overline{\rho\frac{d(A''B'')}{dt}} \equiv \overline{\rho}\frac{D\left(\overline{\rho A''B''}/\overline{\rho}\right)}{Dt} + \partial_j\left(\overline{\rho A''B''V''}\right) =$$

$$= -J^T_{(A)j}\partial_j <B> - J^T_{(B)j}\partial_j <A> + \overline{A''\left(\sigma_{(B)} - \partial_j J_{(B)j}\right)} + \overline{B''\left(\sigma_{(A)} - \partial_j J_{(A)j}\right)},$$

which is the balance equation for the single-point second moments $\overline{\rho A''B''}/\overline{\rho}$ in its substantial form. Hence we finally find the evolutionary transfer equation for the correlations $<A''B''>$:

$$\underbrace{\overline{\rho}\frac{D<A''B''>}{Dt}}_{\text{Convection}} + \underbrace{\partial_j J^{T\Sigma}_{(AB)j}}_{\text{Diffusion}} = \underbrace{-J^T_{(A)j}\partial_j <B> - J^T_{(B)j}\partial_j <A>}_{\text{Reproduction}} +$$

$$+ \underbrace{\overline{A''\sigma_{(B)} + B''\sigma_{(A)}}}_{\text{Redistribution}} \underbrace{- \overline{\rho} <\varepsilon_{(AB)}>}_{\text{Dissipation}}. \tag{4.1.9}$$

Here

$$J^{T\Sigma}_{(AB)j} \equiv \overline{\rho A''B''V''_j} + \overline{A''J_{(B)j}} + \overline{B''J_{(A)j}} = J^T_{(AB)j} + \overline{A''J_{(B)j}} + \overline{B''J_{(A)j}} \tag{4.1.10}$$

is the substantial density of total turbulent flux in the mixed correlation $<A''B''>$, and

$$\overline{\rho} <\varepsilon_{(AB)}> \equiv -\left(\overline{J_{(A)j}\,\partial_j B'' + J_{(B)j}\,\partial_j A''}\right) \tag{4.1.11}$$

is the dissipation rate of the quantity $<A''B''>$ due to molecular transport processes.

The expression for the total flux, $J^{T\Sigma}_{(AB)j}$, may be represented in another form which is more convenient for the subsequent analysis:

$$J^{T\Sigma}_{(AB)j} \equiv \overline{\rho} <A''B''V''_j> + \overline{A'J'_{(B)j}} + \overline{B'J'_{(A)j}} + \overline{\rho} <v''A''> \overline{J_{(B)j}} + \overline{\rho} <v''B''> \overline{J_{(A)j}} =$$

$$= \overline{\rho A'B'V''_j} + \overline{A'J'_{(B)j}} + \overline{B'J'_{(A)j}} + \overline{A''}J^{\Sigma}_{(B)j} + \overline{B''}J^{\Sigma}_{(A)j}, \tag{4.1.10*}$$

where $J^{\Sigma}_{(A)j} \equiv \overline{J_{(A)j}} + J^T_{(A)j}$ is the total flux density (averaged regular and turbulent flux) of property $A(\boldsymbol{r},t)$ in the medium.

When the fluctuating flow properties coincide ($A(\boldsymbol{r},t) = B(\boldsymbol{r},t)$), equation (4.1.9) turns into the balance equation for the mean-square fluctuation ("dispersion") $<A''^2>$ of the thermohydrodynamic parameter $A(\mathbf{r},t)$ (*Kolesnichenko, 1981*):

$$\bar{\rho}\frac{D\langle A''^2/2\rangle}{Dt}+\partial_j\left(\overline{\rho A''^2V_j''/2}+\overline{A''J_{(A)j}}\right)=-J^T_{(A)j}\partial_j\langle A\rangle+\overline{A''\sigma_{(A)}}-\bar{\rho}\langle\varepsilon_{(A)}\rangle, \tag{4.1.12}$$

where

$$\bar{\rho}\langle\varepsilon_{(A)}\rangle\equiv\bar{\rho}\langle\varepsilon_{(AA)}\rangle/2=-\overline{J_{(A)j}\partial_j A''} \tag{4.1.13}$$

is the scalar dissipation rate of the dispersion $\langle A''^2\rangle$.

Thus, the general balance equation for mixed pair correlations (4.1.9), the equation for dispersions (4.1.12) alike, comprises the terms reflecting the influence of various physical processes on the spatial-temporal distribution of the turbulent property $\langle A''B''\rangle$. These are: convective transport; diffusion; energy exchange between the averaged and fluctuation motions; redistribution of turbulent energy between fluctuation motions in various directions; and dissipation of the property $\langle A''B''\rangle$ owing to molecular transfer processes. We will successively consider the various quantities A and B below.

4.2. TURBULENT ENERGY BALANCE EQUATIONS FOR COMPRESSIBLE MULTICOMPONENT MEDIA

Let us derive the transfer equations for the components of the Reynolds stress tensor and the balance equation for the turbulent energy, $\langle e\rangle$, (which follows from the equation for R_{ki} at $i=k$) for compressible multicomponent mixtures. These equations, obtained from the general evolutionary equation (4.1.9) for single-point pair moments, are precise. However, incorporation of the approximating relations with empirical coupling constants (for the sake of modeling a number of the involved unknown correlations) turns them into model equations which are valid only for a particular class of flows. A variety of rather justified closure hypotheses now available has ultimately led to the development of a large number of models of this kind (see, e.g., *Turbulence: Principles and Applications, 1980; Turbulent Flows of Reacting Gases, 1983; Marov and Kolesnichenko, 1987*).

The partial differential equations for R_{ki} are rather complicated from the computational viewpoint, and therefore need to be simplified additionally for their numerical evaluation. At the same time, they are important as a tool for perfection of simpler turbulence models. In particular, under local equilibrium conditions when the convective and diffusion terms mutually balance, the differential equations turn into algebraic equations for R_{ki} preserving the basic properties of the original equations. The solution of these algebraic equations allows one to find, under certain conditions, the local rheological relations for the turbulent stress tensor such as (3.3.21), though the turbulent viscosity coefficient tensor is already known (see section 4.3.9.). The turbulent viscosity coefficients are expressed in this case in terms of empirical constants involved in the equation for R_{ki}, the turbulent energy, $<e>$, and the scale of turbulence, L, (or the turbulent energy dissipation rate, ε_e).

As was already mentioned in Chapter 3, the fundamental transfer equation for the turbulent energy, $<e>$, underlies many modern semi-empirical turbulence models. In particular, if locally-equilibrium changes of the quantity R_{ki} occur, using this equation together with the formula for the turbulence scale, L, allows one, within certain limits, to accommodate an incomplete equilibrium between the turbulent velocity fields and the averaged flow (see, e.g., *Lewellen, 1980*).

Note that deducing the evolutionary transfer equations for R_{ki} (and $<e>$) carried out by conventional methods, i.e. without using weighted averaging, appears cumbersome when applied to compressible mixtures. These equations include a lot of new correlation characteristics of turbulence that incorporate mass density fluctuations and have no clear physical sense (*Monin and Yaglom, 1971*). To reduce such equations to the form of (4.2.9), many additional simplifications are required, in particular, some minor individual terms must be omitted *a priori*, which can be assessed accurately, however, only for specific flow classes (*Ievlev*, 1990).

4.2.1. The Equations for the Turbulent Stress Tensor

Let us assume that $A \equiv V_k(\boldsymbol{r},t)$ and $B \equiv V_i(\boldsymbol{r},t)$ in (4.1.9) and use expression (2.1.22) for the respective fluxes and impulse generation rates:

$$J_{(V_i)j} \equiv -\pi_{ij}, \quad \sigma_{(V_i)} \equiv -\partial_i p + 2\rho\,\varepsilon_{kji} V_k \Omega_j + \rho \sum_{\alpha=1}^{N} Z_\alpha F_{\alpha i}, \qquad (^1)$$

$$J_{(V_k)j} \equiv -\pi_{kj}, \quad \sigma_{(V_k)} \equiv -\partial_k p + 2\rho\,\varepsilon_{jlk} V_j \Omega_l + \rho \sum_{\alpha=1}^{N} Z_\alpha F_{\alpha k}. \qquad (^2) \qquad (4.2.1)$$

This results in the following precise transfer equation for the Reynolds turbulent stress tensor, $R_{kj} = -\overline{\rho V_i'' V_j''}$, in a flow with variable density (*Marov and Kolesnichenko, 1987*):

$$-\bar{\rho}\frac{D(R_{ki})}{Dt} + \partial_j\left(\overline{\rho V_k'' V_i'' V_j''} - \overline{\pi_{ij} V_k''} - \overline{\pi_{kj} V_i''}\right) =$$
$$= R_{kj}\partial_j < V_i > + R_{ij}\partial_j < V_k > + \overline{V_k''\sigma_{(V_i)} + V_i''\sigma_{(V_k)}} - \bar{\rho} < \varepsilon_{(V_k V_i)} >, \qquad (4.2.2)$$

where

$$J^{T*}_{(V_k V_i)j} \equiv \left(\overline{\rho V_k'' V_i'' V_j''} - \overline{\pi_{ij} V_k''} - \overline{\pi_{kj} V_i''}\right) \qquad (4.2.3)$$

is the "diffusion flux" of the Reynolds stress tensor, R_{ki}, and

$$\overline{\rho} < \varepsilon_{(V_k V_i)} > \equiv \left(\overline{\pi_{ij}\partial_j V_k''} + \overline{\pi_{kj}\partial_j V_i''}\right) \tag{4.2.4}$$

is the quantity (a second rank tensor) related to the dissipation rate of the tensor R_{ki} owing to molecular viscosity.

Let us transform equation (4.2.2) rewriting the addends in the penultimate member, in view of the definitions (4.2.1), as follows:

$$\overline{\sigma_{(V_i)} V_k''} = -\overline{V_k''\partial_i p} - 2\varepsilon_{jli} R_{kj}\Omega_l + \sum_{\alpha=1}^{N} J_{\alpha k}^T F_{\alpha i}^*, \qquad (^1)$$

$$\overline{\sigma_{(V_k)} V_i''} = -\overline{V_i''\partial_k p} - 2\varepsilon_{jlk} R_{ij}\Omega_l + \sum_{\alpha=1}^{N} J_{\alpha i}^T F_{\alpha k}^*. \qquad (^2) \tag{4.2.5}$$

Next we apply the same procedure as was used when deriving (3.1.51) to the addends with pressure in (4.2.5); for any parameter A this yields

$$\overline{A''\partial_i p} = \overline{A''}\,\partial_i \overline{p} + \overline{A''\partial_i p'} = \overline{A''}\partial_i \overline{p} + \partial_i\left(\overline{p'A''}\right) - \overline{p'\partial_j A''} =$$

$$= \overline{\rho v'' A''}\partial_i \overline{p} + \partial_j\left(\delta_{ij}\overline{p'A''}\right) - \overline{p'\partial_i A''}, \tag{4.2.6}$$

whence, supposing $A \equiv V_k(\boldsymbol{r},t)$, we have

$$\overline{V_k''\partial_i p} = \overline{V_k''}\,\partial_i \overline{p} + \overline{V_k''\partial_i p'} = J_{(v)k}^T \partial_i \overline{p} + \partial_i\left(\overline{p'V_k''}\right) - \overline{p'\partial_i V_k''} =$$

$$= \overline{\rho v'' V_k''}\,\partial_i \overline{p} + \partial_j\left(\delta_{ij}\overline{p'V_k''}\right) - \overline{p'\partial_i V_k''}, \tag{4.2.7}$$

and similarly,

$$\overline{V_i''\partial_k p} = J_{(v)i}^T \partial_k \overline{p} + \partial_j\left(\delta_{jk}\overline{p'V_i''}\right) - \overline{p'\partial_k V_i''}, \tag{4.2.8}$$

Here δ_{jk} is the *Kronecker delta*. Substituting now the relations (4.2.5)–(4.2.8) into (4.2.2), we rewrite it in the more compact form:

$$-\overline{\rho}\frac{D\left(R_{ki}/\overline{\rho}\right)}{Dt} + \partial_j J_{kij}^{T\Sigma} = P_{ki}^* + \Phi_{ki} + \overline{\rho} G_{ki} - \overline{\rho} < \varepsilon_{ki} >. \tag{4.2.9}$$

The following notation is used here:

$$J_{kij}^{T\Sigma} \equiv J_{(V_k V_i)j}^{T\Sigma} + \overline{p'\left(\delta_{ij}V_k'' + \delta_{kj}V_i''\right)} =$$

$$= \overline{\rho V_k'' V_i'' V_j''} + \overline{\left(\delta_{ij} p' - \pi_{ij}\right) V_k''} + \overline{\left(\delta_{kj} p' - \pi_{kj}\right) V_i''}, \tag{4.2.10}$$

$$P_{ki}^* \equiv R_{kj}\left(\partial_j < V_i > -2\varepsilon_{jli}\Omega_l\right) + R_{ij}\left(\partial_j < V_k > -2\varepsilon_{jlk}\Omega_l\right), \tag{4.2.11}$$

$$\Phi_{ki} \equiv \overline{p'\left(\partial_i V_k'' + \partial_k V_i''\right)}, \tag{4.2.12}$$

$$\overline{\rho} < \varepsilon_{ki} > = \overline{\rho} < \varepsilon_{(V_k V_i)} > \equiv \left(\overline{\pi_{ij} \partial_j V_k''} + \overline{\pi_{kj} \partial_j V_i''}\right) \tag{4.2.13}$$

$$\overline{\rho} G_{ki} \equiv -J_{(v)k}^T \, \partial_i \overline{p} - J_{(v)i}^T \, \partial_k \overline{p} + \sum_{\alpha=1}^{N} \left[J_{\alpha k}^T \, F_{\alpha i}^* + J_{\alpha i}^T \, F_{\alpha k}^*\right]. \tag{4.2.14}$$

Let us discuss the meaning of the individual addends in (4.2.9). They are as follows:

- $J_{kji}^{T\Sigma}$ is the total "diffusion flux" of the stress tensor R_{ki} involving various turbulence transport mechanisms in space, such as: turbulent diffusion (the addend $\overline{\rho V_k'' V_i'' V_j''}$); transport due to molecular viscosity when a pulsating medium entrains in adjacent moving, originally not pulsated gas layers (the addend $-\left(\overline{V_k'' \pi_{ij} + V_i'' \pi_{kj}}\right)$); and transport caused by interaction between pulsating velocity and pressure fields (the addend $\overline{p' \delta_{ij} V_k'' + p' \delta_{kj} V_i''}$);

- P_{ki}^* is the quantity related to turbulence generation (of tensor R_{ki}) due to interaction between turbulent velocity pulsations and an inhomogeneous field of averaged velocities conditioned by wind shear or by rotation of the medium with common angular velocity Ω_j, the latter being the rotation rate of a relative coordinate frame in which the equation of motion is considered;
- Φ_{ki} is the correlation tensor of turbulent velocity and pressure pulsations characterizing the redistribution of turbulent energy among various components of the Reynolds stress tensor;

- $\overline{\rho} G_{ki}$ represents the quantity related to the appearance and disappearance of turbulence depending on the sign of $\overline{\rho} G_{ki}$, when a medium travels in a nonuniform dynamic pressure field (in particular, in a buoyancy force field, when it is possible to put $\partial_j \overline{p} \approx -\delta_{j3} \overline{\rho} g$, see (3.1.39)), or under action of nongravitational mass forces;

- $\overline{\rho} < \varepsilon_{ki} >$ is the addend related to the rate of change of the quantity R_{ki} caused by turbulent energy dissipation due to molecular viscosity.

When the operational relation (3.1.13) is taken into account, the substantial derivative in (4.2.9) can be rewritten as

$$-\overline{\rho}\frac{D(R_{ki}/\overline{\rho})}{Dt}=-\frac{\partial R_{ki}}{\partial t}-\partial_j\left(R_{ki}<V_j>\right).$$

The second term in the right-hand part of this expression is the advective term describing the effects of convective transport of the tensor R_{ki} with the averaged flow velocity.

Equation (4.2.9) for a compressible multicomponent mixture, despite certain analogies, differs substantially from the equation for a flow with constant density (*Monin and Yaglom, 1971*). It concerns the structure of the expression for G_{ki} specifically. Recalling formula (3.3.39) for a flow with specific volume $J^T_{(v)j}$ we can easily transform the expression for G_{ki} as

$$\overline{\rho}G_{ki}=\frac{1}{\overline{p}}\left(\partial_k\overline{p}\,\overline{p'V_i''}+\partial_i\overline{p}\,\overline{p'V_k''}\right)-\frac{<R^*>}{\overline{p}<C_p>}\left(\partial_k\overline{p}J^T_{qi}+\partial_i\overline{p}J^T_{qk}\right)+$$
$$+\sum_{\alpha=1}^{N}J^T_{\alpha k}\left(F^*_{\alpha i}-\mathrm{k}<T_\alpha>\frac{\partial_i\overline{p}}{\overline{p}}\right)+\sum_{\alpha=1}^{N}J^T_{\alpha i}\left(F^*_{\alpha k}-\mathrm{k}<T_\alpha>\frac{\partial_k\overline{p}}{\overline{p}}\right). \quad (4.2.15)$$

The designations

$$<T_\alpha>=\frac{H_\alpha}{<C_p>}=<T>-\frac{C^*_{p\alpha}}{<C_p>^3}\left(\overline{C_p''h''}-\sum_{\beta=1}^{N}<h_\beta>\overline{C_p''Z_\beta''}\right)\cong<T> \quad (4.2.16)$$

are used here for the sake of convenience (see (3.3.31) and (3.3.37)).

Let us note that when addressing, for example, a turbulized atmosphere where mass density changes due to pressure fluctuations are negligible, the first addend in (4.2.15) can be omitted. This is in particular, the case of forced convection in the presence of large-scale atmospheric vortices (see Chapter 3 and *Van Mieghem, 1973*). Besides, the quantity $<T_\alpha>$ in (4.2.15) can be replaced by the temperature $<T>$ provided the influence of heat capacity pulsations ($C_p''\approx 0$) on the averaged motion patterns, pertinent to multicomponent media, is neglected. Apparently, this can be adopted in the case of subsonic motion in a mixture (*Ievlev, 1975*). For turbulent motion in media stratified in a gravitational field, the expression for G_{ki} takes the form

$$\overline{\rho}G_{ki}\cong\frac{\overline{\rho}g<R^*>}{\overline{p}<C_p>}\left(\delta_{k3}J^T_{qi}+\delta_{i3}J^T_{qk}\right)+$$
$$+\sum_{\alpha=1}^{N}J^T_{\alpha k}\left(F^*_{\alpha i}+\delta_{i3}g\frac{\overline{\rho}\mathrm{k}<T>}{\overline{p}}\right)+\sum_{\alpha=1}^{N}J^T_{\alpha i}\left(F^*_{\alpha k}+\delta_{k3}g\frac{\overline{\rho}\mathrm{k}<T>}{\overline{p}}\right). \quad (4.2.15^*)$$

This formula perfectly matches the similar term of equation (4.2.9) obtained for a flow of homogeneous fluids in the Boussinesque approximation (*Turbulence: Principles and*

Applications, 1980), but being modified taking into account the multicomponent nature of the mixture and the compressibility of the medium.

The equations for the Reynolds turbulent stresses noted as (4.2.9) (six equations for the components of the symmetrical tensor R_{ki}) can not be used directly for closing the averaged hydrodynamic equations for a mixture ((3.2.4)–(3.2.8)) because they contain many new uncertain quantities. These are correlations with pressure pulsations, dissipative terms, and the third order moments, which are not expressed in terms of the "known" pair correlations. In case of multicomponent turbulence, modeling these additional quantities substantially depends on having the opportunity to draw analogies to the closure hypotheses approved fairly satisfactory in the turbulence theory for "constant" density flows (*Turbulent shear flows-I, 1982*). Let us emphasize that presently it is not obvious whether it is necessary to modify such approximations for flows of mixtures with variable density.

Next, in order to demonstrate more clearly the opportunities of invariantly modeling turbulized mixtures, we will consider a simplified closure scheme of the second order using a minimum number of arbitrary constants. More sophisticated parametric relations for the modeled correlators in the case of homogeneous fluid can be found elsewhere (e.g., *Turbulence: Principles and Applications*, (*1980*). We will assume that the approximating relations for correlators frequently used in semi-empirical turbulence theories for homogeneous fluids, are also appropriate for multicomponent turbulence when obtained through weighted-mean averaging (*Kolesnichenko and Marov, 1984*). Then we have

$$\bar{\rho} < \varepsilon_{ki} > = \frac{2}{3} K_{e1} \delta_{ki} \frac{\bar{\rho} < e >^{3/2}}{L} - K_{e2} \frac{\nu R_{ki}}{L^2}, \tag{4.2.17}$$

$$\Phi_{ki} = K_{p1} \frac{< e >^{1/2}}{L} \left(R_{ki} + \frac{2}{3} \delta_{ki} \bar{\rho} < e > \right) - K_{p2} \left(P_{ki} - \frac{2}{3} \delta_{ki} P \right), \tag{4.2.18}$$

$$J_{kij}^{T\Sigma} = \left(c_1 L < e >^{1/2} + c_2 \nu \right) \partial_j R_{ki}. \tag{4.2.19}$$

Here $<e> = <V_s'' V_s''>/2$ is the turbulent energy per unit mass (an averaged kinetic energy of turbulent pulsations); ν is the molecular kinematic viscosity coefficient; $L(\boldsymbol{r},t)$ is the external scale of the turbulent velocity field at a point $\boldsymbol{r}$, this scale being mutually related to, though not coincided with, the integral turbulence scale, L, and to the mixing length, Λ, discussed in the previous chapter; $P_{ki} = P_{ki}^* + \bar{\rho} G_{ki}$ is the total appearance (disappearance) rate of turbulent Reynolds stress, R_{ki}; and $P = \frac{1}{2} \delta_{kj} P_{kj} = R_{kj} \partial_j < V_k > + \bar{\rho} G$ is the total turbulent energy generation rate, $<e>$, under action of the mean shear (the first addend) and of buoyancy effects and mutual transformations of turbulent and potential energies in a stratified multicomponent mixture (the second addend). Accounting for (4.2.15) it is easy to transform the latter as

$$\bar{\rho}G = \tfrac{1}{2}\bar{\rho}\delta_{kj}G_{kj} = -J^T_{(v)j}\partial_j\bar{p} + \sum_{\alpha=1}^{N} J^T_{\alpha j}F^*_{\alpha j} =$$
$$= \overline{p'V''_j}\frac{\partial_j\bar{p}}{\bar{p}} - \frac{<R^*>}{\bar{p}<C_p>}J^T_{qj}\partial_j\bar{p} + \sum_{\alpha=1}^{N}\left(F^*_{\alpha j} - \mathrm{k}<T>\frac{\partial_j\bar{p}}{\bar{p}}\right)J^T_{\alpha j}. \quad (4.2.20)$$

Here K_e, K_p, c are free constants determined from experimental evidence (*Kompaniets et al., 1979; Turbulence: Principles and Applications, 1980*), the empirical coefficients K_{e2} and c_2 in the equations (4.2.17)–(4.2.19) being important only for small turbulent Reynolds numbers, $\boldsymbol{Re}^T \equiv L<e>^{1/2}/\nu$, i.e. when $K_{e1}\boldsymbol{Re}^T / K_{e2} \leq 0(1)$ and $c_1\boldsymbol{Re}^{\mathrm{T}} / c_2 \leq 0(1)$ (*Marov and Kolesnichenko, 1987*).

The relations (4.2.17)–(4.2.19) ensure that the minimum requirements which have to be satisfied with any second order closure scheme are fulfilled. Expression (4.2.17) for $<\varepsilon_{ki}>$ describes the effect of viscous dissipation on the structure of the tensor R_{ki}. Occurring small-scale eddies are the main cause of turbulent energy dissipation. For flows with large Reynolds numbers, $\boldsymbol{Re}^T$, this small-scale turbulence is isotropic. Combining the terms in (4.2.17) for large Reynolds numbers ensures isotropy of the viscous dissipation process (the first term), whereas the same combination for small Reynolds numbers takes into account a possible anisotropy of the dissipation processes for every component of the tensor R_{ki} (the second term). As the scale of large vortices, $L(\boldsymbol{r},t)$, does not influence the dissipation process, their destruction rate in (4.2.17) does not depend on ν.

The first addendum in (4.2.18) for the correlations of pressure and velocity pulsations, Φ_{ki}, describes the redistribution of turbulent energy among individual components of the main (averaged) flow velocity fluctuations and a tendency of fluctuating flows to isotropy, which complies to *Rotta's* model (*Rotta, 1951*). The sum in parentheses, ($R_{kl} + \tfrac{2}{3}\bar{\rho}<e>\delta_{ki}$) (or more exactly, the difference, since $\bar{\rho}<e> = -\tfrac{1}{2}R_{lj}\delta_{lj}$), characterizes the degree of anisotropy of the flow and possesses the necessary tensor symmetry. The second addendum in (4.2.18) corresponds to the *Launder* relation (*Launder, 1975; Launder and Morse, 1982*) for many turbulent flows. It balances the total generation of stresses R_{kj} in (4.2.9) and describes how turbulent energy formation impacts on turbulent velocity fluctuations and hence, on pressure fluctuations. Proportionality of the tensors Φ_{ki} and $(P_{ik} - \tfrac{2}{3}\delta_{ik}P)$ is equivalent to the assumption that correlations of pressure and velocity pulsations not only tend to make turbulence as isotropic as the rate, being linearly related to the degree of deviation from isotropy, but also redistribute the turbulence generation at a rate proportional to the anisotropy of such a generation.

The flux defined by (4.2.19) conditions the diffusion transport of the Reynolds stress R_{kj} from one region of the flow to another without generating or attenuating these stresses. The transport prevents the formation of large gradients in the spatial distribution of the quantity R_{kj} (see *Turbulent Shear Flows -I, 1982*). Let us note that, although both parts of this gradient relation are third rank tensors, it does not comply (by symmetry requirements) with the tensor $\left(\overline{\rho V_k'' V_i'' V_j''}\right)$. Besides, it contradicts the so-called *feasibility limit* which is the necessity to satisfy the *Schwarz inequality* for the third order moment (see section 3.2.2.). Nevertheless, we confine ourselves to this elementary approximation of the diffusion term keeping in mind that only locally-equilibrium versions of this model will be used.

The advantage of applicating the transfer equations critically depends on choosing the empirical constants. Basically, they are extracted from studying special turbulent flows depending only on a single unknown coefficient. Ideally, for any closed turbulence model with a chosen approximation method for the unknown terms in the equations, all entered empirical constants should be constant. This circumstance served as a guideline for selecting the numerical constants (*Turbulence: Principles and Applications, 1980*) in the equations (4.2.17)–(4.2.19):

$$K_{e1} = 0.125;\; K_{e2} = 6;\; K_{p1} = 0.275;\; K_{p2} = 0.55;\; c_1 = 0.3 \pm 0.05 . \qquad (4.2.21)$$

At the same time, it was recognized (*Ievlev, 1975*) that the constants may assumed to be invariable only provided some "equilibrium" turbulence spectrum for the considered flow patterns occurs. For some other flow conditions the values of the "constants" can vary strongly. To account for this circumstance, some authors consider the "constants" as simple functions of characteristic dimensionless flow parameters (such as the Reynolds, Richardson, or Rossby numbers) and some other dimensionless turbulence properties. In this case, however, invariant modeling completely loses its advantage over the first order closure schemes.

4.2.2. The Transfer Equation for Turbulent Energy in Compressible Multicomponent Mixtures

Convolving equation (4.2.9) in indexes k and i $(-\frac{1}{2}R_{kj}\delta_{kj} = \overline{\rho} <e>)$ results in the precise equation for the averaged kinetic energy of turbulent pulsations in compressible mixtures (compare with (3.1.68)):

$$\overline{\rho}\frac{D<e>}{Dt} = -\partial_j J_{<e>j} + R_{ij}\partial_j <V_i> + \overline{p'\partial_j V_j''} + \overline{\rho}G - \overline{\rho} <\varepsilon_e> . \qquad (4.2.22)$$

In contrast to the equation for the Reynolds tensor, this equation contains only two unknown correlation terms. The first diffusion term,

$$\begin{aligned} J_{<e>j} &\equiv \overline{\rho(e + p'/\rho)V_j''} - \overline{\pi_{ij} V_i''} = \tfrac{1}{2} J^{T\Sigma}_{(V_i V_i)j} + \overline{p'V_j''} = \\ &= \overline{\rho e V_j''} + \overline{(p'\delta_{ij} - \pi'_{ij})V_i''} - \overline{\pi}_{ij} J^{T}_{(v)i} \end{aligned} \qquad (4.2.23)$$

describes the total substantial turbulence energy flux related to the various transport mechanisms in space. In particular, the quantity $J^T_{(e)j} \equiv \overline{\rho e V_j''}$ is interpreted as the kinetic energy flux of pulsating (vortex) motion, such that the divergence $\partial_j J^T_{(e)j}$ features the mean decreasing rate of vortex kinetic energy per unit volume due to "turbulent diffusion". In its turn, the quantity $-\partial_j\left(\overline{\pi'_{ij} V''_i}\right)$ characterizes the mean increasing rate of vortex kinetic energy at the expense of work effected by fluctuations of the viscous stress tensor, π'_{ij}, at the boundary of a unit volume. The second correlation term,

$$\bar{\rho} < \varepsilon_e > \equiv \overline{\pi_{ij}\partial_j V''_i} = \overline{\pi'_{ij}\partial_j V'_i} + \bar{\pi}_{ij}\partial_j J^T_{(v)i} = \bar{\rho}\varepsilon_e + \bar{\pi}_{ij}\partial_j J^T_{(v)i}, \tag{4.2.24}$$

where

$$\varepsilon_e \equiv \overline{\pi'_{ij}\partial_j V'_i} = \tfrac{1}{2}\nu\overline{\left(\partial_j V'_i + \partial_i V'_j - \tfrac{2}{3}\delta_{ij}\partial_s V'_s\right)^2} > 0, \tag{4.2.25}$$

describes dissipation of the turbulent kinetic energy into heat due to molecular viscosity. This is the mean work per unit time and per unit volume effected by viscous stress tensor fluctuations against turbulent eddies with non-zero velocity shear ($\partial_j V'_i \neq 0$), (see (3.1.68)). The penultimate term in the right hand part of this equation,

$$\bar{\rho} G \cong -\frac{\partial_j \bar{p}}{\bar{p}}\left(\mathrm{k}<T>\sum_{\alpha=1}^{N} J^T_{\alpha j} + \frac{<R^*>}{<C_p>} J^{T*}_{qj}\right) + \sum_{\alpha=1}^{N} J^T_{\alpha j} F^*_{\alpha j} \tag{4.2.20*}$$

is typical for turbulent flows in multicomponent mixtures in a gravitational field. It describes turbulent energy generation due to non-homogeneity of the temperature and/or the stratified mixture composition, as well as the mutual transformation of turbulent and potential energy owing to mass force effects of non-gravitational origin. In this case, the term with pressure fluctuation can be omitted often (see (4.2.20)).

For a self-sustaining turbulent field, the dissipation rate, ε_e, has the same order of magnitude as the turbulence generation rate by a shear flow, $R_{ij}\partial_j < V_i >$. According to (3.1.45), the quantity $R_{ij}\partial_j < V_i >$ relates to the rate of exchange between kinetic energy of mean motion and turbulent (vortex) motion. This energy transition is an exclusively kinematic process depending only on the choice of averaging and the turbulent motion patterns in the mixture.

In case of developed turbulence, when $\boldsymbol{Re} \equiv LV/\nu >> 1$, the inequality $< e >^{1/2} \geq LV / L_{hydr}$ holds (the sign $<$ applies to small $\boldsymbol{Re}$). Here the Reynolds number is determined by the integrated turbulence scale, L, corresponding to large eddies and by the characteristic flow velocity, V, and $L_{hydr} = |\nabla \ln \theta|^{-1}$ is the distance attributed to substantial variations of some averaged thermohydrodynamic parameter θ. For $\boldsymbol{Re} >> 1$, the approximate equality

$$< \varepsilon_e > = \varepsilon_e\left(1 + \bar{\pi}_{ij}\partial_j J^T_{(v)i} / \bar{\rho}\varepsilon_e\right) \approx \varepsilon_e \tag{4.2.26}$$

is valid. Indeed, estimating the order of magnitude of the individual terms in (4.2.26) we have

$$\overline{\rho}\varepsilon_e \propto \left(\overline{\rho}<e>^{3/2}\right)/L; \qquad \overline{\pi}_{ij} \propto \overline{\rho}\nu V/L_{hydr}; \qquad \partial_j J^T_{(v)i} = \overline{\partial_j V''_i} \propto <e>^{1/2}/L_{hydr}.$$

Therefore, the following estimate emerges:

$$\overline{\pi}_{ij}\partial_j J^T_{(v)i}/\overline{\rho}\varepsilon_e \propto \left(\overline{\rho}\nu V/L_{hydr}\right)\times\left(<e>^{1/2}/L_{hydr}\right)\times\left(L/\overline{\rho}<e>^{3/2}\right)=$$

$$=\nu VL/<e>L^2_{hydr} = (\nu/LV)\times\left(V^2L^2/<e>L^2_{hydr}\right)\leq 1/\boldsymbol{Re}.$$

Likewise, it is possible to show that for $\boldsymbol{Re} >> 1$, the last item in (4.2.24) can be omitted:

$$\overline{\partial_j\pi'_i V''_i} + \overline{\pi}_{ij}J^T_{(v)i} = \overline{\pi'_{ij}V''_i}\left(1+\frac{\overline{\pi}_{ij}J^T_{(v)i}}{\overline{\pi'_{ij}V''_i}}\right) \approx \overline{\pi'_{ij}V'_i}. \tag{4.2.27}$$

This is because we can estimate that

$$\overline{\pi}_{ij}J^T_{(v)i}/\overline{\pi'_{ij}V'_i} \propto \left(\overline{\rho}\nu V/L_{hydr}\right)\times\left(L/\overline{\rho}<e>^{3/2}L_{hydr}\right)\times<e>^{1/2}=$$

$$=\nu VL/<e>L_{hydr} \geq 1/\boldsymbol{Re}.$$

Thus, the transfer equation for turbulent energy in an extremely developed turbulent flow of a multicomponent mixture takes the form

$$\begin{aligned}\overline{\rho}\frac{D<e>}{Dt} &+ \partial_j\left(\overline{\rho e V''_j} + \overline{p'V'_j} - \overline{\pi'_{ij}V'_i}\right) = R_{ij}\partial_j<V_i> + \overline{p'\partial_j V'_j} - \\ &- J^T_{(v)j}\partial_j\overline{p} + \sum_{\alpha=1}^{N}J^T_{\alpha j}F^*_{\alpha j} - \overline{\rho}\varepsilon_e = P + \overline{p'\partial_j V'_j} - \overline{\rho}\varepsilon_e,\end{aligned} \tag{4.2.28}$$

where $J^T_{(v)j} = \frac{<R^*>}{<C_p>\overline{p}}J^{T*}_{qj} + \frac{k<T>}{\overline{p}}\sum_{\alpha=1}^{N}J^T_{\alpha j}$ is the turbulent specific volume flux. Using the relations (4.2.23) and (4.2.24), it is easy to show that in case of a moderate $\boldsymbol{Re}$, the turbulent energy equation for mixtures looks like (4.2.28) as well, except for the term containing the pressure gradient which should be replaced by $J^T_{(v)i}\partial_j\left(-\delta_{ij}\overline{p}+\overline{\pi}_{ij}\right)$.

The convolution of the approximating relations (4.2.17) - (4.2.19) leads to the following models for the unknown correlation terms in the balance equation for turbulent energy (4.2.28):

$$\overline{\rho}\varepsilon_e = \tfrac{1}{2}<\varepsilon_{ki}>\delta_{ki} = K_{e1}\frac{\overline{\rho}<e>^{3/2}}{L} + K_{e2}\frac{\overline{\rho}\nu<e>}{L^2}, \tag{4.2.29}$$

$$\overline{p'\partial_j V_j''} = 0, \tag{4.2.30}$$

$$J_{<e>j} = \overline{\rho(e + p'/\rho)V_j''} - \overline{\pi'_{ij}V_i''} = -\left(c_1 L <e>^{1/2} + c_2 \nu\right)\left(\overline{\rho}\partial_j <e>\right). \tag{4.2.31}$$

The relations (4.2.29)–(4.2.31) essentially follow from the dimensionality theory and are regarded as the generalization of the Kolmogorov's hypothesis (*Kolmogorov, 1941;1942*). Recall that the energy dissipation rate, ε_e, at a specified point in the developed turbulent flow is determined only by the local values of the mean turbulent energy, $<e>$, per unit mass and by the turbulence scale, $L(\boldsymbol{r},t)$, while turbulent transport of impulse and fluctuation energy is affected by the gradient type diffusion terms. This form of the equation (4.2.28) is often used in particular calculations of turbulent motions based on the *Kolmogorov-Lounder* and some other models (*Lewellen, 1980*).

Richardson numbers. As is seen from (4.2.28), two additional turbulence generation mechanisms are possible in stratified jet flows of a multicomponent mixture. While the first mechanism is thermal, the second one is diffusive in its nature because it is caused by the concentration gradient of any diffusing component. This is due to the fact that the spatial-temporal non-homogeneity or fluctuations of mass density is conditioned by the non-homogeneity (fluctuations) of both the temperature and concentration fields (see (3.3.27)). Clearly that, according to *Archimedes' principle*, in a gravitational field a local volume of fluid with smaller density is buoyed up by the ambient fluid. Under certain conditions (see section 3.3.2.) such a buoyancy force results in loss of stability and sets the fluid in motion. It is just the quantity $\overline{\rho}G = -J^T_{(v)j}\partial_j \overline{p} + \sum_{\alpha=1}^{N} J^T_{\alpha j}F^*_{\alpha j}$ in (4.2.28) which was repeatedly referred to as the one describing generation of turbulent energy effected by buoyancy forces. Using relation (3.3.39), the term $\overline{\rho}G$ is expressed in terms of turbulent diffusion and heat fluxes (see (4.2.20*)). Thus, the origin (disappearance) of these two kinds of turbulence in mixture jet flows, i.e. thermal turbulence and turbulence caused by non-homogeneity of composition (concentration gradients), is ultimately related to the dual nature of the buoyancy force.

It is a common practice to account for the influence of thermal mixture stratification on the evolution of turbulent flows using the *Richardson dynamic number*, ***Rf***, which, within the developed model, can be denoted in the following general form:

$$\boldsymbol{Rf} = \frac{\left(<R^*>\partial_j\overline{p}/<C_p>\overline{p}\right)J^{T*}_{qj}}{R_{ij}\partial_j<V_i>} \cong -\frac{\left(g\delta_{j3}/<C_p><T>\right)J^{T*}_{qj}}{R_{ij}\partial_j<V_i>} =$$

$$= \frac{\dfrac{g\delta_{j3}}{<C_p><T>}\lambda^T\left(\partial_j<T> + \dfrac{g\delta_{j3}}{<C_p>}\right)}{\left[-\tfrac{2}{3}\overline{\rho}<e>\delta_{ij} + \overline{\rho}\nu^T\left(\partial_j<V_i> + \partial_i<V_j> - \tfrac{2}{3}\delta_{ij}\partial_s<V_s>\right)\right]\partial_j<V_i>} =$$

$$= \frac{\frac{g\delta_{j3}}{<T>}\left(\partial_j <T> + \frac{g\delta_{j3}}{<C_p>}\right)}{\boldsymbol{Pr}^T\left[-\frac{2}{3}\frac{<e>}{\nu^T}\partial_s <V_s> + \left(\partial_j <V_i> + \partial_i <V_j> - \frac{2}{3}\delta_{ij}\partial_s <V_s>\right)^2\right]} = \frac{\boldsymbol{Ri}}{\boldsymbol{Pr}^T}. \tag{4.2.32}$$

Here $\boldsymbol{Ri}$ is the gradient Richardson number and $\boldsymbol{Pr}^T = \overline{\rho}\nu^T <C_p> / \lambda^T$ is the turbulent Prandtl number. The second notation of $\boldsymbol{Rf}$ in (4.2.32) is obtained using the hydrostatic equation (3.1.39) and the averaged equation of state for gas mixtures as in (3.2.2). The final expression for $\boldsymbol{Rf}$ is given for the isotropic case utilizing the rheological relations attributed to the turbulent heat flux (formula (3.3.15)) and to the turbulent stress tensor (formula (3.3.21)).

The Richardson dynamic number is commonly used in atmospheric sciences. In a stratified atmosphere only horizontally homogeneous turbulent shear flows are used to be examined. Then, if one identifies the direction of non-homogeneity with the x_3-axis with the direction of the gravitational force as in the buoyancy problem and if one further assumes that the mean flow velocity, $<V_j>$, has no component along the x_3-axis while the horizontal velocity component is directed along the x_1- axis, the dynamic Richardson number takes the following standard form:

$$\boldsymbol{Rf} = \frac{\frac{g}{<T>}\left(\partial_3 <T> + \frac{g}{<C_p>}\right)}{\boldsymbol{Pr}^T\left(\partial_3 <V_1>\right)^2}. \tag{4.2.33}$$

In order to take into account the effect of composition stratification on the formation (disappearance) of turbulence energy due to buoyancy forces and those of non-gravitational origin (for example, the *Lorentz's ponderomotive force*), we employ the so-called *Kolmogorov's dynamic number*,

$$\boldsymbol{Kf} = -\frac{\sum_{\alpha=1}^{N} J^T_{\alpha j}\left(F^*_{\alpha j} - \mathrm{k} <T> \partial_j \overline{p}/\overline{p}\right)}{R_{ij}\partial_j <V_i>} \cong -\frac{\sum_{\alpha=1}^{N} J^T_{\alpha j}\left(F^*_{\alpha j} + M^* g\delta_{j3}\right)}{R_{ij}\partial_j <V_i>} \cong$$

$$\cong \frac{\sum_{\alpha=1}^{N} F^*_{\alpha j} <Z_\alpha>,_j - g(\ln M^*),_3}{\boldsymbol{Sc}^T\left[-\frac{2}{3}\frac{<e>}{\nu^T} <V_s>,_s + \left(<V_i>,_j + <V_j>,_i - \frac{2}{3}\delta_{ij} <V_s>,_s\right)^2\right]} = \frac{\boldsymbol{Ko}}{\boldsymbol{Sc}^T}, \tag{4.2.34}$$

where $\boldsymbol{Ko}$ is the gradient Kolmogorov number and $\boldsymbol{Sc}^T = \nu^T / D^T$ is the turbulent Schmidt number. Note that in the case under consideration we have that $\boldsymbol{Sc}^T = \boldsymbol{Pr}^T$ as

$\boldsymbol{Le}^T = \chi^T / D^T = \lambda^T / \overline{\rho} D^T < C_P > = 1$ according to the assumption mentioned above. The equality of turbulent diffusivity of different components was supposed in the last notation of $\boldsymbol{Kf}$. This number was first introduced in the turbulence theory of suspension (*Barenblatt, 1978*).

Using (4.2.32) and (4.2.34), we will now combine the terms in the right-hand part of the equations (4.2.22) which express the full generation of turbulence energy. It yields

$$P \equiv R_{ij}\partial_j < V_i > -\left(< R^* > \partial_j \overline{p} / < C_p > \overline{p}\right) J_{qj}^{T*} + \sum_{\alpha=1}^{N} J_{\alpha j}^{T}\left(F_{\alpha j}^* - \mathrm{k} < T > \partial_j \overline{p} / \overline{p}\right) =$$

$$= R_{ij}\partial_j < V_i > \left(1 - \boldsymbol{Rf} - \boldsymbol{Kf}\right). \qquad (4.2.35)$$

Hence it follows that in the case $\boldsymbol{Rf} + \boldsymbol{Kf} < 0$, turbulent energy is generated by both wind shear and Archimedean forces, and mass forces of non-gravitational origin. If $\boldsymbol{Rf} + \boldsymbol{Kf} \to 1$, the respective sum of the terms in the balance equation for turbulent energy becomes zero. This means that turbulent motions are not maintained. In those practically important cases when one of the indicated origins of the buoyancy force is not efficient, it is possible to address either the critical Richardson number, $\boldsymbol{Ri}_c$, or Kolmogorov number, $\boldsymbol{Ko}_c$. These numbers constrain turbulence set up if $\boldsymbol{Ri} < \boldsymbol{Ri}_c$ at constant composition or if $\boldsymbol{Ko} < \boldsymbol{Ko}_c$ at constant temperature. When two origins of the Archimedean forces are effective, these forces may act both in direct and inverse direction. When the depths of the thermal and diffusion mixing layers substantially differ, such a contrary orientation of the turbulent energy sources can result in reversibility inside (external for a thinner layer) some jet flow region. As was shown (*Marov and Kolesnichenko, 1987*), a rather rough criterion for turbulent motion in a mixture to occur, generally can be constrained by the two conditions: $\boldsymbol{Re} > \boldsymbol{Re}_c$ and $(\boldsymbol{Ri} + \boldsymbol{Ko}) < (\boldsymbol{Ri} + \boldsymbol{Ko})_c$.

4.3. TRANSFER EQUATIONS FOR THE PAIR CORRELATIONS OF MIXTURE ENTHALPY AND CONCENTRATIONS OF COMPONENTS

Now we will extend the method of invariant modeling developed until recently for turbulent homogeneous fluids to compressible, multicomponent, and chemically active media. For the purpose, the equation for the Reynolds stress tensor derived in the previous section should be added to the evolutionary transfer equations for the single-point second moments of fluctuating thermohydrodynamic parameters, including turbulent energy dissipation rates. Although the approach employed is rather labor consuming, it allows us to obtain fully justified relations for the indicated correlations and also to reveal limitations inherent to these fairly single-type equations. The ultimate goal is to develop a procedure for modeling the turbulent exchange coefficients involved in the linear rheological relations for turbulent fluxes. Discussion of the local-equilibrium approximation of the derived evolutionary transfer equations and numerical values of the empirical constants appearing in the approximating relations for the simulated unknown correlations is an important part of the study.

4.3.1. The Transfer Equation for Turbulent Heat Fluxes

Let us first identify the parameters A and B in the general evolutionary equation (4.1.9) with the true values of the full enthalpy, h, and the hydrodynamic velocity, V_j, of a turbulized mixture and use also the expressions (2.1.22) and (2.1.52) for the fluxes and origination rate of these quantities:

$$J_{(h)j} \equiv q_j, \qquad \sigma_{(h)} \equiv \frac{dp}{dt} + \pi_{ij}\partial_j V_i + \sum_{\alpha=1}^{N} J_{\alpha j} F^*_{\alpha j},$$

$$J_{(V_i)j} \equiv -\pi_{ij}, \qquad \sigma_{(V_i)} \equiv -\partial_i p + 2\rho\,\varepsilon_{kji} V_k \Omega_j + \rho\sum_{\alpha=1}^{N} Z_\alpha F_{\alpha i}$$

We then come to the strict evolutionary transfer equation for the turbulent heat flux, $q_j^T \equiv \overline{\rho h'' V_j''} = \overline{\rho} < h'' V_j'' >$ (*Kolesnichenko and Marov, 1984*):

$$\overline{\rho}\frac{D(q_j^T/\overline{\rho})}{Dt} + \partial_k\left[\overline{(\rho V_j'' V_k'' + p'\delta_{jk} - \pi_{jk})h''} + \overline{q_k V_j''}\right] = -q_k^T\left(\partial_k < V_j > + 2\varepsilon_{jlk}\Omega_l\right) +$$

$$+ R_{jk}\left(\partial_k < h > - \frac{\partial_k \overline{p}}{\overline{\rho}}\right) + \overline{\rho} G_{hj} + \overline{p'\partial_j h''} + \overline{V_j''\left(\frac{dp'}{dt} + \pi_{lk}\partial_k V_l''\right)} - \overline{\rho} < \varepsilon_{hj} >, \tag{4.3.1}$$

describing the spatial-temporal distribution of the correlations $< h'' V_j'' >$ for a turbulent shear flow. Here

$$\overline{\rho} < \varepsilon_{hj} > \equiv \overline{\pi_{lj}\partial_j h''} - \overline{q_l \partial_l V_j''} \tag{4.3.2}$$

is the rate of destruction of the correlation $< h'' V_j'' >$ under action of molecular viscosity and thermal diffusivity (a vector quantity), and

$$\overline{\rho} G_{hj} \equiv -\overline{h''}\partial_j \overline{p} \tag{4.3.3}$$

is the generation (dissipation) rate of the turbulent heat flux, q_j^T, brought about by the averaged pressure gradient (see (3.3.40)). The term "strict equation" is used by us in the sense that no approximate transformations affected (4.3.1) when deriving it.

By analogy to the equation for the stress tensor (4.2.9), one may argue that left-hand part of the transfer equation for the components of the turbulent heat flux vector (4.3.1) comprises the convective and "diffusion" terms. The right-hand part comprises the terms describing the full generation of q_j^T caused by the gradients of the averaged flow velocity and the averaged temperature (enthalpy), as well as by the buoyancy forces; redistribution of this quantity by the composite correlator of pressure fluctuations and enthalpy gradient in the mixture (this particular term counteracts generation of $< h''V_j''>$ and limits its growth); and flux dissipation of q_j^T, respectively. An estimate of the penultimate addend in (4.3.1) showed that it is small against the dissipative term $\overline{\rho} < \varepsilon_{hj}>$ for all cases in which turbulent heat transport is important (*Ievlev, 1975*). Hence, in what follows this addend will be omitted. We will use the simplified defining relation $q_j \cong -\chi\rho h_j$ for the molecular heat flux, q_j, in (4.3.1), where $\chi = \lambda / \rho C_p$ is the molecular thermal diffusivity coefficient. The deduction of this relation is given below in subsection 7.1.2.

In view of (3.3.37), (3.3.40), (3.1.39), and the approximate relation

$$\Xi_\alpha \equiv < h_\alpha > - \frac{\text{k} < T >< C_p >}{< R^* >} = h_\alpha^0 + \frac{\text{k} < T >}{< R^* >} \sum_{\beta=1}^{N} \left(C_{p\alpha} - C_{p\beta} \right) < Z_\beta > \approx h_\alpha^0 ,$$

the quantity $\overline{\rho} G_{hj}$ for flow stratified in a gravitational field may be rewritten as

$$\overline{\rho} G_{hj} \cong - \frac{< R^* > \partial_j \overline{p}}{< C_p > \overline{p}} \left(\overline{\rho h''^2} - \sum_{\alpha=1}^{N} < h_\alpha > \overline{\rho Z_\alpha'' h''} \right) - \frac{\text{k} < T > \partial_j \overline{p}}{\overline{p}} \sum_{\alpha=1}^{N} \overline{\rho Z_\alpha'' h''} =$$
$$\cong \delta_{j3} \frac{g}{< C_p >< T >} \left(\overline{\rho h''^2} - \sum_{\alpha=1}^{N} h_\alpha^0 \, \overline{\rho Z_\alpha'' h''} \right). \qquad (4.3.4)$$

This expression, describing the generation of the turbulent heat flux due to buoyancy forces, incorporates the mean-square moment ("dispersion") of the enthalpy and the single-point pair correlations of enthalpy and composition fluctuations. The characteristic evolutionary transfer equations for these quantities are required for invariantly modeling multicomponent turbulence.

Other uncertain correlation terms inducing the closure problem are also included in the strict equation (4.3.1) for the turbulent heat flux, q_j^T. In view of the earlier stipulated reasons, we model these additional correlations using the following simple approximations (*Kolesnichenko and Marov, 1984*):

$$\overline{\rho} < \varepsilon_{hj} > = K_{e2} \frac{\nu + \chi}{2L^2} q_j^T , \qquad (4.3.5)$$

$$\overline{p' \partial_j h''} = -K_{s1} \frac{< e >^{1/2}}{L} q_j^T - K_{s2} P_{hj} , \qquad (4.3.6)$$

$$\left[\left(\overline{\rho V_j'' V_k''} + p' \delta_{jk} - \pi_{jk} \right) h'' - \chi \overline{\rho V_j'' \partial_k h} \right] = - \left[c_3 L < e >^{1/2} + c_4 (\nu + \chi) \right] \partial_k q_j^T , \qquad (4.3.7)$$

in which K_{s1}, K_{s2}, c_3, and c_4 are universal empirical constants (*Turbulence: Princi-*

ples and Applications, 1980); and $L = L(\mathbf{r},t)$ is the local turbulence scale defined by (4.2.17)–(4.2.19). The quantity $P_{hj} \equiv q_k^T\left(\partial_k < V_j > + 2\varepsilon_{ljk}\Omega_l\right) + \overline{\rho}\, G_{hj}$ determines the full generation rate of turbulent heat flux, q_j^T, in response to both the mean strain and the Archimedean forces.

As for the relations (4.3.5)–(4.3.7), the following comment is to be made. In case of locally isotropic turbulence for large Reynolds numbers, the correlation moments $\overline{\pi_{lk}\partial_l h''}$ and $\overline{q_l\, \partial_l V_j''}$, and thus the term with the viscous dissipation, $< \varepsilon_{hj} >$, vanish. Hence, the model approximations of the quantity $< \varepsilon_{hj} >$ are needed only for small $\boldsymbol{Re}^T$, as is the case in near-surface layers. The first term in (4.3.6) is most often used for approximating the correlations $\overline{p'\partial_j h''}$. This expression, proposed by Monin (*Monin, 1964*), represents direct analogy to the "tendency-to-isotropy" approach offered for the quantities Φ_{ki} (*Rotta, 1951,b*) which are the strain rate correlations with pressure fluctuations in the model (4.2.18).

Let us recall that to accomplish complete analogy to this model in case of homogeneous turbulent fluids, Launder approximations for the correlations $\overline{p'\partial_j h''}$ with additional terms were supplemented (*Launder and Spalding, 1972*) which are proportional to generating the quantity $< h'' V_j'' >$ due to mean strain and Archimedean forces. This is the term with P_{hj} in (4.3.6). Thus, while the correlation $\overline{p'\partial_j h''}$ partially compensates the term with directly generating the quantity $< h'' V_j'' >$ by buoyancy forces, these forces give rise to either increasing or decreasing vertical turbulent heat fluxes under unstable or stable stratification of the medium, respectively. The diffusion terms in (4.3.1) are modeled like the equation for R_{ij}, using gradient hypotheses with a scalar diffusion coefficient. The values

$$K_{e2} = 6;\ K_{s1} = 0.4;\ K_{s2} = 0.5;\ c_3 = 0.8\,. \tag{4.3.8}$$

can be adopted as free constants in the relations (4.3.5)–(4.3.7) based on the data available on modeling flat mixing layers. These constants are subject, however, to further refinement for other flow types.

4.3.2. A Prognostic Equation for Mean-Square Fluctuations of the Mixture Enthalpy

When using (2.1.51) and (2.1.52) as instantaneous values of the flux and enthalpy generation rate for multicomponent mixtures,

$$J_{(h)j} \equiv q_j, \qquad \sigma_{(h)} \equiv \frac{dp}{dt} + \partial_j \pi_{ij} V_i + \sum_{\alpha=1}^{N} J_{\alpha j} F^*_{\alpha j},$$

the general transfer equation (4.1.12) leads to the following evolutionary equation for the enthalpy variance, $< h''^2 >$:

$$\bar{\rho}\frac{D(<h''^2>/2)}{Dt}+\partial_k\left(\overline{\rho h'' V_k''/2}+\overline{q_k h''}\right)=-q_k^T\left(\partial_k <h>-\frac{\partial_k \bar{p}}{\bar{\rho}}\right)-\bar{\rho}\varepsilon_h - \tag{4.3.9}$$
$$-\overline{h''\left(dp'/dt+\pi_{ij}\partial_j V_i''\right)},$$

where the quantity

$$\bar{\rho}\varepsilon_h \equiv -\overline{q_j \partial_j h''} \cong \chi \overline{\rho\, \partial_j h \partial_j h''} \tag{4.3.10}$$

characterizes the scalar dissipation rate of the correlation $<h''^2>$ owing to molecular thermal diffusivity (see also formula (7.1.7)). According to the estimates available (*Ievlev, 1975*), the last addend in (4.3.9) is small compared to the penultimate dissipative term and thus can be omitted.

Equation (4.3.9) for the dispersion $<h''^2>$ is similar in structure as equation (4.2.22) for kinetic turbulent energy transport. However, it is much simpler than the latter as it does not contain the correlators containing pressure fluctuations and the terms brought about by buoyancy forces. It should be noted that in case of multicomponent chemically active mixtures, equation (4.3.9) is much simpler than the similar equation for the mean square of temperature fluctuations, $<T''^2>$, as well. This is because, unlike the latter, it does not involve a large number of additional pair correlations of temperature and composition fluctuations. The presence of similar correlations in the equation for the dispersion $<T''^2>$ is ultimately due to the explicit involvement of a chemical thermal energy source in equation (2.1.49) for enthalpy noted in terms of temperature. Dissipative and diffusion addends in equation (4.3.9) will be further approximated by the relations

$$\bar{\rho}\varepsilon_h = K_{h1}\frac{\bar{\rho}<e>^{1/2}<h''^2>}{L}+K_{h2}\frac{\chi\overline{\rho h''^2}}{L^2}, \tag{4.3.11}$$

$$J_{<h>j} \equiv \left(\overline{\rho h''^2 V_j''/2}+\overline{q_j h''}\right)=-\left(c_5 L<e>^{1/2}+c_6\chi\right)\partial_j(\bar{\rho}<h''^2/2>), \tag{4.3.12}$$

where K_{h1}, K_{h2}, c_5, c_6 are universal constants. The constants K_{h2} and c_6 are essential for small values of the Reynolds turbulent number, $\boldsymbol{Re}^T$, only, i.e. when $K_{h1}\boldsymbol{Pr}\boldsymbol{Re}^T/K_{h2} \le 0(1)$ and $c_5\boldsymbol{Pr}\boldsymbol{Re}^{\mathrm{T}}/c_6 \le 0(1)$; here $\boldsymbol{Pr}=\nu/\chi$ is the Prandtl molecular number. With reference to *Turbulence: Principles and Applications* (*1980*) it is possible to use the following typical values for these coefficients:

$$K_{h1}=0.45;\quad K_{h2}=3;\quad c_5=0.3\pm 0.05. \tag{4.3.13}$$

Let us note that the dissipation rate, ε_h, of mean-square enthalpy fluctuations, together with the quantities $<h''^2>$ and $<e>$, according to approximation (4.3.11), determines the linear turbulent transport scale of the enthalpy fluctuations, $L_h \propto <e>^{1/2}<h''^2>/\varepsilon_h$.

Therefore, if $L = L_h$, the additional balance equation for the quantity ε_h can be used to assess the turbulence scale, $L(\boldsymbol{r},t)$ (see section 4.3.8.).

4.3.3. Correlations Including Fluctuations of the Substance Production Source

An adequate description of turbulent flows in chemically active media requires knowledge of the spatial-temporal distributions of single-point pair correlations which involve concentration fluctuations, i.e. the correlations $J^T_{\alpha j}$, $<h''Z''_\alpha>$, and $<Z''_\alpha Z''_\beta>$. This follows from considering the averaged equations of motion for multicomponent mixtures (3.2.4)–(3.2.8), the transfer equation (4.2.9) for turbulent Reynolds stresses, equation (4.3.1) for turbulent heat flows, etc. Therefore, an explicit representation of the prognostic equations to clarify the dynamics of these moments in a reacting flow is of key importance from the viewpoint of both general closure problem in the second order and a comprehensive understanding of turbulent substance transport in stratified mixtures.

We will focus now on the in-depth analysis of the transfer equations for the indicated moments (correlations like $<Z''_\alpha A''>$) based on the general equation (4.1.9). As a first step, we obtain a general expression for the quantities $\overline{A''\sigma_\alpha}$ comprising turbulent pulsations of a substance source generated by chemical reactions because such quantities are present in the transfer equations.

Using the weighted-mean averaging properties (3.1.7) and expression (3.1.27), the quantity $\overline{A''\sigma_\alpha}$ may be represented as

$$\overline{A''\sigma_\alpha} = \overline{\sigma}_\alpha \overline{A''} + \overline{\sigma'_\alpha A''} = \left(\sum_{s=1}^{r} \nu_{\alpha s}\overline{\xi}_s\right)\cdot \overline{A''} + \sum_{s=1}^{r} \nu_{\alpha s}\overline{A\xi'_s}, \tag{4.3.14}$$

where for fluctuations of chemical reaction rate s $(s = 1,2,...,r)$ according to (3.2.28) we have

$$\xi'_s = \Lambda_{s0}T' + \sum_{\beta=1}^{N} \Lambda_{s\beta} n'_\beta . \tag{4.3.15}$$

Because in the developed approach all transfer equations include correlations with fluctuations of enthalpy, h'', and/or concentrations, Z''_α, it is convenient to express the chemical reaction rate fluctuations, ξ_s, in (4.3.15) in terms of these particular quantities. This can be done using the following obvious transformations:

$$n'_\beta = n_\beta - \overline{n}_\beta = \rho\, Z_\beta - \overline{\rho} < Z_\beta > =$$

$$= \rho Z_\beta - \rho < Z_\beta > + \rho' < Z_\beta > = \rho\, Z''_\beta + \rho' < Z_\beta >, \qquad (^1)$$

$$T' = T'' - \overline{T''} = \frac{1}{\overline{\rho}}\left(\overline{\rho}T'' + \overline{\rho' T''}\right) = \frac{\rho T''}{\overline{\rho}} - \frac{(\rho' T'')'}{\overline{\rho}}. \qquad (^2)$$

(4.3.16)

In order to eliminate temperature fluctuations, T'', from the right hand part of (4.3.16 (2)) one can take advantage of formula (3.3.29). This yields

$$T' = \frac{1}{\bar{\rho} < C_p >}\left(\rho\, h'' - \sum_{\beta=1}^{N} < h_\beta > \rho\, Z''_\beta - \sum_{\beta=1}^{N} C_{p\beta}\rho\,(Z''_\beta T'')''\right) - \frac{(\rho' T'')'}{\bar{\rho}}. \tag{4.3.17}$$

Applying these relations to define the correlations $\overline{An'_\beta}$ and $\overline{AT'}$ entering (4.3.14), we obtain

$$\overline{An'_\beta} = \overline{\rho A'' Z''_\beta} - \bar{\rho} < Z_\beta > \overline{A''}, \tag{4.3.18}$$

$$\overline{AT'} \cong \frac{1}{\bar{\rho} < C_p >}\left(\overline{\rho h'' A''} - \sum_{\beta=1}^{N} < h_\beta > \overline{\rho Z''_\beta A''}\right), \tag{4.3.19}$$

with only the second order moments being retained in (4.3.19).

Finally, with account for (4.3.15), (4.3.18), and (4.3.19), the correlation term $\overline{A\xi'_s}$ in (4.3.14) can be denoted as

$$\overline{A\xi'_s} = \frac{\Lambda_{s0}}{\bar{\rho} < C_p >}\overline{\rho h'' A''} + \sum_{\beta=1}^{N}\overline{\rho Z''_\beta A''}\left(\Lambda_{s\beta} - \frac{\Lambda_{s0}}{\bar{\rho} < C_p >} < h_\beta >\right) - \left(\bar{\rho}\sum_{\beta=1}^{N}\Lambda_{s\beta} < Z_\beta >\right)\overline{A''}.$$

Hence, by using (3.3.38*) for $\overline{A''}$, we find the required expression for $\overline{A''\sigma_\alpha}$:

$$\overline{A''\sigma_\alpha} = L_{0\alpha}^{(Ch)}\overline{\rho h'' A''} + \sum_{\beta=1}^{N} L_{\beta\alpha}^{(Ch)}\overline{\rho Z''_\beta A''} \tag{4.3.20}$$

with $L_{0\alpha}^{(Ch)}$ and $L_{\beta\alpha}^{(Ch)}$ as

$$L_{0\alpha}^{(Ch)} = \frac{1}{\bar{\rho} < C_p >}\sum_{s=1}^{r}\nu_{\alpha s}\left[\Lambda_{s0} + \frac{k\,\bar{n}}{\bar{p}}\left(\bar{\xi}_s - \sum_{\beta=1}^{N} n_\beta \Lambda_{s\beta}\right)\right], \tag{4.3.21}$$

$$L_{\beta\alpha}^{(Ch)} = \sum_{s=1}^{r}\nu_{\alpha s}\left(\Lambda_{s\beta} - \frac{C_{p\beta} < T >}{\bar{\rho} < C_p >}\cdot\Lambda_{s0}\right) - h_\beta^0\, L_{0\alpha}^{(Ch)}. \tag{4.3.22}$$

Note, that when deriving the relations (4.3.20)–(4.3.22), we neglect the pair correlation products such as $\overline{\rho Z''_\beta A''}$ and $\overline{\rho h'' A''}$. For the same reason, only the first term should be considered when applying formula (3.2.41) for the averaged chemical reaction rates, $\bar{\xi}_s$, contained in (4.3.21). The cases of different quantities A will be examined below.

4.3.4. The Transfer Equation for Turbulent Diffusion Fluxes

Let us assume that $A \equiv Z_\alpha$ and $B \equiv V_j$ in the general transfer equation (4.1.9). Then, using the relations (2.1.10) and (4.2.1) for the instantaneous values of fluxes and sources of α-substance and mixture momentum, respectively, we obtain the following precise evolutionary transfer equations for the turbulent diffusion fluxes, $J^T_{\alpha j} \equiv \overline{\rho Z''_\alpha V''_j}$ $(\alpha = 1,2,\dots,N)$:

$$\overline{\rho}\frac{D(J^T_{\alpha j}/\overline{\rho})}{Dt} + \partial_k\left[\overline{\left(\rho V''_j V''_k + p'\delta_{jk} - \pi_{jk}\right)Z''_\alpha} + \overline{J_{\alpha k}V''_j}\right] = R_{jk}\partial_k < Z_\alpha > +$$

$$+ J^T_{\alpha k}\left(\partial_k < V_j > + 2\varepsilon_{jlk}\Omega_l\right) + \overline{p'\partial_j Z''_\alpha} + \overline{\rho}\ G_{\alpha j} + \sum_{s=1}^{r}\nu_{\alpha s}\overline{\xi_s V''_j} - \overline{\rho} < \varepsilon_{\alpha j} > . \qquad (4.3.23)$$

Here

$$\overline{\rho} < \varepsilon_{\alpha j} > \equiv \overline{\pi_{jk}\partial_k Z''_\alpha - J_{\alpha k}\partial_k V''_j} \qquad (4.3.24)$$

is the destruction rate of the correlation $< Z''_\alpha V''_j >$ due to molecular viscosity and diffusion (a vector quantity), and

$$\overline{\rho}G_{\alpha j} \equiv \overline{\rho' Z''_\alpha}\frac{\partial_j \overline{p}}{\overline{\rho}} \qquad (4.3.25)$$

is the generation rate of the turbulent diffusion flux, $J^T_{\alpha j}$, determined by the averaged pressure gradient. If turbulent gaseous mixture flows stratified in a gravitational field are considered, the quantity $\overline{\rho}\ G_{\alpha j}$ can be rewritten, in view of (3.3.40) for the correlations $\overline{Z''_\alpha}$ and the hydrostatic equation, $\partial_j \overline{p} \approx -\delta_{j3}\overline{\rho}g$, as follows:

$$\overline{\rho}G_{\alpha j} \cong \delta_{j3}\frac{g}{< C_p >< T >}\left(\overline{\rho h'' Z''_\alpha} - \sum_{\beta=1}^{N} h^0_\beta \overline{\rho Z''_\alpha Z''_\beta}\right). \qquad (4.3.25^*)$$

According to (4.3.20)–(4.3.22), the terms in equation (4.3.23) containing correlations with hydrodynamic flow velocity and chemical source fluctuations, $\overline{V''_j \sigma_\alpha}$, take the form

$$\sum_{s=1}^{r}\nu_{\alpha s}\overline{\xi_s V''_j} = L^{(Ch)}_{0\alpha} q^T_j + \sum_{\beta=1}^{N} L^{(Ch}_{\beta\alpha} J^T_{\beta j} . \qquad (4.3.26)$$

To approximate the dissipative and diffusion terms in the evolutionary transfer equation for the diffusion fluxes, $J^T_{\alpha j}$, we use the following simple expressions similar to the relations (4.3.5)–(4.3.7) for the turbulent heat flux:

$$\bar{\rho} < \varepsilon_{\alpha j} > = K_{e2} \frac{\nu + \chi}{2L^2} J^T_{\alpha j}, \tag{4.3.27}$$

$$\overline{p' \partial_j Z''_\alpha} = -K_{\alpha 1} \frac{< e >^{1/2}}{L} J^T_{\alpha j} - K_{\alpha 2} P_{\alpha j}, \tag{4.3.28}$$

$$\overline{\left(\rho V''_j V''_k + p' \delta_{jk} - \pi_{jk}\right) Z''_\alpha + J_{\alpha k} V''_j} = -\left(c_{\alpha 1} L < e >^{1/2} + c_{\alpha 2} (\chi + \nu)\right) \partial_k J^T_{\alpha j}. \tag{4.3.29}$$

Here the quantity $P_{\alpha j} \equiv J^T_{\alpha k} \left(\partial_k < V_j > + 2\varepsilon_{ijk} \Omega_i\right) + \bar{\rho} G_{\alpha j}$ specifies the generation rate of the turbulent diffusion flux, $J^T_{\alpha j}$, owing to the mean strain of a medium and buoyancy effects; and $K_{\alpha 1}$, $K_{\alpha 2}$, $c_{\alpha 1}$, $c_{\alpha 2}$ $(\alpha = 1, 2, \ldots, N)$ are empirical constants. Because *ad hoc* information concerning these model constants is unavailable, we take advantage of their numerical values similar to (4.3.8):

$$K_{e2} = 6; \; K_{\alpha 1} = 0.4; \; K_{\alpha 2} = 0.5; \; c_{\alpha 1} = 0.8 .$$

This is, apparently, acceptable for a passive (non-reacting) mixture. For a chemically active mixture, more accurate values can be furnished based on special experiments to simulate particular cases of reacting turbulence.

It should be noted that the N equations (4.3.23) for the turbulent diffusion fluxes are not independent. Indeed, it is easy to show that their summation after multiplying by M_α results in identity. Therefore, the algebraic integral $\sum_{\alpha=1}^{N} M_\alpha J^T_{\alpha j} = 0$ must be used as an additional relation for defining the fluxes $J^T_{\alpha j}$.

4.3.5. The Evolutionary Transfer Equation For Correlations with Mixture Enthalpy and Composition Pulsations

Assuming that $A \equiv Z_\alpha$ and $B \equiv h$ in the general transfer equation for the second moments (4.1.9) and taking into account the definitions

$$J_{(Z_\alpha) j} \equiv J_{\alpha j}, \quad \sigma_{(Z_\alpha)} \equiv \sigma_\alpha ,$$

$$J_{(h) j} \equiv q_j, \qquad \sigma_{(h)} \equiv \frac{dp}{dt} + \pi_{ij} \partial_j V_i + \sum_{\alpha=1}^{N} J_{\alpha j} F^*_{\alpha j},$$

for the respective fluxes and sources, we come to the exact evolutionary transfer equations for the correlations $< h'' Z''_\alpha >$ $(\alpha = 1, 2, \ldots, N)$:

$$\overline{\rho}\,\frac{D<h''Z''_\alpha>}{Dt}+\partial_j\left(\overline{\rho h''Z''_\alpha V''_j+J_{\alpha j}h''+q_jZ''_\alpha}\right)=-q^T_j\partial_j<Z_\alpha>-$$
$$-J^T_{\alpha j}\left(\partial_j<h>-\frac{\partial_j\overline{p}}{\overline{p}}\right)+\overline{Z''_\alpha\left(p'+\pi_{ij}\partial_jV_i\right)}+\sum_{s=1}^{r}\nu_{\alpha s}\overline{\xi_s h''}-\overline{\rho}<\varepsilon_{h\alpha}> \tag{4.3.30}$$

Here the notation

$$\overline{\rho}<\varepsilon_{h\alpha}>\equiv-\overline{q_j\partial_jZ''_\alpha}-\overline{J_{\alpha j}\partial_jh''} \tag{4.3.31}$$

is introduced for the scalar dissipation rate of the correlation $<h''Z''_\alpha>$, the dissipation being induced by thermal diffusivity and molecular diffusion. The third addend in the right hand part of (4.3.30) is smaller than the dissipative term (*Ievlev, 1975*) and hence in what follows this addend will be omitted.

The addends including the enthalpy fluctuation correlations with chemical reaction rates $\sum_{s=1}^{r}\nu_{\alpha s}\overline{\xi_s h''}$, in view of (4.3.20)–(4.3.22), take the form

$$\sum_{s=1}^{r}\nu_{\alpha s}\overline{\xi_s h''}=L^{(Ch)}_{0\alpha}\overline{\rho}<h''^2>+\sum_{\beta=1}^{N}L^{(Ch)}_{\beta\alpha}\overline{\rho}<Z''_\beta h''>. \tag{4.3.32}$$

Next, we approximate the dissipative and diffusion terms in equation (4.3.30) using the relations (*Kolesnichenko and Marov, 1984*)

$$\overline{\rho}<\varepsilon_{h\alpha}>=K_{h\alpha 1}\frac{\overline{\rho}<e>^{1/2}}{L}<h''Z''_\alpha>+K_{h\alpha 2}\frac{\chi}{L^2}\overline{\rho h''Z''_\alpha}\,, \tag{4.3.33}$$

and

$$\left(\overline{\rho h''Z''_\alpha V''_j+J_{\alpha j}h''+q_jZ''_\alpha}\right)=-\left(c_{h\alpha 1}L<e>^{1/2}+c_{h\alpha 2}\chi\right)\partial_j\left(\overline{\rho\, h''Z''_\alpha}\right). \tag{4.3.34}$$

The empirical constants $K_{h\alpha 2}$ and $c_{h\alpha 2}$ are essential only for small turbulent Reynolds numbers, $\boldsymbol{Re}^T$, i.e. when $K_{h\alpha 1}\boldsymbol{PrRe}^T/K_{h\alpha 2}\le 0(1)$ and $c_{h\alpha 1}\boldsymbol{PrRe}^{\mathrm{T}}/c_{h\alpha 2}\le 0(1)$. By analogy with (4.3.13), it is acceptable to use the following values of the empirical coefficients:

$$K_{h\alpha 1}=0.45;\quad K_{h\alpha 2}=3;\quad c_{h\alpha 1}=0.3\pm 0.05.$$

The equations (4.3.30) are also linearly dependent. For this reason one of them should be omitted. The equality $\sum_{\alpha=1}^{N}M_\alpha<h''Z''_\alpha>=0$ serves as an additional relation for defining the correlations $<h''Z''_\alpha>$.

4.3.6. The Transfer Equation for Correlation Moments of Mixture Composition Pulsations

Now we set $A \equiv Z_\alpha$ and $B \equiv Z_\beta$ in the general transfer equation for the second moments (4.1.9) and take into account the definitions $J_{(Z_\alpha)j} \equiv J_{\alpha j}$ and $\sigma_{(Z_\alpha)} \equiv \sigma_\alpha$ for the true fluxes and sources of a substance (see (2.1.10)). Then, as the result of simple computations, we obtain the following $N(N+1)/2$ evolutionary equations for the mixed pair correlations $< Z''_\alpha Z''_\beta >$ $(\alpha, \beta = 1,2,...N)$ symmetric in indices α and β:

$$\bar{\rho}\frac{D<Z''_\alpha Z''_\beta>}{Dt} + \partial_j\left[\overline{\rho Z''_\alpha Z''_\beta V''_j} - \chi\overline{\rho\left(Z''_\alpha \partial_j Z''_\beta + Z''_\beta \partial_j Z''_\alpha\right)}\right] =$$

$$= -J^T_{\alpha j}\partial_j < Z_\beta > - J^T_{\beta j}\partial_j < Z_\alpha > + \sum_{s=1}^{r}\left(\nu_{\alpha s}\overline{\xi_s Z''_\beta} + \nu_{\beta s}\overline{\xi_s Z''_\alpha}\right) - \bar{\rho} < \varepsilon_{\alpha\beta} >, \tag{4.3.35}$$

where

$$\bar{\rho} < \varepsilon_{\alpha\beta} > \equiv \chi\overline{\rho\left(\partial_j Z_\alpha\, \partial_j Z''_\beta + \partial_j Z_\beta\, \partial_j Z''_\alpha\right)} \tag{4.3.36}$$

is the destruction rate of the correlation $< Z''_\alpha Z''_\beta >$ due to molecular diffusion. The addends containing the pair correlations of composition and chemical reaction rate fluctuations $\sum_{s=1}^{r}\nu_{\beta s}\overline{\xi_s Z''_\alpha}$ and $\sum_{s=1}^{r}\nu_{\alpha s}\overline{\xi_s Z''_\beta}$, in view of (4.3.20)–(4.3.22), become

$$\sum_{s=1}^{r}\nu_{\beta s}\overline{\xi_s Z''_\alpha} = L^{(Ch)}_{0\beta}\overline{\rho h'' Z''_\alpha} + \sum_{\gamma=1}^{N} L^{(Ch)}_{\gamma\beta}\overline{\rho Z''_\gamma Z''_\alpha}\ , \tag{4.3.37}$$

and

$$\sum_{s=1}^{r}\nu_{\alpha s}\overline{\xi_s Z''_\beta} = L^{(Ch)}_{0\alpha}\overline{\rho h'' Z''_\beta} + \sum_{\gamma=1}^{N} L^{(Ch)}_{\gamma\alpha}\overline{\rho Z''_\gamma Z''_\beta}\ . \tag{4.3.38}$$

To approximate the dissipative and diffusion terms in equation (4.3.35), we use the simple expressions (*Kolesnichenko and Marov, 1984*)

$$\bar{\rho} < \varepsilon_{\alpha\beta} > = K_{\alpha\beta 1}\frac{\bar{\rho} < e >^{1/2}}{L} < Z''_\alpha Z''_\beta > + K_{\alpha\beta 2}\frac{\chi}{L^2}\overline{\rho Z''_\alpha Z''_\beta}\ , \tag{4.3.39}$$

and

$$\left[\overline{\rho Z''_\alpha Z''_\beta V''_j} - \chi\overline{\rho\left(Z''_\alpha \partial_j Z''_\beta + Z''_\beta \partial_j Z''_\alpha\right)}\right] = -\left(c_{\alpha\beta 1} L < e >^{1/2} + c_{\alpha\beta 2}\chi\right)\partial_j(\overline{\rho Z''_\alpha Z''_\beta})\ . \tag{4.3.40}$$

The transfer equations (4.3.35) are linearly dependent. The equalities

$$\sum_{\alpha=1}^{N} M_\alpha <Z''_\beta Z''_\alpha> = 0 \qquad (\beta = 1,2,..., N)$$

serve as additional relations for defining the correlations $< Z''_\beta Z''_\alpha >$.

It is worthwhile to emphasize again that, in order to ensure a comprehensive model for multicomponent compressible mixtures, the evolutionary equations for correlations with density pulsations should be incorporated. As follows from the previous analysis, many algebraic relations and differential equations describing the averaged turbulent pulsations of thermohydrodynamic parameters comprise pair moments with mass density pulsations such as $\overline{\rho v'' A''}$ $(= \overline{A''} = -\overline{\rho' A'}/\overline{\rho})$. The quantities $J^T_{(v)j}$, $\overline{Z''_\alpha}$, etc. serve as an example. Proceeding from the general evolutionary equation (4.1.9), the transfer equations similar to those brought above can also be deduced for such correlations. Specifically, the prognostic equation for the turbulent specific volume flux, $J^T_{(v)j}$, takes the form

$$\overline{\rho}\frac{D(J^T_{(v)i}/\overline{\rho})}{Dt} + \partial_j\left[R_{ij} < v > + \overline{v''(p'\delta_{ij} - \pi_{ij})} - J^T_{(v)i} < V_j >\right] = R_{ij}\partial_j < v > -$$

$$- J^T_{(v)j}\left(\partial_j < V_i > + 2\varepsilon_{ilj}\Omega_l\right) + \overline{p'\partial_i v''} + \sum_{\alpha=1}^{N} F_{\alpha i}\overline{Z''_\alpha} - \overline{\rho} < \varepsilon_{(vV_i)} > . \qquad (4.3.41)$$

However, presently the theory of invariantly modeling turbulence for media with variable density is poorly progressed because of some problems related to approximating many unknown terms in such equations (*Ievlev, 1975;1990*). With due regard for this disadvantage we offered an original method to model correlations comprising the density fluctuations $\overline{\rho v'' A''}$. This method is based on systematically applying the approximate algebraic relation (3.3.36). As was already noticed, using this relation is most efficient for studying shear flows of multicomponent mixtures in a stratified atmosphere. This is because in that case it is possible to neglect the relative turbulent pressure pulsations as compared to the relative temperature and concentration pulsations.

We may summarize that the obtained evolutionary transfer equations for the second order moments are capable to close the system of multicomponent hydrodynamic equations (3.2.4)–(3.2.8) subjected to Reynolds averaging and when the turbulence scale, L, is specified. Together with the hydrodynamic equations they form a complicated semi-empirical turbulence model in second approximation, within which rather complex reacting gaseous mixture flows can be described. The proposed systematic deduction of these equations enables to verify whether the original hypotheses and/or assumptions are admissible. This offers an accurate criterion for the exposition completeness of turbulent heat and mass transfer for any particular problem. In addition, based on a general form of the notation underlying their structure (in particular, the retention of non-gravitational mass forces), the equations allow one to obtain modifications needed to describe other turbulized media, for example, wet, fine-grained, or electrically conductive ones.

4.3.7. The Turbulence Scale

The turbulence scale, L, appears in the derived evolutionary transfer equations for the second order correlation moments. For their closure it is necessary to specify how this scale is defined when approximating the unknown correlation terms. Using only one turbulence scale can be insufficient because, in general, the macro scales L_A of turbulent pulsations of the respective thermohydrodynamic characteristics in the flux A (velocity, enthalpy, concentrations) differ from each other.

Deducing the equations for L_A is one of the most complicated problems in the semi-empirical shear turbulence theory. The matter is that the parameters L_A can not be defined, generally speaking, only in terms of the single-point moments of a fluctuating property A. Being a measure of the distance between two points $\boldsymbol{r}_1$ and $\boldsymbol{r}_2$ in a turbulized flow, where the two-point correlators $< A''(\boldsymbol{r}_1)A''(\boldsymbol{r}_2) >$ still noticeably differ from zero, the scales L_A can be deduced from the complex differential equations for these moments through integrating them from $\boldsymbol{r}_1$ to $\boldsymbol{r}_2$ (*Lumley and Panofsky, 1966*; *Lewellen, 1980*). The prognostic equations obtained in such a way, describe convection, generation, and dissipation having a scale L_A though include a large number of proportionality coefficients experimentally defined only poorly. In other words, they are considerably less reliable than, for example, the balance equations for the Reynolds stress tensor components, in which many terms are defined accurately.

For this reason, to ensure the efficiency of practical calculations, the turbulence scale, L, is frequently set either by empirically found functions, or by a simple model. In the latter case, an algebraic formula or a simplified differential equation are used which accounts only for the particular geometry (distance to a wall, the shape of a channel, etc.) and does not depend on special features of the flow. For instance, proceeding from experimental evidence obtained for smooth pipes, Prandtl has offered the following empirical formula for the quantity L:

$$L = R\left[0.14 - 0.08(1 - z/R)^2 - 0.06(1 - z/R)^4\right], \tag{4.3.42}$$

where R is the radius of the pipe, and z is the distance from the wall (see *Nikuradze, 1936*). In case of free convection in stratified layers with velocity shear, it is possible to define the scale L using the following simple differential equation (*Bykova, 1973*):

$$L = -\kappa c^{1/4}\Psi/(1 + a x_3)\partial_3\Psi \,. \tag{4.3.43}$$

It permits one to calculate L in terms of the local mean flow characteristics. Here $\Psi = < e >^{1/2}/L$, $\kappa = 0{,}4$ is Karman constant; a is a coefficient having dimensionality L^{-1} (it is a function of the Rossby number, $\boldsymbol{Ro}$); and c is an empirical constant.

However, both formulas (4.3.42) and (4.3.43) are insufficiently universal and, being suitable for one class of flows, should be considerably modified if being applied to other classes. Besides, formulas such as (4.3.42) can be used only for "equilibrium" turbulence having properties determined by local conditions at every particular point (*Ievlev, 1975*). This is why when dealing with "non-equilibrium" turbulence for which the influence of the pre-history of the flow on its emerging features at every particular point is important, the quantity L should be determined using some equation accounting for all types of energy transformation in the turbulent flow. One aims to use a "universal" evolutionary equation for the turbulence scale, L, when invariantly modeling turbulent transport for homogeneous fluids. Such an approach to some extent eliminates the mentioned disadvantages of finding the turbulence scale.

A similar equation for the turbulence macro scale defined by the formula

$$L = \frac{const}{<e>} \iiint_w < V''(x)V''(x+r) > \frac{dw}{r^2},$$

was obtained by integrating over volume the transfer equation for two-point correlations of the velocity, $< V''(x)V''(x+r) >$ (*Lewellen, 1980*). The equation for L looks like

$$\frac{DL}{Dt} - 0.3\partial_j \left(L < e >^{1/2} \partial_j L \right) = -0.35 \frac{L}{\bar{\rho} < e >} R_{ij} \partial_j < V_i > + 0.8 \frac{L}{<e>} \frac{g\delta_{j3}}{<T>} q_j^T +$$

$$+ 0.6 \frac{\nu L}{\lambda^2} - 0.375 \frac{1}{< e >^{1/2}} \left[\partial_j (< e >^{1/2} L) \right]^2, \qquad (4.3.44)$$

where $\lambda = L/(3 + 0.125\boldsymbol{Re}^T)^{1/2}$ is the *Kolmogorov-Taylor* microscale. The addend comprising the scale λ adapts the coupling of velocity fluctuations at some distance from the wall with the pressure fluctuations at the wall, as well as the difference of vortex sizes along the wall and in the transverse direction. The difficulties encountered when deriving this and other similar equations for L are related to the fact that none of the terms in the initial equation for the correlations $\overline{V_j'(x)V'(x+r)}$ can be integrated. Consequently, all of them should be modeled. On the other hand, for the differential equation (4.3.44) there is a boundary condition problem for free boundaries of the turbulent flow where the scale L does not tend to zero.

For this reason, in case of multicomponent turbulence we will introduce the evolutionary transfer equations for the combinations $(< A''^2 >)^m (L_A)^n$ instead of the balance equations for the macro turbulence scales, L_A. These evolutionary equations determine the scales L_A when they are used together with the transfer equations for the moments $< A''^2 >$ (*Turbulence: Principles and Applications, 1980*). One of these equations is the prognostic equation for the scalar dissipation rate, $< \varepsilon_{(A)} > \equiv \chi_A < \partial_j A \partial_j A'' >$, of some flow feature A, i.e. the quantity which defines the degradation rate of this feature down to the molecular level. Here χ_A is the respective molecular exchange coefficient. Specifically, the equation for the turbulent energy dissipation rate, ε_e, being considered together with the empirical relation $L \propto < e >^{3/2} / \varepsilon_e$ emerging from dimensionality reasons, allows one to close the system of equations for the second order moments. In this case the boundary condition problem becomes much simpler because the quantity ε_e tends to zero at the external boundary.

4.3.8. The Evolutionary Transfer Equation for the Scalar Dissipation Rate

For the first time the evolutionary transfer equation for the turbulent energy dissipation rate, ε_e, was obtained for a flow of homogeneous incompressible fluid (*Davydov, 1959, 1961*). Unfortunately, an attempt to obtain similar equations for turbulized flows with variable density using weighted-mean averaging turned out unsuccessful (*Favre, 1969*).

Let us consider the procedure of deriving the evolutionary transfer equations for compressible gases based on the transfer equation for the scalar dissipation rate of the mean-square fluctuation of feature A (see formula (4.1.13)),

$$\bar{\rho} < \varepsilon_{(A)} > \equiv \overline{\rho\varepsilon_{(A)}} = -\overline{J_{(A)j}\partial_j A''} = \chi_A \overline{\rho \partial_j A \partial_j A''}. \tag{4.3.45}$$

Taking for this purpose the derivatives of the right- and left-hand sides of the substantial balance equation (2.1.1) for some feature A with respect to the spatial coordinate x_i we obtain

$$\frac{d}{dt}(\partial_i A) = -\partial_j A \partial_i V_j + \partial_i\{[\sigma_{(A)} - \partial_j J_{(A)j}]/\rho\}. \tag{4.3.46}$$

Next, using the identity

$$dA''/dt = dA/dt - D<A>/Dt - V_j''\partial_j<A>,$$

and its partial derivative with respect to x_i, after simple transformations we get

$$\frac{d}{dt}(\partial_i A'') = -\partial_j A'' \partial_i V_j + \partial_i\{[\sigma_{(A)} - J_{(A)j},_j]/\rho\} -$$
$$-\partial_i\left(\frac{D<A>}{Dt}\right) - \partial_i[V_j''\partial_j<A>]. \tag{4.3.47}$$

Let us now average the quantity $\rho d\varepsilon_{(A)}/dt$ over an ensemble of probable realizations. Then, taking into account the identity (3.1.12), the relations (4.3.46) and (4.3.47), and the general averaged balance equation (3.1.14),

$$\bar{\rho}\frac{D<A>}{Dt} = \partial_j[\overline{J_{(A)j}} + J_{(A)j}^T] + \overline{\sigma_{(A)}},$$

we obtain the following exact differential transfer equation:

$$\overline{\rho\frac{d}{dt}\varepsilon_{(A)}} = \bar{\rho}\frac{D}{Dt}<\varepsilon_{(A)}> + \partial_j\overline{(\rho\varepsilon_{(A)}V_j'')} = \chi_A \overline{\rho\frac{d}{dt}(\partial_i A \partial_i A'')} =$$

$$= \chi_A \overline{\rho\partial_i A''\{-\partial_j A \partial_i V_j + \partial_i[(\sigma_{(A)} - \partial_j J_{(A)j})/\rho]\}} + \chi_A \overline{\rho\partial_i A\{-\partial_j A''\partial_i V_j +}$$

$$\overline{+\partial_i[(\sigma_{(A)} - \partial_j J_{(A)j})/\rho] - \partial_i[D<A>/Dt] - \partial_i[V_j''\partial_j <A>]}\}. \tag{4.3.48}$$

It is easy to find from (4.3.48) the evolutionary transfer equation for the scalar dissipation rate of the mean-square fluctuation of feature A in the following general form:

$$\bar{\rho}\frac{D}{Dt}<\varepsilon_{(A)}> + \partial_j[\overline{\rho\varepsilon_{(A)}V_j''} + \chi_A\overline{(\partial_j A + \partial_j A'')\partial_i J_{(A)i}}] =$$

$$= \overline{(J_{(A)i}\partial_j A'' + J_{(A)j}\partial_i A'')}\partial_i <V_j> + \overline{J_{(A)i}\partial_i V_j''}\partial_j <A> + \overline{J_{(A)i}V_j''}\partial_j\partial_i <A> +$$

$$+\chi_A\overline{\rho[\partial_i A''\partial_i(\sigma_{(A)}/\rho) + \partial_i A\partial_i(\sigma_{(A)}/\rho)'']} - \chi_A\overline{\rho(\partial_i A\partial_j A'' + \partial_j A\partial_i A'')\partial_i V_j''} +$$

$$+\chi_A\overline{[\partial_j J_{(A)j}\partial_i(\rho\partial_i A + \rho\partial_i A'')]/\rho} - \overline{J_{(A)i}}\,\partial_i(\partial_j J^{\Sigma}_{(A)j}/\bar{\rho}). \tag{4.3.49}$$

A simple estimation shows that the last term in (4.3.49) can be omitted for fairly large $\boldsymbol{Re}^{\mathrm{T}}$ because this does not result in essential loss of information about the flow. Let us notice that new unknown correlation terms, appearing in equation (4.3.49), again give rise to a closure problem. When comparing this equation to the evolutionary transfer equations for the second moments of turbulent pulsations, it is seen that the terms in the right-hand side of (4.3.49) may be interpreted as generation due to averaged flows and dissipation due to molecular transfer processes.

Equation (4.3.49) is appropriate for modeling the unknown correlation terms. Being considered together with the simple empirical relation

$$L_A \propto <e>^{1/2}<A''^2>/<\varepsilon_{(A)}>, \tag{4.3.50}$$

which follows from dimensionality reasons, it allows one to define the turbulent mixing macro scale, L_A, of feature A and to close the system of equations for the second order moments, $<A''^2>$.

The Transfer Equation for the Turbulent Energy Dissipation Rate. As an example of utilizing the general equation (4.3.49) while deriving the transfer equation for the scalar dissipation rate of a particular mean-square moment (for instance, $<e>$, $<Z_\alpha''^2>$, $<h''^2>$, etc.) we will obtain here the prognostic equation for the turbulent kinetic energy dissipation rate. For developed turbulence, equation (4.3.49) written for dissipation of the turbulent kinetic energy, $\varepsilon_e = \frac{1}{2}\nu\overline{(\partial_j V_i' + \partial_i V_j')^2}$, becomes (*Turbulent Shear Flows - I, 1982*)

$$\bar{\rho}\frac{D\varepsilon_e}{Dt} = 0.15\partial_j\left(\frac{<e>}{\varepsilon_e}R_{kj}\partial_k\varepsilon_e\right) + 1.45\frac{\varepsilon_e}{<e>}\left(R_{ij}\partial_j <V_i> + 0.48\bar{\rho}G\right) - 1.92\frac{\bar{\rho}\varepsilon_e^2}{<e>}, \tag{4.3.51}$$

where the quantity $\bar{\rho}G$ is defined by formula (4.2.20*). The right-hand part of equation

(4.3.51) includes the terms retaining for large $\boldsymbol{Re}^T$, which have been modeled by in simplest way: generation of ε_e due to a velocity gradient; turbulent energy dissipation owing to buoyancy forces and non-gravitational mass forces; and molecular destruction of the quantity ε_e. This equation, being considered together with the empirical relation $L \propto \langle e \rangle^{3/2} / \varepsilon_e$ which follows from dimensionality reasons, completely closes the system of equations for the second order moments.

The Transfer Equation for the Scalar Dissipation Rate of Mixture Enthalpy Dispersion. As another application of the general equation (4.3.49) let us obtain the evolutionary transfer equation for the dissipation rate of the enthalpy dispersion,

$$\bar{\rho} \langle \varepsilon_h \rangle = -\overline{q_j \partial_j h''} \approx \chi \overline{\rho \partial_j h \partial_j h''}. \tag{4.3.52}$$

Again, it is necessary to express the unknown correlation members in terms of the "known" second moments and the parameters $\langle \varepsilon_h \rangle$, $\langle e \rangle$. For the most simply modeled terms, we have (*Turbulent Shear Flows -I, 1982*)

$$\bar{\rho} \frac{D \langle \varepsilon_h \rangle}{Dt} = k_1 \partial_j \left[\frac{\langle e \rangle}{\langle \varepsilon_e \rangle} R_{ij} \partial_i \langle \varepsilon_h \rangle \right] + k_2 \frac{\langle \varepsilon_h \rangle}{\langle h''^2 \rangle} R_{ij} \partial_i \langle V_j \rangle +$$

$$-0.97 \frac{\bar{\rho} \langle \varepsilon_h \rangle}{\langle h''^2 \rangle} q_j^T \partial_j \langle h \rangle + 30.0 \frac{\bar{\rho} \langle \varepsilon_e \rangle^2}{\langle e \rangle^3} (q_j^T)^2 - 3.0 \frac{\bar{\rho} \langle \varepsilon_h \rangle^2}{\langle h''^2 \rangle} - 0.75 \frac{\bar{\rho} \varepsilon_e \varepsilon_h}{\langle e \rangle}. \tag{4.3.53}$$

In the right hand side of equation (4.3.53), the correlation terms are retained which dominate for large Reynolds numbers, namely in the following processes: diffusion; origination of the quantity $\langle \varepsilon_h \rangle$ due to shear wind velocity and temperature; origination of the quantity $\langle \varepsilon_h \rangle$ due to buoyancy forces; and, finally, dissipation of the mean square of enthalpy fluctuations due to molecular destruction. Assuming that the time scale defined by the quantities $\langle h''^2 \rangle$ and $\langle \varepsilon_h \rangle$ is also applicable to the turbulent energy dissipation rate, $\langle e \rangle$, i.e.

$$\langle e \rangle / \langle \varepsilon_e \rangle = const \langle h''^2 \rangle / \langle \varepsilon_h \rangle, \tag{4.3.54}$$

equation (4.3.53) can be used also for defining the turbulence scale (4.3.50) instead of (4.3.51).

4.3.9. Algebraic Closure Models

The complicated phenomenological model for multicomponent turbulence we have discussed, incorporates the averaged hydrodynamic equations for mixtures together with the additional prognostic equations (4.2.9), (4.3.1), (4.3.9), (4.3.23), (4.3.30), and

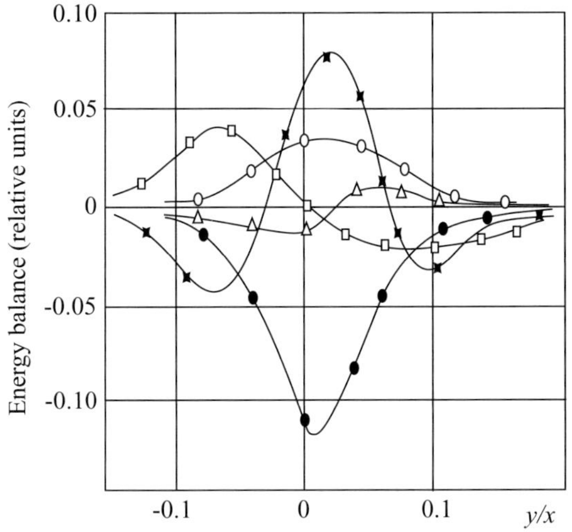

Fig. 4.3.1. Turbulent energy balance in the flat mixing zone reduced to a dimensionless form multiplying by factor $x/<u>^3$: ■ – turbulent energy generation; △ – turbulent energy generation due to normal stresses; ∘ – energy dissipation; □ – turbulent energy transfer by the mean flow; × – diffusion. *y*, *x* – longitudinal and transversal coordinates, respectively; *u* – transversal velocity. According to *Townsend, 1956*.

(4.3.35) for the pair correlations R_{ij}, q_j^T, $<h''^2>$, $J_{\alpha j}^T$, $<h''Z''_\alpha>$, $<Z''_\alpha Z''_\beta>$ of the fluctuating thermohydrodynamic parameters. This model corresponds to the most realistic approach when describing turbulent transport in a reacting mixture. For this reason such a model potentially overcomes simpler models based on using the gradient relations (3.3.3**), (3.3.15**), and (3.3.19**) for turbulent flows. However, the prognostic equations are rather difficult to evaluate and as a consequence, currently they are not quite appropriate for practical applications. Nevertheless, they are important as a tool for perfecting simpler models. Specifically, it is possible to reduce the differential equations for the second correlation moments to algebraic relations for the same quantities. These relations can be used to model the turbulent transport coefficients D_{ij}^T, χ_{ij}^T and ν_{ij}^T involved in the gradient relations.

The Local-Equilibrium Approach. Assume there is an internal equilibrium within the turbulence structure (though not a complete equilibrium with the mean velocity field) such that the convective and diffusion terms in the evolutionary transfer equations (4.2.9), (4.3.1), (4.3.9), (4.3.23), (4.3.30), and (4.3.35) are mutually balanced because they are approximately equal in magnitude and opposite in sign (see Figures 4.3.1 and 4.3.2). Then the second order correlation moments R_{ij}, q_j^T, $<h''^2>$, $J_{\alpha j}^T$, $<h''Z''_\alpha>$, $<Z''_\alpha Z''_\beta>$ are in local equilibrium, in other words, they do not vary in time and/or space. In this case, the differential equations (4.2.9), (4.3.1), (4.3.9), (4.3.23), (4.3.30), and (4.3.35) are transformed into the following algebraic equations:

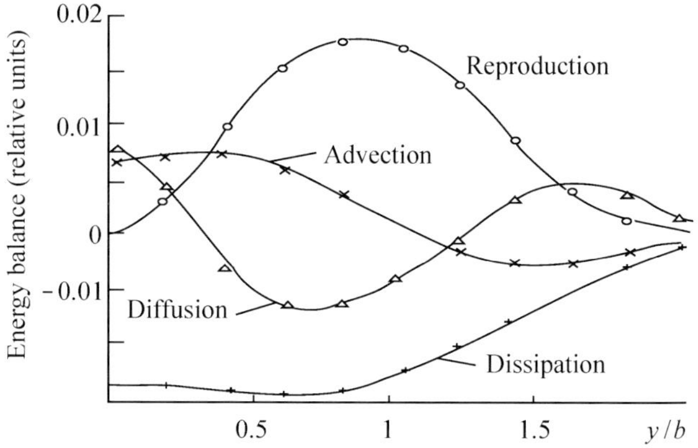

Fig. 4.3.2. Turbulent energy balance in the cross-section of a flat stream (b – half-width of a stream, y – longitudinal coordinate). Like the previous case, reproduction and diffusion components are approximately equal in magnitude and opposite in sign and hence they are mutually balanced. According to *Bradbury, 1965*.

1) The equation for the Reynolds stress tensor

$$K_{p1}\frac{<e>^{1/2}}{L}\left(R_{ki}+\frac{2}{3}\delta_{ki}\bar{\rho}<e>\right)+(1-K_{p2})P_{ki}+\frac{2}{3}K_{p2}\delta_{ki}P-$$
$$-\frac{2}{3}K_{e1}\delta_{ki}\frac{\bar{\rho}<e>^{3/2}}{L}+K_{e2}\frac{\nu R_{ki}}{L^2}=0, \tag{4.3.55}$$

2) The equation for the turbulent heat flux

$$-\left(K_{s1}\frac{<e>^{1/2}}{L}+K_{e2}\frac{\nu+\chi}{2L^2}\right)q_j^T+R_{jk}\left(\partial_k<h>-\frac{\partial_k\bar{p}}{\bar{\rho}}\right)+(1-K_{s2})P_{hj}=0, \tag{4.3.56}$$

3) The equation for the enthalpy dispersion

$$K_{h1}\frac{\bar{\rho}<e>^{1/2}<h''^2>}{L}+K_{h2}\frac{\chi\overline{\rho h''^2}}{L^2}+q_k^T\left(\partial_k<h>-\frac{\partial_k\bar{p}}{\bar{\rho}}\right)=0, \tag{4.3.57}$$

4) The equations for the turbulent diffusion fluxes

$$-\left(K_{\alpha 1}\frac{<e>^{1/2}}{L}+K_{e2}\frac{\nu+\chi}{2L^2}\right)J_{\alpha j}^T+(1-K_{\alpha 2})P_{\alpha j}+R_{jk}\partial_k<Z_\alpha>+$$
$$+L_{0\alpha}^{(Ch)}q_j^T+\sum_{\beta=1}^{N}L_{\beta\alpha}^{(Ch}J_{\beta j}^T=0, \qquad (\alpha=1,2,\ldots,N) \tag{4.3.58}$$

5) The equations for the correlations of enthalpy and composition pulsations

$$-K_{h\alpha 1}\frac{\overline{\rho}<e>^{1/2}}{L}<h''Z''_\alpha>-K_{h\alpha 2}\frac{\chi}{L^2}\overline{\rho h''Z''_\alpha}-q^T_j\partial_j<Z_\alpha>-$$

$$-J^T_{\alpha j}\left(\partial_j<h>-\frac{\partial_j\overline{p}}{\overline{p}}\right)+L^{(Ch)}_{0\alpha}\overline{\rho}<h''^2>+\sum_{\beta=1}^{N}L^{(Ch)}_{\beta\alpha}\overline{\rho}<Z''_\beta h''>,\qquad(\alpha=1,2,...,N)\qquad(4.3.59)$$

6) The equations for the correlations of mixture composition pulsations

$$-K_{\alpha\beta 1}\frac{\overline{\rho}<e>^{1/2}}{L}<Z''_\alpha Z''_\beta>-K_{\alpha\beta 2}\frac{\chi}{L^2}\overline{\rho Z''_\alpha Z''_\beta}-J^T_{\alpha j}\partial_j<Z_\beta>-J^T_{\beta j}\partial_j<Z_\alpha>+$$

$$+L^{(Ch)}_{0\beta}\overline{\rho h''Z''_\alpha}+L^{(Ch)}_{0\alpha}\overline{\rho h''Z''_\beta}+\sum_{\gamma=1}^{N}L^{(Ch)}_{\gamma\beta}\overline{\rho Z''_\gamma Z''_\alpha}+\sum_{\gamma=1}^{N}L^{(Ch)}_{\gamma\alpha}\overline{\rho Z''_\gamma Z''_\beta}=0.\qquad(4.3.60)$$

$$(\alpha,\beta=1,2,...,N)$$

Here the quantities P_{ki}, P, P_{hj}, and $P_{\alpha j}$ describing the generation of turbulence energy, are defined by the relations

$$P_{ki}=R_{kj}\left(\partial_j<V_i>-2\varepsilon_{jli}\Omega_l\right)+R_{ij}\left(\partial_j<V_k>-2\varepsilon_{jlk}\Omega_l\right)+\frac{\overline{\rho}g<R^*>}{\overline{p}<C_p>}\left(\delta_{k3}J^T_{qi}+\delta_{i3}J^T_{qk}\right)+$$

$$+\sum_{\alpha=1}^{N}J^T_{\alpha k}\left(F^*_{\alpha i}+\delta_{i3}g\frac{\overline{\rho}\mathrm{k}<T>}{\overline{p}}\right)+\sum_{\alpha=1}^{N}J^T_{\alpha i}\left(F^*_{\alpha k}+\delta_{k3}g\frac{\overline{\rho}\mathrm{k}<T>}{\overline{p}}\right),\qquad(4.3.61)$$

$$P=R_{kj}\partial_j<V_k>+\frac{g<R^*>\delta_{j3}}{\overline{p}<C_p>}\left(q^T_j-\sum_{\alpha=1}^{N}<h_\alpha>J^T_{\alpha j}\right)+\sum_{\alpha=1}^{N}J^T_{\alpha j}\left(F^*_{\alpha j}+\delta_{j3}g\frac{\mathrm{k}<T>}{\overline{p}}\right)=$$

$$=R_{ij}\partial_j<V_i>(1-\boldsymbol{Rf}-\boldsymbol{Kf}),\qquad(4.3.62)$$

$$P_{hj}\equiv q^T_k\left(\partial_k<V_j>+2\varepsilon_{ljk}\Omega_l\right)+\delta_{j3}\frac{g}{<C_p><T>}\left(\overline{\rho h''^2}-\sum_{\alpha=1}^{N}h^0_\alpha\overline{\rho Z''_\alpha h''}\right),\qquad(4.3.63)$$

$$P_{\alpha j}\equiv J^T_{\alpha k}\left(\partial_k<V_j>+2\varepsilon_{ijk}\Omega_i\right)+\delta_{j3}\frac{g}{<C_p><T>}\left(\overline{\rho h''Z''_\alpha}-\sum_{\beta=1}^{N}h^0_\beta\overline{\rho Z''_\alpha Z''_\beta}\right).\qquad(4.3.64)$$

The equations (4.3.56)–(4.3.60) form a closed system of algebraic equations relating the second order correlation moments R_{ij}, q^T_j, $<h''^2>$, $J^T_{\alpha j}$, $<h''Z''_\alpha>$, $<Z''_\alpha Z''_\beta>$ to the averaged flow parameters, as well as the turbulence scale, L, and the turbulent energy, $<e>$, meeting the basic principles of the K-theory of turbulence. Therefore, the general system should also involve the turbulent energy balance equation (4.2.28) and an expression for the scale L. However, the system incorporating the equations for R_{ij} (for all i and j) and for $<e>$ becomes overdetermined, that is, one equation is superfluous. Hence, one of the whatever equations (4.3.56) may be omitted.

Generally speaking, it would be necessary to neglect the terms in (4.3.56)–(4.3.60) and (4.2.28) that describe effects inherent to a laminar flow because small Reynolds numbers are incompatible with the hypothesis of local-equilibrium turbulence. However, for the approximate approach we are pursuing, these terms are retained and furthermore, the convective and diffusion terms are inserted into the turbulent energy balance equation (4.2.28):

$$\bar{\rho}\,\frac{D<e>}{Dt}-\partial_j\left[\left(c_1 L<e>^{1/2}+c_2\nu\right)\left(\bar{\rho}\partial_j<e>\right)\right]=P-K_{e1}\frac{\bar{\rho}<e>^{3/2}}{L}-K_{e2}\frac{\bar{\rho}\nu<e>}{L^2}. \tag{4.3.65}$$

This approach (the so-called *quasi-equilibrium approximation*) appears to be more accurate than the *super-equilibrium approach*. This is because it allows, to some extent, to account for the effects of the "non-equilibrium" nature of the turbulent field where the turbulence features at every particular point in the flow are related to the entire field of various defining parameters. Applicating the complete equation (4.3.65) to define the turbulent energy, $<e>$, can be justified by the fact that the relaxation time for the "locally equilibrium" turbulent structures is much shorter compared to the time needed for the total intensity of the turbulent velocity field to reach a level where production and dissipation of turbulent energy become equal.

Thus, using (4.3.56)–(4.3.60) and assuming that the gradient relations

$$J^T_{\alpha j}=-\bar{\rho}\,D^T_{jk}\,\partial_k<Z_\alpha>, \tag{4.3.66}$$

$$q^T_j=-\bar{\rho}\,D^T_{jk}\left(\partial_k<h>-\frac{\partial_k\bar{p}}{\bar{\rho}}\right), \tag{4.3.67}$$

and

$$R_{ij}=-\tfrac{2}{3}\bar{\rho}<e>\delta_{ij}+\tfrac{1}{2}\bar{\rho}\left(\nu^T_{is}\delta_{jl}+\nu^T_{js}\delta_{il}\right)\left(\partial_j<V_i>+\partial_i<V_j>-\tfrac{2}{3}\delta_{ij}\partial_k<V_k>\right) \tag{4.3.68}$$

are valid, it is possible, in general, to find expressions for the turbulent viscosity and diffusion coefficients. Particular formulas for the turbulent exchange can be obtained based on the quasi-equilibrium turbulence theory in case of a horizontal turbulent flow in a stratified medium with a velocity shear. This approach is discussed in the following chapters.

Finally, we have to acknowledge that invariantly modeling in the second closure order is, nevertheless, incapable to calculate accurately flows in which transport of any quantity contrarily to its gradient occurs. Such a phenomenon arises, for example, in the boundary layer of the Earth's atmosphere, which is neutrally stratified in temperature, for developed convection when the heat flux is directed upwards against the potential temperature gradient. This leads to a negative turbulent heat conductivity coefficient in (4.3.67), that is, to negative thermal conduction. Accordingly, an adequate theory for counter-gradient transport can be advanced, apparently, only based on models of the third closure order (see *Mellor and Yamada, 1974;1982*; *Lykosov, 1991* for further discussion).

Summary

- *A comprehensive turbulence model for multicomponent chemically active continua was developed. The model permits to examine gas flows with the account for compressibility, variability of thermal properties, and stratification in the gravity field. As a basis, the model comprises the evolutionary transfer equations for the single-point second moments of fluctuating thermohydrodynamic flow parameters, in addition to the hydrodynamic equations for the mean motion of the mixture.*

- *The system of equations for the pulsating components of the thermohydrodynamic parameters was utilized in order to accommodate a missing information on the turbulent flow when applying a statistical averaging operator to the equations describing instantaneous motions. This system was also used as a basis for deriving the transfer equations for the single-point pair correlations of the pulsating thermohydrodynamic parameters.*

- *The following model equations were derived based on the general balance equation for the second moments: the evolutionary transfer equations for the components of the Reynolds stress tensor; the transfer equation for the turbulent energy of a multicomponent mixture; the evolutionary equations for the turbulent heat flux vector; the transfer equation for the dispersion of enthalpy pulsations; the transfer equations for the turbulent diffusion vectors; the transfer equations for the second moments of pulsating enthalpy and mixture composition; and the transfer equations for the pair correlations of the pulsating mixture component concentrations. For multicomponent mixtures with variable density, the evolutionary transfer equation for the scalar dissipation rate were obtained, which allows one to close completely the system of hydrodynamic equations for the mean motion at the second moment level.*

- *The considered model for second order multicomponent turbulence can be used effectively for evaluating complex flows of multicomponent reacting gases with variable density when convective and "diffusion" turbulence transport (the pre-history of the flow) is important. Simpler models based on the gradient closure hypothesis appear inadequate for such flows. Semi-empirical expressions for the turbulent exchange coefficients obtained within the developed approach are generally equivalent to those used in the first order closure procedure.*

CHAPTER 5

THE STEFAN-MAXWELL RELATIONS AND THE HEAT FLUX FOR TURBULENT MULTICOMPONENT CONTINUUM MEDIA

Convective and diffusive turbulence transport are of specific interest for the numerous problems of motion of the multicomponent gaseous mixtures. We have shown that these problems can be solved by means of second order closure models, when evolutionary transfer equations for the second correlation moments are incorporated and a number of mechanisms which are responsible for generating these moments are taken into account in a fairly accurate fashion.

Nonetheless, one must admit that the respective second order models distinguished by their mathematical complexity are not much advantageous compared to the simpler first order models discussed in section 3.3. Indeed, because the system of model equations for the correlations $<A''B''>$, deduced from the general evolutionary equation (4.1.9) for the single-point pair moments is not closed, it should be supplemented with one or several differential equations for the statistical characteristics of the turbulent motion. The latter are equivalent, to certain extent, to the spatial turbulence scale, L. With such an approach, the model expressions for some higher order terms should be added to these equations. However, the relevant approximating expressions in the form of gradient relations with some universal (for a given class of problems) proportionality constants are often insufficiently correct.

Moreover, assessing the overall state of the first order closure problem, it should be recognized that, currently, there is virtually no general phenomenological turbulent theory for heat conduction and turbulent diffusion for multicomponent mixtures. The known gradient relations (see, e.g., *Monin and Yaglom 1971; Van Miegham, 1973; Lapin and Strelets, 1989*) lack sufficient generality. They have mainly been obtained for a homogeneous fluid either for turbulent flows with a pronounced dominant direction, or under strong and poorly justifiable assumptions. One such assumption, discussed in section 3.3, suggests equality of the mixing lengths for turbulent transport of momentum, heat, and substance of a passive admixture. Anyway, the closure problem of the hydrodynamic equations for the mean motion at the level of the first order models deserves more in-depth study.

For example, the problem can be treated in the framework of the thermodynamic approach to the turbulence theory for compressible gaseous continua. In particular, the Onsager formalism of non-equilibrium thermodynamics allows one to obtain the most general structure of rheological relations for turbulent diffusion and heat fluxes in multicomponent mixtures. The generalized Stefan-Maxwell relations for turbulent multicomponent diffusion and the respective expression for the total heat flux (see (2.3.69) and (2.3.75)) are of specific interest. In that case, deducing such relations in detail for small-scale turbulence in flows exhibiting a tendency to establishing a local statistical isotropy must include turbulent heat and mass transport most comprehensively described with the respective rheological relations at the accepted closure level (*Kolesnichenko and Marov, 1985*).

5.1. THE BALANCE EQUATION FOR THE AVERAGED ENTROPY IN TURBULENT FLOWS OF GASEOUS MIXTURES

First, we will recall the basic postulates which underlay the formalism of non-equilibrium thermodynamics for continuum media. These can be practically used for the thermodynamic treatment of any irreversible process, including heat and mass transfer in turbulent flows of multicomponent compressible mixtures (*de Groot and Mazur, 1962*). They are as follows:

- the principle of quasi-local thermodynamic equilibrium is valid;
- the inequality $\sigma_{(S)} \geq 0$ expressing the second law of thermodynamics is true for the entropy production related to irreversible processes in the very system;
- the phenomenological relations (2.2.1) are valid for the fluxes and thermodynamic forces entering expression (2.2.4) for entropy production, if it is constrained to the linear range;
- the relations of symmetry (2.2.2) are valid for the kinetic coefficients entering the linear laws (2.2.1).

Here the thermodynamic analysis of turbulent multicomponent media is performed based on the assumption that the single-point correlations, $< A''B''>$, are small compared to the first order terms, $< A >< B >$. Then they may be omitted for any fluctuating thermodynamic parameters A and B excluding those not equal to the hydrodynamic flow velocity, V_j, i.e.

$$< A''B'' > / < A >< B > \ll 1 \qquad (A \neq V_j, B \neq V_j). \tag{5.1.1}$$

Thereby, a turbulized multicomponent continuum is represented as a thermodynamic complex composed of two subsystems (*Nevzglyadov, 1945 a, b*): the subsystem of mean motion (averaged molecular and turbulent chaos), and the subsystem of pulsating motion (turbulent chaos itself, or turbulent superstructure). This allows one to obtain the necessary rheological relations for the turbulent fluxes of diffusion, heat, and the turbulent stress tensor, which extend the hydrodynamic results of homogeneous fluids over the case of multicomponent mixtures (*Kolesnichenko and Marov, 1984*).

5.1.1. The General Form of the Evolutionary Equation for the Weighted-Mean Entropy

Let us obtain the balance equation for the weighted-mean specific entropy, $<S>\equiv \overline{\rho S}/\overline{\rho}$, of the turbulent multicomponent mixture by means statistically averaging (see 3.1.5) the evolutionary equation (2.2.5) for the instantaneous value of the pulsating parameter S:

$$\overline{\rho}\frac{D<S>}{Dt}+\partial_j\left(\overline{J}_{(S)j}+J^T_{(S)j}\right)=\sigma_{<S>}. \tag{5.1.2}$$

Here $\sigma_{<S>}\equiv\overline{\sigma}_{(S)}$ is the local production of the averaged entropy, $<S>$, i.e. the origination of the quantity $<S>$ per unit time and per unit volume; $\overline{J}_{(S)j}$ and $J^T_{(S)j}\equiv\overline{\rho S''V''_j}$ are the averaged substantial density of instantaneous molecular entropy flux and the turbulent entropy flux density of the averaged motion, respectively.

There are two possible ways to obtain an explicit form of the expressions for the averaged quantities $J^T_{(S)j}$, $\overline{J}_{(S)j}$, and $\sigma_{<S>}$ in (5.1.2): the first is to average their respective instantaneous analogs (for example, over an ensemble of probable realizations); and the second is to associate equation (5.1.2) with the one obtained from the averaged Gibbs identity (2.2.6) after eliminating the respective substantial derivatives of the mean parameters $<v>$, $<Z_\alpha>$, $<\varepsilon>$. We will take advantage of the second option.

The Averaged Gibbs Identity. The fundamental Gibbs identity (2.2.6) is assumed to be valid for micro-movements within a mixture. Being averaged and written in reference to the trajectory of the mass center of a physical elementary volume, it results in the following equation for the weighted-mean specific entropy, $<S>$, and specific internal energy, $<\varepsilon>$ (*Marov and Kolesnichenko, 1983*):

$$\overline{\rho}<T>\frac{D<S>}{Dt}=\overline{\rho}\frac{D<\varepsilon>}{Dt}+\overline{\rho}\overline{p}\frac{D<v>}{Dt}-\overline{\rho}\sum_{\alpha=1}^{N}<\mu_\alpha>\frac{D<Z_\alpha>}{Dt}+\Im. \tag{5.1.3}$$

Here the following notation was introduced:

$$\Im\equiv-\overline{T''\rho dS/dt}-<T>\partial_j\left(\overline{\rho S''V''_j}\right)+\partial_j\left(\overline{\rho\varepsilon''V''_j}\right)+\overline{p'\partial_jV''_j}-$$
$$-\sum_{\alpha=1}^{N}\overline{\mu''_\alpha\rho\, dZ_\alpha/dt}-\sum_{\alpha=1}^{N}<\mu_\alpha>\partial_j\left(\overline{\rho Z''_\alpha V''_j}\right). \tag{5.1.4}$$

It is possible to show that $\Im\equiv 0$ provided the same thermodynamic relations hold for both the averaged and the corresponding instantaneous values of the thermodynamic parameters, and in particular, if the thermodynamic key identities

$$<G>=\sum_{\alpha=1}^{N}<\mu_\alpha><Z_\alpha>=<\varepsilon>+\overline{p}<v>-<T><S>, \qquad ({}^1)$$

$$\langle T\rangle\delta\langle T\rangle+\sum_{\alpha=1}^{N}\langle\mu_\alpha\rangle\delta\langle Z_\alpha\rangle=\delta\langle\varepsilon\rangle+\bar{p}\,\delta\langle v\rangle \qquad (^2) \qquad (5.1.5)$$

are valid (δ is the increment symbol). Note that this holds when the conditions (5.1.1) are fulfilled. In this case the fundamental Gibbs identity (2.2.6) in its substantial form retains its shape for a subsystem of averaged molecular and turbulent chaos as well.

Indeed, by averaging over an ensemble of probable realizations the identity

$$\delta(\rho A\,\varepsilon)-T\delta(\rho\, AS)+p\,\delta A\;-\sum_{\alpha=1}^{N}\mu_\alpha\delta(\rho AZ_\alpha)\equiv 0,$$

which holds for any field quantity A, we have that

$$\delta(\bar\rho\langle A\rangle\langle\varepsilon\rangle)-\langle T\rangle\,\delta(\bar\rho\langle A\rangle\langle S\rangle+\bar{p}\delta\langle A\rangle-\sum_{\alpha=1}^{N}\langle\mu_\alpha\rangle\,\delta(\bar\rho\langle A\rangle\langle Z_\alpha\rangle)=$$

$$=-\,\delta\,(\overline{\rho A''\varepsilon''})+\langle T\rangle\,\delta(\overline{\rho\, A''S})+\overline{T''\delta(\rho\, S\, A)}-\overline{p\,\delta A''}+$$

$$+\sum_{\alpha=1}^{N}\langle\mu_\alpha\rangle\,\delta(\overline{\rho Z''_\alpha A''})+\sum_{\alpha=1}^{N}\overline{\mu_\alpha\delta(\rho Z_\alpha A)}. \qquad (5.1.6)$$

By virtue of assumption (5.1.5), the left-hand part of this equality vanishes for any A. Setting sequentially $A=1$ and $A=V_j$ in (5.1.6), we obtain, respectively, two identities:

$$-\overline{T''\partial_t(\rho\, S)}-\sum_{\alpha=1}^{N}\overline{\mu''_\alpha\partial_t(\rho Z_\alpha)}=0\,, \qquad \text{and} \qquad (^1)$$

$$-\sum_{\alpha=1}^{N}\langle\mu_\alpha\rangle\,\overline{\partial_j(\rho\, Z_\alpha V''_j)}+\overline{\partial_j(\rho\,\varepsilon'' V''_j)}-\langle T\rangle\,\overline{\partial_j(\rho\, S'' V''_j)}-$$

$$-\overline{T''\partial_j(\rho S V_j)}+\overline{p\partial_j V''_j}-\sum_{\alpha=1}^{N}\overline{\mu''_\alpha\,\partial_j(\rho Z_\alpha V_j)}=0\,, \qquad (^2) \qquad (5.1.7)$$

from which follows obviously that $\Im\equiv 0$.

A Formula for the Production of Weighted-Mean Mixture Entropy. Let us exclude from the right hand part of the averaged Gibbs relation (5.1.3) the substantial time derivatives of the averaged state variables $\langle v\rangle$, $\langle\varepsilon\rangle$ and $\langle Z_\alpha\rangle$ $(\alpha=1,2,\ldots,N)$. For the purpose we take advantage of the hydrodynamic equations for the mean motion scale in the following form:

$$\bar\rho\frac{D\langle v\rangle}{Dt}=\partial_j\langle V_j\rangle, \qquad (5.1.8)$$

$$\bar{\rho}\frac{D<Z_\alpha>}{Dt}=-\partial_j(\bar{J}_{\alpha j}+J^T_{\alpha j})+\sum_{s=1}^{r}\nu_{\alpha s}\bar{\xi}_s\,, \qquad (\alpha=1,2,...,N) \tag{5.1.9}$$

$$\bar{\rho}\frac{D<\varepsilon>}{Dt}=-\partial_j(\bar{q}_j+q^T_j-\overline{p'V''_j})-\bar{p}\partial_j<V_j>+\bar{\pi}_{ij}\partial_j<V_i>-\overline{p'\partial_j V''_j}+$$

$$+J^T_{(v)j}\partial_j\bar{p}+\sum_{\alpha=1}^{N}\bar{J}_{\alpha j}F^*_{\alpha j}+\bar{\rho}<\varepsilon_e>, \tag{5.1.10}$$

where $J^T_{(v)j}\equiv\overline{\rho v''V''_j}$, $J^T_{\alpha j}\equiv\overline{\rho Z''_\alpha V''_j}$, $q^T_j\equiv\overline{\rho h''V''_j}$ are the turbulent fluxes of specific volume, diffusion, and heat, respectively, and $<\varepsilon_e>\equiv\overline{\pi_{ij}\partial_j V''_i}/\bar{\rho}$ is the mean specific dissipation rate of turbulent kinetic energy into heat due to molecular viscosity. This gives rise to an explicit substantial balance form of the weighted-mean entropy, $<S>$, for the subsystem of averaged motion of a turbulized multicomponent continuum:

$$\bar{\rho}<T>\frac{D<S>}{Dt}=-\partial_j(\bar{q}_j+q^T_j-\overline{p'V''_j})+\bar{\pi}_{ij}\partial_j<V_i>-\overline{p'\partial_j V''_j}+J^T_{(v)j}\partial_j\bar{p}+$$

$$+\sum_{\alpha=1}^{N}\bar{J}_{\alpha j}F_{\alpha j}-\sum_{\alpha=1}^{N}<\mu_\alpha>\partial_j(\bar{J}_{\alpha j}+J^T_{\alpha j})+\sum_{s=1}^{r}<A_s>\bar{\xi}_s+\bar{\rho}<\varepsilon_e>. \tag{5.1.11}$$

Here we introduced the averaged chemical affinity, $<A_s>$, for reaction s in the turbulent medium that is expressed in terms of the averaged chemical potential per a molecule (see (3.2.14)):

$$<A_s>=-\sum_{\alpha=1}^{N}\nu_{\alpha s}<\mu_\alpha> \qquad (s=1,2,...,r). \tag{5.1.12}$$

The quantity $<A_s>$ is the scalar thermodynamic force conjugated with the scalar $\bar{\xi}_s$ which is the s-th chemical reaction rate (*Prigogine and Defay, 1954*).

Using the simple transformations

$$\frac{\partial_j q^\Sigma_j}{<T>}=\partial_j\left(\frac{q^\Sigma_j}{<T>}\right)-q^\Sigma_j\partial_j\left(\frac{1}{<T>}\right)=\partial_j\left(\frac{q^\Sigma_j}{<T>}\right)+q^\Sigma_j\frac{\partial_j<T>}{<T>^2},$$

$$\frac{<\mu_\alpha>}{<T>}\partial_j J^\Sigma_{\alpha j}=\partial_j\left(\frac{<\mu_\alpha>J^\Sigma_{\alpha j}}{<T>}\right)-J^\Sigma_{\alpha j}\partial_j\left(\frac{<\mu_\alpha>}{<T>}\right),$$

where

$$q^\Sigma_j\equiv\bar{q}_j+q^{T*}_j, \qquad (q^{T*}_j\equiv q^T_j-\overline{p'V''_j}), \quad J^\Sigma_{\alpha j}\equiv J_{\alpha j}+J^T_{\alpha j} \tag{5.1.13}$$

are the total heat and diffusion fluxes in the multicomponent turbulent continuum, respectively, it is easy to reduce equation (5.1.11) to the form (5.1.2) of the balance equation for the averaged mixture entropy. It yields

$$\bar{\rho}\frac{D<S>}{Dt}+\partial_j\left\{\frac{q_j^{\Sigma}-\sum_{\alpha=1}^{N}<\mu_\alpha>J_{\alpha j}^{\Sigma}}{<T>}\right\}=-q_j^{\Sigma}\frac{\partial_j<T>}{<T>^2}-$$

$$-\frac{1}{<T>}\left\{\sum_{\alpha=1}^{N}\bar{J}_{\alpha j}\left[<T>\partial_j\left(\frac{<\mu_\alpha>}{<T>}\right)-F_{\alpha j}\right]-\sum_{\alpha=1}^{N}J_{\alpha j}^{T}<T>\partial_j\left(\frac{<\mu_\alpha>}{<T>}\right)-\bar{\pi}_{ij}\partial_j<V_i>+\right.$$

$$\left.-\sum_{s=1}^{r}<A_s>\bar{\xi}_s+\overline{p'\partial_jV_j''}+\overline{\rho'V_j''}\frac{\partial_j\bar{p}}{\bar{\rho}}-\bar{\rho}<\varepsilon_e>\right\}. \qquad (5.1.14)$$

Comparing now (5.1.14) to (5.1.2), we easily obtain the expressions for the averaged molecular flux, $\bar{J}_{(S)j}$, the turbulent flux, $J_{(S)j}^{T}$, and the entropy production, $\sigma_{<S>}$, in the subsystem of averaged (molecular plus turbulent) chaos. They are as follows (see (3.3.9)):

$$\bar{J}_{(S)j}=\frac{1}{<T>}\left[\bar{q}_j-\sum_{\alpha=1}^{N}<\mu_\alpha>\bar{J}_{\alpha j}\right]=\frac{1}{<T>}\bar{J}_{qj}+\sum_{\alpha=1}^{N}<S_\alpha>\bar{J}_{\alpha j}, \qquad (5.1.15)$$

$$J_{(S)j}^{T}=\frac{1}{<T>}\left[q_j^{T*}-\sum_{\alpha=1}^{N}<\mu_\alpha>J_{\alpha j}^{T}\right]=\frac{1}{<T>}J_{qj}^{T*}+\sum_{\alpha=1}^{N}<S_\alpha>J_{\alpha j}^{T}, \qquad (5.1.16)$$

$$\sigma_{<S>}\equiv\sigma_{<S>}^{i}+\sigma_{<S>}^{e}, \qquad (5.1.17)$$

where

$$0\le\sigma_{<S>}^{i}<T>\equiv-q_j^{\Sigma*}\frac{\partial_j<T>}{<T>}+\sum_{\alpha=1}^{N}\left[-<T>\partial_j\left(\frac{<\mu_\alpha>}{<T>}\right)+F_{\alpha j}\right]J_{\alpha j}^{\Sigma}+$$

$$+\bar{\pi}_{ij}\partial_j<V_i>+\sum_{s=1}^{r}<A_s>\bar{\xi}_s, \qquad (5.1.18)$$

$$\sigma_{<S>}^{e}<T>\equiv-\overline{p'\partial_jV_j''}+J_{(v)j}^{T}\partial_j\bar{p}+\bar{\rho}<\varepsilon_e>-\sum_{\alpha=1}^{N}J_{\alpha j}^{T}F_{\alpha j}\equiv\Delta. \qquad (5.1.19)$$

Recalling also (3.3.10 [(a)]) and (3.3.10 [(b)]), we have

$$J_{qj}^{T} \equiv q_j^{T} - \sum_{\alpha=1}^{N} < h_\alpha > J_{\alpha j}^{T}, \qquad (^1)$$

$$J_{qj}^{T*} \equiv q_j^{T*} - \sum_{\alpha=1}^{N} < h_\alpha > J_{\alpha j}^{T} = q_j^{T} - \overline{p' V_j''} - \sum_{\alpha=1}^{N} < h_\alpha > J_{\alpha j}^{T}, \qquad (^2) \qquad (5.1.20)$$

where $\sigma^{i}_{<S>}$ is the local production rate of the averaged entropy, $< S >$, determined by irreversible processes within the subsystem of averaged motion of the turbulized multi-component continuum.

The quantity $\sigma^{e}_{<S>}$ specifies the entropy exchange between the subsystems of fluctuating and averaged motions and will be further clarified. Its sign varies depending on particular conditions of the turbulent flow (*Nikolaevsky, 1984*). For instance, the energy transition rate, $\overline{p' \partial_j V_j''}$, representing the work done against eddies by the surrounding medium per unit time and per unit volume, can be different in sign due to pressure fluctuations, p', and expansion $(\partial_j V_j'' > 0)$ or compression $(\partial_j V_j'' < 0)$ of the eddies. Similarly, the quantity $J_{(v)j}^{T} \partial_j \overline{p} \approx g \overline{\rho' V_3}$ representing the turbulence energy generation rate by buoyancy forces is positive in case of small-scale turbulence and negative for large eddies (*Van Miegham, 1973*). The transformation rate of mean motion potential energy into turbulent energy, $\sum_{\alpha=1}^{N} J_{\alpha j}^{T} F_{\alpha j}^{*}$, related to the work done by non-gravitational mass forces, can be different in sign as well. Thus, as follows from (5.1.14), the entropy, $< S >$, of the subsystem of averaged chaos can either grow or decrease, which is a general property of thermodynamically open systems.

Let us note once again that assigning the individual terms in equation (5.1.14) to the turbulent flow or to the average entropy production is ambiguous because a number of alternative statements are possible, based on definitions of the turbulent heat flux differing from (5.1.20). These ideas are discussed in more detail elsewhere (*de Groot and Mazur, 1962; Dyarmati, 1974*).

5.1.2. The Balance Equation for Entropy and Entropy Production for the Turbulent Chaotic Subsystem

Such a fundamental feature as the mixture entropy, $< S >$, averaged according to Favre still does not determine completely the thermodynamic state of the whole system because it is independent of the *state variables* describing the turbulent *overstructure*, for example, the turbulent energy, $< e > \equiv \overline{\rho V_j'' V_j''} / 2\overline{\rho}$. Following (*Blackadar, 1955*), we define the thermodynamic structure (model) of a turbulent chaotic subsystem by setting the *caloric equation of state*, i.e. the functional relation $< e > = < e > \left(\overline{\rho}, S_F\right)$ between state variables $< e >$, S_F and $\overline{\rho}$, where S_F is the so-called *turbulization entropy*. Then the fundamental Gibbs identity postulated (see section 2.2) as

$$\frac{D<e>}{Dt}=T_F\frac{DS_F}{Dt}-p_F\frac{D<v>}{Dt}, \tag{5.1.21}$$

introduces the so-called *turbulization temperature*, $T_F\equiv\{\partial<e>/\partial S_F\}_{<v>}$, and *turbulent (or pulsation) pressure*, $p_F\equiv\{\partial<e>/\partial<v>\}_{S_F}$, for a turbulent chaotic subsystem, or turbulent overstructure. It is supposed that the respective derivatives are positive, i.e. $\{\partial<e>/\partial S_F\}_{<v>}>0$ and $\{\partial<e>/\partial<v>\}_{S_F}>0$.

Various functional relations between the parameters $<e>$, S_F, T_F and p_F, obtained in an ordinary way from the formulas given above, can be interpreted as distinct equations of state for a turbulent chaotic subsystem. Let us further assume that the function $<e>=<e>(\overline{\rho},S_F)$ set *a priori* allows us to obtain similar relations which are analogous in their structure to the various equations of state for an ideal monatomic gas. In particular, we assume that the formulas

$$<e>=\tfrac{3}{2}<v>p_F=\tfrac{3}{2}R^*T_F,\quad S_F=\tfrac{3}{2}R^*\ln\left(p_F<v>^{5/3}\right)+const \tag{5.1.22}$$

are valid for $<e>$ and S_F.

We will obtain the respective evolutionary transfer equation for the turbulization entropy, S_F, from (5.1.21) using the mentioned method and equation (5.1.8) for the specific volume, $<v>$, and the balance equation (4.2.22) for the turbulent energy, $<e>$. Let us rewrite the latter as

$$\overline{\rho}\frac{D<e>}{Dt}=-\partial_jJ_{<e>j}+R_{ij}\partial_j<V_i>+\overline{p'\partial_jV_j''}-J^T_{(v)j}\partial_j\overline{p}+\sum_{\alpha=1}^{N}J^T_{\alpha j}F_{\alpha j}-\overline{\rho}<\varepsilon_e>,$$

where $J_{<e>j}\equiv\overline{\rho(e+p'/\rho)V_j''}-\overline{\pi_{ij}V_i''}$ is the total substantial turbulence energy flux (see (4.2.23)) and $R_{ij}\equiv-\overline{\rho V_i''V_j''}$ is the Reynolds stress tensor. We then obtain the following balance equation for the turbulization entropy, S_F:

$$\overline{\rho}\frac{DS_F}{Dt}+\partial_jJ_{(S_F)j}=\sigma_{(S_F)}\equiv\sigma^i_{(S_F)}+\sigma^e_{(S_F)}, \tag{5.1.23}$$

where

$$J_{(S_F)j}\equiv\frac{1}{T_F}\left[\overline{\rho(e+p'/\rho)V_j''-\pi_{ij}V_i''}\right]=\frac{1}{T_F}J_{<e>j} \tag{5.1.24}$$

is the substantial flux of S_F for a turbulent chaotic subsystem. The energy dissipation, being a local measure of the non-equilibrium state of a turbulent chaotic subsystem, is written as

$$0 \le T_F \sigma^i_{(S_F)} = -J_{<e>j} \frac{\partial_j T_F}{T_F} + R_{ij}\partial_j <V_i> + p_F \partial_j <V_j>, \tag{5.1.25}$$

$$T_F\, \sigma^e_{(S_F)} = \overline{p'\partial_j V''_j} - J^T_{(v)j}\partial_j \overline{p} + \sum_{\alpha=1}^{N} J^T_{\alpha j} F^*_{\alpha j} - \overline{\rho} < \varepsilon_e > \equiv -\Delta\,. \tag{5.1.26}$$

These look like the sum of pair products of thermodynamic forces and fluxes while the quantities $\sigma^i_{(S_F)}$ and $\sigma^e_{(S_F)}$ can be regarded as the local production and sinking rates of S_F.

We will now transform the quantity $T_F \sigma^i_{(S_F)}$ into a form convenient for further analysis. For this purpose, we divide the velocity gradient, $\partial_j <V_i>$, into its symmetric and anti-symmetric parts, $\partial_j <V_i> = \frac{1}{2}\left[\partial_j <V_i> + \partial_i <V_j>\right] + \frac{1}{2}\left[\partial_j <V_i> - \partial_i <V_j>\right]$. Then we take into account the "equation of state" (5.1.22) and introduce the designations

$$\overset{\circ}{e}_{ij} \equiv e_{ij} - \tfrac{1}{3}\left(e_{kl}\delta_{kl}\right)\delta_{ij} = e_{ij} - \tfrac{1}{3}\delta_{ij}\partial_k <V_k> \tag{5.1.27}$$

for the symmetric zero-trace part of the strain rate tensor, $e_{ij} \equiv \frac{1}{2}\left[\partial_j <V_i> + \partial_i <V_j>\right]$, of an averaged continuum, and

$$\overset{\circ}{R}_{ij} \equiv R_{ij} - \tfrac{1}{3}\left(R_{kl}\delta_{kl}\right)\delta_{ij} = R_{ij} + \tfrac{2}{3}\delta_{ij}\overline{\rho} <e> = R_{ij} + p_F \delta_{ij} \tag{5.1.28}$$

for the zero-trace symmetrical Reynolds tensor, R_{ij}. This yields

$$T_F \sigma^i_{(S_F)} = -J_{(S_F)}\partial_j T_F + \overset{\circ}{R}_{ij}\, \overset{\circ}{e}_{ij}\,. \tag{5.1.29}$$

Here we took advantage of the condition that the scalar product of symmetric and anti-symmetric tensors is always equal to zero.

For the turbulent flux in a specific volume

$$J^T_{(v)j} \cong -\frac{1}{\overline{p}}\overline{p'V_j} + \frac{<R^*>}{\overline{p}<C_p>} J^T_{qj} + \frac{\mathrm{k}<T>}{\overline{p}} \sum_{\alpha=1}^{N} J^T_{\alpha j}\,,$$

we can write, in view of (3.3.39), the energy dissipation, $T_F \sigma^e_{(S_F)}$, as

$$T_F \sigma^e_{(S_F)} = \overline{p'\partial_j V''_j} + \frac{\partial_j \overline{p}}{\overline{p}} \overline{p'V''_j} - \frac{\partial_j \overline{p}}{\overline{\rho}<T><C_p>} J^T_{qj} - \overline{\rho}<\varepsilon_e> +$$

$$\sum_{\alpha=1}^{N}\left(F_{\alpha j} - \frac{\partial_j \overline{p}}{\overline{n}}\right) J^T_{\alpha j} = \overline{p'\partial_j V''_j} + \frac{\partial_j \overline{p}}{\gamma^T \overline{p}} \overline{p'V''_j} - \frac{\partial_j \overline{p}}{\overline{\rho}<T><C_p>} J^{T*}_{qj} +$$

$$+\sum_{\alpha=1}^{N}\left(F_{\alpha j}-\frac{\partial_j \bar{p}}{\bar{n}}\right)J^T_{\alpha j}-\bar{\rho}<\varepsilon_e>, \tag{5.1.30}$$

where $\gamma^T \equiv <C_p>/<C_V>$ is *Poisson's constant* for the averaged motion of the multi-component turbulized continuum.

Let us examine now the evolutionary transfer equation (5.1.23), which with regard to the transformations (5.1.29) and (5.1.30), can be rewritten as

$$\bar{\rho}\frac{DS_F}{Dt}+\partial_j J_{(S_F)j}=\sigma_{(S_F)}\equiv\frac{1}{T_F}\left[-J_{(S_F)j}\partial_j T_F+\mathring{R}_{ij}\,\mathring{e}_{ij}(1-\boldsymbol{Rf}-\boldsymbol{Kf})+\right.$$

$$\left.+\overline{p'\partial_j V''_j}+\frac{\partial_j\bar{p}}{\gamma^T\bar{p}}\overline{p'V''_j}-\bar{\rho}<\varepsilon_e>\right], \tag{5.1.31}$$

where $\boldsymbol{Rf}$ and $\boldsymbol{Kf}$ are the Richardson and Kolmogorov dynamic numbers defined by (4.2.32) and (4.2.34), respectively. They characterize how stratification of temperature and mixture composition affects the evolution of the average kinetic energy of turbulent fluctuations. From (5.1.31) it follows that:

- If $(\boldsymbol{Rf}+\boldsymbol{Kf})<1$, the turbulization entropy is increased at the rate of $\mathring{R}_{ij}\,\mathring{e}_{ij}$ under influence of Reynolds stresses. The reason for this is that the mechanism to maintain turbulence in a shear flow operates faster than the mechanism to consume energy for work done against the buoyancy force and other, non-gravitational mass forces. This is the case, for instance, for forced convection developed in a planetary atmosphere. The term $\mathring{R}_{ij}\,\mathring{e}_{ij}$ plays a key role in this medium and controls the transition of kinetic energy from averaged into turbulent motion (*Marov and Kolesnichenko, 1987*).

- If $(\boldsymbol{Rf}+\boldsymbol{Kf})=1$, the energy determined by the gradient of the averaged hydrodynamic velocity is consumed to overcome the Archimedean forces and the non-gravitational mass forces, and thus turbulence can not be maintained.

- If $(\boldsymbol{Rf}+\boldsymbol{Kf})>1$, the main role in forming both thermal and concentrational turbulence is played by the buoyancy force (the term $J^T_{(v)j}\partial_j\bar{p}$), responsible for transforming the system's internal energy into averaged mean flow kinetic energy. This case matches, for example, free convection in a planetary atmosphere.

It should be noted that the formulas (4.2.32) and (4.2.34), also resulting from analyzing the balance equation (5.1.23) for the turbulization entropy, S_F, define the Richardson and Kolmogorov dynamic numbers most comprehensively. Other approaches for defining $\boldsymbol{Rf}$ and $\boldsymbol{Kf}$ are less valuable and can be accepted in some particular cases only under additional assumptions.

5.1.3. The Balance Equation for the Total Turbulized Continuum Entropy

The two entropies we have introduced, $<S>$ and S_F, renders concrete our idea of turbulent flows as the thermodynamic complex which includes two mutually open subsystems: the turbulent overstructure and the averaged (turbulent and molecular) chaos. From (5.1.14) and (5.1.23) the summarizing evolutionary transfer equation for the total turbulized continuum entropy, $<S>+S_F$, follows:

$$\bar{\rho}\frac{D(<S>+S_F)}{Dt}=\partial_j\left\{\overline{J_{(S)j}}+J^T_{(S)j}+J_{(S_F)j}\right\}+\sigma, \tag{5.1.32}$$

where

$$0\le\sigma\equiv\sigma^i_{<S>}+\sigma^i_{(S_F)}+\sigma^e_{<S>}+\sigma^e_{(S_F)}=\sigma^i_{<S>}+\sigma^i_{(S_F)}+\sigma_{(<S>,S_F)}, \tag{5.1.33}$$

$$\sigma_{(<S>,S_F)}\equiv\sigma^e_{<S>}+\sigma^e_{(S_F)}=\left[(T_F-<T>)/T_F<T>\right]\Delta\,. \tag{5.1.34}$$

It is seen from (5.1.18), (5.1.25), and (5.1.34) that the production per unit volume of the total entropy, σ, in a turbulized multicomponent chemically active gas-phase continuum represents a bilinear form. It is composed of the generalized thermodynamic fluxes and their conjugated thermodynamic forces having a substantially different physical nature. Transport of heat, substance, momentum, and chemical reactions contribute jointly to this production. Thus, this entropy production, $\sigma_{(<S>,S_F)}\equiv\left[(T_F-<T>)/T_F<T>\right]\Delta$, describes irreversible dissipative processes caused by the exchange between the averaged and pulsating turbulent motions. The source of entropy related to chemical reactions is defined as the sum of the bilinear forms of the scalars $\bar{\xi}_s$ and $<A_s>$:

$$\sigma_{ch}=\frac{1}{<T>}\sum_{s=1}^{r}<A_s>\bar{\xi}_s\,. \tag{5.1.35}$$

The entropy production brought about by the averaged molecular and turbulent diffusion, is the sum of the scalar products of the following polar vectors:

$$\sigma^i_{dif}=\sum_{\alpha=1}^{N}\partial_j\left[\left(\frac{<\mu_\alpha>}{<T>}\right)-\frac{F_{\alpha j}}{<T>}\right](J^T_{\alpha j}+\bar{J}_{\alpha j})\,. \tag{5.1.36}$$

Similarly, the entropy production related to the transport of thermal energy by heat conduction and *turbulent diffusion*, is determined by the following product of polar vectors:

$$\sigma^i_{h.c.}=-(q^{T*}_j+\bar{q}_j)\frac{\partial_j<T>}{<T>^2}-J_{<e>j}\frac{\partial_j T_F}{T_F^2}\,. \tag{5.1.37}$$

Finally, the entropy production due to the viscous forces,

$$\sigma^{i}_{vis} = \frac{1}{T_F}(R_{ij}\partial_j\langle V_i\rangle + p_F\partial_j\langle V_j\rangle) + \frac{1}{\langle T\rangle}\overline{\pi}_{ij}\partial_j\langle V_i\rangle, \qquad (5.1.38)$$

is determined by the bilinear form of the viscous and turbulent stress tensors and the gradient of the mean mass flow velocity.

5.2. DEFINING RELATIONS FOR MULTICOMPONENT TURBULIZED MEDIA: THE THERMODYNAMIC APPROACH

In order to find the rheological relations (the *defining relations*) between the generalized thermodynamic fluxes and forces, it is possible to take advantage of the *Onsager formalism* of non-equilibrium thermodynamics beginning at the total entropy growth of a multicomponent turbulized continuum. It also allows to account for a mutual impact of the averaged molecular and specific turbulent features of the flow.

5.2.1. Linear Kinematic Constitutive Relations

Recall (see (3.2.12)) that the gradient of the averaged chemical potential $\langle\mu_\alpha\rangle$ in (5.1.14) has the following form for turbulent ideal mixtures:

$$\langle\mu_\alpha\rangle = \mu^0_\alpha(\overline{p}, \langle T\rangle) + \mathrm{k}\langle T\rangle \ln(\overline{n}_\alpha/\overline{n}) \quad (\alpha = 1,2,\ldots,N). \qquad (5.2.1)$$

where k is the Boltzmann constant. Let us express it in terms of the averaged thermodynamic parameters $\langle T\rangle$, $\overline{p}$, $x^*_\alpha(=\overline{n}_\alpha/\overline{n})$, where μ^0_α is the chemical potential of the proper α-component with temperature $\langle T\rangle$ and pressure $\overline{p}$. Then, the following bilinear form can be obtained for the volume origination rate σ of the total entropy, $(\langle S\rangle + S_F)$, in the thermodynamic complex under consideration:

$$0 \le \sigma \equiv \frac{1}{\langle T\rangle}\left\{J^{\Sigma*}_{qj}X^T_{0j} - \sum_{\alpha=1}^{N} J^{\Sigma}_{\alpha j}X^{T*}_{\alpha j} + p^v\partial_j\langle V_j\rangle + \overline{\overset{o}{\pi}_{ij}}\,\overset{o}{e}_{ij} + \sum_{s=1}^{r}\langle A_s\rangle\overline{\xi}_s\right\} +$$

$$+\frac{1}{T_F}\left\{-J_{\langle e\rangle j}\left(\frac{\partial_j T_F}{T_F}\right) + \overset{o}{R}_{ij}\,\overset{o}{e}_{ij}\right\} + \frac{T_F - \langle T\rangle}{T_F\langle T\rangle}\Delta. \qquad (5.2.2)$$

The following designations are introduced here for the generalized thermodynamic fluxes and their conjugated thermodynamic forces:

$$J^{\Sigma*}_{qj} \equiv q^{\Sigma*}_j - \sum_{\alpha=1}^{N}\langle h_\alpha\rangle J^{\Sigma}_{\alpha j}, \quad q^{\Sigma*}_j \equiv \overline{q}_j + q^{T*}_j = \overline{q}_j + q^T_j - \overline{p'V''_j}, \qquad (5.2.3)$$

$$\overline{\pi}_{ij} \equiv \overline{\overset{o}{\pi}_{ij}} - p^v \delta_{ij}, \quad p^v \equiv \tfrac{1}{3}\overline{\pi}_{kl}\delta_{kl}, \tag{5.2.4}$$

$$\overset{o}{R}_{ij} \equiv R_{ij} - \tfrac{1}{3}\left(R_{kl}\delta_{kl}\right)\delta_{ij} = R_{ij} + \tfrac{2}{3}\overline{\rho} < e > \delta_{ij}, \tag{5.2.5}$$

$$\overset{o}{e}_{ij} \equiv \tfrac{1}{2}\left(\partial_j < V_i > + \partial_i < V_j >_{,i}\right) - \tfrac{1}{3}\partial_k < V_k > \delta_{ij}, \quad e_{ij} \equiv \tfrac{1}{2}\left(\partial_i < V_j > + \partial_j < V_i >\right), \tag{5.2.6}$$

$$X^T_{0j} \equiv -\partial_j < T > / < T >, \tag{5.2.7}$$

$$X^{T*}_{\alpha j} \equiv -F_{\alpha j} + < T > \partial_j\left(\frac{< \mu_\alpha >}{< T >}\right) + < h_\alpha > \left(\frac{\partial_j < T >}{< T >}\right) =$$

$$= -F_{\alpha j} + \frac{\partial_j \overline{p}}{\overline{n}} + \mathrm{k} < T > \partial_j (\ln x^*_\alpha), \tag{5.2.8}$$

$$\Delta \equiv -\overline{p'\partial_j V''_j} + J^T_{(v)j}\partial_j \overline{p} - \sum_{\alpha=1}^{N} J^T_{\alpha j} F_{\alpha j} + \overline{\rho} < \varepsilon_e > . \tag{5.2.9}$$

The Onsager formalism of non-equilibrium thermodynamics allows us to find the linear constitutive relations between the thermodynamic fluxes and forces for three basic regions of a turbulent flow: the laminar sublayer; the buffer zone (an intermediate region where the effects of molecular and turbulent transport are comparable); and the region with a developed turbulent flow where $R_{ij} >> \overline{\pi}_{ij}$, $q^T_j >> \overline{q}_j$, etc.

The fluxes may be represented (see formula (2.2.1)) as the symmetric functions of the thermodynamic forces, $\Im_{\alpha i} = \sum_\beta \Lambda^{ij}_{\alpha\beta} X_{\beta j}$ $(\alpha, \beta = 1,2,..., N)$, when the situation is close to local equilibrium. For a turbulized continuum, in general, the matrix of the coefficients $\Lambda^{ij}_{\alpha\beta}$ depends not only on the averaged state variables of the medium (temperature, pressure, and concentrations) and the parameters describing its geometrical symmetry, but also on the correlation features of the turbulent velocity field, i.e. on the quantities $< e >$, $< \varepsilon_e >$, etc. As is seen from (5.2.2), the spectrum of possible cross effects expands for turbulent flows in multicomponent media as compared to laminar flows. For example, the averaged molecular diffusion fluxes, $\overline{J}_{\alpha j}$, are influenced additionally by the thermodynamic forces, $\partial_j(\ln T_F)$, responsible for the "diffusion" transport of turbulent kinetic energy by the flux $J_{<e>j}$.

However, at present there is no experimental evidence which could verify and describe quantitatively such cross effects. Besides, the contribution of any cross effect to the common thermodynamic process rate is an order of magnitude less as compared to the direct effects (*de Groot and Mazur, 1962*; *Dyarmati, 1974*). Keeping this in mind, in what follows we will require positiveness of the local production rates of the entropies $\sigma^i_{(S_F)}$, $\sigma^i_{<S>}$, $\sigma_{(<S>+S)}$ independent from each other. In addition, we assume that the turbulent chaotic subsystem does not render a noticeable influence on the respective linear relations inherent to the averaged molecular chaotic subsystem, such as, for instance, the relations between the symmetric part of the averaged viscous stress tensor, $\overline{\overset{o}{\pi}_{ij}}$, with zero trace and the viscous force tensor, $\overset{\circ}{e}_{ij}$. Also, a number of cross effects in the following rheological relations will be omitted, which is not stated explicitly.

As an example of the thermodynamic approach to deducing such kind of relations, we will focus on the detailed deduction of defining relations for small-scale turbulence. This type of turbulent flow usually has a tendency to relaxation to local statistical isotropy when the statistical properties of the turbulent flow are directionally independent. This systematic approach is easily extended to non-isotropic (large-scale) turbulence (*Marov and Kolesnichenko, 1987*).

Proceeding from the general tensor function theory (*Sedov, 1984*), the symmetry properties of isotropic media are totally characterized by the metric tensor, g^{ij}. This means that all tensors will be tensor functions of the metric tensor $\Lambda^{ij}_{\alpha\beta}=L_{\alpha\beta}g^{ij}$ $(\alpha,\beta=1,2,...N)$ only. Besides, owing to the absence of interference between fluxes and thermodynamic forces with various tensor dimensionalities in an isotropic system (the *Curie principle*), one may consider, for example, phenomena such as thermal conduction or diffusion as being described by polar vectors, irrespective of scalar and tensor phenomena. Two additional hypotheses inherent to non-equilibrium thermodynamics can be also adopted: first, the *Markovian behavior* of the system when instantaneous fluxes depend only on the generalized forces taken at the same instant, and second, the linearity of transfer processes when the fluxes are proportional to the thermodynamic forces. Then we obtain from (5.2.2) the following phenomenological relations in a rectangular coordinate frame $(g^{ij}\equiv\delta_{ij})$:

$$J^{\Sigma*}_{qj}=-L^{\Sigma}_{00}\partial_j(\ln<T>)-\sum_{\beta=1}^{N}L^{\Sigma}_{0\beta}X^{T*}_{\beta j}, \tag{5.2.10}$$

$$J^{\Sigma}_{\alpha j}=-L^{\Sigma}_{\alpha 0}\partial_j(\ln<T>)-\sum_{\beta=1}^{N}L^{\Sigma}_{\alpha\beta}X^{T*}_{\beta j}\quad(\alpha=1,2,...,N), \tag{5.2.11}$$

$$\overline{\overset{o}{\pi}_{ij}}=2\mu\overset{o}{e}_{ij},\quad p^{v}=\eta^{v}\partial_k<V_k>, \tag{5.2.12}$$

$$\overset{o}{R}_{ij}=2\mu^{T}\overset{o}{e}_{ij}, \tag{5.2.13}$$

$$\xi_s=\sum_{k=1}^{r}L^{Ch}_{sk}<A_k>\quad(s=1,2,...,r), \tag{5.2.14}$$

$$J_{<e>j}=-L^{F}_{00}\left(\partial_j T_F/T_F\right),\quad \Delta=L^{F}\frac{T_F-<T>}{T_F<T>}. \tag{5.2.15}$$

Here the kinetic coefficients L, η, μ are functions of the local state variables. As soon as the linear relations (5.2.10) and (5.2.11) are postulated, the *Onsager-Kasimir* theorem (*de Groot and Mazur, 1962*) asserts that

$$L^{\Sigma}_{\alpha\beta} = L^{\Sigma}_{\beta\alpha} \quad (\alpha,\beta = 1,2,\ldots,N)\,. \tag{5.2.16}$$

In addition, the relations

$$\sum_{\alpha=1}^{N} M_{\alpha} L^{\Sigma}_{\alpha 0} = 0, \qquad \sum_{\alpha=1}^{N} M_{\alpha} L^{\Sigma}_{\alpha\beta} = 0 \qquad (\beta = 1,2,\ldots,N) \tag{5.2.17}$$

follow from linear independence of the thermodynamic forces (see (2.3.11)). Now, taking into account the relations (5.2.4) and (5.2.12), for the averaged viscous stress tensor we obtain

$$\overline{\pi}_{ij} = \mu\left(\partial_j < V_i > + \partial_i < V_j >\right) - \left(\mu^{v} - \tfrac{2}{3}\mu\right)\delta_{ij}\partial_k < V_k >, \tag{5.2.18}$$

where μ and μ^{v} are the coefficients of viscosity and dilatational viscosity, respectively, which depend on the averaged parameters of the turbulent medium $<T>$, $\overline{\rho}$, $<Z_{\alpha}>$ $(\alpha = 1,2,\ldots,N)$.

We see that the defining relation for $\overline{\pi}_{ij}$ is obtained here directly using the methods of non-equilibrium thermodynamics, without the necessity to invoke respective analogues for the instantaneous values of the corresponding quantities. Similarly, using (5.2.13) and (5.1.28) the following, most complete form of the Reynolds tensor valid for isotropic turbulence is obtained:

$$R_{ij} = -\tfrac{2}{3}\overline{\rho} < e > \delta_{ij} + \mu^{T}\left(\partial_j < V_i > + \partial_i < V_j > - \tfrac{2}{3}\delta_{ij}\partial_k < V_k >\right), \tag{5.2.19}$$

where $\mu^{T} (= \overline{\rho}\nu^{T})$, ν^{T} are the scalar turbulent viscosity coefficient and the kinematic turbulent viscosity coefficient, respectively. However, let us emphasize that the rheological relation (5.2.19) for the Reynolds stresses is not the only one possible; indeed, if anisotropy of the turbulent field is taken into account, this relation becomes more complicated because the turbulent viscosity coefficient, μ^{T}, should be replaced by a 4-th rank tensor (see formula (3.3.19) and *Monin and Yaglom, 1971; Libby, 1975*).

5.2.2. Turbulent Diffusion and Heat Fluxes in Developed Turbulent Flows

The expression for the origination rate per unit volume of the total entropy, $\sigma_{(<S>+S_F)}$, can be simplified for the locally stationary state of the developed turbulent field. In this important case there is some intrinsic equilibrium in the turbulence structure when production and sink of the turbulization entropy are approximately balanced. As measurements of the turbulence energy budget in a homogeneous flow with a velocity gradient showed, this situation is observed, for instance, under the condition of forced and free convection in an atmospheric layer adjacent to the surface (*Lumley and Panofski, 1966; Van Miegham, 1973*).

Let $\sigma_{(S_F)} \approx 0$; then, by virtue of (5.1.25) and (5.1.26), we have

$$-J_{(S_F)}\partial_j T_F \approx -\overset{o}{R}_{ij}\overset{o}{e}_{ij} + J^T_{(v)j}\partial_j\overline{p} - \sum_{\alpha=1}^{N} J^T_{\alpha j}F_{\alpha j} - \overline{p'\partial_j V''_j} + \overline{\rho}<\varepsilon_e> \equiv -\overset{o}{R}_{ij}\overset{o}{e}_{ij} + \Delta \tag{5.2.20}$$

and expression (5.2.2) for the energy dissipation ($<T>\sigma$) in the developed turbulized flow where $R_{ij} >> \overline{\pi_{ij}}$, $J^T_{\alpha j} >> \overline{J_{ij}}$, etc., takes the form

$$0 \leq \sigma <T> \cong J^{T*}_{qj}X^T_{0j} - \sum_{\alpha=1}^{N} J^T_{\alpha j}X^{T*}_{\alpha j} - J_{(S_F)j}\partial_j T_F + \overset{o}{R}_{ij}\overset{o}{e}_{ij} + \sum_{s=1}^{r} <A_s> \overline{\xi}_s. \tag{5.2.21}$$

In this case it is possible to write the following constitutive kinematic relations for the turbulent diffusion flows, $J^T_{\alpha j}$ $(\alpha = 1,2,...,N)$, and heat flows, J^{T*}_{qj}, as well as for the turbulization entropy flux, $J_{(S_F)j}$:

$$J^{T*}_{qj} = L^T_{00}X^T_{0j} - \sum_{\beta=1}^{N} L^T_{0\beta}X^{T*}_{\beta j}, \tag{5.2.22}$$

$$J^T_{\alpha j} = L^T_{\alpha 0}X^T_{0j} - \sum_{\beta=1}^{N} L^T_{\alpha\beta}X^{T*}_{\beta j}, \quad (\alpha = 1,2,...,N), \tag{5.2.23}$$

$$J_{(S_F)j} = -L_{FF}T_{F,j}. \tag{5.2.24}$$

Here the scalar phenomenological coefficients L^T, depending on both the averaged state parameters $\overline{\rho}$, $<T>$, $<Z_\alpha>$ $(\alpha = 1,2,...,N)$ and the parameter $<e>$, describing the physical nature of the turbulent medium, comply with the Onsager-Kazimir symmetry conditions (5.2.16) and the conditions (5.2.17).

In order to maintain some further analogy with the respective expressions for the regular (laminar) flow patterns (see section 2.3), we introduce a new set of linearly dependent vectors, $d^T_{\beta j}$ $(\beta = 1,2,...,N)$, instead of the linearly independent diffusion forces, $X^{T*}_{\beta j}$. For this purpose we put

$$d^T_{\beta j} \equiv \frac{\overline{n}_\beta}{\overline{p}}(X^{T*}_{\beta j} + M_\beta R_j) \quad (^1), \qquad \sum_{\beta=1}^{N} d^T_{\beta j} = 0 \quad (^2), \tag{5.2.25}$$

where R_j is a certain unknown vector common for all components of the system, which is defined by formula (5.2.25 (^{2})). Then, in view of (5.2.8), defining the vectors $X^{T*}_{\beta j}$, we obtain the following expression for the generalized diffusion forces, $d^T_{\beta j}$, in a turbulized continuum:

$$d^T_{\beta j} \equiv \partial_j x^*_\beta + \left(x^*_\beta - M_\beta < Z_\beta >\right)\frac{\partial_j \overline{p}}{\overline{p}} + \frac{\overline{n}_\beta}{\overline{p}}\left[- F_{\beta j} + M_\beta \sum_{\alpha=1}^{N} < Z_\alpha > F_{\alpha j}\right], \qquad (5.2.26)$$

where $x^*_\alpha \equiv \overline{n}_\alpha / \overline{n}$. Comparing this with formula (2.3.15) shows that it is quite similar in its structure to the respective expression for the laminar regime of mixture motion.

Substituting now (5.2.25) into (5.2.22) and taking into account the relations (5.2.17), we obtain

$$J^{T*}_{qj} = L^T_{00} X^T_{0j} - \overline{p}\sum_{\beta=1}^{N}(1/\overline{n}_\beta)\, L^T_{0\beta} d^T_{\beta j}, \qquad (5.2.27)$$

$$J^T_{\alpha j} = L^T_{\alpha 0} X^T_{0j} - \overline{p}\sum_{\beta=1}^{N}\left(1/\overline{n}_\beta\right) L^T_{\alpha\beta} d^T_{\beta j} \quad (\alpha = 1,2,\ldots, N). \qquad (5.2.28)$$

Finally, using (5.1.20) for the reduced heat flux, J^{T*}_{qj}, and (5.2.7) for the thermodynamic force, X^T_{0j}, from the relations (5.2.27) and (5.2.28) we find

$$q^T_j = \overline{p' V''_j} + \sum_{\beta=1}^{N} < h_\beta > J^T_{\beta j} - \lambda^T_0 \partial_j < T > - \overline{p}\sum_{\beta=1}^{N} D^T_{t\beta} d^T_{\beta j}, \qquad (5.2.29)$$

$$J^T_{\alpha j} = -\overline{n}_\alpha D^T_{t\alpha} \partial_j (\ln < T >) - \overline{p} n_\alpha \sum_{\beta=1}^{N} D^T_{\alpha\beta} d^T_{\beta j} \quad (\alpha = 1,2,\ldots, N). \qquad (5.2.30)$$

In analogy to the formulas for laminar fluid flows (see (2.3.21)), the symmetric multicomponent eddy diffusion coefficients, $D^T_{\alpha\beta}$ $(\alpha, \beta = 1,2,\ldots, N)$, the thermal eddy diffusion coefficients, $D^T_{t\beta}$ $(\beta = 1,2,\ldots, N)$, and the turbulent heat conduction coefficient, λ^T_0, for multicomponent gases are introduced in (5.2.29) and (5.2.30) using the following definitions:

$$\begin{cases} \lambda^T_0 = L^T_{00} / < T >, \quad D^T_{t\beta} = L^T_{0\beta} / \overline{n}_\beta & (\beta = 1,2,\ldots, N), \\ D^T_{\alpha\beta} = D^T_{\beta\alpha} = \overline{p} L^T_{\alpha\beta} / \overline{n}_\alpha \overline{n}_\beta & (\alpha, \beta = 1,2,\ldots, N). \end{cases} \qquad (5.2.31)$$

The turbulent transport coefficients, $D^T_{\alpha\beta}$ and $D^T_{t\alpha}$, by virtue of (5.2.17), satisfy the relations

$$\sum_{\alpha=1}^{N} M_\alpha <Z_\alpha> D_{t\alpha}^T = 0, \ (^1) \quad \sum_{\alpha=1}^{N} M_\alpha <Z_\alpha> D_{\alpha\beta}^T = 0, \quad (\beta = 1,2,..., N). \ (^2) \qquad (5.2.32)$$

The derived rheological equations (5.2.29) and (5.2.30) for the fluxes of diffusion, $J_{\alpha j}^T$, and heat, q_j^T, describe heat and mass transfer in developed isotropic turbulent flows in a multicomponent gas mixture most completely. However, because of insufficient experimental data on the turbulent exchange coefficients, simpler models (see section 3.3) are often used at the current stage of development of a semi-empirical theory for multicomponent turbulence.

We end this section with a few notes. The rheological relations for other flow patterns in multicomponent turbulized mixtures can also be obtained in the framework of the developed thermodynamic approach. For example, the defining relations for both the total thermal fluxes, $q_j^{\Sigma*}$, and the diffusion fluxes, $J_{\alpha j}^{\Sigma}$, (specifically, in the form of the Stefan-Maxwell relations) and the total friction force stress, π_{ij}^{Σ}, (describing the total transport of heat, various substances, and momentum) can be deduced for flows in a buffer zone where the molecular and turbulent fluxes are of the same order of magnitude and the equality $<T> = T_F$ is fulfilled. Furthermore, modeling the proposed turbulent exchange coefficients, particularly the multicomponent turbulent diffusion coefficients, can be carried out within the domain of the *K-theory for multicomponent turbulence* (see section 4.3.9) with writing the complete system of additional evolutionary transfer equations for pair correlations of fluctuating thermohydrodynamic mixture parameters. A particular case of such an approach is examined in detail in Chapter 8.

5.3. THE STEFAN-MAXWELL RELATION AND THE HEAT FLUX FOR TURBULENT MIXTURES.

Like in the case of laminar mass and heat transfer patterns, it is convenient to reduce the defining relations for turbulent diffusion and heat fluxes to the form of the generalized Stefan-Maxwell relations comprising the binary turbulent diffusion coefficients. This is possible, in particular, due to the fact that, generally speaking, it is easier to take advantage of the empirical data available for the binary coefficients than using the data available for the multicomponent turbulent transport coefficients.

5.3.1. The Stefan-Maxwell Relation

To derive the generalized Stefan-Maxwell relations (see subsection 2.3.3) acceptable for turbulent multicomponent mixtures, let us first rewrite the equations (5.2.27) and (5.2.28) into the form

$$J_{qj}^{T*} - L_{00}^T X_{0j}^T = \sum_{\beta=1}^{N-1} L_{0\beta}^T \left[X_{\beta j}^T - \frac{M_\beta}{M_N} X_{Nj}^T \right], \qquad (5.3.1)$$

$$J^T_{\alpha j} - L^T_{\alpha 0} X^T_{0j} = \sum_{\beta=Б}^{N-1} L^T_{\alpha\beta}\left[X^T_{\beta j} - \frac{M_\beta}{M_N} X^T_{Nj}\right], \quad (\alpha = 1,2,..., N-1) \tag{5.3.2}$$

and resolve them with respect to the thermodynamic forces in terms of the fluxes J^{T*}_{qj} and $J^T_{\alpha j}$ $(\alpha = 1,2,..., N)$. Here $X^T_{\beta j}$ $(\beta = 1,2,..., N)$ are the thermodynamic forces, defined by the expressions

$$X^T_{\beta j} \equiv -\frac{\bar{p}}{\bar{n}_\beta} d^T_{\beta j}, \quad (^1) \qquad \sum_{\beta=1}^{N} <Z_\beta> X^T_{\beta j} = 0. \quad (^2) \tag{5.3.3}$$

This procedure is quite similar to the one performed for laminar flows in multicomponent mixtures when the relations (2.3.31) and (2.3.32) were resolved with respect to the thermodynamic forces in terms of the diffusion fluxes, $J_{\alpha j}$, and heat fluxes, J_{qj}. Continuing this analogy, we bring the final result:

$$X^T_{0j} = (1/A^T_{00}) J^{T*}_{qj} - \sum_{\alpha=1}^{N} (A^T_{0\alpha} / A^T_{00}) J^T_{\alpha j}, \tag{5.3.4}$$

$$-X^T_{\beta j} = A^T_{\beta 0} X^T_{0j} + \sum_{\alpha=1}^{N} A^T_{\beta\alpha} J^T_{\alpha j}, \qquad (\beta = 1,2,..., N). \tag{5.3.5}$$

Here the coefficients A^T are determined by the expressions

$$A^T_{00} = L^T_{00} - \sum_{\beta=1}^{N-1}\sum_{\gamma=1}^{N-1} L^T_{0\beta} \mathcal{M}^T_{\beta\gamma} L^T_{\gamma 0}, \tag{5.3.6}$$

$$A^T_{0N} = A^T_{N0} = a^T_0 M_N = -M_N \sum_{\alpha=1}^{N-1}\sum_{\gamma=1}^{N-1} < Z_\gamma > \mathcal{M}^T_{\gamma\alpha} L^T_{\alpha 0}, \tag{5.3.7}$$

$$A^T_{0\alpha} = A^T_{\alpha 0} = a^T_0 M_\alpha + \sum_{\beta=1}^{N-1} L^T_{0\beta} \mathcal{M}^T_{\beta\alpha} = \sum_{\beta=1}^{N-1} L^T_{\beta 0}\left[\mathcal{M}^T_{\beta\alpha} - M_\alpha \sum_{\gamma=1}^{N-1} < Z_\gamma > \mathcal{M}^T_{\gamma\beta}\right], \quad (\alpha = 1,2,..., N-1) \tag{5.3.8}$$

$$A^T_{\beta\alpha} = A^T_{\alpha\beta} = -\frac{M_\alpha M_\beta}{M_N} a^T_N + \sum_{\gamma=1}^{N-1} <Z_\gamma> \left(M_\alpha \mathcal{M}^T_{\gamma\beta} + M_\beta \mathcal{M}^T_{\gamma\alpha}\right) - \mathcal{M}^T_{\beta\alpha} \quad (\alpha, \beta = 1,2,..., N-1) \tag{5.3.9}$$

$$A^T_{N\alpha} = A^T_{\alpha N} = -a^T_N M_\alpha + M_N \sum_{\gamma=1}^{N-1} < Z_\gamma > \mathcal{M}^T_{\gamma\alpha}, \tag{5.3.10}$$

$$A^T_{NN} = -a^T_N M_N, \tag{5.3.11}$$

while the coefficients a^T are

$$a_0^T = -\sum_{\alpha=1}^{N-1}\sum_{\gamma=1}^{N-1} <Z_\gamma> \mathcal{M}_{\gamma\alpha}^T L_{\alpha 0}^T, \tag{5.3.12}$$

$$a_N^T = M_N \sum_{\beta=1}^{N-1}\sum_{\gamma=1}^{N-1} <Z_\beta><Z_\gamma> \mathcal{M}_{\beta\gamma}^T. \tag{5.3.13}$$

The elements $\mathcal{M}_{\beta\alpha}^T$ of the inverse matrix comply with the relation

$$\sum_{\alpha=1}^{N-1} \mathcal{M}_{\beta\alpha}^T L_{\alpha\gamma}^T = \delta_{\beta\gamma} = \begin{cases} 1, & \beta = \gamma \\ 0, & \beta \neq \gamma \end{cases}, \tag{5.3.14}$$

and symmetry of the coefficients $\mathcal{M}_{\beta\alpha}^T$ follows from the symmetry of the phenomenological coefficients $L_{\alpha\beta}^T$:

$$\mathcal{M}_{\alpha\beta}^T = \mathcal{M}_{\beta\alpha}^T \qquad (\beta = 1,2,\ldots, N-1). \tag{5.3.15}$$

In addition, the coefficients $A_{0\alpha}^T$ and $A_{\alpha\beta}^T$ satisfy the identities (compare to (2.3.49) and (2.3.52))

$$\sum_{\alpha=1}^{N} <Z_\alpha> A_{\alpha 0}^T = 0, \quad (^1) \qquad \sum_{\beta=1}^{N} <Z_\beta> A_{\beta\alpha}^T = 0, \quad (\alpha = 1,2,\ldots, N). \quad (^2) \tag{5.3.16}$$

Now let us transform the equations (5.3.5) into the form of the generalized Stefan-Maxwell relations for multicomponent diffusion in turbulent flows. For this purpose, we subtract expression (5.3.16 (^{2})) multiplied by $J_{\beta j}^T / <Z_\beta>$ from (5.3.5). This yields

$$-X_{\beta j}^T = A_{\beta 0}^T X_{0j}^T + \sum_{\substack{\alpha=1 \\ \alpha\neq\beta}}^{N} A_{\beta\alpha}^T \left(J_{\alpha j}^T - \frac{<Z_\alpha>}{<Z_\beta>} J_{\beta j}^T \right) \qquad (\beta = 1,2,\ldots, N), \tag{5.3.17}$$

or, using the conventional designations,

$$d_{\beta j}^T = \sum_{\substack{\alpha=1 \\ \alpha\neq\beta}}^{N} \frac{A_{\beta\alpha}^T}{\bar{p}} \left(\bar{n}_\beta J_{\alpha j}^T - \bar{n}_\alpha J_{\beta j}^T \right) - \frac{A_{\beta 0}^T \bar{n}_\beta}{\bar{p}} \frac{\partial_j <T>}{<T>} \qquad (\beta = 1,2,\ldots, N). \tag{5.3.18}$$

After that, it remains to show that

$$A_{\beta 0}^{T}=-\sum_{\alpha=1}^{N}\overline{n}_{\alpha}A_{\beta\alpha}^{T}\left[\frac{L_{0\alpha}^{T}}{\overline{n}_{\alpha}}-\frac{L_{0\beta}^{T}}{\overline{n}_{\beta}}\right],\quad (\beta=1,2,\ldots,N). \tag{5.3.19}$$

Using the relations (5.3.8) and (5.3.9) and identity (5.3.16), we find

$$\sum_{\alpha=1}^{N}\overline{n}_{\alpha}A_{\beta\alpha}^{T}\left[\frac{L_{0\alpha}^{T}}{\overline{n}_{\alpha}}-\frac{L_{0\beta}^{T}}{\overline{n}_{\beta}}\right]=\sum_{\alpha=1}^{N}A_{\beta\alpha}^{T}L_{0\alpha}^{T}-\frac{L_{0\beta}^{T}}{\overline{n}_{\beta}}\sum_{\alpha=1}^{N}\overline{n}_{\alpha}A_{\beta\alpha}^{T}=A_{\beta N}^{T}L_{0N}^{T}+\sum_{\alpha=1}^{N-1}A_{\beta\alpha}^{T}L_{0\alpha}^{T}=$$
$$=\sum_{\alpha=1}^{N-1}L_{0\alpha}^{T}\left[A_{\beta\alpha}^{T}-\frac{M_{\alpha}}{M_{N}}A_{\beta N}^{T}\right]=\sum_{\alpha=1}^{N-1}L_{0\alpha}^{t}\left[-\mathcal{M}_{\beta\alpha}^{T}+M_{\beta}\sum_{\gamma=1}^{N-1}<Z_{\gamma}>\mathcal{M}_{\gamma\alpha}^{T}\right]=-A_{0\beta}^{T}. \tag{5.3.20}$$

Then, substituting (5.3.19) into (5.3.18) and using the designations (5.2.31), we finally obtain

$$d_{\beta j}^{T}=\sum_{\substack{\alpha=1\\ \alpha\neq\beta}}^{N}\frac{A_{\beta\alpha}^{T}}{\overline{p}}\left(\overline{n}_{\beta}J_{\alpha j}^{T}-\overline{n}_{\alpha}J_{\beta j}^{T}\right)+\frac{\partial_{j}<T>}{<T>}\sum_{\substack{\alpha=1\\ \alpha\neq\beta}}^{N}\frac{\overline{n}_{\beta}\overline{n}_{\alpha}}{\overline{p}}A_{\beta\alpha}^{T}\left(D_{t\alpha}^{T}-D_{t\beta}^{T}\right), \tag{5.3.21}$$
$$(\beta=1,2,\ldots,N).$$

Further introducing the binary eddy diffusion coefficients, $\mathcal{D}_{\alpha\beta}^{T}$, and the thermal eddy diffusion ratios, $K_{t\beta}^{T}$, and using the formulas

$$\mathcal{D}_{\alpha\beta}^{T}=\mathrm{k}<T>/\overline{n}A_{\alpha\beta}^{T},\qquad (\alpha,\beta=1,2,\ldots,N),\quad (^{1})$$
$$K_{t\beta}^{T}=\overline{n}_{\beta}A_{\beta 0}^{T}/\overline{p},\qquad (\beta=1,2,\ldots,N).\quad (^{2}) \tag{5.3.22}$$

the following generalized form of the Stefan-Maxwell relations for developed turbulent multicomponent flows emerges from (5.3.21):

$$(1/\overline{n}^{2})\sum_{\substack{\alpha=1\\ \alpha\neq\beta}}^{N}\frac{\overline{n}_{\beta}J_{\alpha j}^{T}-\overline{n}_{\alpha}J_{\beta j}^{T}}{\mathcal{D}_{\alpha\beta}^{T}}=d_{\beta j}^{T}+K_{t\beta}^{T}\frac{\partial_{j}<T>}{<T>},\quad (\beta=1,2,\ldots,N), \tag{5.3.23}$$

where

$$d_{\beta j}^{T}\equiv\partial_{j}x_{\beta}^{*}+\left(x_{\beta}^{*}-M_{\beta}<Z_{\beta}>\right)\frac{\partial_{j}\overline{p}}{\overline{p}}+\frac{\overline{n}_{\beta}}{\overline{p}}\left[-F_{\beta j}+M_{\beta}\sum_{\alpha=1}^{N}<Z_{\alpha}>F_{\alpha j}\right].$$

These relations are similar in their structure to the Stefan-Maxwell relations (2.3.69), deduced with using thermodynamic (*Kolesnichenko and Tirskiy, 1976*) and gas-kinetic (*Marov and Kolesnichenko, 1987*) approaches for laminar fluids of plasma flows.

In analogy to the relations (2.3.64) and (2.3.65), it is also possible to show that the formulas

$$K^T_{t\beta} = \sum_{\alpha=1}^{N} \frac{x^*_\beta x^*_\alpha}{\mathcal{D}^T_{\alpha\beta}} \left(D^T_{t\beta} - D^T_{t\alpha}\right), \qquad (\beta = 1,2,\ldots,N) \tag{5.3.24}$$

are true. They link the thermal eddy diffusion ratios, $K^T_{t\beta}$, (see 5.3.23) and the thermal eddy diffusion coefficients, $D^T_{t\beta}$, introduced by formula (5.2.31) for multicomponent mixtures, and the formulas

$$\sum_{\alpha=1}^{N} K^T_{t\alpha} = 0, \quad (^1) \qquad \sum_{\beta=1}^{N} D^T_{\alpha\beta} K^T_{t\beta} = D^T_{t\alpha}, \quad (\alpha = 1,2,\ldots,N), \quad (^2) \tag{5.3.25}$$

which are appropriate for finding the parameters $K^T_{t\beta}$ in terms of the parameters $D^T_{\alpha\beta}$ and $D^T_{t\alpha}$ in case of turbulized media. Thus, the formulas (5.3.24) and (5.3.25) are universal in their character, i.e. they can be used when describing both laminar and turbulized regimes of multicomponent mixture motion.

At last, using the turbulent thermal diffusion ratios, $K^T_{t\beta}$, the expressions (5.2.30) for the turbulent diffusion fluxes, $J^T_{\alpha j}$, may be rewritten as

$$J^T_{\alpha j} = -\bar{n}_\alpha \sum_{\beta=1}^{N} D^T_{\alpha\beta} \left[d^T_{\beta j} + K^T_{t\beta} \frac{\partial_j <T>}{<T>} \right], \qquad (\alpha = 1,2,\ldots,N). \tag{5.3.26}$$

5.3.2. The Heat Flux in Turbulent Multicomponent Media

According to (5.1.20) and (5.2.29), the total heat flux in a mixture due to turbulent heat conduction and eddy diffusion, $q^T_j = q^{T*}_j + \overline{p'V''_j}$, is

$$q^{T*}_j = J^{T*}_{qj} + \sum_{\alpha=1}^{N} <h_\alpha> J^T_{\alpha j} = -\lambda^T_0 \partial_j <T> - \bar{p} \sum_{\alpha=1}^{N} D^T_{t\alpha} d^T_{\alpha j} + \sum_{\alpha=1}^{N} <h_\alpha> J^T_{\alpha j}. \tag{5.3.27}$$

From the equations (5.3.4), taking into account (5.3.22 (^{2})), it is possible to obtain another expression for the reduced turbulent heat flux, J^{T*}_{qj}, which complies with the Stefan-Maxwell relations (5.3.23), namely

$$J^{T*}_{qj} = A^T_{00} X^T_{0j} + \sum_{\alpha=1}^{N} A^T_{0\alpha} J^T_{\alpha j} = -\lambda^T \partial_j <T> + \bar{p} \sum_{\alpha=1}^{N} (K^T_{t\alpha} / \bar{n}_\alpha) J^T_{\alpha j}. \tag{5.3.28}$$

Here the proper turbulent heat conduction coefficient is designated in terms of

$$\lambda^T = A_{00} / <T>. \tag{5.3.29}$$

It is bound to the coefficient λ_0^T entered above (see (5.2.31)) by the relation

$$\lambda^T = \lambda_0^T - \mathrm{k} <T> \sum_{\alpha=1}^{N} K_{t\alpha}^T D_{t\alpha}^T = \lambda_0^T - \mathrm{k} <T> \sum_{\alpha=1}^{N} \sum_{\beta=1}^{N} K_{t\alpha}^T D_{\alpha\beta}^T K_{t\beta}^T. \tag{5.3.30}$$

The procedure for deriving (5.3.30) is similar to that for obtaining formula (2.3.74). Thus, the total heat flux in the turbulent multicomponent gaseous medium can finally be written as

$$q_j^T = \overline{p'V_j''} - \lambda^T \partial_j <T> + \mathrm{k} <T> \sum_{\alpha=1}^{N} (K_{t\alpha}^T / x_\alpha^*) J_{\alpha j}^T + \sum_{\alpha=1}^{N} <h_\alpha> J_{\alpha j}^T. \tag{5.3.31}$$

Except for the first addend, this relation, like for multicomponent diffusion in turbulent flows (5.3.23), is quite similar in its structure to the respective expression (2.3.75) for laminar mixture motion obtained by both thermodynamic and kinetic approaches for multicomponent gases. Let us emphasize that the proper turbulent heat conduction coefficient, λ^T, unlike the coefficient λ_0^T, can be measured experimentally in the steady state condition of mixture motion because in that case (and when the gas as a whole is at rest) all turbulent diffusion fluxes, $J_{\alpha j}^T$, vanish. In addition, it is worth to note that, in analogy to the system (2.3.80), it is also possible to write down the system of equations permitting to define the multicomponent eddy diffusion coefficients, $D_{\alpha\beta}^T$, in terms of the binary eddy diffusion coefficients, $\mathcal{D}_{\alpha\beta}^T$.

Summary

- *Based on the phenomenological turbulence theory for multicomponent chemically active gaseous continua, the thermodynamic approach was applied to close the hydrodynamic equations for the averaged motion at the level of the first order models. This approach allowed us to reveal more general expressions for the turbulent fluxes in multicomponent media as compared to those derived using the concept of mixing length. Representing a turbulized continuum as a thermodynamic complex composed of two subsystems: mean motion (the averaged molecular and turbulent chaos) and pulsative motion (the turbulent overstructure), enabled us to obtain rheological relations for the turbulent fluxes of diffusion, heat, and momentum using the methods of non-equilibrium thermodynamics. This approach extends the respective hydrodynamic results for homogeneous fluids to the case of multicomponent mixtures.*

- *The substantial form of the weighted-mean specific entropy balance for the subsystem of averaged motion in a turbulized continuum was obtained by means of averaging the fundamental Gibbs identity valid for micromotions in multicompo-*

nent mixtures. An explicit form for the averaged molecular and turbulent entropy fluxes, related to the respective diffusion and heat fluxes, was found. A similar form was also obtained for the local production rate of the averaged entropy, brought about by irreversible processes within the subsystem of averaged molecular chaos, and for the entropy exchange rate between the subsystems of pulsation and mean motion. The state parameters for the subsystem of turbulent chaos, such as temperature and turbulization pressure, were introduced by postulating the Gibbs identity. The evolutionary balance equation for turbulization entropy was analyzed and the expressions for the pulsation entropy fluxes, as well as for the local production and sink of entropy for the subsystem of turbulent chaos, were found. The refined rheological relations for the turbulent thermodynamic fluxes in multicomponent media were deduced using the evolutionary transfer equation for the total entropy of the turbulized continuum.

- *For small-scale multicomponent turbulence with a tendency to local statistical isotropy, when the statistical properties of the turbulent flow, and hence the turbulent exchange coefficients, are direction independent, the rheological relations were obtained. The developed procedure is advantageous in terms of the proposed thermodynamic approach to the first order closure problem and may be extended to non-isotropic large-scale turbulence and to the case when the influence of the integrated flow properties on the turbulence features at a point must be taken into account.*

- *For a locally stable turbulent field distinguished by some inner equilibrium in the turbulence structure, the Stefan-Maxwell relations for multicomponent diffusion and the respective expression for the heat flux in the turbulized continuum were derived. These relations ensure the most complete description of heat and mass transfer in the multicomponent medium, although due to the lack of experimental evidence simpler models are to be utilized.*

PART II

SOME MODEL PROBLEMS OF MULTICOMPONENT TURBULENCE

Explosive advancement of computer science involving the development of efficient numerical techniques provided broad opportunities for handling numerous complicated physical problems. These problems involve, in particular, the diverse models of mechanics of fluids including turbulized multicomponent gas mixtures where complex physicochemical processes occur. The outer gaseous envelopes of the planets, subjected to the direct impact of EUV and X-ray solar radiation, and primordial accretion discs giving birth to planets, serve as representative examples.

Here we will address the problems of setting and evaluating some planetary aeronomy and cosmogony models based on the principal concepts developed in the previous chapters. Special focus will be given to the problems of modeling the composition, dynamics, and thermal regime of those areas in the Earth's upper/middle atmosphere whose structure and variations are strongly affected, in addition to the radiation transfer, by turbulent mixing. We will also deal with the processes of formation and evolution of gas-dust subdisc around solar-type star involving particles coagulation and transfer in the heterogeneous turbulized medium. Unfortunately, because of the great complexity of the natural phenomena under study, the treatment of the drawn examples is rather limited. Nevertheless, they allow one to clarify ideas about specific approaches to solving the relevant modeling problems and the trend in the development of these particular fields.

CHAPTER 6

DIFFUSION PROCESSES IN THE THERMOSPHERE

The thermosphere is the very dynamical region of the planetary environment. The main problem posed by planetary aeronomy is how to describe most adequately the parameters responsible for the structure and variations of this region in the upper atmosphere (temperature, mass density, abundance of chemical components, wind velocity, etc.) depending on solar activity and various factors of solar-planetary interaction.

Two main paths of modeling are used to be pursued: developing semi-empirical models based on the experimental data available, and theoretical modeling based on the analysis of major physicochemical mechanisms responsible for the observed spatial-temporal distributions and pronounced variations of the atmospheric parameters. The latter approach incorporates in-depth study of photolysis, chemical kinetics, molecular and turbulent heat and mass transfer. Its advantage is the principal capability as to provide a physical interpretation of the solar and magnetic impact on the structure and composition of such a gaseous medium, and to forecast its behavior depending on the level of external disturbances is an ultimate goal. Great importance of the latter is clearly manifested, for example, by the necessity to target back to the ground properly large constructions deployed in near-space (like the *MIR* orbital complex or the *International Space Station*) after completing their operational lifetime.

Naturally, the main progress has been accomplished in studying the upper atmosphere of the Earth gearing its extension over other planets in the solar system. Numerous empirical models were developed and tested throughout about four 11-years cycles of solar activity within a broad range of heights, specifically those where orbits of satellites and manned orbital stations reside. The most comprehensive and world recognized are the models developed under auspice of the Committee on Space Research (*COSPAR*), referred to as *COSPAR International Reference Atmospheres* (*CIRA*) based on the continuously updating data storage.

In parallel, diverse attempts to describe the Earth's upper atmosphere as a specific gaseous medium based on both hydrodynamic and gas-kinetic approaches have been undertaken, involving detailed investigation of photolysis mechanisms and accompanying aeronomic reactions (see, e.g., *Nicolet, 1963; Banks and Kockarts, 1973; McEwan and Phillips, 1975; Whitten and Poppoff, 1975; Koshelev et al., 1983; Chamberlain and Hunten, 1987; Brasseur and Solomon, 1987; 1990; Marov and Kolesnichenko, 1987; Namgaladze et al., 1990; Marov et al., 1996;1997*). Since the data on the inner and outer planets accumulated, the basic ideas on both physical and chemical aeronomy, rooted in originally terrestrial studies, were significantly extended (see, e.g., *Krasnopolsky, 1987; Nagy et al., 1983; Atreya et al., 1984; Russell, 1991; Barth et al., 1992; Bishop et al. 1995; Kasprzak et al. 1997; Marov et al., 1997; Marov and Grinspoon, 1998*).

One problem of specific importance that we will address is coupling the temporal-spatial distribution of atmospheric species and the key chemical processes with account for dynamic transport. Historically, a simplified approach to solving this problem was one-dimensional modeling with averaging in latitude and longitude, the vertical component of the hydrodynamic velocity vector affecting the atmospheric composition as strongly as diffusion and horizontal transport. Therefore, the vertical transport of a substance driven by wind (advection) and turbulence was described as vertical turbulent diffusion with some effective eddy diffusion coefficient D^T that accounts for atmospheric dynamics. As was earlier mentioned, it was introduced as a fitting parameter to ensure the best agreement of calculations with the data available.

Critical assessment of such an approach seems expedient when applying the results of studying multicomponent turbulent media we obtained in Part 1 to the pertinent regions of a planetary upper atmosphere. In particular, a poor support is revealed for some assumptions adopted in numerically modeling mass transfer (*Atmosphere. A Handbook, 1991*). More specifically, with the utilized expressions for the turbulent diffusion fluxes the total substance flux does not become zero, in other words, condition (3.1.29) is not fulfilled. On the other hand, the calculations of molecular mass transfer are performed using either the simplest formulas (first suggested by *Collegrove et al., 1966*) or defining relations for the molecular diffusion fluxes with asymmetrical diffusion coefficients, $D_{\alpha\beta}$ (unsuccessfully introduced, as we claimed earlier, by *Hirschfelder et al., 1961*). However, asymmetry of the coefficients $D_{\alpha\beta}$ does not comply with the fundamental Onsager reciprocity relation in non-equilibrium thermodynamics (see section 2.2), though such a concordance is of key importance for modeling heat and mass transfer in the actual polyatomic, chemically active mixtures of atmospheric gases (*Curtiss, 1968*). Unfortunately, the results of the kinetic theory for monoatomic nonreacting gases are often imprudently used for these purposes. This is why it is necessary to study in more detail diffusive transfer in a stratified atmosphere, which covers, in addition to the diffusion itself, also heat conduction and thermal diffusion.

6.1. DIFFUSIVE TRANSFER IN ATMOSPHERIC MULTICOMPONENT GAS MIXTURES

As an example, we will focus here on our results of modeling the terrestrial atmosphere in the height range of 70–400 km allowing to describe quite accurately heat and mass transfer. The model is based on the hydrodynamic equations for mixtures and also incorporates Stefan-Maxwell relations for multi-component molecular diffusion (obtained thermodynamically in section 2.3) and rheological relations for the turbulent diffusion and heat fluxes (obtained in section 3.3).

6.1.1. Molecular Heat and Mass Transfer

Molecular diffusion considerably contributes to transferring a substance in a planetary thermosphere where a positive temperature lapse rate occurs and lighter components eventually begin to dominate due to diffusive separation of gases in the gravitational field (see Chapter 1). In multicomponent gaseous media molecular diffusion can be described on the basis of either the generalized Fick law (2.3.19) or the Stefan-Maxwell relations (2.3.69). Here we will consider these two interdependent approaches taking into account some restrictions imposed by the specificity of the aeronomic problems.

In accordance with the basic hypothesis of non-equilibrium thermodynamics (the existence of local linear dependence between the fluxes and their conjugated thermodynamic forces, see subsection 2.2.1), the macroscopic theory of heat and mass transfer leads to the following defining relations for the diffusion rates, $w_{\alpha j} = V_{\alpha j} - V_j \quad (= J_{\alpha j} / n_\alpha)$, of the individual mixture components:

$$w_{\alpha j} = -\sum_{\beta=1}^{N} D_{\alpha\beta}\left[d_{\beta j} + K_{T\beta}\left(\frac{\partial_j T}{T}\right)\right], \qquad (\alpha = 1,2,\ldots,N). \tag{6.1.1}$$

Here, as earlier (see (2.3.23*), (2.3.65) and (2.3.67))

$$d_{\beta j} = \partial_j x_\beta + (x_\beta - c_\beta)\frac{\partial_j p}{p} + \frac{\rho c_\beta}{p}\left[-\frac{1}{M_\beta}F_{\beta j} + \sum_{\alpha=1}^{N}\frac{c_\alpha}{M_\alpha}F_{\alpha j}\right] \tag{6.1.2}$$

are the thermodynamic diffusion force vectors; $K_{T\beta}$ are the thermal diffusion ratios, defined uniquely from solving the system of algebraic equations

$$\sum_{\alpha=1}^{N} K_{T\alpha} = 0, \qquad \sum_{\beta=1}^{N} D_{\alpha\beta}K_{T\beta} = D_{T\alpha}, \qquad (\alpha = 1,2,\ldots,N); \tag{6.1.3}$$

$D_{\alpha\beta}$, $D_{T\alpha}$ are the multicomponent and thermal diffusion coefficients, respectively; $x_\beta = n_\beta / n$, $c_\beta = M_\beta Z_\beta (= M_\beta n_\beta / \rho)$ are the mole fraction and mass concentration of any mixture component β ($\sum_{\beta=1}^{N} c_\beta = 1$); $V_{\beta j}$ is the mean particle velocity (of species β); and $n = \sum_{\beta=1}^{N} n_\beta$, $p = \mathrm{k}Tn$, V_j are the total number density, pressure and hydrodynamic velocity of the N-component continuum, respectively.

The multicomponent diffusion coefficients, $D_{\alpha\beta}$, and thermal diffusion coefficients, $D_{T\alpha}$, are linearly dependent (see (2.3.24)), i.e.

$$\sum_{\beta=1}^{N} c_\beta D_{T\beta} = 0, \quad (^1) \qquad \sum_{\beta=1}^{N} c_\beta D_{\alpha\beta} = 0, \quad (\alpha = 1,2,\ldots,N). \quad (^2) \tag{6.1.4}$$

In addition, the coefficients $D_{\alpha\beta}$ are defined as symmetric quantities with respect to rearranging indices: $D_{\alpha\beta} = D_{\beta\alpha}$ $(D_{\alpha\alpha} > 0)$ (see (2.3.21)). Therefore, there are only $\frac{1}{2}N(N+1)$ independent multicomponent diffusion coefficients for the N-component mixture.

We also defined the mean mass hydrodynamic flow velocity as $V_j = \sum_{\beta=1}^{N} c_\beta V_{\beta j}$ and therefore, the diffusion rates, $w_{\beta j}$, satisfy the condition

$$\sum_{\beta=1}^{N} c_\beta w_{\beta j} = 0. \tag{6.1.5}$$

This means that there is no total momentum in the system due to mutual molecular diffusion.

In view of the hydrostatic equation $\partial p / \partial z = -\rho g$, the formulas (6.1.1) and (6.1.2) can be transformed into the following form for stratified atmospheres (compare to (2.3.96)):

$$w_{\alpha z} = -\frac{1}{n}\sum_{\beta=1}^{N} D_{\alpha\beta}\left[\frac{\partial n_\beta}{\partial z} + (1+\alpha_{T\beta})\frac{n_\beta}{T}\frac{\partial T}{\partial z} + \frac{n_\beta}{H_\beta}\right], \tag{6.1.6}$$

where $\alpha_{T\beta} \equiv nK_{T\beta}/n_\beta$ is the thermal diffusion factor; $H_\beta \equiv \mathrm{k}T/M_\beta g$ is the local scale height for the β-component, g is the acceleration due to gravity, and $z(\equiv x_3)$ is the height. Let us note that the expressions for the diffusion rates, $w_{\alpha z}$, used in some models for the thermospheric composition (*Atmosphere. A Handbook, 1991; Vlasov and Davydov, 1982; Bryunelli and Namgaladze, 1988*) are incorrect because the factor $(1/n_\alpha)$ instead of $(1/n)$ is entered before the summation sign in (6.1.6).

Using the methods from kinetic theory for heat and mass transfer we have obtained (*Marov and Kolesnichenko, 1987*) the expressions for the thermal diffusion ratios, $K_{T\alpha}$, as the ratios of determinants with composite elements. These expressions permit to calculate the ratios in the first approximation of the *Chapman-Enskog* theory in terms of the full bracket integrals Λ – the key quantities in the kinetic theory for rarefied gases, i.e. without preliminary evaluating the molecular exchange coefficients, $D_{\alpha\beta}$ and $D_{T\alpha}$. The expressions for Λ, being algebraic functions of the reduced integrals of all pair collisions of particles (Ω^*-integrals, see formula (2.3.88)), are available up to the 4-th approximation (*Hirschfelder et al., 1961*).

The multicomponent diffusion coefficients, $D_{\alpha\beta}$, in (6.1.1) can be found, in general, from solving the system of algebraic equations (see (2.3.80))

$$\sum_{\substack{\beta=1\\ \beta\neq\alpha}}^{N}\left[\frac{x_\alpha x_\beta}{\mathcal{D}_{\alpha\beta}} + \sum_{\substack{\delta=1\\ \delta\neq\alpha}}^{N}\frac{M_\beta}{M_\alpha}\frac{x_\alpha x_\delta}{\mathcal{D}_{\alpha\delta}}\right]D_{\gamma\beta} = c_\alpha - \delta_{\gamma\alpha}, \quad (\alpha,\gamma = 1,2,\ldots,N), \tag{6.1.7}$$

linking the coefficients $D_{\alpha\beta}$ with the diffusion coefficients, $\mathcal{D}_{\alpha\beta}$, of binary mixtures. In the particular case of a three component mixture, equation (6.1.7) allows one to define the coefficients $D_{\alpha\beta}$ as

$$D_{11} = \frac{n^2}{n_1\rho^2}\frac{n_1 n_3 M_3^2 \mathcal{D}_{23}\mathcal{D}_{31} + n_1 n_2 M_2^2 \mathcal{D}_{12}\mathcal{D}_{23} + (\rho-\rho_1)^2 \mathcal{D}_{31}\mathcal{D}_{12}}{n_1\mathcal{D}_{23} + n_2\mathcal{D}_{31} + n_3\mathcal{D}_{12}},$$

$$D_{22} = \frac{n^2}{n_2\rho^2}\frac{n_2 n_3 M_3^2 \mathcal{D}_{13}\mathcal{D}_{32} + n_1 n_2 M_1^2 \mathcal{D}_{12}\mathcal{D}_{13} + (\rho-\rho_2)^2 \mathcal{D}_{21}\mathcal{D}_{32}}{n_1\mathcal{D}_{23} + n_2\mathcal{D}_{31} + n_3\mathcal{D}_{12}},$$

$$D_{33} = \frac{n^2}{n_3\rho^2}\frac{n_1 n_3 M_1^2 \mathcal{D}_{21}\mathcal{D}_{31} + n_3 n_2 M_2^2 \mathcal{D}_{12}\mathcal{D}_{23} + (\rho-\rho_3)^2 \mathcal{D}_{32}\mathcal{D}_{13}}{n_1\mathcal{D}_{23} + n_2\mathcal{D}_{31} + n_3\mathcal{D}_{12}},$$

$$D_{12} = \frac{n^2}{\rho^2}\frac{n_3 M_3^2 \mathcal{D}_{23}\mathcal{D}_{31} - M_2(\rho_1+\rho_3)\mathcal{D}_{12}\mathcal{D}_{23} - M_1(\rho_2+\rho_3)\mathcal{D}_{31}\mathcal{D}_{12}}{n_1\mathcal{D}_{23} + n_2\mathcal{D}_{31} + n_3\mathcal{D}_{12}},$$

$$D_{13} = \frac{n^2}{\rho^2}\frac{n_2 M_2^2 \mathcal{D}_{23}\mathcal{D}_{21} - M_3(\rho_1+\rho_2)\mathcal{D}_{13}\mathcal{D}_{23} - M_1(\rho_1+\rho_3)\mathcal{D}_{21}\mathcal{D}_{13}}{n_1\mathcal{D}_{23} + n_2\mathcal{D}_{31} + n_3\mathcal{D}_{12}},$$

$$D_{32} = \frac{n^2}{\rho^2}\frac{n_1 M_1^2 \mathcal{D}_{13}\mathcal{D}_{21} - M_3(\rho_1+\rho_2)\mathcal{D}_{32}\mathcal{D}_{13} - M_2(\rho_1+\rho_3)\mathcal{D}_{21}\mathcal{D}_{32}}{n_1\mathcal{D}_{23} + n_2\mathcal{D}_{31} + n_3\mathcal{D}_{12}}. \qquad (6.1.8)$$

Obviously, these formulas for the coefficients $D_{\alpha\beta}$ considerably differ from the asymmetrical expressions used for the same coefficients elsewhere (*Vlasov and Davydov, 1982*).

However, in general, in particular when the mixture is composed of more than four or five components, the system of combined equations (6.1.7) is difficult to use for determining the multicomponent diffusion coefficients, $D_{\alpha\beta}$, because direct matrix inversion in (6.1.7) requires about N^3 arithmetical operations per difference node (*Oran and Boris 1990*). In addition, the system of differential equations obtained after substituting the rates $w_{\alpha j}$ from (6.1.1) into the respective diffusion equations (2.1.58) for species abundance, proves to be non-resolvable with respect to higher derivatives. As is known, numerical realization of such a system poses additional difficulties. Therefore, when dealing with aeronomic problems, it is more convenient to have the defining relations (6.1.1) solved with respect to the diffusion rates, $w_{\alpha j}$, via thermodynamic diffusion forces, $d_{\alpha j}$. They are written in the form of Stefan-Maxwell relations including the diffusion binary gaseous mixture coefficients, $\mathcal{D}_{\alpha\beta}$, (2.3.86) instead of the multicomponent diffusion coefficients, $D_{\alpha\beta}$.

The Stefan-Maxwell Relations for multicomponent diffusion. Stefan-Maxwell relations (2.3.96) for mixtures stratified in a gravitational field can be rewritten as

$$\frac{\partial n_\beta}{\partial z} = -\frac{n_\beta}{H_\beta} - \left(1+\alpha_{T\beta}\right)\frac{n_\beta}{T}\frac{\partial T}{\partial z} + \sum_{\substack{\alpha=1\\ \alpha\neq\beta}}^{N}\frac{n_\alpha J_{\beta z} - n_\beta J_{\alpha z})}{n\mathcal{D}_{\alpha\beta}}, \quad (\beta = 1,2,\ldots,N). \qquad (6.1.9)$$

These are the basic relations to study multicomponent molecular diffusion in a planetary upper atmosphere. Because the resistance coefficients matrix, $R_{\alpha\beta} = (x_\alpha x_\beta / \mathcal{D}_{\alpha\beta})$, defined by the equations (6.1.9) is degenerate (see (2.3.52) and (2.3.59)), only a part of the N different diffusion fluxes are independent. The condition

$$\sum_{\beta=1}^{N} M_\beta J_{\beta z} = 0 \tag{6.1.5*}$$

serves as the necessary supplementary equation.

Basically, directly solving the Stefan-Maxwell relations (6.1.9) for the diffusion fluxes, $J_{\alpha z}$, leads to the same difficulties as solving the algebraic equations (6.1.7) for the coefficients $D_{\alpha\beta}$. The problem is that the number of relevant arithmetical operations increases proportionally to $(N-1)^3$ because of the necessity to inverse matrices with dimensionality $(N-1)^2$. This virtually excludes the possibility to use the relations (6.1.9) without additional modification when dealing with a mixture of atmospheric gases composed of numerous species $(N \geq 10)$.

Therefore, to determine the diffusion fluxes from (6.1.9) computational procedures which do not require matrix inversion (for example, the method of successive approximations) are the most convenient (*Tirskiy, 1963*). Similar iterative procedures, being applied to the problem of calculating the thermospheric composition, can be performed using the following form of Stefan-Maxwell relations (6.1.9):

$$J_{\alpha z} = -\rho D_\alpha \frac{\partial}{\partial z}\left(\frac{n_\alpha}{\rho}\right) + \delta J_{\alpha z}, \qquad \frac{1}{D_\alpha} = \sum_{\beta=1}^{N} \frac{x_\beta}{\mathcal{D}_{\alpha\beta}},$$

$$\delta J_{\alpha z} \equiv n_\alpha D_\alpha \left[\frac{1}{n} \sum_{\substack{\beta=1 \\ \beta\neq\alpha}}^{N} \left(\frac{1}{\mathcal{D}_{\alpha\beta}} - \frac{M_\beta}{M_\alpha \mathcal{D}_{\alpha\alpha}} \right) J_{\beta z} - \frac{\partial \ln \mathrm{M}}{\partial z} + \left(1 - \frac{M_\beta}{\mathrm{M}}\right)\frac{1}{H} - \alpha_{T\beta} \frac{\partial \ln T}{\partial z} \right], \tag{6.1.10}$$

where $H \equiv \mathrm{k}T/\mathrm{M}g$ is the local scale height for the mixture as a whole, $\mathrm{M} \equiv \rho/n$ is the mean molecular mass of the mixture, obtained from (6.1.9) using the condition of the total mixture component momentum being zero (6.1.5*). According to (6.1.10), the diffusion flux of the α-component depends on the composition and diffusion fluxes of other components; however, this circumstance is not essential because methods of successive approximations are frequently used for solving the problem. This is followed by the numerical calculation with formula (6.1.10) using the *Gauss-Seidel* method; then only one iteration at each time step is performed for solving such a stationary problem using the relaxation method (*Samarsky, 1977, 1978; Lapin et al., 1984*).

Let us note that this method is essentially equivalent to calculating the diffusion fluxes, $J_{\alpha z}$, based on the widely used generalized *Fick law*. The latter is true, however, only in the approximation of minor admixture diffusion with the effective *Wilkey* diffusion coefficient,

$$J_{\alpha z} = -\rho D_\alpha \frac{\partial}{\partial z}\left(\frac{n_\alpha}{\rho}\right), \qquad \frac{1}{D_\alpha} = \sum_{\beta=1}^{N} \frac{x_\beta}{\mathcal{D}_{\alpha\beta}}, \qquad (\alpha = 1,2,...,N). \tag{6.1.11}$$

In contrast, our procedure is more advantageous because it allows us, in particular, to keep the fulfillment of equality (6.1.5*) and therefore, to avoid a noticeable violation of the integral conditions of mass balance for the individual mixture components.

Molecular Transport of Energy by Thermal Conduction. Molecular thermal conduction is an important mechanism to redistribute thermal energy in the thermosphere where its role in forming the temperature height profile is of particular importance. The total molecular heat flux in a multicomponent mixture defined by relation (2.3.75), can be written for a stratified atmosphere as

$$q_z = -\lambda \frac{\partial T}{\partial z} + p \sum_{\alpha=1}^{N} K_{T\alpha} w_{\alpha z} + \sum_{\alpha=1}^{N} h_\alpha n_\alpha w_{\alpha z}, \tag{6.1.12}$$

where $\lambda(T, n_1, ..., n_N)$ is the molecular heat conductivity coefficient that depends on the temperature and gas composition. The coefficient λ may be evaluated for a mixture of polyatomic atmospheric gases with rather good accuracy using the following formulas (*Maison and Saxena, 1958*):

$$\lambda = \sum_{\alpha=1}^{N} \lambda_\alpha \boldsymbol{Eu}_\alpha \left(1 + 1.065 \sum_{\substack{\beta=1 \\ \beta \neq \alpha}}^{N} \frac{x_\beta}{x_\alpha} G_{\alpha\beta}^{\lambda} \right)^{-1}, \tag{6.1.13}$$

$$\lambda_\alpha = \frac{25\sqrt{\pi M_\alpha \mathrm{k} T}}{32\pi\sigma_\alpha^2 \Omega_\alpha^{(2,2)*}} \frac{3\mathrm{k}}{2M_\alpha}, \qquad \boldsymbol{Eu}_\alpha = 0.115 + 0.354 \frac{C_{p\alpha}}{\mathrm{k}}, \tag{6.1.14}$$

$$G_{\alpha\beta}^{\lambda} = \frac{1}{4}\left(\frac{2M_\beta}{M_\alpha + M_\beta}\right)^{1/2} \left[1 + \left(\frac{M_\alpha}{M_\beta}\right)^{1/4} \left(\frac{\lambda_\alpha}{\lambda_\beta}\right)^{1/2}\right]^2. \tag{6.1.15}$$

Here $\boldsymbol{Eu}_\alpha$ is the *Eiken* correction for the heat conductivity coefficient for a pure λ_α – gas taking into account the internal energy of real multicomponent molecules; note that the same designations are used in the formulas (6.1.14) and subsection 2.3.6.

The derivative

$$Q_T \equiv -\frac{\partial}{\partial z}\left(q_z - \sum_{\alpha=1}^{N} h_\alpha n_\alpha w_{\alpha z}\right) = -\frac{\partial}{\partial z}\left(\lambda \frac{\partial T}{\partial z} - p \sum_{\alpha=1}^{N} K_{T\alpha} w_{\alpha z}\right), \tag{6.1.16}$$

defining the input or sinking of thermal energy due to molecular heat conductivity (see the energy equation (3.1.78)), is positive in the lower terrestrial thermosphere at heights of 90–120 km. At the same time, the molecular heat conductivity promotes cooling of the thermosphere ($Q_T < 0$) at heights larger than 160 km.

6.1.2. Turbulent Multicomponent Fluxes of Mass and Energy

Turbulent mixing along with molecular diffusion plays an important role in the mesosphere and the lower thermosphere of the Earth below about 120 km. The mechanisms of turbulent diffusion and heat conduction operating here impact the height profiles of composition and temperature in the entire upper atmosphere. Turbulence is influenced by various factors specific to the upper atmosphere, such as its multicomponent gas composition, the mass density drop with height by several orders of magnitude, multiple aeronomic reactions affecting the energy budget, etc.. This limits the opportunity to take advantage of the results obtained in studying turbulent flows of homogeneous fluid in a gravitational field (*Monin and Yaglom, 1971*) and to apply them to modeling mass and heat transfer in the upper atmosphere.

In Chapter 3 the system of hydrodynamic equations for the averaged values of the thermohydrodynamic parameters has been obtained using weighted-mean Favre averaging, which is pertinent for adequately modeling planetary gas envelopes. The rheological relations for the turbulent fluxes of diffusion, $J^T_{\alpha z}$, and heat , q^T_z, derived in section 3.3 allow to close the system.

When utilizing the hydrodynamic approach to describe the dynamic and thermal state of such a medium, the spatial modeling scale, $L_{hydr} = |\nabla \ln \theta|^{-1}$, for any structural parameter θ, is usually taken about one kilometer in the vertical and hundreds of kilometers in the horizontal direction. Hence the motions at considerably smaller spatial scales are treated as diffusion resulting in turbulent mixing, rather than ordered motions within the respective scales.

Turbulent Diffusion Fluxes. In the one-dimensional approximation, expression (3.3.3) for the turbulent diffusion flux, $J^T_{\alpha z}$, becomes

$$J^T_{\alpha z} = -\bar{\rho} D^T \frac{\partial}{\partial z}\left(\frac{\overline{n}_\alpha}{\bar{\rho}}\right), \quad (\alpha = 1,2,\ldots, N), \tag{6.1.17}$$

where the fluxes $J^T_{\alpha z}$ are linearly dependent (see (3.1.29)):

$$\sum_{\alpha=1}^{N} M_\alpha J^T_{\alpha z} = 0. \tag{6.1.18}$$

Here D^T is the eddy diffusion coefficient in the vertical direction. The expression for $J^T_{\alpha z}$ may be transformed in view of the averaged equation of state for multicomponent mixtures (3.2.2) and hydrostatics equation (3.3.4) to the following form convenient for numerical calculations:

$$J^T_{\alpha z} = -\overline{\rho} D^T \frac{\partial}{\partial z}\left(\frac{\overline{n}_\alpha}{\overline{\rho}}\right) = -D^T \left\{ \frac{\partial \overline{n}_\alpha}{\partial z} + \overline{n}_\alpha \left(\frac{1}{<H>} + \frac{1}{<T>}\frac{\partial <T>}{\partial z} - \frac{1}{M^*}\frac{\partial M^*}{\partial z} \right) \right\}, \tag{6.1.19}$$

where $<H> \equiv \mathrm{k} <T>/M^* g = \overline{p}/\overline{\rho} g$ is the weighted-mean value of the local scale height for the atmosphere as a whole. Obviously, the last term in (6.1.19) is of great importance for the turbopause where the averaged molecular mass, $M^* = \overline{\rho}/\overline{n}$, strongly varies with height, though this term was omitted elsewhere (*Lettau, 1951; Atmosphere. A Handbook, 1990*). Recall that the eddy diffusion coefficient D^T, unlike its molecular analog D_α, is not an inherent feature but depends both on the choice of the spatial averaging scale of the turbulent field and on the vertical hydrodynamic transport.

Turbulent Heat Fluxes. Below about 105 km, heating of atmospheric gas by absorbed solar radiation and the chemical processes initiated by this absorption, is indemnified by turbulent heat conduction. The total thermal energy flux in the multicomponent mixture, arising due to the correlation between the specific enthalpy fluctuations and the mean-mass-flow velocity and transferred by turbulence, can be written for the stratified atmosphere as (see (3.3.15*))

$$q^T_z = \overline{p' V''_z} + \sum_{\alpha=1}^{N} <h_\alpha> J^T_{\alpha z} - \lambda^T \left(\frac{\partial <T>}{\partial z} + \frac{g}{<C_p>} \right), \tag{6.1.20}$$

where $\lambda^T = \overline{\rho} <C_p> D^T$ is the turbulent heat conduction coefficient in the vertical direction. According to equation (3.1.78) for the internal energy of an averaged turbulized continuum, the heating (cooling) rate at the expense of heat transport by turbulent heat conduction is described as

$$Q^T_T \equiv -\frac{\partial}{\partial z}\left(q^T_z - \overline{p' V''_z} - \sum_{\alpha=1}^{N} <h_\alpha> J^T_{\alpha z} \right) = \frac{\partial}{\partial z}\left[\overline{\rho} <C_p> D^T \left(\frac{\partial <T>}{\partial z} + \frac{g}{<C_p>} \right) \right]. \tag{6.1.21}$$

A similar formula was first used in early models of the terrestrial thermosphere (*Chandra and Sinha, 1974; Johnson, 1975*). Let us note that the adiabatic gradient $(\partial <T>/\partial z + g/<C_p>) > 0$ and the integral of Q^T_T with respect to height is always negative when the atmospheric stratification is stable; therefore, turbulent heat conduction lowers the thermosphere as a whole. Nevertheless, the quantity Q^T_T can be arbitrary in its sign at some heights in the thermosphere. As follows from (6.1.21), because near the mesopause $|\partial <T>/\partial z| << g/<C_p>$, in this region the sign of Q^T_T is determined by the sign of the derivative $\partial(\overline{\rho} D^T)/\partial z$.

Turbulent Gas Heating. As discussed in Chapter 1, internal gravity waves are regarded as an important mechanism to generate turbulence in the mesosphere and the lower thermosphere. Viscous dissipation of turbulent energy should result in dynamically heating these regions, the heating rate being comparable to that supplied by the incident solar radiation.

Let us obtain the algebraic expression for the local heating rate, $Q^T_{<e>}$, of a multicomponent turbulent flow determined by dissipation of turbulent energy due to both the molecular viscosity and the work done by turbulent pressure fluctuations against the buoyancy forces. To this end, we will consider the local-equilibrium one-dimensional approximation of the evolutionary transfer equation for turbulent energy (4.2.28), which takes the following form:

$$\bar{\rho}\frac{D<e>}{Dt}+\frac{\partial}{\partial z}\left(\overline{\rho e V_z''}+\overline{p'V_z'}-\overline{\pi_{iz}'V_i'}\right)\approx 0 =$$
$$= R_{iz}\frac{\partial}{\partial z}<V_i>+\overline{p'\partial V_z'/\partial z}-J^T_{(v)z}\frac{\partial \bar{p}}{\partial z}-\bar{\rho}\varepsilon_e. \qquad (6.1.22)$$

Here $J^T_{(v)z}$ is the turbulent flux in a specific volume in the vertical direction defined by relation (4.2.28)

$$J^T_{(v)z}=\frac{<R^*>}{<C_p>\bar{p}}J^{T*}_{qz}+\frac{\mathrm{k}<T>}{\bar{p}}\sum_{\alpha=1}^{N}J^T_{\alpha z}, \qquad (6.1.23)$$

where

$$J^{T*}_{qz}=q^T_z-\overline{p'V_z''}-\sum_{\alpha=1}^{N}<h_\alpha>J^T_{\alpha z}=-\lambda^T\left(\frac{\partial<T>}{\partial z}+\frac{g}{<C_p>}\right)$$

is the reduced heat flux (see (3.3.10)). From (6.1.22) it follows that within the local-equilibrium approximation which fulfills at some critical value of the dynamic Richardson number, $\boldsymbol{Rf}_c$, the energy supply rate from a shear wind, $R_{iz}\partial<V_i>/\partial z$, is equal to the turbulent energy dissipation rate, $\bar{\rho}\varepsilon_e$, plus the work done by turbulent pressure fluctuations against the buoyancy forces, $J^T_{(v)z}\,\partial\bar{p}/z$. It also follows from (6.1.22) that the turbulent heating rate of the atmospheric gas, $Q^T_{<e>}\equiv\bar{\rho}\varepsilon_e+J^T_{(v)z}\partial\bar{p}/\partial z-\overline{p'\partial V_z''/\partial z}$, (see the balance equation (3.1.78) for the averaged internal energy of a medium) can be written under stationary conditions as

$$Q^T_{<e>}\approx R_{iz}\partial<V_i>/\partial z, \qquad (i=1,2). \qquad (6.1.24)$$

When using the critical value of the dynamic Richardson number, $\boldsymbol{Rf}_c$, (formula (4.2.32))

$$\boldsymbol{Rf} = \frac{\left(\dfrac{<R^*>}{<C_p>\,\overline{p}}\right)\dfrac{\partial \overline{p}}{\partial z}J^{T*}_{qz}}{R_{iz}\dfrac{\partial}{\partial z}<V_i>} \cong \frac{\dfrac{g}{<T>}\left(\dfrac{\partial <T>}{\partial z}+\dfrac{g}{<C_p>}\right)}{\boldsymbol{Pr}^T\left[(\partial <V_x>/\partial z)^2+(\partial <V_y>/\partial z)^2\right]} = \frac{\boldsymbol{Ri}}{\boldsymbol{Pr}^T}, \tag{6.1.25}$$

the formula (6.1.24) can be rewritten as

$$Q^T_{<e>} = \frac{1}{\boldsymbol{Rf}_c}D^T\frac{g}{<T>}\left(\frac{\partial <T>}{\partial z}+\frac{g}{<C_p>}\right). \tag{6.1.26}$$

The quantities $\boldsymbol{Pr}^T$ and $\boldsymbol{Ri}_c$ needed to calculate $Q^T_{<e>}$ should be taken from measurements. For the terrestrial thermosphere $1/\boldsymbol{Pr}^T = 0.3$ (*Justus, 1967*) and $\boldsymbol{Ri}_c$ is estimated to vary from 0.66 to 2 (*Gordiets et al., 1976*). Similarly, when the influence of stratification of the atmospheric mixture composition on turbulence is taken into account and the critical value $\boldsymbol{Ko}_c$ of the dynamic Kolmogorov number (formula (4.2.34))

$$\boldsymbol{Kf} = \frac{M^* g\sum_{\alpha=1}^{N} J^T_{\alpha z}}{R_{iz}\partial <V_i>/\partial z} \cong \frac{g\dfrac{\partial}{\partial z}(\ln M^*)}{\boldsymbol{Sc}^T\left[(\partial <V_x>/\partial z)^2+(\partial <V_y>/\partial z)^2\right]} = \frac{\boldsymbol{Ko}}{\boldsymbol{Sc}^T} \tag{6.1.27}$$

is used, the expression for $Q^T_{<e>}$ becomes

$$Q^T_{<e>} = \frac{1}{\boldsymbol{Kf}_c}D^T\frac{g}{M^*}\left(\frac{\partial M^*}{\partial z}\right). \tag{6.1.28}$$

Recall that in (6.1.27), $\boldsymbol{Ko}$ is the Kolmogorov gradient number and $\boldsymbol{Sc}^T = \nu^T/D^T$ is the turbulent Schmidt number. Let us note that in the case under consideration $\boldsymbol{Sc}^T = \boldsymbol{Pr}^T$ because it is usually supposed that $\boldsymbol{Le}^T = \chi^T/D^T = \lambda^T/\overline{\rho}D^T<C_p>=1$.

When using the critical values of the Richardson and Kolmogorov numbers ($\boldsymbol{Rf}_c$ and $\boldsymbol{Kf}_c$) and taking into account (6.1.25), (6.1.27), and (6.1.22) it is possible to evaluate the volume dissipation rate, ε_e, of turbulent energy due to molecular viscosity:

$$\varepsilon_e = R_{iz}\frac{\partial <V_i>}{\partial z}(1-\boldsymbol{Rf}_c-\boldsymbol{Kf}_c) = \frac{g\,D^T}{<T>}\left(\frac{\partial <T>}{\partial z}+\frac{g}{<C_p>}\right)\frac{1-\boldsymbol{Rf}_c-\boldsymbol{Kf}_c}{\boldsymbol{Rf}_c} =$$

$$= gD^T\frac{\partial(\ln M^*)}{\partial z}\frac{1-\boldsymbol{Rf}_c-\boldsymbol{Kf}_c}{\boldsymbol{Kf}_c}. \tag{6.1.29}$$

This formula can serve as a basis for experimental assessing the turbulent diffusion coefficient D^T by measuring the quantity ε_e.

6.2. MODELING THE TERRESTRIAL LOWER THERMOSPHERE

In Chapter 1 we defined the homopause, located in the lower planetary thermosphere, as the quite narrow height range where the competing processes of molecular and turbulent heat and mass transfer are equally important. The treatment of this range entails certain difficulties because the numerous processes involved must be addressed and evaluated thoroughly.

To benefit the modeling, the theoretical prerequisites summarized in the previous section should be complemented with the valuable experimental information. In particular, an approximation of the effective turbulent exchange coefficient, D^T, is introduced to parametrize the dynamic effects. Until recently, it was used as the only turbulent transport coefficient in modeling multicomponent turbulence at a given turbulent Prandtl-Schmidt number, $\boldsymbol{Pr}^T = \nu^T / D^T = \nu^T / \chi^T$, assuming the latter to be approximately constant within a wide height range. With such an approach, the coefficient D^T (together with the dynamic Richardson number, $\boldsymbol{Rf}$) serves as fitting parameter for matching the results of calculations with the measurement data available.

Historically, some other empirical and semi-empirical techniques to determine both D^T and $\boldsymbol{Pr}^T$ with application to the upper atmosphere were developed (see, e.g., *Lloyd at al., 1972; Rees et al., 1972; Roper, 1974; Izakov, 1978; Gordiets and Kulikov, 1981; Sassi and Visconti, 1990*). The wind shear of the averaged motion was postulated as the main mechanism responsible for maintaining turbulence, and variations in wind shear at the height of the turbopause was associated with dissipation of the outward propagating acoustic-gravitational waves, their energy and momentum being transformed into heat and kinetic energy of the mean flow (*Hines, 1965; Justus, 1969; Gavrilov, 1974; Gavrilov and Shved, 1975; Ginzburg and Kuzin, 1981*). To accommodate such a mechanism, a specific approach for evaluating the turbulent exchange coefficients for the stratified jet flows in multicomponent gaseous mixtures was developed (*Kolesnichenko and Marov, 1980, 1985; Kolesnichenko and Vasin, 1984*).

For an in-depth study of the upper atmosphere in the framework of a global model capable to describe self-consistently the large-scale dynamic processes, the gas composition, and the heat budget, it is necessary to know both the molecular and turbulent exchange coefficients entering the defining relations (3.3.3*), (3.3.15*), and (5.2.19). Indeed, the height dependence of these coefficients affects the atmospheric dynamics, the distribution of individual gaseous components, and the temperature. In particular, the spatial profiles of the turbulent exchange coefficients λ^T and ν^T determine the relative efficiency of two earlier mentioned basic manifestations of flow turbulization in the lower thermosphere: cooling due to turbulent heat conduction or heating due to dissipation of turbulent energy at the expense of molecular viscosity.

The advancement in the development of global thermospheric models was mainly achieved through utilization of the greater dimensionality, the more complete account for the elementary processes involved, and the energy balance evaluation. Nonetheless, the basic approach to approximating the effects of turbulent heat and mass exchange remains essentially unchanged. In other words, in view of the obvious difficulties of modeling large- and small-scale dynamics at the same level of accuracy, the method of parametrizing the dynamic effects in terms of the effective eddy diffusion coefficient (or tensor), D^T, continues to be a common tool. Here we first illustrate such an approach based on our early study. It does not claim to be complete but rather aims to clarify the concept of modeling the turbulent heat and mass exchanges in the upper atmosphere. We the proceed to a more comprehensive model involving the direct semi-empirical evaluation of the turbulent exchange coefficients, based on the results of Part 1. Such an attempt is undertaken in the next chapter.

6.2.1. The System of Differential Equations of the Model

In order to model numerically the homopause region, we will take advantage of the system of averaged hydrodynamic equations for mixtures. This system comprises the continuity equation (3.2.4) for the continuum as a whole; the diffusion equations (3.2.5) for the individual chemical components in the medium with regard for the aeronomic reactions and molecular and turbulent diffusion; the rheological Stefan-Maxwell relations (5.3.23) alike for the averaged molecular diffusion fluxes; the equation for internal energy of the averaged turbulized continuum (3.1.78); the hydrostatic equation (3.3.4); and the averaged equation of state for the pressure (3.2.2).

In the limit of one-dimensional approximation, sufficient for the purpose we pursue, these equations take the following form:

1) the continuity equation for the gas as a whole

$$\overline{\rho}\left(\frac{\partial(1/\overline{\rho})}{\partial t}+<V_z>\frac{\partial(1/\overline{\rho})}{\partial z}\right)=\frac{\partial<V_z>}{\partial z}, \tag{6.2.1}$$

where $<V_z>$ is the vertical component of the hydrodynamic velocity;

2) the equation for concentrations of neutral gas species

$$\overline{\rho}\left(\frac{\partial(\overline{n}_\alpha/\overline{\rho})}{\partial t}+<V_z>\frac{\partial(\overline{n}_\alpha/\overline{\rho})}{\partial z}\right)+\frac{\partial}{\partial z}\left(\overline{J}_{\alpha z}+J^T_{\alpha z}\right)=P_\alpha-\overline{n}_\alpha L_\alpha, \tag{6.2.2}$$

$$(\alpha=1,2,\ldots,N-1)$$

where the terms P_α and $\overline{n}_\alpha L_\alpha$ designate the generation and disappearance rates for α-particles, respectively, owing to photochemical and chemical reactions;

3) the averaged Stefan-Maxwell relations for multicomponent molecular diffusion

$$\frac{\partial\overline{n}_\alpha}{\partial z}=-\frac{\overline{n}_\alpha}{\overline{H}_\alpha}-\left(1+\overline{\alpha}_{T\alpha}\right)\frac{\overline{n}_\alpha}{<T>}\frac{\partial<T>}{\partial z}+\sum_{\substack{\alpha=1\\ \alpha\neq\beta}}^{N}\frac{\overline{n}_\beta\overline{J}_{\alpha z}-\overline{n}_\alpha\overline{J}_{\beta\ \ z})}{\overline{n}\mathcal{D}_{\alpha\beta}}, \tag{6.2.3}$$

$$(\alpha=1,2,\ldots,N-1),$$

where the index α relates to the components $O,\ O_2,\ N_2,\ H,\ He,\ Ar$;

4) the hydrostatic equation

$$\frac{\partial \bar{p}}{\partial z} = -\bar{\rho} g\,; \tag{6.2.4}$$

5) the equation of state for ideal gas mixtures

$$\bar{p} = \mathrm{k} < T > \sum_{\beta=1}^{N} \bar{n}_\beta\,; \tag{6.2.5}$$

6) the heat balance equation

$$\bar{\rho} < C_p > \left(\frac{\partial < T >}{\partial t} + < V_z > \frac{\partial < T >}{\partial z} \right) =$$

$$= -\frac{\partial}{\partial z}\left(\bar{q}_z + q_z^{T*} - \sum_{\beta=1}^{N} < h_\beta > (\bar{J}_{\beta z} + J_{\beta z}^{T}) \right) + \left(\frac{\partial \bar{p}}{\partial t} + < V_z > \frac{\partial \bar{p}}{\partial z} \right) + \bar{\pi}_{iz} \frac{\partial < V_i >}{\partial z} -$$

$$-\frac{\partial < T >}{\partial z} \sum_{\beta=1}^{N} C_{p\alpha} (\bar{J}_{\beta z} + J_{\beta z}^{T}) + Q_{\odot}^{uv} + Q_{<e>}^{T} + Q_L + Q_{\sup}\,, \tag{6.2.6}$$

where the term $q_z^{T*} \equiv q_z^T - \overline{p' V_z''}$ determines the turbulent heat flux in the vertical direction (see (3.1.54)); $Q_{\odot}^{uv}$ is the heating rate of neutral gas by the solar *EUV* radiation due to the joint contributions of the radiative processes and chemical reactions into the energy budget of the medium; Q_L is the cooling rate of neutral gas due to *IR*-radiation; $Q_{<e>}^{T}$ is the heat source at the expense of dissipation of turbulent energy (see (6.1.26)); $Q_{\sup}$ incorporates the auxiliary heat sources and sinks. The procedure of concretely defining the term $Q_{\odot}^{uv}$ (also called the heating function) for various gas mixtures and all other terms responsible for heat sources and sinks, are examined in detail elsewhere (*Marov and Kolesnichenko, 1987*). Of particular importance is the term Q_L accounting for the heat loss by radiation emitted at wavelengths of 1–20 microns, specifically in the molecular *NO* and CO_2 rotation-vibration bands at 5.3 and 15 microns.

The fluxes $\bar{q}_z,\ J_{\alpha z}^{T},\ q_z^{T}$ are determined by the formulas (6.1.12), (6.1.19), and (6.1.20), respectively. Calculating the quantity $Q_{<e>}^{T}$ (see (6.1.26)) incorporates the height profile of the critical dynamic Richardson number (6.1.25), obtained with using the parameter $(\boldsymbol{Pr}^T)$ averaged in height; it is adopted that $(\boldsymbol{Pr}^T)^{-1} = 0.3$. The module of the horizontal wind velocity, $\sqrt{(\partial < V_x > / \partial z)^2 + (\partial < V_y > / \partial z)^2}$, based on experimental data serves as an external parameter of the problem.

6.2.2. Boundary and Initial Conditions

The system of hydrodynamic equations for the mean motion scale (6.2.1)–(6.2.6) is a closed system of quasi-linear partial differential equations. It is appropriate for finding the averaged vertical component of the mean mass velocity, $<V_z>$, the averaged number densities of the chemical mixture components, $\overline{n}_\beta$, the averaged mean mass density, $\overline{\rho}$, and the averaged temperature, $<T>$, of the reacting mixture of atmospheric gases. The particular solution of the system is determined by the given initial and boundary conditions, as well as by the set of chemical components, together with their gas-dynamical, thermophysical, and chemical properties.

The boundary conditions are defined as follows:

a) At the lower boundary (z =70 km) the values of the temperature and the number concentrations are taken from the standard atmospheric model *MSISE90* (*Hedin, 1991*):

$$<T>\big|_{z=70} = T_{MSISE}, \quad \overline{n}_\beta\big|_{z=70} = n_\beta\big|_{MSISE}, \quad \overline{p}\big|_{z=70} = p\big|_{MSISE}, \tag{6.2.7}$$

while the atomic oxygen values are found from the photochemical equilibrium condition (see (2.1.58*)):

$$\frac{\partial}{\partial t}\overline{n}(0) = P(0) - \overline{n}(0)L(0) = 0; \tag{6.2.8}$$

b) At the upper boundary (z = 400 km) the vertical temperature gradient is assumed to vanish due to the high thermal diffusivity of the medium, $\chi = \lambda / \overline{\rho} < C_p >$, while the vertical wind velocity vanishes in view of the requirement that the mass flux across the upper boundary is zero. To model the condition $\lim_{z\to\infty} \overline{\rho} < V_z > = 0$, it is necessary to know the asymptotic behavior of the parameters $\overline{\rho}$ and $<V_z>$ at infinity. Because $\overline{\rho}$ decreases exponentially with height, while $<V_z>$ should aim linearly to infinity, the second derivative of the velocity with respect to height should vanish (*Izakov et al., 1972; Morozov and Krasitskiy, 1978*). Thus, the following boundary conditions are set at $z = 400$ km:

$$\partial^2 < V_z > / \partial z^2\big|_{z=400} = 0, \qquad \partial < T > / \partial z\big|_{z=400} = 0. \tag{6.2.9}$$

It was additionally supposed that all gases are in the state of gravitational-diffusion equilibrium at the upper boundary (see (2.3.96)):

$$\frac{\partial \overline{n}_\alpha}{\partial z} + \frac{\overline{n}_\alpha}{H_\alpha} = 0, \qquad \alpha = O,\ O_2,\ N_2,\ H,\ He,\ Ar\,. \tag{6.2.10}$$

It should be noted that the requirement of no diffusion flux of a substance across the upper boundary is not quite correct for light species such as hydrogen and helium because of their dissipation from the atmosphere from about this level. Describing this phenomenon in detail is bound to the necessity to define the fluxes of these components across the upper boundary. An estimate of the dissipation rate for hydrogen amounts to $2\cdot10^7$ $cm^{-2}s^{-1}$ (*Chamberlain and Hunten, 1987*). As was shown by the calculations of height distributions of atmospheric components (*Shimazaki, 1972*), the flux accuracy is better only less than 10% as compared to zero diffusion flux (for instance, for *n(H)* at 140 km altitude).

At last, the initial conditions are defined by also using the *MSISE* model, although it should be noted that the solution obtained by the relaxation method does not depend on the initial conditions.

6.2.3. The Transfer Coefficients

In the calculations the semi-empirical expressions for the binary diffusion coefficients describing variations of $D_{\alpha\beta}$ in a wide range of thermospheric temperatures (*Oran and Boris 1990*) were employed. They are summarized in Table 6.2.1. The following factors of thermodiffusion, slightly varying in the height range $120 \le z \le 250$ km, were accepted (*Banks and Kockarts, 1973*): $\alpha_T(0_2) = 0.12$; $\alpha_T(N_2) = 0.08$; $\alpha_T(0) = -0.08$; $\alpha_T(H) = -0.38$; $\alpha_T(He) = -0.27$; $\alpha_T(Ar) = 0.17$.

The Turbulent Diffusion Coefficient. There are various experimental estimates of the turbulent diffusion coefficient, D^T. As was mentioned, they are usually obtained using relations such as (6.1.29) through directly measuring the turbulence energy dissipation rate, ε_e, in the atmosphere. In the model under consideration, the following height dependence of the turbulent diffusion coefficient, $D^T(z)$, and the respective numerical values of the parameters involved were used (*Shimazaki, 1971; Chandra and Sinha, 1976*):

$$\begin{cases} D^T(z) = D_m^T \exp\left[-S_1(z-z_m)^2\right], & z \ge z_m \\ D^T(z) = \left(D_m^T - D_0^T\right)\exp\left[-S_2(z-z_m)^2\right] + D_0^T \exp\left[S_3(z-z_m)\right], & z \le z_m \end{cases} \qquad (6.2.11)$$

where D_m^T, D_0^T, z_m, S_1, S_2, S_3 are the empirical constants; z is the height; and D_m^T is the maximum value of D^T at height z_m. Here $D_m^T = 10^6 \div 10^7$ cm^2 s^{-1}, (z_m is varied from 85 to 105 km; $D_0^T = 2\times10^6$ cm^2 s^{-1}; $S_1 = S_2 = 0.05$ km^{-1}; and $S_3 = 0.07$ km^{-1}.

Apparently, the tentative values of z_m and D_m^T need to be clarified additionally with due regard for the continuously improving data base. In particular, as the measurements show, the coefficient D^T undergoes substantial variations depending on the solar activity and the season. Its values are in the range $10^5 \div 10^7$ cm^2 s^{-1} at the heights of the mesosphere and the lower thermosphere, the maximum values being attributed to higher levels of solar activity and winter. At the same time, there is some evidence that D^T exhibits an inverse seasonal trend reaching the maximum ($D_m^T = 10^6 - 10^7$ cm^2 s^{-1}) in the moderate and high latitudes at $z_m \approx 90 - 110$ km (*Danilov and Kalgin, 1992*). One

Table 6.2.1 Binary diffusion coefficients

Mixture	$\mathcal{D}_{\alpha\beta}$, cm^2 s^{-1}	Mixture	$\mathcal{D}_{\alpha\beta}$, cm^{2}s^{-1}
O_2-N_2	$0.829\times10^{17}\,T^{0.724}\,n^{-1}$	O-He	$3.43\times10^{17}\,T^{0.749}\,n^{-1}$
O_2-He	$3.21\times10^{17}\,T^{0.710}\,n^{-1}$	O-Ar	$0.551\times10^{17}\,T^{0.841}\,n^{-1}$
O_2-Ar	$0.717\times10^{17}\,T^{0.736}\,n^{-1}$	N_2-He	$1.16\times10^{20}\,T^{0.524}\,n^{-1}$
O-N_2	$0.969\times10^{17}\,T^{0.774}\,n^{-1}$	N_2-Ar	$0.663\times10^{17}\,T^{0.752}\,n^{-1}$
O-O_2	$0.969\times10^{17}\,T^{0.774}\,n^{-1}$	N_2-H	$6.10\times10^{17}\,T^{0.732}\,n^{-1}$

may assume that meridional circulation renders a pronounce influence on this seasonal trend being responsible for the stronger development of turbulence in winter than in summer.

The Molecular Heat Conduction Coefficient. The following semi-empirical expression (*Banks and Kockarts, 1973*) was adopted for the molecular heat conduction coefficient for gaseous mixture in the thermosphere:

$$\lambda = \frac{1}{\overline{n}} \sum_{\alpha=1}^{N} b_\alpha \overline{n}_\alpha < T >^{0,69}, \tag{6.2.12}$$

where $b(O_2) = b(N_2) = 56, \quad b(O) = 75.9, \quad b(H) = 379, \quad b(He) = 299.$

6.2.4. Ionization and Dissociation of the Atmospheric Components

As was said, absorption of the incident solar *EUV* radiation is responsible for the elementary processes in the upper atmosphere, first of all photodissociation and photoionization of atmospheric components. The relative share of the solar energy ultimately transformed into heating of neutral gas, or the so-called solar heating efficiency (*Marov and Kolesnichenko, 1987*), may be expressed as

$$Q_\circ^{uv} \approx \varepsilon^{dis} \sum_{\lambda<\lambda_{d\alpha}} h\nu q_{d\alpha}^{\lambda} + \varepsilon^{ion} \sum_{\lambda<\lambda_{i\alpha}} h\nu q_{i\alpha}^{\lambda} . \tag{6.2.13}$$

Here $h\nu$ is the photon energy; $q_{d\alpha}^{\lambda}$ and $q_{i\alpha}^{\lambda}$ are the dissociation and ionization rates, respectively, of the α-component by solar radiation with wavelength λ defined by the expressions

$$q_{d\alpha}^{\lambda} = I_{\lambda\infty} \sigma_{d\alpha}^{\lambda} \overline{n}_\alpha \exp(-\tau_\lambda), \qquad q_{i\alpha}^{\lambda} = I_{\lambda\infty} \sigma_{i\alpha}^{\lambda} \overline{n}_\alpha \exp(-\tau_\lambda); \tag{6.2.14}$$

$\lambda_{d\alpha}$, $\lambda_{i\alpha}$ are the respective threshold wavelengths; $I_{\lambda\infty}$ is the intensity of the incident solar radiation at "an upper boundary" of the atmosphere; $\sigma^{\lambda}_{d\alpha}$, $\sigma^{\lambda}_{i\alpha}$ are the ionization and dissociation cross-sections, respectively, for α-particles; and ε^{dis} and ε^{ion} are the dissociation and ionization efficiencies, respectively.

The optical depth, τ_{λ}, is expressed as

$$\tau_{\lambda} = \sum_{\alpha=1}^{N} \sigma^{\lambda}_{a\alpha} \int_{z}^{\infty} \bar{n}_{\alpha} ds, \tag{6.2.15}$$

where $\sigma^{\lambda}_{a\alpha}$ is the cross-section of the incident radiation absorbed by the α-component, the integration being taken along the light beam, s, from infinity to a specified point at some height z. The wavelength dependence of the incident solar radiation intensity and the respective cross-sections of absorption, ionization, and dissociation by the atmospheric species are summarized elsewhere (see, e.g., *Meteorology and Atomic Energy, 1971; Torr et al., 1979; Torr and Torr, 1985; Tobiska, 1990, 1991, 1993*).

If atmospheric components are distributed in height according to the barometric formula, the following approximate relation for the optical depth, τ_{λ}, is true:

$$\tau_{\lambda} \cong \sum_{\alpha=1}^{N} \sigma^{\lambda}_{a\alpha} \bar{n}_{\alpha} H_{\alpha} Ch(X_{\alpha}, \chi_{\odot}), \tag{6.2.16}$$

where $Ch(X_{\alpha}, \chi_{\odot})$ is the Chapman function; $X_{\alpha} \equiv (R_{pl} + z)/H_{\alpha}$; $H_{\alpha} \equiv kT/M_{\alpha}g$ is the local scale height for the α-component; R_{pl} is the radius of the planet; $\chi_{\odot}$ is zenith angle of the Sun, which is a function of local time, latitude, and season,

$$\cos\chi_{\odot} = \cos\delta_{\odot} \cos\varphi \cos\Theta_{t} + \sin\delta_{\odot} \sin\varphi\ ; \tag{6.2.17}$$

$\delta_{\odot}$ and $\alpha_{\odot}$ are Sun's right ascension and declination, respectively; φ is the latitude; Θ_{t} is hour angle defined by the formula

$$\Theta_{t} = S + \Omega t - \alpha_{\odot}. \tag{6.2.18}$$

Here S is the sidereal time at Greenwich midnight; Ω is the angular velocity of the Earth's rotation, $(2\pi/86400)$; t is local time. The sidereal time, S, the declination, $\delta_{\odot}$, and the right ascension, $\alpha_{\odot}$, of the Sun were calculated using the respective formulas (*Marov et al., 1989*).

Solar radiation with wavelengths shorter than 1000 $\mathring{A}$ triggers ionization of the basic atmospheric components at heights of 100–300 km and is responsible for forming the ionosphere. Photo-electrons originating from ionization lose their excess energy in elastic and inelastic collisions and in addition excite molecules. As this occurs, the energy released partially transforms into heat and forms the basic heat source of the terrestrial thermosphere (*Marov and Kolesnichenko, 1987; Marov et al, 1997*). In turn,

Table 6.2.2 Dissociation Cross-sections and Fluxes of Solar UV radiation

$\Delta\lambda, \overset{o}{A}$	$I_{\lambda\infty}$, cm^{-2} s^{-1}	$\sigma_a(O_2) = \sigma_d(O_2)$, cm^2	$\Delta\lambda, \overset{o}{A}$	$I_{\lambda\infty}$, cm^{-2}·s^{-1}	$\sigma_a(O_2) = \sigma_d(O_2)$, cm^2
1200 – 1250	1.0+10	4.7 – 18	1800 – 1850	6.7+11	3.3 – 20
1215.67	2.7+11	1.5 – 20	1850 – 1900	1.0+12	4.7 – 21
1250 – 1300	4.7+09	4.4 – 19	1900 – 1950	1.6	6.0 – 22
1300 – 1350	1.5+10	2.0 – 18	1950 – 2000	2.8	6.7 – 23
1350 – 1400	9.5+09	1.1 – 17	2000 – 2050	4.1	7.4 – 24
1400 – 1450	1.2+10	1.4	2050 – 2100	7.4	7.0
1450 – 1500	2.0	1.3	2100 – 2150	1.8+13	6.2
1500 – 1550	3.6	1.0	2150 – 2200	2.2	5.3
1550 – 1600	4.0	6.6 – 18	2200 – 2250	3.1	4.2
1600 – 1650	6.5	3.5	2250 – 2300	3.4	3.3
1650 – 1700	1.3+11	1.5	2300 – 2350	3.4	2.4
1750 – 1750	2.4	5.4 – 19	2350 – 2400	3.4	1.7
1750 – 1800	4.3	1.3	2400 – 2450	4.2	1.2

Note. The dissociation threshold for O_2 is located at 2423.7 $\overset{o}{A}$.

solar radiation with wavelengths $1000 < \lambda < 2200\,\overset{o}{A}$ is absorbed mainly by molecular oxygen while ozone absorbs radiation with longer wavelengths. With regard for the absorption cross-sections of these two components, they mainly determine the photodissociation of the Earth's neutral atmosphere.

Moreover, the absorption cross-section of molecular oxygen widely varies from 10^{-24} cm^2 at $\lambda = 2420\,\overset{o}{A}$ up to the maximum value 10^{-17}cm^2 at $\lambda = 1450\,\overset{o}{A}$. This circumstance is intrinsically related to the existence of characteristic absorption in the *Schumann-Runge* continuum ($1310 < \lambda < 1750\,\overset{o}{A}$) in the lower thermosphere, in the *Schumann-Runge* bands ($1750 < \lambda < 2100\,\overset{o}{A}$) in the mesosphere, and in the *Herzberg* continuum ($2100 < \lambda < 2400\,\overset{o}{A}$) in the stratosphere. As an example, the values of the dissociation cross-sections for O_2 and the respective fluxes of solar *UV* radiation are given in Table 6.2.2.

It should be emphasized that the radiative transfer pattern within the *Schumann-Runge* bands is rather complicated. To simplify calculations of the photo-dissociation rate, J_{SRB}, approximating formulas (*Nicolet, 1984*) can be used which provide an accuracy of $\pm 10\%$ and $\pm 15\%$ for O_2 column abundance molecules less or greater than 10^{19} cm^{-2}, respectively. Thus, $J_{SRB} = 1.1\times10^{-7}$ $\exp\left[-1.97\times10^{-10} N(O_2)^{0,522}\right]$s^{-1} for $N(O_2) \leq 1\times10^{19}$cm^{-2} in the thermosphere and $J_{SRB} = 1.45\times10^{-8} N(O_2)^{-0.83}$ s^{-1} for $N(O_2) \geq 1\times10^{19}$cm^{-2} in the mesosphere.

The sequence of photochemical and accompanying chemical processes used in the model is as follows (J and k are the reaction rate coefficients):

- *Dissociation of O_2 by solar radiation with wavelengths 2200–2400 $\overset{o}{A}$*

$$O_2 + h\nu \rightarrow O + O; \qquad (J_1)$$

- *Recombination of atomic oxygen due to triple collisions*

$$O + O + M \rightarrow O_2 + M, \qquad (k_1)$$

$$(k_1 = 4.7\times10^{-33}\left(<T>/300\right)^{-2} cm^6 s^{-1});$$

- *Recombination of O and O_2 due to triple collisions*

$$O + O_2 + M \rightarrow O_3 + M, \qquad (k_2)$$

$$(k_2 = 6.0\times10^{-34}\left(T/300\right)^{-2,3} cm^6 s^{-1});$$

- *Dissociation of O_3 by solar radiation*

$$O_3 + h\nu \rightarrow O_2 + O; \qquad (J_2)$$

- *Decomposition of ozone*

$$O_3 + O \rightarrow O_2 + O_2 \qquad (k_3)$$

$$(k_3 = 8.8\times10^{-12} \exp\left(-2060/T\right) cm^3 s^{-1}).$$

The values of k_1, k_2, and k_3 can be found elsewhere (*Jet Propulsion Laboratory, 1987*). A crude estimate shows that at the height where the production rate of atomic oxygen is maximal (at about 100 km) the lifetime of O -atoms is more than one day and exponentially grows with height.

6.2.5. The Modeling Results

The system of hydrodynamic equations (6.2.1)–(6.2.6) with the specified energy sources and sinks allows us to evaluate the simultaneous behavior of temperature and composition in the lower thermosphere. It was solved numerically using the relaxation

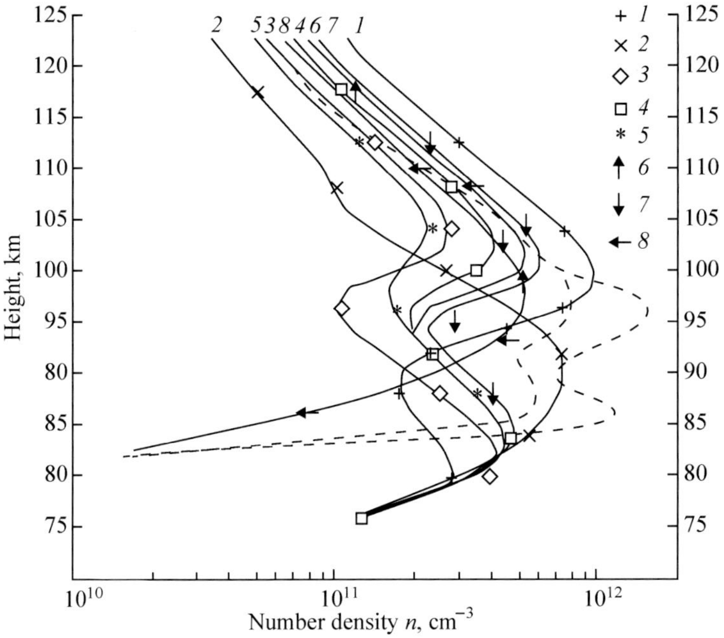

Fig. 6.2.1. Number density height profiles of atomic oxygen, $n_o(z)$, calculated with formula (6.2.1) for different values of maximal eddy diffusion coefficient, D_m^T at height z_m, and empirical constants, S_1 and S_2: 1. $D_m^T = 10^7 cm^2 s^{-1}$, $z_m = 85$ km, $S_1 = S_2 = 0.05$ km^{-2};
2. $D_m^T = 10^7 cm^2 s^{-1}$, $z_m = 100$ km, $S_1 = S_2 = 0.05$ km^{-2};
3. $D_m^T = 3 \times 10^7 cm^2 s^{-1}$, $z_m = 90$ km, $S_1 = S_2 = 0.05$ км$^{-2}$;
4. $D_m^T = 10^7 cm^2 s^{-1}$, $z_m = 90$ km, $S_1 = S_2 = 0.05$ km^{-2};
5. $D_m^T = 10^7 cm^2 s^{-1}$, $z_m = 90$ km, $S_1 = 0.025$ km^{-2};
6. $D_m^T = 10^7 cm^2 s^{-1}$, $z_m = 90$ km, $S_1 = 0.075$ km^{-2};
7. $D_m^T = 10^7 cm^2 s^{-1}$, $z_m = 90$ km, $S_1 = 0.1$ km^{-2};
8. *MSISE 90* model (*Hedin, 1991*); dashed curves - measurements (*Murtagh et al., 1990*).

method. The results correspond to the Sun's position at 15^h *LT* at $45°$ *N* during the vernal equinox and to the solar activity index $F_{10,7} = 150$. The solar zenith angle under these conditions is $\chi_\odot = 60.6°$.

As an example, Figure 6.2.1 shows the height profiles of the number density of the atomic oxygen $n_O(z)$ computed for different eddy diffusion coefficients, D^T. It is clearly seen that both the height z_m, where D^T is maximal ($D^T = D_m^T$) and the magnitude of the parameter S_1, defining the pattern of D^T, decreasing for $z > z_m$, strongly affect the *O*-atom distribution. Only one maximum is formed when $z_m \geq 100$ km (version 2), but if $z_m \leq 90$ km (versions 1, 3-7), two maxima occur. Of special interest is the fact that in version 8, corresponding to the *MSISE* model, also only one maximum, $n_O(z)$, at

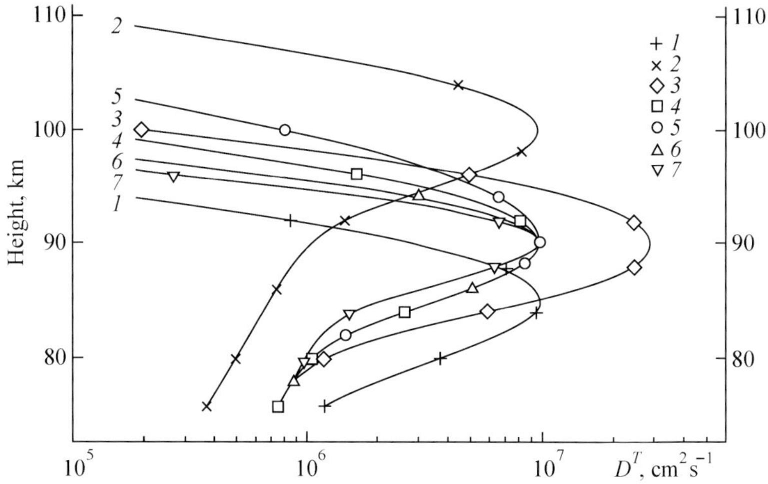

Fig. 6.2.2. Height profiles of the eddy diffusion coefficient, D^T, calculated with formula (6.2.11) for different values of D^T_m, z_m, and S_1 (see Fig. 6.2.1). The area between the shaded curves – the value D^T according to *Hocking, 1990.*

the height of 98 km is pronounced. However, when z_m = 85–90 km, the first maximum $n_O(z)$ is located at 100–104 km, i.e. a bit higher compared to the *MSISE* model, and the second maximum occurs at 80–84 km.

The height profiles of the eddy diffusion coefficient, D^T, in Figure 6.2.2 depending on the parameters z_m and S_1, correspond to the versions shown in Figure 6.2.1. In version 1, $D^T(h)$ perfectly matches the one deduced from the *MSISE* model though $n_O(z)$ at 100–400 km turns out to be about twice as much as compared to this model. Coincidence of the computed $n_O(z)$ values with the respective model profile in this height range can be achieved at the expense of increasing D^T. Then, the position of D^T_m should be lowered, however, to 85–90 km which is substantially lower than the usually accepted value z_m =105 km. Also, when z < 90 km, the computed O number densities considerably exceed both the model (*MSISE*) and experimental values.

On the other hand, the amplitude and the height distribution of $n_O(z)$ are determined by the parameter S_1, the double maximum formation enhancing with its growth. The measurement data (*Murtagh et al., 1990*), shown as dashed curves in Figure 6.2.1, confirm the presence of a double maximum and fit the modeling results. It is also evident that number density of atomic oxygen in the lower thermosphere can vary twofold depending on S_1 and by a factor of four depending on z_m.

Figure 6.2.3 shows the characteristic time scales of the various physical-chemical processes determining the height distribution of the basic thermospheric components for version 1 in Figures 6.2.1 and 6.2.2. Curve 1 featuring the chemical relaxation time of atomic oxygen reveals a sharp decrease in the characteristic time with height. It means that the condition of photochemical equilibrium is met for atomic oxygen for heights

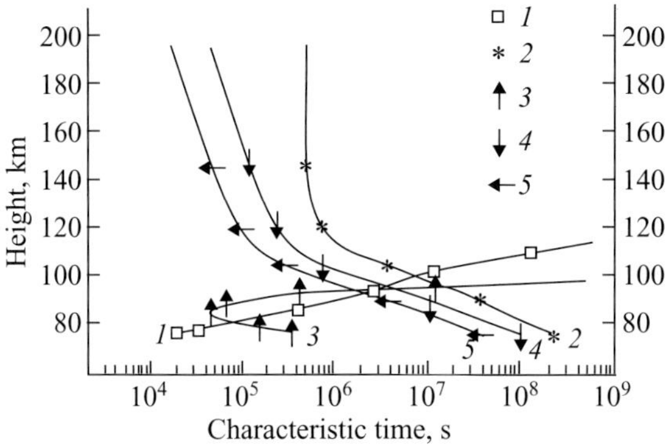

Fig. 6.2.3. Characteristic times of physical-chemical processes: 1 – chemical relaxation of O ; 2 – chemical relaxation of O_2 ; 3 – eddy diffusion; 4 – molecular diffusion for O ; 5 – molecular diffusion for O_2 .

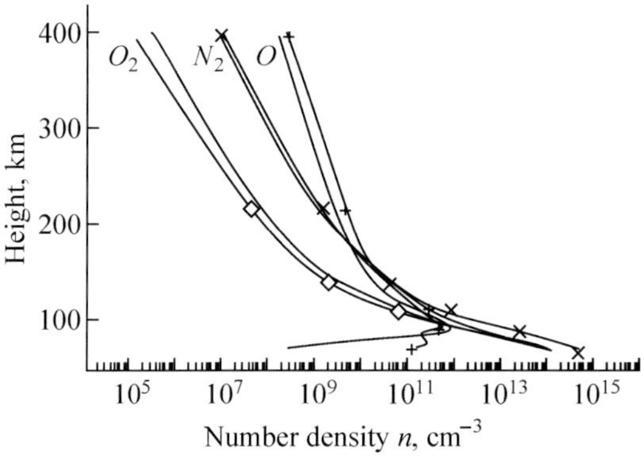

Fig. 6.2.4. Comparison of the computed profiles of the O (+), N_2 (×), and O_2 (□) number densities with the *MSISE 90* model (solid curves without markers).

below 80 km. At the same time, eddy diffusion and recombination of O-atoms are the dominant processes at 80–100 km, whereas molecular diffusion begins to contribute significantly into substance transport above 100 km.

The height profiles of the O, N_2, and O_2 number densities (computed with the parameters used for version 1 in Figure 6.2.1) are compared with the *MSISE* model in Figure 6.2.4. As is seen, for the selected profile of D^T (version 1 in Figure 6.2.2), the number density of O is approximately twice as high as the model density at 100 to 400 km, the number density of O_2 at 400 km being three times smaller than predicted by the model. The deviation of the computed curve from the barometric distribution can be caused by the poorly known efficiency of O_2 - dissociation. In turn, the molecular nitrogen distribution virtually coincides with the model profile. Possible minor diversions can be attributed to the errors related to difference approximations in height.

As was already mentioned, the temperature profile particularly depends on the efficiencies of turbulent heat conduction, Q_T^T, (see 6.1.21) and molecular turbulent energy dissipation, $Q_{<e>}^T$, (see 6.1.26) into the energy budget of the lower thermosphere. In other words, its pattern reflects the balance of heat released due to turbulent energy dissipation against the energy drained through turbulent heat conduction, positive or negative. The modeling results with a typical height profile for D^T and $D_m^T \approx 110$ km support earlier inferences (*Marov and Kolesnichenko, 1987*) that turbulence cools the lower thermosphere at $z \geq$ 100–105 km and heats it at $z \leq$ 100–105 km.

Let us note that in the thermospheric model under consideration the functional links between basic features of the turbulent field such as $D^T = D^T(\boldsymbol{Rf}_c, \boldsymbol{Kf}_c)$, were ignored. Unfortunately, an exact pattern of such links presently cannot be ascertained because of scarce experimental information. This prevents us from using similar models to compute the height profile of the eddy diffusion coefficient in the lower thermosphere self-consistently with the temperature and the composition. For the same reason, in view of poorly correlated and considerably scattered experimental data, estimates of turbulent cooling and/or heating efficiencies, being ultimately dependent on the height distribution of the external parameters $D^T(z)$, $\boldsymbol{Rf}_c(z)$, $\boldsymbol{Kf}_c(z)$, are too crude. In this connection, the theoretical methods for finding functional relations such as $D^T = D^T(\boldsymbol{Rf}_c, \boldsymbol{Kf}_c)$ become especially important because they open the opportunity to self-consistently evaluate the heat balance in multicomponent turbulent natural media such as the lower planetary thermospheres.

Summary

- *A diffusive-photochemical model of the upper terrestrial atmosphere in the height range 70–400 km was developed. The model is focused on evaluating the composition and the heat budget and accounts for the main sources and sinks of heat. Diffusion processes in turbulent multicomponent mixtures were studied based on the Stefan-Maxwell relations for multicomponent molecular diffusion and on gradient relations for turbulent heat and substance fluxes.*

- *The dynamic processes in the lower thermosphere including the homopause were examined using the time-averaged eddy diffusion coefficient,* D^T*, commonly utilized as a fitting parameter in modeling aeronomic problems. This coefficient was estimated quantitatively by comparing model and experimental height distributions of atomic oxygen.*

- *The developed model serves to illustrate the available approach to evaluating diffusion processes in a turbulent reactive mixture such as the planetary lower thermosphere. This approach poses further improvement through self-consistently calculating the temperature, the composition, and the turbulent heat and mass transfer coefficients for the shear flows in multicomponent gaseous mixtures.*

CHAPTER 7

TURBULENT TRANSFER COEFFICIENTS IN PLANETARY UPPER ATMOSPHERES: A SEMI-EMPIRICAL DETERMINATION

The evolutionary transfer equations for the turbulent energy, $<e>$, and for the mean-square fluctuations of the enthalpy, $<h''^2>$, for multicomponent gas mixtures deduced in Chapter 4, allows us to develop a technique for modeling the turbulent exchange coefficients in a horizontally homogeneous flow with a transverse hydrodynamic velocity shear.

The concept of quasi-local equilibrium in the theory for turbulent transport in multicomponent mixtures (subsection 4.3.9) serves as a basis for a semi-empirical procedure to derive the algebraic formulas for calculating the coefficients D^T, χ^T, and ν^T. These coefficients are a part of the defining relations (3.3.3*), (3.3.15*), and (5.2.19) for the turbulent flows of diffusion, heat, and the Reynolds stress tensor. An advanced approach to model the turbulent exchange coefficients is considered here, taking the terrestrial upper atmosphere as an example of a turbulized natural multicomponent medium. For this region, comprising both the mesosphere and the lower thermosphere, numerical estimates of the turbulent viscosity coefficient, ν^T, and thermal conduction coefficient, χ^T, are obtained.

7.1. THE ORIGINAL EQUATIONS AND THEIR TRANSFORMATIONS

We have seen that in order to model self-consistently the atmospheric gas temperature, the composition, and the dynamics involving the turbulent exchange coefficients, functional relations like $D^T(z) = D^T[\boldsymbol{Rf}_c(z), \boldsymbol{Kf}_c(z)]$ are needed. These relations should meet some physical criteria and permit to calculate adequately the complicated heat and mass transfer processes.

The procedure for finding such relations represents a particular application of the general K-theory for multicomponent turbulence discussed in subsection 4.3.9. As an example, below we will consider an elementary approach to develop a semi-empirical theory for the turbulent exchange coefficients. A more comprehensive approach is given in Chapter 8 where the modeling of tensors of the turbulent exchange coefficients is performed with the account for the different intensities of turbulent fluctuations of thermo-hydrodynamic parameters in different coordinate axes.

7.1.1. The Initial Evolutionary Transfer Equations

It was shown in section 4.1. that the balance transfer equations for the mean-square fluctuations $<A''^2>$ of the thermohydrodynamic parameters $A(\boldsymbol{r},t)$ (whose instantaneous values comply with a conservation law such as (2.1.1)), can be obtained from the following general equation:

$$\bar{\rho}\frac{D<A''^2/2>}{Dt}+\partial_j\left(\overline{\rho A''^2 V_j''/2}+\overline{A'' J_{(A)j}}\right)=-J_{(A)j}^T\,\partial_j<A>+\overline{A''\sigma_{(A)}}-\bar{\rho}<\varepsilon_{(A)}>, \tag{7.1.1}$$

This equation comprises, along with the convective and diffusion terms, the terms responsible for the formation, redistribution, and dissipation of the respective turbulent properties of the flow. Here $\bar{\rho}<\varepsilon_{(A)}>\equiv-\overline{J_{(A)j}\,\partial_j A_j''}$ is the scalar dissipation rate of the variance $<A''^2>$, and $J_{(A)j}^T\equiv\overline{\rho A'' V_j''}$ is the turbulent flow density of property $A(\boldsymbol{r},t)$.

When quantity A is identified with the hydrodynamic flow velocity, V_k, in (7.1.1) and in addition, the expressions $J_{(V_k)j}=-\pi_{kj}$, $\sigma_{(V_k)}=-\partial_k p$ (see (2.1.22)) for instantaneous values of the fluxes and origination rates of momentum are used, the balance equation for the turbulent energy, $<e>$, can be written as (cf. with (4.2.22))

$$\bar{\rho}\frac{D<e>}{Dt}=-\partial_j[\overline{\rho(e+p'/\rho)V_j''}-\overline{\pi_{ij}V_i''}]+R_{ij}\partial_j<V_i>+\overline{p'\partial_j V_j''}+\bar{\rho}G-\bar{\rho}<\varepsilon_e>, \tag{7.1.2}$$

where

$$\bar{\rho}G=\partial_j\bar{p}\left[\frac{J_{qj}^T}{\bar{\rho}<T><C_p>}-\frac{1}{n}\sum_{\alpha=1}^{N}J_{\alpha j}^T\right]$$

is the generation rate of turbulent energy, $<e>$, due to buoyancy effects;

$$J_{qj}^T\equiv q_j^T-\sum_{\alpha=1}^{N}<h_\alpha>J_{\alpha j}^T$$

is the reduced turbulent heat flux (see (3.3.10)); and $<\varepsilon_e>$ is the of turbulent energy into heat due to molecular viscosity (all designations correspond to those used in sections 3.3 and 4.2).

In general, for chemically active gaseous mixtures, the balance equation for enthalpy variance, $<h''^2>$, is much simpler than the analogous equation for the mean square of temperature fluctuations, $<T''^2>$, because it does not involve a large number of additional temperature and concentrations correlations related to the chemical reactions occurring in the medium. When deriving the transfer equation for the mean-square enthalpy fluctuations, $<h''^2>$, in a turbulized multicomponent flow, we will identify the quantity A with the specific enthalpy, h, in equation (7.1.1). We will also take into account the expressions $J_{(h)j} \equiv q_j$ and $\sigma_{(h)} \equiv dp/dt + \pi_{ij}\partial_j V_i$ for the instantaneous values of the substantial heat flux and the enthalpy source density (see (2.1.51) and (2.1.52)). It yields

$$\bar{\rho}\frac{D(<h''^2>/2)}{Dt} + \partial_k\left(\overline{\rho h'' V_k''/2} + \overline{q_k h''}\right) = -q_k^T\left(\partial_k < h > - \frac{\partial_k \bar{p}}{\bar{\rho}}\right) - \bar{\rho} < \varepsilon_h >, \qquad (7.1.3)$$

where the quantity $\bar{\rho} < \varepsilon_h > \equiv -\overline{q_j \partial_j h''}$ (see (4.3.10)) characterizes the scalar dissipation rate of the correlation $< h''^2 >$ due to molecular thermal diffusivity.

Let us now make an important simplification in equation (7.1.3) for $< h''^2 >$. The total molecular heat flux vector (see (2.3.75)),

$$q_j = -\lambda \partial_j T + p\sum_{\alpha=1}^{N} K_{T\alpha} w_{\alpha j} + \sum_{\alpha=1}^{N} h_\alpha n_\alpha w_{\alpha j}$$

can be transformed in the following way. From (2.1.45) for the mixture enthalpy it follows that

$$\partial_j h = C_p \partial_j T + \sum_{\alpha=1}^{N} h_\alpha \partial_j Z_\alpha,$$

whence it is possible to obtain

$$q_j = -\frac{\lambda}{C_p}\partial_j h + \sum_{\beta=1}^{N} h_\beta\left[\frac{\lambda}{C_p}\partial_j Z_\beta + J_{\beta j}\right] + \mathrm{k}T\sum_{\beta=1}^{N} \alpha_{T\beta} J_{\beta j}. \qquad (7.1.4)$$

Obviously, the Stefan-Maxwell relations (2.3.69) should be used to quantitatively analyze multicomponent molecular diffusion. For the sake of simplicity, we confine ourselves to considering gaseous mixtures comprising components with similar molecular weights and assume that the binary diffusion coefficients, $D_{\alpha\beta}$, are approximately equal to each other ($D_{\alpha\beta} \approx D$). As is known, in this case the cross-effects of molecular heat transfer (thermo- and baro-diffision) are also identical. Then, as follows from the adopted assumptions, relation (2.3.69) takes the form of Fick's law:

$$J_{\beta j} = -n\mathcal{D}\partial_j(n_\beta / n) \cong -\rho\mathcal{D}\partial_j Z_\beta, \quad (\rho \propto n), \qquad (7.1.5)$$

and the formula for the molecular heat flux, q_j, can be rewritten as

$$q_j = -\frac{\lambda}{C_p}\partial_j h + \sum_{\beta=1}^{N} h_\beta\left[\frac{\lambda}{C_p} - \rho\mathcal{D}\right]\partial_j Z_\beta. \qquad (7.1.6)$$

If, in addition, the intermolecular force potentials (see (2.3.90)) for different particles are rather similar, then the binary diffusion coefficients, D, for all mixture components are approximately equal to the heat conduction coefficient, $\chi = \lambda / \rho C_p$ within the whole temperature range typical for the upper atmosphere. Then for the molecular heat flux, q_j, we finally have

$$q_j = -\rho\chi\,\partial_j h. \tag{7.1.7}$$

In view of this relation, the transfer equation for the variance $<h''^2>$ (see (7.1.3)) takes the form

$$\bar{\rho}\frac{D(<h''^2>)}{Dt} + \partial_k\left(\overline{\rho h'' V_k''} - 2\chi\overline{\rho\, h''\,\partial_k h}\right) = -2q_k^T\left(\partial_k <h> - \frac{\partial_k \bar{p}}{\bar{\rho}}\right) - 2\bar{\rho} <\varepsilon_h>, \tag{7.1.3*}$$

where $\bar{\rho} <\varepsilon_h> \cong \chi\overline{\rho\,\partial_j h\,\partial_j h''}$.

7.1.2. Modeling the Correlation Terms

In order to model the turbulent energy flux, $J^T_{<e>k}$, and the enthalpy variance flux, $J^T_{<h>k}$ (including the third order moments), as well as the dissipative terms $<\varepsilon_e>$ and $<\varepsilon_h>$ entering the evolutionary transfer equations (7.1.2) and (7.1.3*), we will use the following elementary approximating expressions:

a) the turbulent kinetic energy flux is defined by the relation (cf. with (4.2.31))

$$J_{<e>j} = \overline{\rho(e + p'/\rho)V_j''} - \overline{\pi'_{ij}V_i''} = -\left(c_1\nu^T + c_2\nu\right)\partial_j\left(\bar{\rho} <e>\right); \tag{7.1.8}$$

b) the enthalpy variance diffusion flux, $J^T_{<h>k}$, entering (7.1.3*) is defined as (cf. with (4.3.12))

$$J_{<h>j} \equiv \left(\overline{\rho h''^2 V_j''/2} + \overline{q_j h''}\right) = -\left(c_5\chi^T + c_6\chi\right)\partial_j\left(\bar{\rho} < h''^2/2 >\right); \tag{7.1.9}$$

c) the turbulent energy dissipation rate, $<\varepsilon_e>$, depending on the turbulent Reynolds number, $\boldsymbol{Re}^T \equiv L <e>^{1/2}/\nu$, is assumed to be proportional to:

- the quantity $\nu <e>/L^2$, where L is the external turbulence scale, for small $\boldsymbol{Re}^T$;

- the quantity $\nu_m <e>/L^2$, where ν_m is the effective kinematic viscosity coefficient defining the impact of small turbulent eddies on large-scale velocity fluctuations, for large $\boldsymbol{Re}^T$;

- the quantity $(\nu+\nu_m)<e>/L^2$ for intermediate (arbitrary) values of $\boldsymbol{Re}^T$;

d) similarly, we have $<\varepsilon_h> \propto (\chi+\chi_m)<h''^2>/L^2$ for the mean destruction rate of the variance $<h''^2>$.

It is natural to assume that the coefficients ν_m and χ_m are proportional to the respective turbulent transfer coefficients determined by large vortices. In other words, these coefficients are proportional to the quantities ν^T and χ^T, such that $\nu_m=\alpha\nu^T$ and $\chi_m=\beta\chi^T$, where α, β are constants (α, β < 1). Then we finally obtain (cf. with (4.2.29) and (4.3.11))

$$<\varepsilon_e>=\frac{1}{a_1^4}\left(\nu^T+\frac{\nu}{\alpha}\right)\frac{<e>}{L^2}, \tag{7.1.10}$$

$$<\varepsilon_h>=\frac{1}{a_2^4}\left(\chi^T+\frac{\chi}{\beta}\right)\frac{\overline{h''^2}}{L^2}, \tag{7.1.11}$$

where a_1, a_2 are empirical constants.

The balance equations (7.1.2) and (7.1.3*) for the correlations $<e>$ and $<h''^2>$ we have deduced, serve as the basis for approximately (semi-empirically) determining the turbulent exchange coefficients, ν^T and χ^T, in what follows. In the specific case of isotropic turbulence, the rheological relations for the turbulent fluxes can be written as (see (3.3.3*), (3.3.15*), and (5.2.19))

$$J_{\alpha j}^T=-\bar{\rho}\,D^T\partial_j<Z_\alpha>, \tag{7.1.12}$$

$$R_{ij}=-\tfrac{2}{3}\bar{\rho}<e>\delta_{ij}+\bar{\rho}\nu^T\left[\partial_j<V_i>+\partial_i<V_j>\right], \tag{7.1.13}$$

$$q_j^T=-\lambda^T\left[\partial_j<T>-\frac{\partial_j\bar{p}}{\bar{\rho}<C_p>}\right]+\sum_{\beta=1}^{N}<h_\beta>J_{\beta j}^T\cong-\bar{\rho}\chi^T\partial_j<h>+\chi^T\partial_j\bar{p}, \tag{7.1.14}$$

where $\chi^T=\lambda^T/\bar{\rho}<C_p>$ is the turbulent heat conduction coefficient. The second (approximate) expression for q_j^T was written in view of an assumption commonly used in turbulence theory, which asserts that the turbulent Lewis number $\boldsymbol{Le}^T(\equiv\chi^T/D^T)=1$.

7.2. THE METEOROLOGICAL APPROXIMATION

The Boussinesque approximation is pertinent for studying a major part of the atmospheric dynamic systems, specifically, for the analytically treating both natural and forced convective motions of atmospheric gases. As was shown earlier, in the case of mass density changes mainly due to concentrations (temperature) variations in a gravitational field, the hydrodynamic equations can be simplified provided the temperature variations do not exceed several degrees and the volume expansion coefficient, $\rho'/\overline{\rho}T''$, (see (3.3.27)) is smaller than ~ $10^{-3}\ K^{-1}$. For such media, the relative mass density deviation $\rho'/\rho \approx \rho'/\overline{\rho}$ can be neglected in all terms of the hydrodynamic equations of motion except for the terms expressing the influence of a gravitational field on the density fluctuations, ρ', like the Archimedean forces in the equation of motion (3.1.38).

7.2.1. The Oberbeck-Boussinesque Approximation

For this approach the following approximate relations are valid: $\rho \approx \overline{\rho}$, $\rho' \approx 0$, $<V_j> \approx \overline{V}_j$, $V_j'' \approx V_j'$, $<h> \approx \overline{h}$, $h'' \approx h'$, $Z_\alpha'' \approx Z_\alpha'$, etc. Let us rewrite the evolutionary transfer equations (7.1.2) and (7.1.3*) for the correlations $<e>$ and $<h''^2>$ for an averaged horizontally homogeneous flow in the plane $z = const$. Entering the designations

$$V_j' = u_j, \quad h' = H, \quad T' = \Theta, \quad n_\alpha' = \gamma_\alpha \tag{7.2.1}$$

and for simplicity keeping intact the averaging symbols only in the fluctuation terms we will have

$$\frac{\partial b}{\partial t} + \frac{1}{\rho}\frac{\partial}{\partial z} J^T_{<e>z} = -\overline{u_x u_y}\frac{\partial V}{\partial z} + \frac{g}{T}\overline{\Theta u_z} + \frac{g}{n}\sum_{\alpha=1}^{N}\overline{\gamma_\alpha u_z} - \varepsilon_e, \tag{7.2.2}$$

$$\frac{\partial \overline{(H^2/2)}}{\partial t} + \frac{1}{\rho}\frac{\partial}{\partial z} J^T_{<h>z} = -\overline{Hu_z}\left(g + \frac{\partial h}{\partial z}\right) - \varepsilon_h . \tag{7.2.3}$$

Here $<V_j> = V(z)\delta_{3j}$; $V(z)$ is horizontal velocity component; and $b \equiv \overline{e}$ is used for designating the kinetic energy of turbulent fluctuations.

Now let us take advantage of the definitions of the dynamic Richardson and Kolmogorov numbers, ***Rf*** and ***Kf***, respectively, for a multicomponent mixture (see (4.2.33) and (4.2.34)), which specify the relation between the stabilizing effects of buoyancy and stratification and the destabilizing effect of wind shear. Accordingly,

$$\boldsymbol{Rf} + \boldsymbol{Kf} = g\left[\frac{1}{T}\overline{\Theta u_z} + \frac{1}{n}\sum_{\alpha=1}^{N}\overline{\gamma_\alpha u_z}\right]\cdot\left[\overline{u_x u_y}\frac{\partial V}{\partial z}\right]^{-1} =$$

$$= g\frac{\chi^T}{\nu^T}\left[\frac{g}{C_p T}+\frac{1}{T}\frac{\partial T}{\partial z}-\frac{1}{M^*}\frac{\partial M^*}{\partial z}\right]\left(\frac{\partial V}{\partial z}\right)^{-2}=\frac{\boldsymbol{Ri}+\boldsymbol{Ko}}{\boldsymbol{Pr}^T}, \tag{7.2.4}$$

where $\boldsymbol{Pr}^T=\chi^T/\nu^T$ is the turbulent Prandtl number; and $\boldsymbol{Ri}$, $\boldsymbol{Ko}$ are the gradient Richardson and Kolmogorov numbers, respectively. Then the transfer equation for turbulent energy, (7.2.1), takes the form

$$\frac{\partial b}{\partial t}+\frac{1}{\rho}\frac{\partial}{\partial z}J^T_{<e>z}=\nu^T(1-\boldsymbol{Rf}-\boldsymbol{Kf})\left(\frac{\partial V}{\partial z}\right)^2-\varepsilon_e \tag{7.2.5}$$

suitable for numerical simulating the shear flow.

7.2.2. The Turbulent Exchange Coefficients

Later on, we will confine ourselves to stratified flows in local equilibrium in a gravitational field when the second order correlation moments do not vary with time and space (see section 4.3.9). Such a situation arises for certain critical values of the Richardson and Kolmogorov numbers, $\boldsymbol{Rf}_{cr}$ and $\boldsymbol{Kf}_{cr}$, respectively. Then the equations (7.2.3) and (7.2.5) will take the following form (below the index *cr* in $\boldsymbol{Rf}_{cr}$ and $\boldsymbol{Kf}_{cr}$ is omitted):

$$-\overline{u_x u_y}(1-\boldsymbol{Rf}-\boldsymbol{Kf})\frac{\partial V}{\partial z}=\left(\frac{\nu}{\alpha}+\nu^T\right)\frac{b}{a_1^4 L^2}, \tag{7.2.6}$$

$$-\overline{Hu_z}\left(g+\frac{\partial h}{\partial z}\right)=\left(\frac{\chi}{\beta}+\chi^T\right)\frac{\overline{H^2}}{a_2^4 L^2}. \tag{7.2.7}$$

An important feature of the equations (7.2.6) and (7.2.7) is that they comprise only local values of the quantities $\nu(z)$, $\chi(z)$, $\partial V(z)/\partial z$, $\partial h(z)/\partial z$, $b(z)$, and $L(z)$, which ultimately control the desired turbulent transport coefficients, ν^T and χ^T. This embodies the so-called *principle of local similarity* in turbulent transfer based on the hypothesis of the occasional occurrence of *local equilibrium* structures in the turbulent field. Such an approach to modeling the turbulent motions was used in the studies of high-temperature boundary layers (*Ievlev, 1975;1990*).

The semi-empirical expressions for the coefficients ν^T and χ^T involving the algebraic equations (7.2.6) and (7.2.7) can be obtained either by a simple or a more complicated method. Obviously, the latter approach is hardly justified because of the approximate nature of the expressions for ν^T and χ^T. Therefore, it seems more reasonable to focus on the simple semi-empirical method.

Let us assume that the correlation $-\overline{u_x u_y}$ is proportional to the turbulent energy, b, and the correlation $-\overline{Hu_z}$ is proportional to the quantity $b^{1/2}\left(\overline{H^2}\right)^{1/2}$. In this case, it is possible to write

$$-\overline{u_x u_y} = k_1 b(\partial V / \partial z)/(|\partial V / \partial z|) = \nu^T \partial V / \partial z\,, \tag{7.2.8}$$

$$-\overline{Hu_z} = k_2 b^{1/2} (\overline{H^2})^{1/2} = \chi^T (g + \partial h / \partial z)\,, \tag{7.2.9}$$

where k_1, k_2 are the so-called correlation coefficients. Excluding now the quantities $\partial V / \partial z$ and $(\overline{H^2})^{1/2}$ from the equations (7.2.6) and (7.2.7) with the use of the relations (7.2.8) and (7.2.9), we obtain

$$\frac{\nu^T}{L\sqrt{b}} = -\frac{\nu}{2\alpha L\sqrt{b}} + \sqrt{1 - \boldsymbol{Rf} - \boldsymbol{Kf} + \frac{\nu^2}{4\alpha^2 b L^2}}\,, \tag{7.2.10}$$

and

$$\frac{\chi^T}{L\sqrt{b}} = -\frac{1}{\boldsymbol{Pr}}\frac{\nu}{2\beta L\sqrt{b}} + \frac{1}{\boldsymbol{Pr}}\sqrt{\frac{\nu^2}{4\beta^2 b L^2} + \left(\frac{\boldsymbol{Pr}}{\boldsymbol{Pr}^{T*}}\right)^2}\,, \tag{7.2.11}$$

where $\boldsymbol{Pr}$ is the molecular Prandtl number. Here the designation $\boldsymbol{Pr}^{T*} = 1/k_2 a_2^2$ is introduced for the set of empirical constants. The quantities $\boldsymbol{Pr}^{T*}$, α and β, contained in (7.2.10) and (7.2.11), are the additional empirical constants. As we can always enter one of the constants into the linear scale, L, it will be further assumed that $a_1^4 k_1^2 = 1$. Thus, the coefficients ν^T and χ^T prove to be expressed only in terms of the parameters b and L and the local properties of the medium. For $\boldsymbol{Rf} + \boldsymbol{Kf} = 1$, we obtain from (7.2.10) that $\nu^T = 0$; the same value for ν^T should be also adopted for $\boldsymbol{Rf} + \boldsymbol{Kf} > 1$ since in this case the simplified balance equation for turbulent energy, (7.2.5), can not be satisfied for any positive value of ν^T. Additionally, it follows from (7.2.10) and (7.2.11) that the parameter $\boldsymbol{Pr}^{T*}$ represents the turbulent Prandtl number in the domain ($\nu^T \gg \nu$) when both the buoyancy forces and composition stratification do not exert any influence on the turbulent energy balance, i.e. when $\boldsymbol{Rf} + \boldsymbol{Kf} \to 0$.

It would be not superfluous to note that, strictly speaking, in the local-equilibrium approximation the turbulence energy, b, should not be involved as an independent variable in the turbulent transport coefficients because this very energy is determined by the same flow parameters as the coefficients ν^T and χ^T. Nevertheless, the quantity b is entered deliberately in the formulas (7.2.10) and (7.2.11) with the purpose to account indirectly for the "non-equilibrium" nature of the turbulent velocity field (see section 4.3.9).

The formulas

$$\nu^T / \nu = -1/\alpha + a_1^2 (1 - \boldsymbol{Rf} - \boldsymbol{Kf}) \boldsymbol{Re}^{T*}\,, \tag{7.2.12}$$

$$\chi^T / \nu = -1/\beta \boldsymbol{Pr} + \boldsymbol{Re}^{T*} \boldsymbol{Pr}^T / (\boldsymbol{Pr}^{T*})^2 \tag{7.2.13}$$

are easy to obtain also from (7.2.5)–(7.2.9). They allow us to calculate the coefficients ν^T and χ^T for equilibrium conditions from the known values of $\boldsymbol{Rf}_{cr}$ and $\boldsymbol{Kf}_{cr}$. Here $\boldsymbol{Re}^{T*} = L^2(\partial V/\partial z)/\nu$ is the local turbulent Reynolds number.

Finally, from (7.2.12) and (7.2.13) the algebraic equation for defining the turbulent Prandtl number, $\boldsymbol{Pr}^T$, in the local-equilibrium approximation follows:

$$\frac{1}{(\boldsymbol{Pr}^{T*})^2}(\boldsymbol{Pr}^T)^3 + \frac{1}{a_1^2\beta \boldsymbol{Pr}\,\boldsymbol{Re}^{T*}}(\boldsymbol{Pr}^T)^2 + \left(\frac{1}{a_1^2\alpha\,\boldsymbol{Re}^{T*}} - 1\right)\boldsymbol{Pr}^T + \boldsymbol{Ri} + \boldsymbol{Ko} = 0. \quad (7.2.14)$$

In context of the above discussion, let us make some essential remarks. Strictly speaking, experiments provide evidence that the local equilibrium condition is not fulfilled for every turbulent flow pattern. In particular, this is observed in tube flows, in boundary layers, and in mixing layers when the production of turbulence energy is approximately equal to the dissipation in basic or intermediate flow regions (where the largest total velocity change occurs). However, this condition is violated, firstly, in a thin surface layer (near the wall) where the "diffusion" transport of turbulence (mainly caused by molecular viscosity, heat conduction, and pressure fluctuations), is important; and, secondly, in a broad external domain of a boundary layer where both turbulence diffusion and convective terms play a substantial role.

As far as the motions in the gaseous medium of a planetary atmosphere stratified in a gravitational field are concerned, it is important to model the transport processes in free turbulent flows with transverse shears, such as jets and mixing layers. It is known (*Townsend, 1959*) that the energy balance in the mixing layer of a jet strongly differs from the balance in a channel or in a boundary layer (see Figures 4.3.1–4.3.2). There are virtually no domains in a jet where the production of turbulence energy would be balanced by dissipation. In such a case, the most important processes become convective transport and turbulence diffusion. Therefore, when modeling these flows, one may suppose by analogy to homogeneous fluids, that there is a certain *inner equilibrium* in the turbulence structure in reference to the mean velocity and temperature fields (*Ievlev, 1975*).

Accordingly, the turbulent exchange coefficients, ν^T and χ^T, can be expressed in terms of the parameters b and L and local properties of the medium (for example, the parameters $\boldsymbol{Rf}(z)$ and $\boldsymbol{Kf}(z)$). In other words, this approach can be extended also to cases for which the distribution of turbulent energy, b, is not an equilibrium one but depends on the turbulization pre-history and should be determined from the complete differential equation for turbulent energy transfer (7.2.5). In these conditions, the relaxation time for inner equilibrium is less than the time needed for the total turbulence intensity to reach the level corresponding to the equality of production and dissipation of the turbulent energy. This is why this equilibrium occurs often.

Based on this perception, we can accept the following dependencies,

$$\nu^T/bL = f(\nu/bL, \boldsymbol{Rf}, \boldsymbol{Kf}), \quad (7.2.15)$$

$$\chi^T/bL = g(\nu/bL, \boldsymbol{Rf}, \boldsymbol{Kf}), \quad (7.2.16)$$

as universal. They are appropriate when turbulence production equals its dissipation and, hence, are valid for non-equilibrium cases. Therefore, the relations (7.2.12) and (7.2.13) derived for local-equilibrium conditions, can be used also as the desired relations between the individual correlation properties of the turbulent flow when the turbulent energy, b, is determined from the complete evolutionary transfer equation (7.2.5). Let us recall that in order to close the model equations (7.2.12), (7.2.13), (7.2.14), and (7.2.5) it is necessary to specify a procedure to determine the turbulence scale, L. The latter can be defined as an empirical function taking into account the geometry of the flow, or it can be found from solving the semi-empirical model equation, as it was shown in subsection 4.3.7.

7.3. NUMERICAL CALCULATION OF THE TURBULENT EXCHANGE COEFFICIENTS

Now, let us apply the results of the previous study to model the turbulent viscosity and heat conduction coefficients in a natural medium of our specific interest. To pursuit the idea to success the content of Chapter 6, we take the upper terrestrial atmosphere in the height range $90 \le z \le 130$ km comprising the homopause region as an example of such a medium.

7.3.1. Basic Equations and Boundary Conditions

In order to obtain the turbulent exchange coefficients, the system of equations (7.2.5), (7.2.10), (7.2.11), and (7.2.14) is to be solved numerically. For the turbulence scale, L, one may use the one defined by the simplified differential equation (4.3.43)

$$L = -\kappa c^{1/4} \Psi / (1 + az) \partial \Psi / \partial z . \tag{7.3.1}$$

This equation allows us to calculate L for shear flows in terms of the mean local properties of the medium for free convection or various modes of flow stratification. Here $\Psi = b^{1/2} / L$, $\kappa = 0{,}4$ is the Karman constant; the coefficient a having the dimension L^{-1} is a function of the Rossby number, $\boldsymbol{Ro}$; c is an empirical constant. The correction factor $(1+az)$ to the *Laikhtman* formula in expression (7.3.1) originally proposed in (*Bykova, 1973*), permits to obtain a more realistic height distribution of the turbulent exchange coefficients, ν^T and χ^T (*Vager and Nadezhina, 1979*). Generally speaking, a simplified formula like (7.3.1) fails to be valid when all terms in the evolutionary transfer equation (7.2.5) for the turbulence energy are taken into account. Unfortunately, more sophisticated differential equations for L were not confirmed to be superior in terms of their universality and accuracy (*Ievlev, 1975; Turbulence: Principles and Applications, 1980*).

The following boundary conditions for the system of equations (7.2.5), (7.2.10), (7.2.11), and (7.2.14) were assigned:

- $\overline{\rho}\partial b/\partial z = 0$ at $z = z_0$, i.e. the requirement of "impermeability" of the thermosphere for turbulent energy flows at its lower boundary $L(z_0) = L_0$; this means that the turbulent diffusion at the boundary is slowed down;

- $b = 0$ at $z = z_0 + d$, i.e. the requirement of turbulence attenuation at the upper boundary of the turbopause; it was taken $d = 40$ km.

The stated initial-boundary problem for the system (7.2.5), (7.2.10), (7.2.11), (7.2.14), and (7.3.1), together with the specified time-independent boundary conditions and arbitrary initial conditions can be solved by the relaxation method (*Samarsky, 1977, 1978*; *Marov and Kolesnichenko, 1987*). In other words, the quantities b, ν^T, χ^T are subjected to determination at $t \to \infty$.

The numerical value of the coefficient a_1 in (7.1.10) was determined as follows. As the formula (7.2.10) is universal, it can be used in regions with developed turbulence as well as when buoyancy forces do not affect the turbulent energy balance ($\boldsymbol{Rf} + \boldsymbol{Kf} \to 0$). Then (7.2.10) evolves to the well known Kolmogorov relation $\nu^T \cong b^{1/2}L$. On the same assumptions, the Richardson-Obukhov law, $\nu^T = a_1^{4/3}\varepsilon_e^{1/3}L^{4/3}$, follows from (7.1.10) (see section 1.1.1). The latter has a good accuracy for the atmosphere for $a_1 = 0.18$.

The parameter c_1 in (7.1.8) is usually of order unity (see Chapter 4). For a mixture flow having the parameters ρ, ν = *const*, it is possible to take advantage of the transfer equation (7.2.5) for finding the coefficient c_2 in the immediate proximity to the upper boundary of the turbopause where $\nu^T << \nu$; then $b/\alpha a_1^4 L^2 = -c_2 \partial^2 b/\partial z^2$. Hence, bearing in mind that close to the "wall" $\sqrt{b} \propto z_*$ and $a_1^2 L \propto \kappa z^*$ $(z^* = z - d)$, we find that $c_2 = 1/2\alpha\kappa^2$. The scale $L \sim 0.5$ km adopted at 90 km height conforms both to the sizes of globules observed experimentally in smoke traces and to the turbulence scales evaluated from the structure functions.

The molecular kinematic viscosity and thermal diffusivity coefficients were determined from the formulas (6.1.13) for multicomponent gas composed of N_2, O_2 and O. The enthalpy gradient for the mixture was computed from the formula

$$\partial < h > / \partial z = < C_p > \partial < T > / \partial z + \sum_{\alpha=1}^{N} < h_\alpha > \partial < Z_\alpha > / \partial z$$

using the height profiles of temperature, $< T(z) >$, and mixture species concentrations, $< Z_\alpha(z) >$, taken from the standard atmosphere. The heat production of a substance, h_α^0, under standard conditions were adopted (*Shchetinkov, 1965*).

Figure 7.3.1 shows the gradient Richardson number, $\boldsymbol{Ri}$, depending on height for the mentioned mixture of atmospheric gases. Besides the model's temperature height and number density profiles, the vertical wind velocity gradient was used, based on numerous experimental data (see *Middle Atmosphere Program, Handbook for MAP, 1985*

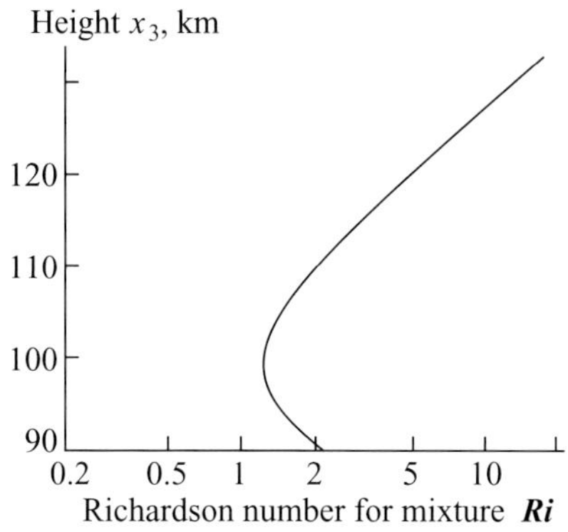

Fig. 7.3.1. Gradient Richardson number, ***Ri***, depending on height in the terrestrial lower thermosphere.

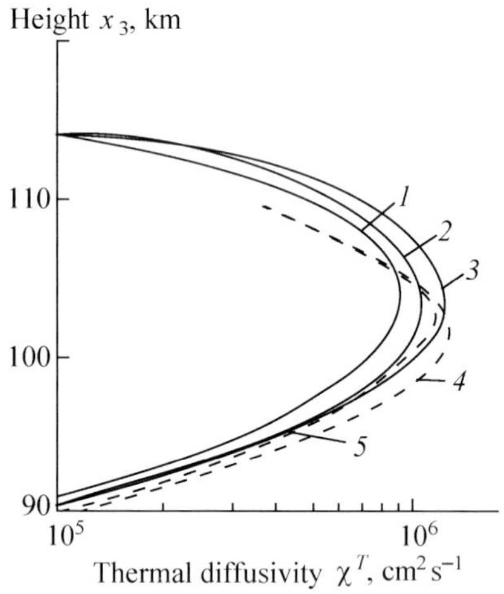

Fig. 7.3.2. The thermal turbulent diffusivity factor, χ^T, depending on height in the terrestrial lower thermosphere: Curves 1 to 3 – solutions of the turbulent energy equation (7.2.5) with $\boldsymbol{Pr}^{T*} = 3$, 2.6, and 2.2, respectively; $\alpha = \beta = 0.5$; Curves 4 and 5 – calculations based on the algebraic relations (7.2.12)–(7.2.13) with $\boldsymbol{Pr}^{T*} = 2.2$ and 2.6, respectively; $\alpha = \beta = 1$.

for references) and averaged over a considerable time interval. The free parameters of the problem, α and β, varied from 0.125 to 1 and $\boldsymbol{Pr}^{T*}$ changed from 0.9 to 4.

The numerical values of the computed turbulent thermal diffusivity coefficients are shown as an example in Figures 7.3.2 and 7.3.3. In order to choose the most correct values of the fitting parameters α, β and $\boldsymbol{Pr}^{T*}$, the computed viscous dissipation rate profiles for the turbulent energy, ε_e, with regard to the modeling results using (7.2.5), (7.2.10), (7.2.11) for various values of these parameters, were compared with the aver-

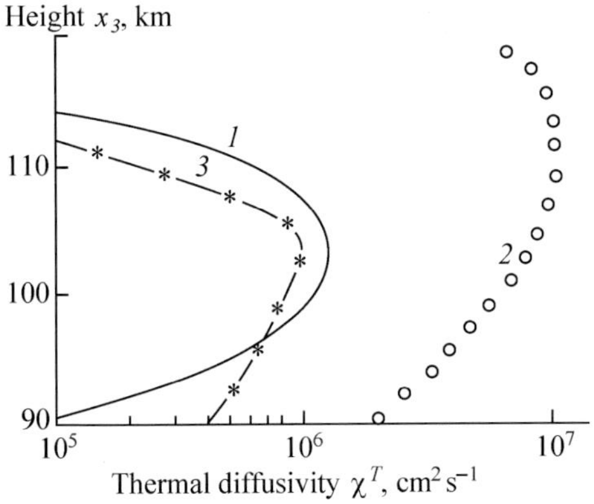

Fig. 7.3.3. The thermal turbulent diffusivity factor, χ^T, depending on height in the terrestrial lower thermosphere: 1 – our calculation for $\boldsymbol{Pr}^{T*} = 2.2$, $\alpha = \beta = 0.5$; 2 – calculation of *Gordiets et al., 1979*; 3 – the profile corresponding to the standard atmosphere *CIRA-72* model.

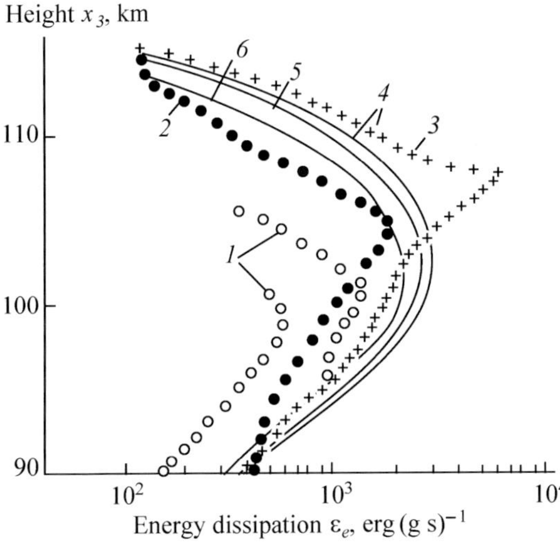

Fig. 7.3.4. Specific viscous dissipation rate of turbulent energy, ε_e, depending on height in the terrestrial lower_thermosphere: curves 1, 2, and 3 – the measurements;
4, 5, and 6 – the calculations for $\alpha = \beta = 0.5$ and $\boldsymbol{Pr}^{T*} = 3$, 2.6 , and 2.2, respectively.

aged measured values of ε_e. The energy dissipation profiles from the model, ε_e, corresponding to $\boldsymbol{Pr}^{T*} = 2.2$ and $\alpha = \beta = 0.5$ are distinguished as the best fit for the experimental curves (Figure 7.3.4).

The peak values in the height distribution of the free parameters α and β are responsible for the height profiles of the coefficients ν^T and χ^T. These parameters were

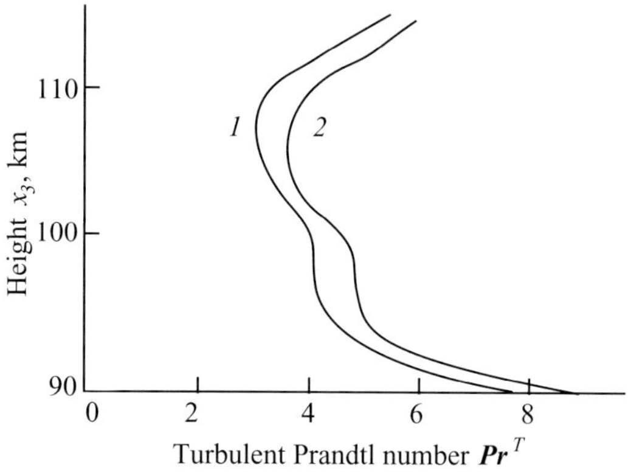

Fig. 7.3.5. Turbulent Prandtl number depending on height in the terrestrial lower thermosphere; the calculations for $Pr^{T*} = 2.2$, $\alpha = \beta = 0.5$ (curve 1);
$Pr^{T*} = 2.6$, $\alpha = \beta = 0.5$ (curve 2).

finally chosen taking into account the results of the computer modeling summarized in Chapter 6. It was shown, among others, that the theoretical height profiles of gas temperature, obtained with regard to the IR emission, coincide best with the empirical data when χ^T does not exceed $(1 \div 2) \times 10^6$ cm^2/s and has a maximum at about 105 km. Obviously, this result is not consistent with some other semi-empirical models (the curves 2 and 3 in Figure 7.3.3). Although the procedure developed for evaluating the parameter χ^T is addressed as the most accurate, we may admit that the discrepancy can be also partially caused by an inaccurate assessment for the relative contribution of the *IR* emission in the overall heat budget of the lower thermosphere.

The computed height distribution of the turbulent Prandtl number, is shown in Figure 7.3.5. Analysis of the turbulent energy balance throughout the whole lower thermosphere based on the transfer equation (7.2.5), has shown that up to 110 km the turbulence energy production and dissipation are approximately balanced. Above 110 km, where turbulence diffusion becomes substantial, such a balance is broken. Calculations of the turbulent coefficients using the algebraic relations (7.2.12)–(7.2.13) showed that the simplified balance equations (7.2.6) and (7.2.7) are applicable only below this height. Comparing Figure 7.3.1 to Figure 7.3.5, it follows that the turbulent Prandtl number, $\boldsymbol{Pr}^T$, grows with the Richardson number, $\boldsymbol{Ri}$. In fact, the buoyancy forces hamper the vertical heat (mass) transport stronger than the vertical transport being created by pressure fluctuations.

Special attention was also paid to the relative contribution of the quantities Q_T^T and $Q_{<e>}^T$ (formulas (6.1.21) and (6.1.26)) into the total heat balance of the lower thermosphere. These quantities are responsible for the turbulent heat conduction and the turbulent energy dissipation due to molecular viscosity. The calculations performed testify that heating caused by dissipation occurs in the height range from 90 to 105 km, while above 105 km turbulent heat conduction predominates and the atmospheric gas cools, in support to the earlier assertion.

In conclusion, let us notice that the lower thermosphere, and specifically the homopause, is poorly covered with experimental data though it has a large impact on the structure and thermal conditions of the entire near-Earth space. Obviously, for the other planets the situation is even worse. This is why the developed method of semi-empirical modeling the turbulent exchange coefficients is extremely valuable being focused on regions with a potential to improve the prognostic models. At the same time, for the general 3D case, the anisotropy of these coefficients is to be taken into account and therefore more universal approaches including the precise differential equations for defining the external turbulence scale should be developed.

Summary

- *The procedure of semi-empirically modeling the isotropic turbulent exchange coefficients was developed for a turbulized multicomponent gaseous flow stratified in a gravitational field and with a transverse hydrodynamic velocity shear. This approach is based on the evolutionary transfer equations for the turbulent energy and the mean square enthalpy fluctuations of a mixture.*

- *The universal algebraic expressions for the turbulent viscosity heat conductivity coefficients of a mixture in the vertical direction were obtained. These expressions depend on the local values of parameters such as the kinetic energy of turbulent fluctuations, the dynamic Richardson and Kolmogorov numbers, and the external turbulence scale. The algebraic equation for the turbulent Prandtl number was also derived. It was shown that when turbulent energy in the expressions for the turbulent exchange coefficients is used as an argument and an additional differential equation is incorporated, the non-equilibrium nature of turbulence with respect to the mean velocity and temperature fields can be considered approximately. The free flows within the transverse velocity shear layers in the planetary atmosphere serve as an example of a medium where non-equilibrium turbulence patterns occur.*

- *The developed procedure was applied to numerically modeling the turbulent viscosity and thermal diffusivity coefficients in the lower terrestrial thermosphere and the results were compared with the available experimental data. In the general three-dimensional case, however, the anisotropy of the turbulent exchange coefficients should be taken into account, which requires to develop a refined approach to model them.*

CHAPTER 8

STATISTICAL PARAMETERS OF TURBULENCE: MODELING FROM FLUCTUATIONS OF THE REFRACTIVE INDEX

Studying dynamic properties of planetary atmospheres based on the measurements of fluctuations of refractive index is one of the important applications of the turbulence theory for multicomponent gases.

Probing the atmosphere at optical and radio wavelengths is done widely to investigate the atmospheric structural parameters and their variations. Various experimental methods are employed. Among them regular space monitoring using the occultation technique becomes increasingly important. This technique rests on the models of probing beam interaction with a medium and can be used for real time data processing and interpretation. A specific feature of such an interaction is electromagnetic wave propagation through a turbulent atmosphere which results in very complicated patterns in the recorded signal. It hampers the data analysis, but at the same time contains valuable information on the turbulence properties and atmospheric dynamics.

The basic principles on how to extract information on statistical properties of turbulence from the patterns of light wave interaction with turbulized gaseous media (*Rytov, 1937; Obukhov, 1953*) were developed successfully and extended for particular applications (see, e.g., *Tatarskiy, 1967; Gurvich, 1968; Gurvich et al., 1976*). The middle atmosphere of a planet, and in particular the Earth's, is an example. Unlike the boundary layer adherent the surface, atmospheric layers located close and above the stratopause are rather poorly studied. Although it is known that the vertical and horizontal turbulence structure in the free atmosphere is non-uniform and that there are layers with sharp wind velocity and temperature gradients, the relationship of these features with the macro scale meteorological phenomena is difficult to reveal. Moreover, such layers cause regular internal hydrodynamic waves and sometimes may be a source of turbulent heating (*Aleksandrov et al., 1990; Gavrilov et al, 1974*).

Unfortunately, there is no comprehensive information on the fluctuation spectrum of the refractive index of atmospheric gases and its variations taking into account the layered structure of the atmosphere, coupled with meteorological phenomena. A reasonable assumption based on the measured microstructure of the wind velocity and the temperature in such layers, is that the respective spectra follow a power law. Then, when considering the influence of atmospheric turbulence on the propagation of probing radiation, it is possible to use the theory of locally-uniform and locally-isotropic turbulence for the domains with much smaller sizes than the external turbulence scale, L, related to the characteristic scale of large, anisotropic, energy carrying eddies (*Tatarskiy, 1967*).

Indeed, Kolmogorov's law, $\Phi_n(\kappa) \propto C_n^2 \kappa^{-11/3}$, (*Kolmogorov, 1941*) fairly well approximates the fluctuation spectrum of the refractive index in the middle atmosphere within the inertial wave number range 10^{-5} cm$^{-1} \leq \kappa \leq 10^{-1}$ cm^{-1}. Then the structure characteristic of the refractive index, C_n^2, proves to be the only undefined parameter describing the intensity of turbulent fluctuation field.

Basically, experimental determination of the statistical parameters of electromagnetic waves penetrating a turbulized layer, enables to define the quantity C_n^2. The key element in the theory for fluctuations caused by wave propagation through a turbulized medium, is the structure characteristic of the refractive index entering, for example, the dispersion formula for the logarithm of radiation intensity fluctuations. This allows us to find the parameter C_n^2 from the measured optical spectra. On the other hand, the structure characteristic, C_n^2, is functionally coupled with two fundamental quantities in the turbulence theory of the averaged velocity and temperature fluctuation fields, namely, the dissipation rate of turbulence energy, ε_e, (see equation (4.2.25)) and the dispersion rate of the potential temperature fluctuations, ε_h, (4.3.2). This gives a different (hydrodynamic) method for finding the functional relation of the quantity C_n^2 to the averaged thermohydrodynamic parameters of the atmosphere and their gradients.

Obviously, such an approach opens the opportunity to evaluate some important statistical characteristics of atmospheric turbulence. In particular, these are the external and internal turbulence scales, the turbulent exchange coefficients, etc., which can be found from measurements of fluctuations in the refractive index of air. This basic concept was in support of the idea to monitor continuously from space the ozone depletion in the middle terrestrial atmosphere by the star occultation technique (*Bertaux et al., 1988; Ioltuhovski and Marov, 1998*) in the framework of the space project *GOMOS* (Global Ozone Monitoring by Occultation of Stars). In the following sections we will consider some feasible approaches to model the external turbulence scale, L, based on the known height distribution of the structure characteristic of the refractive index, C_n^2. In turn, C_n^2 is going to be determined from the measured fluctuation dispersion of the logarithm of the optical light beam intensity emitted from a satellite to probe the atmosphere.

8.1. ALGEBRAIC EQUATIONS FOR MODELING THE TURBULENT EXCHANGE COEFFICIENTS

The system of hydrodynamic equations of mean motion for mixtures (3.2.4)–(3.2.8) is appropriate to model the middle atmosphere. The defining relations (3.3.3), (3.3.15), and (3.3.19) for the turbulent fluxes of diffusion, heat, and the turbulent stress tensor comprise the turbulent exchange coefficients (tensors, in the general case), which should be set *a priori.* The Kolmogorov hypothesis is commonly employed as a basic principle of local similarity in the theory of semi-empirically modeling the turbulent coefficients for homogeneous fluids. This means that the turbulent exchange coefficients, such as ν^T, χ^T, and the turbulent energy dissipation rate, ε_e, at a point in the developed turbulent flow depend only on the properties of the medium, the volume forces at that point (see section 3.3.2), the turbulence energy, $<e>$, and the local external turbulence scale, $L(\boldsymbol{r})$.

Accordingly, the quantities ν^T and ε_e in the case of quasi-equilibrium turbulence will be expressed as $\nu^T = c_\mu L\sqrt{<e>}$, and $\varepsilon_e = <e>^{3/2}/L$. The evolutionary transfer equation (4.2.28) for turbulent energy was used to define the parameter $<e>$. However, it should be emphasized that the external turbulence scale, L, remains undetermined. In the case of multicomponent turbulence more general semi-empirical equations for the turbulent exchange coefficients are obtained using the derived quasi-equilibrium algebraic transfer equations (see subsection 4.3.9) for the single-point second moments of the thermohydrodynamic pulsating parameters of the flow.

8.1.1. Modeling the Turbulent Exchange Coefficients

Let us consider a quasi-equilibrium approximation of the multicomponent turbulence model when the convection and diffusion terms in the differential transfer equations (4.2.9), (4.3.1), and (4.3.9) for the second moments of the pulsating thermohydrodynamical flow parameters are omitted. Such a model can be used to find algebraic relations between the correlations R_{ij}, $<e>$, q_j^T and $<h''^2>$. We have

$$-\bar{\rho}\frac{\partial}{\partial t}\left(\frac{R_{ij}}{\bar{\rho}}\right) \cong 0 \cong (1-K_{p2})[R_{ik}\partial_k<V_j> + R_{jk}\partial_k <V_i> + \bar{\rho}G_{ij}] +$$

$$+K_{p1}^*\left(\frac{<e>^{1/2}}{L}\right)[R_{ij} + \tfrac{2}{3}\delta_{ij}\bar{\rho}<e>] - \tfrac{2}{3}\delta_{ij}\{ K_{p2}[R_{sk}\partial_k<V_s> -$$

$$-\bar{\rho}G] - K_{e1}^*\bar{\rho}\frac{<e>^{3/2}}{L}\Big\}, \tag{8.1.1}$$

$$\bar{\rho}\frac{\partial}{\partial t}\left(\frac{q_j^T}{\bar{\rho}}\right) \cong 0 \cong R_{jk}\left[\partial_k <h> - \frac{\partial_k \bar{p}}{\bar{\rho}}\right] - (1-K_{s2})q_k^T\partial_k <V_j> +$$

$$+ (1-K_{s2})\bar{\rho}G_j - K_{s1}^*\frac{<e>^{1/2}}{L}q_j^T, \tag{8.1.2}$$

$$\bar{\rho}\frac{\partial <e>}{\partial t} \cong 0 \cong R_{ik}\partial_k <V_j> + \bar{\rho}G - K_{e1}^*\bar{\rho}\frac{<e>^{3/2}}{L}, \tag{8.1.3}$$

$$\bar{\rho}\frac{\partial}{\partial t}\frac{<h''^2>}{2} \cong 0 \cong -q_k^T\left[\partial_k <h> - \frac{\partial_k \bar{p}}{\bar{\rho}}\right] - K_{h1}^*\bar{\rho}\frac{<e>^{1/2}}{L}<h''^2>. \tag{8.1.4}$$

The closed system of algebraic equations (8.1.1)–(8.1.4) links the second order correlation moments to the gradients of the defining parameters of the averaged mixture flow. It represents the *K-theory* for multicomponent turbulence we earlier discussed. Generally speaking, the system permits to obtain the semi-empirical expressions for the turbulent exchange coefficients.

The following designations were used:

$$K_{p1}^{*}=K_{p1}\left(1+\frac{1}{\boldsymbol{Re}^{T}}\frac{K_{e2}}{K_{p1}}\right),\qquad K_{e1}^{*}=K_{e1}\left(1+\frac{1}{\boldsymbol{Re}^{T}}\frac{K_{e2}}{K_{e1}}\right),$$

$$K_{s1}^{*}=K_{s1}\left(1+\frac{1+1/\boldsymbol{Pr}}{2\boldsymbol{Re}^{T}}\frac{K_{e2}}{K_{s1}}\right),\qquad K_{h1}^{*}=K_{h1}\left(1+\frac{1}{\boldsymbol{Pr}\,\boldsymbol{Re}^{T}}\frac{K_{h2}}{K_{h1}}\right). \tag{8.1.5}$$

The coefficients K^{*} in (8.1.5) depend on the turbulent Reynolds number, $\boldsymbol{Re}^{T}(\equiv \nu^{-1}L<e>^{1/2})$, and the molecular Prandtl number, $\boldsymbol{Pr}\equiv\nu/\chi$, , where $\chi=\lambda/\rho C_{p}$ is the molecular thermal diffusivity. Note that the case in which $\boldsymbol{Re}^{T}$ is large and all coefficients are $K^{*}\equiv K$, is known as the *super-equilibrium* approximation (*Donaldson, 1972*). Although small Reynolds numbers are incompatible with the hypothesis of locally-equilibrium turbulence, the terms describing the effects inherent to a laminar flow were retained in the equations (8.1.1)–(8.1.4). This is because in the following analysis it is convenient to use the formulas for the turbulent exchange coefficients obtained for locally-equilibrium conditions and the quantity $<e>$ determined from the complete differential transfer equation (7.2.5).

As an elementary example of semi-empirically modeling the turbulent exchange coefficients, let us consider a free horizontal two-dimensional shear flow when $<V_{j}>=\{<V(z)>,0,0\}$, $g_{j}=(0,0,-g)$, and $<T>=<T(z)>$. Here *x*-coordinate is horizontally aligned with the main flow and the *z*-coordinate is directed vertically. In this case the initial system of equations (8.1.1)–(8.1.4) for the correlations R_{ij}, $<e>$, q_{j}^{T} and $<h''^{2}>$ allows us to obtain the following rheological relations for the pair correlations (the turbulent transport fluxes) if the medium is stratified in a gravitational field:

$$<e>=\frac{1}{K_{e1}^{*}}L^{2}\boldsymbol{Pr}^{T}\left(\frac{\partial<V>}{\partial z}\right)^{2}\Psi(1-\boldsymbol{Rf}-\boldsymbol{Kf}), \tag{8.1.6}$$

$$R_{xz}=\overline{\rho}\,L<e>^{1/2}\boldsymbol{Pr}^{T}\Psi\frac{\partial<V>}{\partial z}, \tag{8.1.7}$$

$$R_{zz}=-\frac{2}{3}\overline{\rho}<e>+\frac{4}{3}\overline{\rho}<e>E_{2}\frac{0{,}5+\boldsymbol{Rf}+\boldsymbol{Kf}}{\boldsymbol{1}-\boldsymbol{Rf}-\boldsymbol{Kf}}, \tag{8.1.8}$$

$$q_{x}^{T}=\Psi L^{2}\overline{\rho}\frac{1}{K_{s1}^{*}}(1-\boldsymbol{Pr}^{T}-K_{s2})\frac{\partial<V>}{\partial z}\left(g+\frac{\partial<h>}{\partial z}\right), \tag{8.1.9}$$

$$q_z^T = -\bar{\rho} L < e >^{1/2} \Psi \left(g + \frac{\partial < h >}{\partial z} \right), \tag{8.1.10}$$

$$< h''^2 > = L^2 \Psi \frac{1}{K_{h1}^*} \left(g + \frac{\partial < h >}{\partial z} \right)^2. \tag{8.1.11}$$

The dynamic Richardson and Kolmogorov numbers, $\boldsymbol{Rf}$ and $\boldsymbol{Kf}$, respectively, account for temperature and chemical composition stratification for turbulence maintenance. These numbers, as well as the turbulent Prandtl number, $\boldsymbol{Pr}^T$, and the characteristic function Ψ for the shear flow conditions, are defined by the expressions

$$\boldsymbol{Rf} \equiv - \frac{g q_z^T}{< C_p >< T > R_{xz} \dfrac{\partial < V >}{\partial z}} = \frac{\chi_{zz}^T}{\nu_{xz}^T} \frac{g \left[g + \dfrac{\partial < h >}{\partial z} \right]}{< C_p >< T > \left(\dfrac{\partial < V >}{\partial z} \right)^2} = \frac{\boldsymbol{Ri}}{\boldsymbol{Pr}^T}, \tag{8.1.12}$$

$$\boldsymbol{Kf} \equiv \frac{g \sum_{\alpha=1}^{N} h_\alpha J_{\alpha z}^T}{< C_p >< T > R_{13} \dfrac{\partial V}{\partial z}} = - \frac{\chi_{zz}^T}{\nu_{xz}^T} \frac{g \sum_{\alpha=1}^{N} \dfrac{\partial < Z_\alpha >}{\partial z}}{< C_p >< T > \left(\dfrac{\partial < V >}{\partial z} \right)^2} = \frac{\boldsymbol{Ko}}{\boldsymbol{Pr}^T}, \tag{8.1.13}$$

$$\boldsymbol{Pr}^T \equiv \frac{\nu_{xz}^T}{\chi_{zz}^T} = \gamma \frac{\boldsymbol{Rf}_2 - \boldsymbol{Rf} - \boldsymbol{Kf}}{\boldsymbol{Rf}_1 - \boldsymbol{Rf} - \beta \boldsymbol{Kf}}, \qquad \Psi = \Psi_0 \frac{\boldsymbol{Rf}_0 - \boldsymbol{Rf} - \alpha \boldsymbol{Kf}}{1 - \boldsymbol{Rf} - \boldsymbol{Kf}}. \tag{8.1.14}$$

In turn, the parameters α, β, γ, Ψ_0 and $\boldsymbol{Rf}_0$, $\boldsymbol{Rf}_1, \boldsymbol{Rf}_2$ depending on the Reynolds number, $\boldsymbol{Re}^T$, are defined in terms of the empirical constants of the turbulence model as follows:

$$\alpha = \frac{1 + 2E_2}{1 + 2E_2 + 1{,}5E_1}, \qquad \beta = \frac{2 + E_2}{2 + E_2 + 3E_1}, \qquad \gamma = \frac{K_{s1} E_2}{K_{e1}} \cdot \frac{2 + 4E_2 + 3E_3}{2 + E_2 + 3E_1}, \tag{8.1.15}$$

$$\boldsymbol{Rf}_0 = \frac{1 - E_2}{1 + 2E_2 + 1{,}5E_1}, \quad \boldsymbol{Rf}_1 = \frac{1 - E_1}{1 + 0{,}5E_2 + 1{,}5E_1}, \quad \boldsymbol{Rf}_2 = \frac{1 - E_2}{1 + 2E_2 + 1{,}5E_3}, \tag{8.1.16}$$

$$\Psi_0 = \frac{2 + 4E_2 + 3E_1}{3K_{s1}}, \tag{8.1.17}$$

where

$$E_1 = \frac{K_{e1}^* (1 - K_{s2})}{K_{h1}^*}, \qquad E_2 = \frac{K_{e1}^* (1 - K_{p2})}{K_{p1}^*}, \qquad E_3 = \frac{K_{e1}^* (1 - K_{s2})}{K_{s2}^*}. \tag{8.1.18}$$

Table 8.1.1. Model Values of the Parameters $\alpha, \beta, \ldots$,

E_1	E_2	E_3	α	β	γ	Ψ_0	$\boldsymbol{Rf}_0$	$\boldsymbol{Rf}_1$	$\boldsymbol{Rf}_2$	$\boldsymbol{Pr}^*$	a	b	c	D
2.25	0.12	0.11	0.28	0.24	0.34	0.76	0.18	0.19	0.55	0.99	2.29	0.65	-0.87	1.16

Comparative analysis of the parameters (8.1.16) calculated for various semi-empirical turbulence models (see, e.g., *Deardorff, 1973; Andre et al., 1976; Levellen, 1980*), has shown that the parameters $\boldsymbol{Rf}_0$, $\boldsymbol{Rf}_1$ and $\boldsymbol{Rf}_2$ are positive definite in case of developed turbulence (i.e., for $\boldsymbol{Re}^T \gg 1$) where the inequality $\boldsymbol{Rf}_1 < \boldsymbol{Rf}_2$ is always correct and the condition $\boldsymbol{Rf}_0 < \boldsymbol{Rf}_1$ is valid if $E_1 > E_2 + E_3$. The last inequality holds for all known models (see Table 8.1.1). Thus there is some critical value of the Richardson number, $\boldsymbol{Rf}_c = \boldsymbol{Rf}_0 < 1$, limiting the combination $\boldsymbol{Rf} + \alpha\boldsymbol{Kf}$ in the developed turbulent flow. When $\boldsymbol{Rf} + \alpha\boldsymbol{Kf} > \boldsymbol{Rf}_0$, the characteristic function $\Psi < 0$, and therefore the formulas (8.1.6)–(8.1.11) loose the physical sense which corresponds to turbulence decay.

Moreover, using (8.1.14) it is possible to find the critical values of the Prandtl $\boldsymbol{Pr}_c^T$ and Richardson $\boldsymbol{Ri}_c = \boldsymbol{Rf}_c\boldsymbol{Pr}_c^T$ numbers which also represent limiting values in the turbulent flow. For $\boldsymbol{Ri} > \boldsymbol{Ri}_c$, these formulas make physical sense provided that the characteristic function is $\Psi \equiv 0$. The critical values of these parameters computed for the model from (*Mellor, 1973*) are $\boldsymbol{Ri}_c = 2.2$ and $\boldsymbol{Pr}_c^T = 12.6$.

8.1.2. Turbulent Exchange Coefficients

For the turbulent Prandtl number, $\boldsymbol{Pr}^T$, it is convenient to use the expression linking the $\boldsymbol{Pr}^T$ with the gradient Richardson $\boldsymbol{Ri}$ and Kolmogorov $\boldsymbol{Ko}$ numbers. The following formula for $\boldsymbol{Pr}^T$ can be easily found from (8.1.14):

$$\boldsymbol{Pr}^T = 0.5\boldsymbol{Pr}^* + (a\boldsymbol{Ri} + b\boldsymbol{Ko}) + [0.25\boldsymbol{Pr}^{*2} + (a\boldsymbol{Ri} + b\boldsymbol{Ko})^2 - c\boldsymbol{Ri} - d\boldsymbol{Ko}]^{1/2}, \qquad (8.1.19)$$

where

$$a = \frac{2 + E_2 + 3E_1}{4(1 - E_2)}, \qquad b = \frac{2 + E_2}{4(1 - E_2)}, \qquad c = \boldsymbol{Pr}^* \frac{2 + 7E_2 + 6E_3 - 3E_1}{4(1 - E_2)},$$

$$d = \boldsymbol{Pr}^* \frac{2 + 7E_2 + 6E_3}{4(1 - E_2)}, \qquad \boldsymbol{Pr}^* = \frac{K_{s1}^* E_2}{K_{e1}^*}. \qquad (8.1.20)$$

The radicand in (8.1.19) is positive for any $\boldsymbol{Ri}$ and $\boldsymbol{Ko}$ provided that $E_1 > E_2 + E_3$. The combination of the model parameters, entering in (8.1.19) and designated as $\boldsymbol{Pr}^*$, has the meaning of the turbulent Prandtl number for the special flow condition when the buoyancy forces do not affect the turbulent energy balance, i.e. when $\boldsymbol{Ri} \to 0$ and

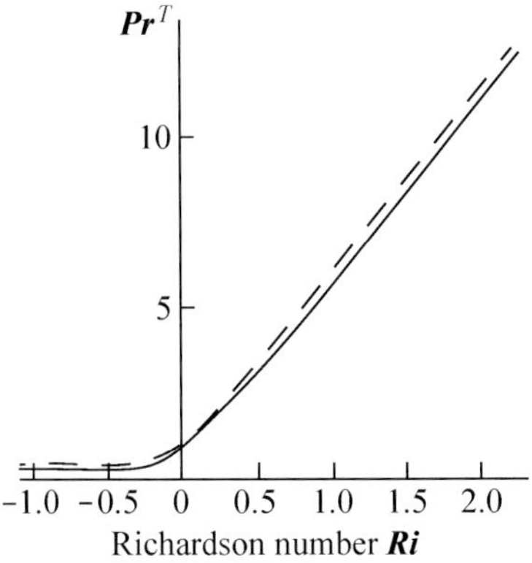

Fig. 8.1.1. Functional dependence of the Prandtl number, Pr^T, on the gradient Richardson number, Ri, derived on the basis of the "super-equilibrium" approximation: $Ko = 0$ (solid curve) and $Ko = 0.1$ (dashed curve).

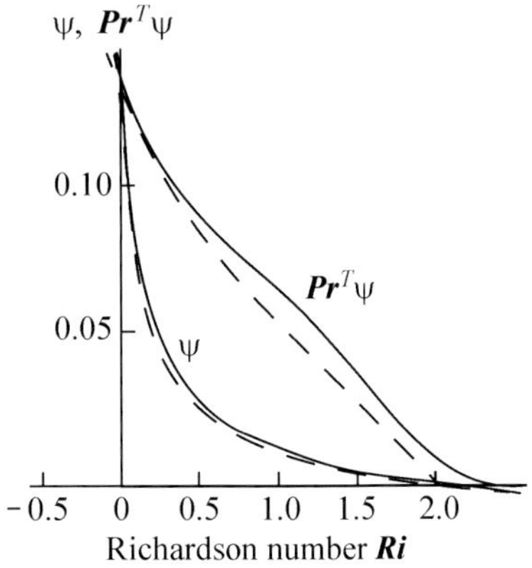

Fig. 8.1.2. Functional dependence of the functions Ψ and $\Psi\, Pr^T$ on the gradient Richardson number, Ri, derived on the basis of the "super-equilibrium" approximation: $Ko = 0$ (solid curve) and $Ko = 0.1$ (dashed curve).

$Ko \to 0$. Calculating the Pr^T number for various turbulence models shows that $Pr^* \sim 1$. Obviously, as follows from equation (8.1.19), for atmospheric stratification close to neutral, $Pr^T \sim Pr^*$, whereas the quantity Pr^T decreases with growing instability and conversely, it sharply increases for stabilizing stratification, that is, with growing Ri and Rf (see Figure 8.1.1). The functional dependence of the most important characteristics of multicomponent turbulent media, such as the characteristic function, Ψ, and the combination $\Psi\, Pr^T$, on the main flow gradients represented by Ri and Ko, is shown in Figure 8.1.2. This dependence was obtained from the equations (8.1.1)–(8.1.5) in the "super-equilibrium" approximation. In that case, the Kolmogorov number for the main flow is the only free fitting parameter.

Using the relations (8.1.6)–(8.1.10) for the correlation moments, we can obtain the following expressions for the turbulent exchange coefficients:

$$\nu^T_{xz} \equiv \frac{R_{xz}}{\bar{\rho}\partial < V > / \partial z} = \frac{L^2 Pr^T \Psi^{3/2}}{K_{e1}^{*1/2}} \sqrt{Pr^T - Ri - Ko}\, \frac{\partial < V >}{\partial z}, \tag{8.1.21}$$

$$\nu^T_{zz} \equiv \frac{R_{zz} + 2/3\,\bar{\rho} < e >}{\bar{\rho}\partial < V > / \partial z} = \frac{4}{3} L^2 \Psi) \frac{1 - K_{p2}}{K^*_{p1}} (0,5 Pr + Ri + Ko) \frac{\partial < V >}{\partial z}, \tag{8.1.22}$$

$$\chi^T_{xz} \equiv -\frac{J^T_{qx}}{\bar{\rho} \frac{\partial < V >}{\partial z}\left(g + \frac{\partial < h >}{\partial z}\right)} = -\frac{L^2 \Psi^{3/2}}{K_{e1}^{*1/2}} (1 + Pr^t - K_{s2}) \frac{\partial < V >}{\partial z}, \tag{8.1.23}$$

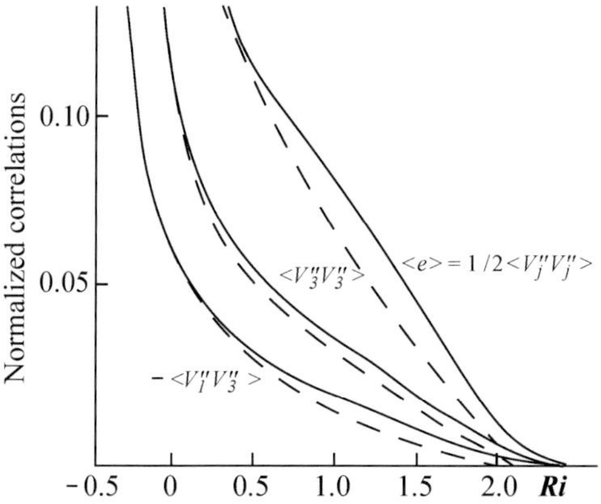

Fig. 8.1.3. Functional dependence of the components of the Reynolds stress tensor, normalized to $L^2(\partial_z <V>)^2$, on the gradient Richardson number, $\boldsymbol{Ri}$, for $\boldsymbol{Ko} = 0$ (solid curve) and $\boldsymbol{Ko} = 0.1$ (dashed curve).

$$\chi^t_{zz} \equiv -\frac{J^T_{qz}}{\bar{\rho}\left(g + \dfrac{\partial <h>}{\partial z}\right)} = -\frac{L^2 \Psi^{3/2}}{K^{*1/2}_{el}} (\boldsymbol{Ri} + \boldsymbol{Ko} - \boldsymbol{Pr}^T) \frac{\partial <V>}{\partial z}. \tag{8.1.24}$$

When $\boldsymbol{Rf} + \alpha \boldsymbol{Kf} = \boldsymbol{Rf}_0$, it follows from (8.1.21)–(8.1.24) that $\nu^T = \chi^T = 0$; the same values for the coefficients ν^T and χ^T should be adopted when $\boldsymbol{Rf} + \alpha \boldsymbol{Kf} > \boldsymbol{Rf}_0$, since the simplified equations (8.1.6)–(8.1.10) can not be satisfied also for any positive values of ν^T and χ^T.

Figures 8.1.3 and 8.1.4 show the functional dependence of the correlation moments on the main flow gradients (the $\boldsymbol{Ri}$ number) in the super-equilibrium approximation. The profiles of the Reynolds stress tensor components and the enthalpy fluctuation correlations were normalized to $L^2(\partial_z <V>)^2$ and $L^2 \partial_z <V> (g + \partial_z <h>)$, respectively, to become dimensionless.

Proceeding from the necessity to take into account the non-equilibrium nature of the real turbulence fields, we assume, as in Chapter 7, that some *inner equilibrium* occurs in the turbulence structure with respect to the mean velocity and temperature fields. Then the turbulent exchange coefficients can be expressed in terms of the parameters $<e>$ and L and the local properties of the medium by the relations

$$\nu^T_{zz} = L <e>^{1/2} \boldsymbol{Pr}^T \Psi, \quad \chi^T_{zz} = L <e>^{1/2} \Psi, \tag{8.1.25}$$

where the spatial-temporal distribution of the quantity $<e>$ is determined by the following evolutionary transfer equation for turbulent energy:

$$\bar{\rho}\frac{D<e>}{Dt}-\frac{\partial}{\partial z}\left(c_1\bar{\rho}L<e>^{1/2}\frac{\partial<e>}{\partial z}\right)=\nu^T_{zz}\left(\frac{\partial<V>}{\partial z}\right)^2(1-\boldsymbol{Rf}-\boldsymbol{Kf})-K_{e1}\bar{\rho}\frac{<e>^{3/2}}{L}. \quad (8.1.26)$$

Let us note that this approach is more accurate than those developed earlier for shear turbulence in a homogeneous medium (*Mellor and Yamada, 1982*), because it takes into account both convective and diffusion transport and the pre-history of the flow in the equation for turbulent energy transfer. Therefore, this approach is more advantageous than the simple super-equilibrium approximation, especially for free atmosphere modeling when these factors are important.

Turbulence Scale and Systematic Application. The external turbulence scale, L, has to be set in order to close the model under consideration. Let us recall that the scale L, characterizing the dimensions of large energy containing vortices, appears in the evolutionary transfer equations for the second moments when parametrizing the unknown correlations. In general, it depends on convective transport, generation and dissipation of turbulence, as well as on the pre-history of the process.

As was shown in Chapter 7, the scale L in free shear layers can be defined by a simple model equation (see formula (7.3.1)). Deducing more general differential equations for L is one of the challenging problems of the semi-empirical theory for multicomponent turbulence. In section 4.3 we already emphasized that the external turbulence scale, L, can not be determined only in terms of the single-point moments of fluctuating quantities. Instead, being a measure of the distance between two points in the flow, $\boldsymbol{r}_1$ and $\boldsymbol{r}_2$, where the correlations $<V''(\boldsymbol{r}_1)V''(\boldsymbol{r}_2)>$ are still not vanishing, this parameter should be found by integrating over the path from $\boldsymbol{r}_1$ to $\boldsymbol{r}_2$, the respective differential equation being derived from the equation for $<V''(\boldsymbol{r}_1)V''(\boldsymbol{r}_2)>$. Meanwhile, similar equations for defining the scale L, published elsewhere, usually include numerous proportionality coefficients poorly justified experimentally. Therefore, such equations should be regarded as less reliable than, for example, the balance equation for the Reynolds stress tensor where many terms are defined quite accurately.

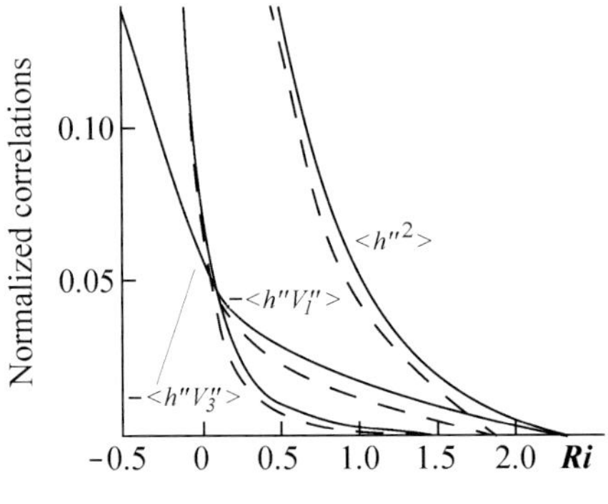

Fig. 8.1.4 Functional dependence of the enthalpy and velocity correlations, normalized to $L^2\partial_z<V>(g+\partial_z<h>)$, on the gradient Richardson number, $\boldsymbol{Ri}$, for $\boldsymbol{Ko}=0$ (solid curve) and $\boldsymbol{Ko}=0.1$ (dashed curve).

In view of the limits discussed and to pursuit the idea to develop a suitable and operational technique in support for the *GOMOS* space project, we have proposed another way to define the scale L (*Kolesnichenko and Marov, 1996*). It permits us to evaluate this parameter from the known structure characteristic of the refractive index of air, C_n^2, measured by probing the atmosphere by light from a reference star. A high-resolution spectrometer on board of a satellite is used to record the specific features in the attenuated spectra during occultation of standard stars (*Bertaux et al., 1988*), these features being caused by light absorption of ozone and other atmospheric minor species affecting their chemical equilibrium and variations. Obviously, the main objective to reconstruct the content of the atmospheric components is hampered by the distortions of light propagation (the random refraction) due to atmospheric turbulence because the light beam, before hitting a receiver, passes through layers of various spatial inhomogeneities and thus possessing a different refractive index. This results in stellar scintillations, or random temporal brightness changes. Measuring the so-called scintillation index, it is possible, for example, to evaluate the amplitude fluctuation dispersion and to use it conveniently as a basic statistical parameter of the light wave.

Therefore, complex analysis of the experimental data is necessary to implement the project successfully, including filtering out the effects caused by turbulence, which should be correlated with the results of numerical calculations of the spatial distributions of minor atmospheric species. In turn, the measured patterns of these very effects (scintillations) allows us to determine the turbulence characteristics, in particular, the turbulence transfer coefficients. This enables one to develop more representative models of sophisticated media such as the middle/upper atmosphere of a planet determined by complex, interacting processes as dynamics, heat and mass transfer, and chemical kinetics.

8.2. THE REFRACTIVE INDEX AND THE EXTERNAL TURBULENCE SCALE

Basically, in non-uniformly heated, horizontal atmospheric layers gravitational mechanical equilibrium is impossible and therefore the originating Archimedean buoyancy forces give rise to eddy motions (thermal and/or concentration turbulence) tending to mix the gaseous components. Large pressure (and temperature) gradients promote global displacement of the air layers with a vertical wind shear, which provokes the generation of additional eddies (mechanical turbulence) and shares the energy. A broad spectrum of eddy motions typical for the terrestrial atmosphere ranges from microscales, molecular thermal motion being an extreme, up to macroscales, the largest corresponding to zonal flow with horizontal linear scales up to 10^4 km.

Specifically the small-scale turbulence with eddies extending from millimeters to hundreds of meters, which are caused mainly by temperature fluctuations, are responsible for the fluctuating refractive index of air. Thus, small thermal inhomogeneities not only ensure turbulent heat transport but also significantly influence the propagation of probing radiation, causing light wave distortions. In this section we examine the influence of random inhomogeneities in the refractive index of air, $n(\boldsymbol{r})$, on the propagation of radiation at optical wavelengths used for the remotely sensing the atmosphere by the stellar occultation technique. First of all, we focus on the kinematics of turbulence and the microstructure in the $n(\boldsymbol{r})$, which are discussed in more detail elsewhere (*Kolesnichenko and Marov, 1996*).

8.2.1. Kinematics of Turbulence and Microstructure in Atmospheric Parameters

It is known that turbulent inhomogeneities arise for large Reynolds numbers, $\boldsymbol{Re} >> \boldsymbol{Re}_c = 2500... 5000$, in flows in which large-scale eddies have characteristic sizes close to the turbulence scale, L, and a characteristic velocity, V, defined as the difference of the hydrodynamic velocities over a distance L. The microstructure in such inhomogeneities is determined by cascade splitting of large eddies into progressively smaller eddies due to inertial forces.

This process accompanies the redistribution of the kinetic energy (per unit mass), V^3/L, of the carrying flow from large-scale to small-scale eddies, having characteristic parameters L_k and V_k, and further on up to the smallest eddies with characteristic sizes in the order of the internal turbulence scale, l. For a certain turbulent structure this cascade mechanism of the kinetic energy transfer of atmospheric eddies, $V^3/L \approx V_1^3/L_1 \approx V_2^3/L_2 \approx ... \approx V_n^3/L_n \equiv v^3/l$ (where $L > L_1 > L_2 > ... > L_n \equiv l$), does not damp with time only because there are external energy sources continuously feeding the shear flow.

As was indicated in Chapter 1, the kinetic energy acquired by rather large eddies which experience cascade transformation down to smaller eddies, finally dissipates into heat. The energy dissipation per unit time and per unit mass, being of the order of $\nu V_k^2/L_k^2$ for eddies with parameters L_k and V_k, grows when the eddy size decreases, continuing until the minimum eddy size, l, is attained and its kinetic energy has become of the same order of magnitude as the energy dissipation due to molecular viscosity, $V^3/L \approx V_1^3/L_1 \approx ... \approx V_n^3/L_n \approx \nu v^2/l^2 \approx \varepsilon_e$. Thus, the quasi-stationary regime of such a cascade mechanism is characterized, on the one hand, by some nearly stationary quantity ε_e describing the transfer rate of rotational kinetic energy from large to small eddies, and on the other hand, by molecular kinematic viscosity, ν, for the smallest eddies.

Note that the quantity ε_e introduced earlier by formula (4.2.25) corresponds to the dissipation rate of turbulent energy into heat. Breaking up of large eddies stops when viscosity forces in the fluid begin to play a decisive role. This occurs for eddies with Reynolds numbers $\boldsymbol{Re} = v_0 l_0/\nu \sim 1$. The domain of turbulent energy dissipation owing to molecular viscosity is determined by the Kolmogorov turbulence scale, $l_k = \nu^{3/4}\varepsilon_e^{-1/4}$, which is of the same order of magnitude as the internal turbulence scale.

In Chapter 1 it was also noted that only a statistical approach enables to describe adequately the turbulent pulsation field of the thermohydrodynamic parameters of the turbulized atmosphere, such as wind velocity, temperature, air refractive index, etc. If $\boldsymbol{Re} = VL/\nu$ is large enough, then, despite the anisotropy and inhomogeneity of the initial large eddies, their random breaking up eventually results in a substantially homogeneous and isotropic structure of developed turbulence (i.e. small scale eddy fields of the thermohydrodynamic parameters) within the scale range $l_k < r < L$. In this case the so-called structure functions for random parameters and their respective spectra serve as the most suitable tool for studying turbulence.

8.2.2. Structure and Spectral Functions of Random Fields

In general, when dealing with statistically modeling atmospheric turbulence, one utilizes the random functions for fluctuating thermohydrodynamic parameters depending on three spatial coordinates and time.

Let us first of all consider the time-independent random fields, $f(\boldsymbol{r})$, ($f(\boldsymbol{r})$ being real), which are locally homogeneous in the domain G and thus, invariant with respect to shifting a pair of points, $\boldsymbol{r}_0$ and $\boldsymbol{r}_0 + \boldsymbol{r}$. The structure functions, $D_f(\boldsymbol{r}) = <[f(\boldsymbol{r}_0 + \boldsymbol{r}) - f(\boldsymbol{r}_0)]^2>$, for the random fields $f(\boldsymbol{r})$ of some macroscopic parameter (*Kolmogorov, 1941*) are convenient to introduce as statistical characteristics of the locally homogeneous fluctuating atmosphere. We will confine ourselves to locally homogeneous and isotropic random fields: those invariant with respect to rotation and specular reflections of the vector $\boldsymbol{r}_0$. Then the structure function depends only on the radius-vector, **r**, but not on its direction: $D_f(\boldsymbol{r}) = D_f(r)$, where $r = |\boldsymbol{r}|$ is the distance between the points of observation. The spectral expansion of $D_f(r)$ in the three-dimensional (or one-dimensional) Fourier integrals can be represented as (see, e.g., *Tatarskiy, 1967*)

$$D_f(r) = 8\pi \int_0^\infty \left(1 - \frac{\sin \kappa r}{\kappa r}\right) \Phi_f(\kappa)\kappa^2 d\kappa \quad \left(= 2\pi \int_{-\infty}^\infty (1 - \cos \kappa r) V_f(\kappa) d\kappa\right), \tag{8.2.1}$$

where κ is the wave number module, $\kappa = |\boldsymbol{\kappa}|$; $\Phi_f(\kappa)$ (≥ 0) is the three-dimensional spectral density of the isotropic random field, $f(r)$, related to the one-dimensional spectral density, $V_f(\kappa)$, by the relations

$$\Phi_f(\kappa) = -\frac{1}{2\pi\kappa} \frac{dV_f(\kappa)}{d\kappa}, \tag{8.2.2}$$

$$V_f(\kappa) = \frac{1}{8\pi\kappa} \int_0^\infty \sin \kappa r \frac{dD_f(r)}{dr} dr. \tag{8.2.3}$$

Learning the structure function, $D_f(r)$, allows us to calculate the correlation function, $B_f(r) = < f(r_0 + r) f(r_0) >$, of the random isotropic field, $f(r)$, (provided such a function exists) with the help of the simple formula $B_f(r) = \frac{1}{2}[D_f(\infty) - D_f(r)]$. In this case, the corresponding three-dimensional spectral density, $\Phi_f(\kappa)$, conjugated by a Fourier integral to the correlation function $B_f(r)$, can be found from (8.2.2) and (8.2.3).

When analyzing the random fluctuation parameters of light, emitted by a star and penetrating the Earth's atmosphere, and the statistical characteristics of these parameters in a plane normal to the light propagation are of special interest. The two-dimensional Fourier spectra, $F_f(0,\kappa_2,\kappa_3)$, of the structure function, $D_f(0,\eta,\xi)$, are convenient to use for the describing the properties of a random field, $f(r)$, homogeneous and isotropic in the plane x = const. Here $x = x'$, $y - y' = \eta$, $z - z' = \xi$ are the coordinates differences of two points in the plane x = const. When $f(\mathbf{r})$ is locally uniform in the plane x = const, it is possible to write (*Tatarskiy, 1967*)

$$D_f(0,\eta,\xi) = 2\int\int_{-\infty}^{\infty}[1-\cos(\kappa_2\eta+\kappa_3\xi)]F(0,\kappa_2,\kappa_3)\,d\kappa_2 d\kappa_3. \qquad (8.2.4)$$

In addition, if $f(\mathbf{r})$ is locally isotropic in the plane $x = const$, the two-dimensional spectral density, $F_f(|x|,,\kappa_2,\kappa_3)$, depends only on the complex $\kappa = (\kappa_2^2+\kappa_3^2)^{1/2}$. We can then in this plane obtain the following expressions in polar coordinates for the structure and correlation functions, $D_f(\rho)$ and $B_f(\rho)$, respectively:

$$D_f(\rho) = 4\pi\int_0^{\infty}[1-J_0(\kappa\rho)]\,F_f(\kappa,0)\kappa d\kappa\,, \qquad (8.2.5)$$

$$B_f(\rho) = 2\pi\int_0^{\infty}J_0(\kappa\rho)F_f(\kappa,0)\kappa d\kappa\,, \qquad (8.2.6)$$

where $J_0(x)$ is the Bessel function; and $\rho = (\eta^2+\xi^2)^{1/2}$ are the distances between the points (y,z) and (y',z') in the plane $x = const$. The two-dimensional spectral density, $F_f(\kappa,x)$, of the locally isotropic random field can be expressed in terms of its three-dimensional spectral density, $\Phi_f(\kappa)$, with the help of the following conjugated Fourier relations:

$$F_f(\kappa_2,\kappa_3,x) = \int_{-\infty}^{\infty}\cos(\kappa_1 x)\,\Phi_f(\kappa_1,\kappa_2,\kappa_3)d\kappa_1, \qquad (^1)$$

$$(8.2.7)$$

$$\Phi_f(\kappa_1,\kappa_2,\kappa_3) = \frac{1}{2\pi}\int_{-\infty}^{\infty}F(\kappa_1,\kappa_2,x)\cos(\kappa_1 x)dx. \qquad (^2)$$

Frozenness Hypothesis. When studying how atmospheric turbulence affects stellar light propagation (in the satellite's coordinate system), the random fields $f(\mathbf{r},t)$, such as, for example, the radiation intensity field, $I_\lambda(y,z,t)$, play an important role. Their time variations are caused by a simple displacement of a "frozen" spatial structure at a constant velocity, $\mathbf{w}$, in our case $\mathbf{w}$ being the satellite velocity component normal to the direction of light propagation. The “frozenness” condition is expressed mathematically as $f(\boldsymbol{r},\tau)=f(\boldsymbol{r}-\boldsymbol{w}\tau,0)$.

It is possible to set the relation between the spatial-temporal structure functions, $D_f(\boldsymbol{r},\tau)$, and the purely spatial functions, $D_f(\boldsymbol{r})$, for such invariable fields:

$$D_f(\boldsymbol{r},\tau) = D_f(\boldsymbol{r}-\boldsymbol{w}\tau).$$

From here we may obtain the formula linking the time (frequency) spectrum of the random scalar field, $W_f(\omega)$, defined by the relation

$$B_f(\tau)=2\int_0^{\infty} W_f(\omega)\cos\omega\tau\, d\omega, \qquad (8.2.8)$$

with its three-dimensional spatial spectral density, Φ_f, where $B_f(\tau)$ is the temporal auto-correlation function. For locally isotropic random fields, $f(\boldsymbol{r},t)$, the relation between $W_f(\omega)$ and Φ_f is expressed as

$$\Phi_f(\omega/|\boldsymbol{w}|)=-\frac{|\boldsymbol{w}|^3}{2\pi\omega}\frac{dW_f(\omega)}{d\omega}, \qquad (8.2.9)$$

$$W_f(\omega)=\left(\frac{2\pi}{|\boldsymbol{w}|}\right)\int_{|\omega|/|\boldsymbol{w}|}^{\infty}\Phi_f(\kappa)\kappa d\kappa=\frac{1}{|\boldsymbol{w}|}V_f(\omega/|\boldsymbol{w}|). \qquad (8.2.10)$$

Here $\omega\,(=2\pi\mathrm{f})$ is the circular frequency (f is the frequency). These conversion formulas are convenient to use for processing experimental data because just the frequency spectra are recorded when remotely sensing the atmosphere, while the spatial structure of the random field of the structure parameter under study, $f(\boldsymbol{r})$, is easier to obtain from the theory.

8.2.3. The Structure Function of the Refractive Index

Small-scale inhomogeneities in the refractive index, $n(\boldsymbol{r})$, at optical wavelengths are mainly caused by random spatial-temporal temperature variations. In turn, micropulsations of the temperature field results from the turbulent mixing in the thermally stratified atmosphere. Numerous observations of light refraction from the outer space (see, e.g., *Grechko et al., 1981*) have shown that small-scale temperature inhomogeneities representing strongly anisotropic layered formations, are permanently present in the upper troposphere and stratosphere. The occurrence of anisotropic inhomogeneities of the refractive index was also confirmed by radar studies of the stratosphere with which a considerable excess of echo signals at vertical sensing against those at oblique sensing was found (*Rottger et al., 1981*).

As is known, the structure function, $D_n(r) = \langle [n''(r_0 + r) - n''(r_0)]^2 \rangle$, for fluctuations in the refractive index, $n'' = n - \langle n \rangle$, in the domain of locally isotropic atmospheric turbulence can be described by the Kolmogorov-Obukhov *two-thirds law* for the refractive index random field,

$$D_n(r) = \begin{cases} C_n^2 r^{2/3}, & l_0 << r << L, \\ C_n^2 l_0^{2/3} (r / l_0)^2, & r << l_0, \end{cases} \tag{8.2.11}$$

where $l_0 = 3.8\ l_k$ is the internal turbulence scale; C_n^2 is the structure characteristic of the refractive index of air describing the intensity of turbulent fluctuations of $n(\mathbf{r})$ in the atmosphere. In the boundary atmospheric layer, the internal turbulence scale, l_0, is 0.1–1 mm, while the external scale of turbulence is $L = 0.4z$ (see subsection 3.3.2).

Having applied formula (8.2.3), it is possible to calculate the one-dimensional spectral density corresponding to any structure function with the general form $D_f(r) = C_f^2 r^p$, where $0 < p < 2$. This yields

$$V_f(\kappa) = \frac{\Gamma(p+1)}{2\pi} \sin\left(\frac{\pi p}{2}\right) C_f^2 \kappa^{-(p+1)}. \tag{8.2.12}$$

Using (8.2.3) and (8.2.12) enables to find the respective three-dimensional spectral density, $\Phi_f(\kappa)$, of any random process, $f(r)$:

$$\Phi_f(\kappa) = \frac{\Gamma(p+1)}{4p^2} \sin\left(\frac{\pi p}{2}\right) C_f^2 \kappa^{-(p+3)} \approx 0.033 C_f^2 \kappa^{-11/3}. \tag{8.2.13}$$

The relations (8.2.11)–(8.2.13) yield the formula for the Kolmogorov fluctuation spectrum of the refractive index in the atmosphere for locally homogeneous and isotropic turbulent fields:

$$\Phi_n(\kappa) = \begin{cases} 0.033 C_n^2 \kappa^{-11/3}, & 2\pi / L << \kappa << 2\pi / l_0, \\ 0, & 2\pi / l_0 << \kappa, \end{cases} \tag{8.2.14}$$

where ($2\pi / L << \kappa << 2\pi / l_0$) and ($2\pi / l_0 << \kappa$) are the inertial and viscous wave number intervals, respectively (see Chapter 1). In the wave number range $0 < \kappa < 2\pi / L$ (the energy interval), the three-dimensional spectrum, $\Phi_n(\kappa)$, generally is anisotropic, and its form depends on the particular turbulence formation mechanism. In other words, the spectrum $\Phi_n(\kappa)$ is not universal.

Inhomogeneities inherent to the inertial wave number interval are mostly responsible for fluctuations in the optical radiation parameters. Sometimes, however, it is necessary to take into account how larger turbulent inhomogeneities than those of the inertial interval affect the fluctuation properties of the passing radiation. It is common in these rare cases to use various model descriptions of the turbulence spectrum. Below, we will use the Karman spectrum for calculating the desired statistical characteristics of the fluctuating optical wave field within the whole wave number range being guided exclusively by mathematical convenience. It has the form

$$\Phi_n(\kappa) = 0.033 C_n^2 \left(\kappa^2 + \frac{1}{L^2} \right)^{-11/6} \exp\left(-\frac{\kappa^2}{\kappa_m^2} \right), \qquad (8.2.15)$$

where the number κ_m is connected to the internal turbulence scale, l_0, by the relation $\kappa_m = 5.92 / l_0$. This spectrum, while being poorly justified from the physical viewpoint, coincides with the formula (8.2.14) for the three-dimensional spectrum, $\Phi_n(\kappa)$, in the inertial wave number interval and accounts for the finiteness of the external ($L = 2\pi / \kappa_0$) and internal ($l_0 = 5.92 / \kappa_m$) atmospheric turbulence scales.

The properties of the terrestrial atmosphere as a turbulized medium are characterized by a broad range of variations in the structure characteristics, C_n^2, of the refractive index of air. For example, in the boundary atmospheric layer, C_n^2 varies from 10^{-13} cm$^{-2/3}$ to 10^{-17} cm$^{-2/3}$ depending on the meteorological conditions. As will be shown in section 8.3, the statistical parameters of an optical wave probing the atmosphere, essentially depend on C_n^2.

8.2.4. The Structure Characteristics of the Inhomogeneous Turbulent Atmosphere

Evidently, real atmospheric turbulence can be treated as locally homogeneous and isotropic only within a domain G having the size of the external turbulence scale, L. When $l_0 << r << L$, the structure function for the random field $f(r)$ of any thermohydrodynamic parameter is $D_f(r) = C_n^2 r^{2/3}$. The field $f(r)$ in any other domain G_1 of the same characteristic size L at a distance L from G will be locally homogeneous and isotropic as well, and the structure function will also be expressed by the two-thirds law though, generally speaking, with some other value for the constant C_n^2.

Hence, the structure function, $C_n^2(\boldsymbol{r})$, may be thought of as a spatially very smooth function that substantially varies over distances of order L. Therefore, when studying light waves propagating through the real atmosphere when the key role is played by small inhomogeneities, atmospheric turbulence should be described by the following structure functions (*Tatarskiy, 1967*):

$$D_f(\boldsymbol{r}_0, \boldsymbol{r}_0 + \boldsymbol{r}) = < [f(\boldsymbol{r}_0) - f(\boldsymbol{r}_0 + \boldsymbol{r})]^2 > = C_n^2(\boldsymbol{r}_0 + \boldsymbol{r}/2) r^{2/3}, \quad l_0 << r << L \qquad (8.2.16)$$

The scale $r << r_0$ and the value of the refractive index structure characteristic in formula (8.2.16), $C_n^2(\boldsymbol{r}_0 + \boldsymbol{r}/2)$, should be computed halfway the segment connecting the points of observation. Physically this approach means that the values of the parameters $\varepsilon_e(\boldsymbol{r}_0 + \boldsymbol{r}/2)$ and $\varepsilon_h(\boldsymbol{r}_0 + \boldsymbol{r}/2)$ are assumed to be constant inside a volume with characteristic size of about r centered at the point $(\boldsymbol{r}_0 + \boldsymbol{r}/2)$ (see the formula (8.2.21) below).

It follows from the foregoing discussion that we can write immediately the local expression for the three-dimensional spectral density in the inhomogeneous atmosphere, $\Phi_n(\boldsymbol{r}, \kappa)$, having replaced the constant quantity C_n^2 in (8.2.14) by the function $C_n^2(\boldsymbol{r})$. This yields

$$\Phi_n(\boldsymbol{r}, \kappa) = C_n^2(\boldsymbol{r}) \Phi_n^0(\kappa), \tag{8.2.17}$$

where $\Phi_n^0(\kappa) = 0.033\kappa^{-11/3}$ for $\kappa < \kappa_m$; $\Phi_n^0(\kappa) = 0$ for $\kappa > \kappa_m$. Specifically, when the light beam travels through the atmosphere, it is necessary to account for the changes in the refractive index structure characteristic, C_n^2, along the entire propagation path. The numerical value of the quantity C_n^2 decreases with height. The formula (*Zuev et al., 1988*)

$$C_n^2(z) = C_n^2(z_0)\exp(-z/z_e), \quad z < z_0, \tag{8.2.18}$$

serves as a good approximation of $C_n^2(\boldsymbol{r})$ in the atmosphere at heights exceeding 10 km. This formula fits to the experimental data satisfactorily when the values of "effective thickness" of the atmosphere, z_e ($z_e \approx 3200$), are chosen properly. According to the C_n^2 measurements in the microwave range, this quantity decreases with height proportionally to z^{-2}:

$$C_n^2(z) = C_n^2(z_0)/[1 + (z/H_0)^2] \tag{8.2.19}$$

where H_0 is the scale height.

8.2.5. Fluctuations in the Refractive Index of Air

The formula

$$n_\lambda = 1 + 10^{-6} N_\lambda, \quad N_\lambda = (77.6p/T) - (0.584p/T\lambda^2) - 0{,}06 p_{wv}, \tag{8.2.20}$$

allows to calculate the refractive index of the terrestrial atmosphere, $n(r)$, in the wavelength range from 0.2 through 20 μm. Here N accounts for the dependence of the refractive index on temperature (at the *Kelvin scale*), atmospheric pressure, p, and *water vapor pressure* (both in mbars). Formula (8.2.20) is valid only in the *transparency windows* in the terrestrial atmosphere and can noticeably differ from (8.2.20) in the absorption bands and lines.

Fluctuations in N are mainly generated by pulsations of the temperature and pressure fields, whereas humidity pulsations renders no substantial effects on $n(r)$ at optical wavelengths. As is known, in the case of turbulence, wind velocity pulsations have a

Kolmogorov's spectrum (as in (8.2.13)). At the same time, the parameters p and T fluctuate chaotically and do not necessarily follow the turbulent motion. It is also known (see Chapter 1) that the combination of these parameters (like that giving the potential temperature, $\Theta = (h + gz)/C_p$), is transferred in turbulent velocity fields without noticeable changes. This means that the quantity Θ formally can be considered as a passive admixture and for this reason, like the flow velocity, it has a Kolmogorov's spectrum like (8.2.13). Particularly, for the structure function for the potential temperature fluctuation, Θ'', the two-thirds law (formula (8.2.11)) also holds when C_n^2 is replaced by C_h^2.

The structure characteristic of the potential temperature field, C_h^2, relates to two quantities, fundamental in the macroscopic turbulence theory: the turbulent energy dissipation rate, $\varepsilon_e(\boldsymbol{r})$, and the smoothing rate of enthalpy inhomogeneities, $\varepsilon_h(\boldsymbol{r})$. The latter relation is

$$C_h^2(\boldsymbol{r}) = \alpha\varepsilon_h(\boldsymbol{r})/[\varepsilon_e(\boldsymbol{r})]^{1/3}, \qquad l_0 << r << L \tag{8.2.21}$$

where α is a universal constant ranging from 1.5 to 3.5, the recommended value being α ~2.8, as a corollary of the two-thirds law for the temperature field (*Monin and Yaglom, 1971; Tatarskiy, 1967*).

It should be emphasized that the quantities $\varepsilon_e(\boldsymbol{r})$ and $\varepsilon_h(\boldsymbol{r})$ are simultaneously regarded as the characteristics of the microstructure in turbulent fields, which determine the transfer rate of kinetic energy and temperature inhomogeneities, respectively, for the eddy cascade from large to progressively smaller proportions. Therefore, these parameters connect the microstructure in turbulent fields with the averaged properties of turbulence introduced in Chapters 3 and 4. This is why it is possible to ultimately determine the large scale turbulence parameters (for example, L) from measuring the fluctuation properties of electromagnetic waves with remotely sensing planetary atmospheres.

As was said, fluctuations of the refractive index in the atmosphere are mainly caused by temperature fluctuations, while pressure fluctuations have less influence and can virtually be neglected. Using this approximation, it is easy to obtain from (8.2.20) the formula linking fluctuations of the refractive index, n'', to potential temperature fluctuations, Θ'':

$$n'' = 10^{-6}(N - \langle N \rangle) = 10^{-6}\left[(77.6 - 0.584\lambda^{-2})\bar{p}\Theta''\right]/(\gamma - 1)\langle T \rangle^2. \tag{8.2.22}$$

Because the fluctuations n'' and Θ'' are proportional, the statistical characteristics of these quantities may be considered as similar. Then the structure characteristic of the refractive index of dry air, $C_n^2(\boldsymbol{r},\lambda)$, at optical wavelengths is linked to $C_h^2(\boldsymbol{r})$ by the following important relation:

$$C_n^2(\boldsymbol{r},\lambda) = \mathrm{M}^2(\lambda)C_h^2(\boldsymbol{r}) = \alpha\mathrm{M}^2(\lambda)\varepsilon_h(\boldsymbol{r})/[\varepsilon_e(\boldsymbol{r})]^{1/3}, \tag{8.2.23}$$

where the parameter

$$M(\lambda) = \frac{10^{-6}}{(\gamma - 1) < T >} \left(\frac{77.6 \overline{p}}{< T >} - \frac{0.584 \overline{p}}{\lambda^2 < T >} \right) = \frac{10^{-6} < N >}{(\gamma - 1) < T >} \tag{8.2.24}$$

accounts, among others, for the dependence of $C_n^2(\boldsymbol{r}, \lambda)$ on the wavelength of the sensing beam, λ.

8.2.6. The External Turbulence Scale and the Structure Characteristic of the Refractive Index

According to (8.2.23), the structure characteristic of the refractive index of air, $C_n^2(\boldsymbol{r})$, can be evaluated given the turbulent energy dissipation rate, $\varepsilon_e(z)$, and the dissipation rate of potential temperature fluctuations, $\varepsilon_h(z)$, are known. This allows us to relate $C_n^2(\boldsymbol{r})$ to the external turbulence scale, L.

Indeed, according to (8.1.3) and (8.1.4) the following relations can be used for evaluating the dissipative quantities $\varepsilon_e(z)$ and $\varepsilon_h(z)$:

$$\varepsilon_e = K_{e1} < e >^{3/2} / L, \tag{8.2.25}$$

$$\varepsilon_h = K_{h1} < e >^{1/2} < h''^2 > / L = < e >^{1/2} L \Psi \left(g + \frac{\partial < h >}{\partial z} \right), \tag{8.2.26}$$

where the characteristic function is $\Psi = \Psi_0(\boldsymbol{Rf}_0 - \boldsymbol{Rf} - \alpha \boldsymbol{Kf})(1 - \boldsymbol{Rf} - \boldsymbol{Kf})^{-1}$ (see formula (8.1.14)). This enables us to derive hydrodynamically with the known turbulence scale the expression for height distribution of $C_n^2(\boldsymbol{r})$ in the free atmosphere:

$$C_n^2(\boldsymbol{r}, \lambda) = \alpha M^2(\lambda) \varepsilon_h(\boldsymbol{r}) / [\varepsilon_e(\boldsymbol{r})]^{1/3} = \frac{\alpha M^2(\lambda)}{(K_{e1})^{1/3}} \Psi L^{4/3} \left(g + \frac{\partial < h >}{\partial z} \right)^2. \tag{8.2.27}$$

This formula extends to the expression for the boundary layer of a homogeneous atmosphere for the free flow of a multicomponent gas mixture (*Ishimaru, 1981*).

On the other hand, an important relation follows from (8.2.27) for the external turbulence scale, L:

$$L(z) = [C_n(z, \lambda)]^{3/2} \delta / \Psi^{3/4} M^{3/2} (g + \partial < h > / \partial z)^{3/2}, \quad \delta \equiv (K_{e1})^{1/4} / \alpha^{3/4}. \tag{8.2.28}$$

It permits to link L to the averaged hydrodynamic parameter gradients of the gaseous mixture flow (i.e. the Richardson, $\boldsymbol{Ri}$, and Kolmogorov, $\boldsymbol{Ko}$, numbers and the structure characteristic of refractive index fluctuations, C_n^2.

Let us note that the relations (8.2.27) and (8.2.28) account for the influence of Archimedean forces only on the properties of the mean motion. However, these forces render influence not only on the averaged motion parameters, but also on the microstructure of the turbulent field (*Obukhov, 1959; Monin, 1962*). With regard for this circumstance, formula (8.2.11) for the structure function, $D_n(r)$, acquires the form

$$D_n(r) \propto \varepsilon_h^{4/5}(g/<T>)^{-2/5} r^{2/5}. \tag{8.2.29}$$

Then the linear scale, $L_k \propto \varepsilon_e^{5/4}(g/<T>)^{-3/2}\varepsilon_h^{-3/4}$, obtained through the dimensionality analysis, which defines the inhomogeneity dimensions which are affected significantly by the buoyancy forces, can be related to L. For the particular application under consideration, for free horizontal two-dimensional shear flows, it is possible to obtain

$$L_k \propto L\Psi^{3/4}\left(\frac{\boldsymbol{Pr}^T - \boldsymbol{Ri} - \boldsymbol{Ko}}{|\boldsymbol{Ri}|}\right)^{3/2} \tag{8.2.30}$$

using (8.2.25) and (8.2.26). Obviously, if $|\boldsymbol{Ri}| << 1$, the scale L_k can be several times the external turbulence scale, L, in the free atmosphere. In this case the influence of Archimedean forces on the microstructure of the turbulent field appears only outside of the inertial turbulence interval. But if $|\boldsymbol{Ri}| >> 1$, then L_k falls within the inertial interval, dividing it into two parts where either the common formulas for the structure function, $D_n(\boldsymbol{r})$, (for $l_0 << r << L_k$) are valid, or the more sophisticated formulas (for $L_k << r << L$) should be used (see, e.g., (*Ishimaru, 1981*)).

We may conclude that for real time space monitoring, the averaged structure parameters of the medium are needed in order to evaluate the minor atmospheric species in the turbulized middle atmosphere. Accordingly, when modeling atmospheric dynamics and chemical kinetics, it is necessary to solve numerically both the system of hydrodynamic equations for the mean motion scales and the evolutionary equation for turbulent energy transport. The latter should be supplemented by expression (8.2.28) for the external turbulence scale, L, and the data on the spatial distribution of the structure characteristic of the refractive index of air, C_n^2. These particular data can be obtained, for example, from a remote sensing experiment.

8.3. THE STRUCTURE CHARACTERISTIC OF THE REFRACTIVE INDEX: DETERMINATION FROM ATMOSPHERIC REMOTE SENSING

It was indicated that, while the structure characteristic, $C_n^2(\boldsymbol{r})$, can be evaluated using formula (8.2.27), it can be also derived independently based on the analysis of the sensing radiation parameters. Basically, any experiment on determining statistical properties of an optical wave that passed through a layer of turbulized atmosphere, enables to reveal the intensity of turbulent fluctuations described by $C_n^2(\boldsymbol{r})$. We will focus here on how to find this quantity from analyzing the measured fluctuation dispersion of the intensity of the propagating monochromatic optical wave through a turbulent atmosphere. Basic concepts of the theory on electromagnetic wave propagation through a turbulized medium, discussed at the beginning of this chapter, underlie the developed method in application to monitoring the middle atmosphere from space.

8.3.1. The Wave Equation and the *SSP* Method

Let us assume that a flat, monochromatic electromagnetic wave incidents on some volume of the turbulized medium. Chaotic fluctuations in the refractive index arising due to turbulent mixing inside this volume result in scattering of the incident wave. The problem is how to determine the fluctuations of the propagating wave parameters in order to calculate finally the value of $C_n^2(\boldsymbol{r})$.

For this purpose we first of all connect the statistical characteristics of the turbulent field (for example, the spectral densities of the structure or the correlation functions of the refractive index) to the respective characteristics of an individual light wave. Next, leaning on the additional assumptions on the distribution probabilities of the intensity fluctuations, we transit to the averaged radiation intensity, i.e. to finding the spectral characteristics of the quantities measured directly in the experiment.

To account for diffraction effects while solving the posed problem, we will start from the stochastic wave equation $\Delta u + k^2 n^2(\boldsymbol{r})u = 0$, where $k = 2\pi/\lambda$ is the wave number and $u(\boldsymbol{r})$ is any of the three components of the electric field, $\boldsymbol{E}$. The refractive index in this equation, $n(\boldsymbol{r})$, is a spatially random function with the mean value $< n(\boldsymbol{r}) > = 1$. Designating $n''(\boldsymbol{r}) = n(\boldsymbol{r}) - 1$ as the refractive index fluctuation and assuming that $|n''(\boldsymbol{r})| << 1$, we search the solution of the wave equation as $u = A\exp(i\Psi)$, where A is the amplitude and Ψ is the phase of the electromagnetic wave. The function $u(\boldsymbol{r})$ is equal to the sum $(u = u_0 + u_1)$ of the unperturbed solution, $u_0 = A_0 \exp(i\Psi_0)$, satisfying the equation $\Delta\, u_0 + k^2 u_0 = 0$, for small u_1. Assuming that $\Psi = \Psi_0 + \Psi_1$ and $\ln A = \ln A_0 + X$, where $\Psi_1 = \mathrm{Im}(u_1/u_0)$ and $X = \mathrm{Re}(u_1/u_0)$ are small perturbations, $X = \ln(A/A_0)$ being the logarithmic "level" of optical wave amplitude fluctuations, we will look for plane wave fluctuations with amplitude X and phase Ψ_1. The smallness of the refractive index fluctuations allows us to apply the small and smooth perturbations (*SSP*) method to the solution of the initial wave equation (*Rytov, 1937*). Let us underline that the contemporary methods of atmospheric remote sensing almost entirely lean on the weak fluctuation theory.

8.3.2. Plane Wave Amplitude Fluctuations in a Locally Isotropic Turbulent Medium

Assume that propagation of a monochromatic light wave is aligned with the x-axis, the wave's source is located in the plane $x = 0$, and the point of observation has the coordinates (S, y, z), i.e. it is at the distance S from the source. The complete solution of the considered wave equation allows us to relate the two-dimensional spectral density (see (8.2.7)) of amplitude fluctuations, $F_A(\kappa, 0) = F_A(0, \kappa_2, \kappa_3)$, (or phase $F_\Psi(\kappa, 0)$) in the plane $x = S$ to the three-dimensional spectral density of correlation function of the refractive index $\Phi_n(\kappa) = \Phi_n(0, \kappa_2, \kappa_3)$ of the medium.

For a plane wave propagating through a non-uniform turbulent air flow whose characteristics such as $C_n^2(\boldsymbol{r})$ experience smooth variations (*Tatarskiy, 1967*), the relation for the quantity $F_A(\kappa,0)$ looks as follows:

$$F_A(\kappa,0) \cong 2\pi k^2 \int_0^S C_n^2(\boldsymbol{r})\Phi_n^0(\kappa)\sin^2[\kappa^2(S-\eta)/2k]\,d\eta \tag{8.3.1}$$

where, as earlier, $k = 2\pi/\lambda$ is the wave number; λ is the wavelength, $\kappa^2 = \kappa_2^2 + \kappa_3^2$, S is the path length. Using (8.2.5) and (8.2.6), it is possible to pass from the spectral density, $F_A(\kappa,0)$, to the structure and correlation functions of the amplitude fluctuations, $D_A(\rho)$ and $B_A(\rho)$, respectively, in the plane $x = S$.

So, according to the general formula (8.2.6), for the amplitude fluctuation correlation function in the plane $x = S$, we can write

$$B_A(\rho) = 4\pi^2 k^2 \int_0^S C_n^2(\boldsymbol{r})d\eta \int_0^\infty J_0(\kappa\rho)\Phi_n^0(\kappa)\sin^2\left(\frac{\kappa^2(S-\eta)}{2k}\right)\kappa d\kappa, \tag{8.3.2}$$

whence

$$B_A(0) \equiv \langle[\ln(A/A_0)]^2\rangle = 4\pi^2 k^2 \int_0^S C_n^2(\boldsymbol{r})d\eta \int_0^\infty \Phi_n^0(\kappa)\sin^2\left(\frac{\kappa^2(S-\eta)}{2k}\right)\kappa\, d\kappa, \tag{8.3.3}$$

thus solving the amplitude fluctuation problem for monochromatic plane waves propagating through the turbulent medium. Similar relations are obtained for the correlation function of the phase fluctuations.

In the particular case when the Karman formula (8.2.15) for the fluctuation density spectrum of the refractive index of air, $\Phi_n^0(\kappa)$, is applicable, the following basic relation for the mean square of the logarithm of the amplitude fluctuations in a locally homogeneous turbulent medium can be derived:

$$\begin{aligned}\sigma_x^2 &\equiv \langle X^2\rangle \equiv \langle[\ln(A/A_0)]^2\rangle \equiv B_A(0) \equiv D_x(\infty)/2 = \\ &= 4\pi^2 0.033 k^2 \int_0^S C_n^2(\boldsymbol{r})d\eta \int_0^{\kappa_m} \kappa^{-11/3}\sin^2[\kappa^2(S-\eta)/2k]\kappa d\kappa\,.\end{aligned} \tag{8.3.4}$$

For developing the technique further it is important to derive simpler asymptotic formulas from (8.3.4). Because for a rather short path S the inequality $\kappa_m^2 S/k \ll 1$, i.e. $\sqrt{\lambda S} \ll l_0$, corresponding to the ray approximation, this yields

$$\sigma_x^2 = 4\pi^2 0.033k^2 \int_0^S C_n^2(\boldsymbol{r})d\eta \int_0^{\kappa_m} \kappa^{-11/3}\left(\frac{\kappa^4 (S-\eta)^2}{4k^2}\right)\kappa d\kappa =$$
$$= 7.37 l_0^{-7/3} \int_0^S C_n^2(\boldsymbol{r})(S-\eta)^2 d\eta. \tag{8.3.5}$$

Integration in (8.3.4) can be extended up to infinity for $\kappa_m^2 S/k >> 1$, i.e. $\sqrt{\lambda S} >> l_0$, when diffraction effects become substantial, which virtually always occurs for a light wave sensing the atmosphere. Then we obtain

$$\sigma_x^2 = D_x(r)/2 = 0.563k^{7/6}\int_0^S C_n^2(\boldsymbol{r})(S-\eta)^{5/6}d\eta = 0.56k^{7/6}\int_0^S C_n^2(\boldsymbol{r})x^{5/6}dx. \quad (8.3.5^*)$$

The integration in the formulas (8.3.5) and (8.3.5*) is performed along the line of sight from the point of observation, S, to the light source, the origin of the coordinates for the second integral being located at the point of observation. If $C_n^2(\boldsymbol{r})$ = const, corresponding to totally homogeneous turbulence along the integration path (such a situation is feasible, for example, when probing the atmosphere along a horizontal path), the following relations result from (8.3.5) for the logarithm of the mean square amplitude fluctuations:

$$\sigma_x^2 = 2.46\, C_n^2 S^3 l_0^{-7/3}, \quad \text{at} \quad \sqrt{\lambda S} << l_0, \quad (^1)$$
$$\sigma_x^2 = 0.31 C_n^2 k^{7/6} S^{11/6}, \quad \text{at} \quad \sqrt{\lambda S} >> l_0. \quad (^2) \tag{8.3.6}$$

Allowing for path inhomogeneities does not result in any calculation difficulties provided a specific model dependence of the structure characteristic, $C_n^2(\boldsymbol{r})$, on the distance is set. For example, it is necessary to model an extremely non-uniform height distribution of refractive index fluctuations caused by scintillations and trembling of stellar images. Obviously, the strongest fluctuations occur closely to the Earth's surface, and they weaken upward. Therefore, lower atmospheric layers are the main contributors to the integral $\int_0^\infty C_n^2(z)z^{5/6}dz$ related to amplitude fluctuations. Assuming that $C_n^2(\boldsymbol{r})$ is determined by formula (8.2.18) and considering $\sqrt{\lambda z_0} >> l_0$, we have that

$$\sigma_x^2 = \langle X^2 \rangle = 0.56k^{7/6}C_{n0}^2 z_0^{11/6}\Gamma\left(\tfrac{11}{6}\right) = 0.53C_{n0}^2 k^{7/6} z_0^{11/6}. \tag{8.3.7}$$

Otherwise, if $C_n^2(\boldsymbol{r})$ is determined by formula (8.2.19), then

$$\langle X^2 \rangle = 0.56\pi C_{n0}^2 k^{7/6} H_0^{11/6}/2\sin(\pi/12) \cong 3.4\, C_{n0}^2 k^{7/6} H_0^{11/6}. \tag{8.3.8}$$

Evidently, the expressions (8.3.5)–(8.3.8) are identical in structure but differ strongly in the numerical coefficient values. The structure characteristic, C_n^2, is included in the expressions as a co-factor. This allows us to extract its value from an experiment, for example, using the measured dispersion of the logarithm of amplitude fluctuations, σ_x^2 , at optical wavelengths. Naturally, in every particular case an appropriate height profile for $C_n^2(\boldsymbol{r})$ should be introduced.

8.3.3. The Evaluation of Intensity Fluctuations with the *SSP* Method

Let us consider the procedure of evaluating the quantity σ_x^2. Assume that the intensity, $I_\lambda(y,z)$, is distributed in the plane of observation, $x = S$, where the properties of the light waves are measured. Obviously, the quantity $I_\lambda(y,z)$ may be represented as

$$I_\lambda(y,z) = I_{\lambda 0}\exp\left[2\ln(A/A_0)\right] = I_{\lambda 0}\exp[2X(y,z)]. \tag{8.3.9}$$

As follows from the theory, when the first approximation of the *SSP* method is applicable, the amplitude, A, (or the complex phase, Ψ_1) of an optical wave looks like an integral of some determined function $F(\boldsymbol{r})$ and a random field of refractive index fluctuations, n'', over the volume D occupied by a turbulent medium along the propagation path: $\ln(A/A_0) = \iiint_D F(\boldsymbol{r})n''dv$. The linear dimensions of this volume in this particular case are far in excess of the correlation scale, L, of the random field of the refractive index, n. Dividing the integration domain into volumes having dimensions of the order of the external turbulence scale, L, we can represent this triple integral as the sum of a large number of independent addends. The resulting formula,

$$\ln(A/A_0) = \sum_j \iiint_{D_j} F(\boldsymbol{r})n''dv_j$$

, thus expresses $\ln(A/A_0)$ as the sum of uncorrelated quantities. Hence, in accordance with the central limit theorem, it is possible to conclude that $\ln(A/A_0)$ in the *SSP* approximation is distributed normally and that the intensity, I_λ, follows a log-normal probability distribution.

Numerous measurements of probing light radiation (including laser beam) were carried out for real atmospheric air routes under conditions of both weak intensity fluctuations and *saturation* (see below). In most cases they revealed distributions close to the *log-normal law*, with some deviations occuring in the domains of profound fading and big intensity cast relative to the mean level. With this circumstance in mind, we will consider below some features of the log-normal law pertinent to the atmospheric remote sensing data analysis.

8.3.4. Plane Light Wave Intensity Fluctuations and the Scintillation Index

The density of the log-normal probability distribution, $P(I_\lambda)$, depends on the mean intensity level, $<\ln I_\lambda>$, and the level of dispersion, $\sigma^2_{\ln I} = <[\ln I_\lambda - <\ln I_\lambda>]^2>$, and is expressed as

$$P(I_\lambda) = \frac{1}{\sqrt{2\pi} I_\lambda \sigma_{\ln I}} \exp\left(-\frac{(\ln I_\lambda - <\ln I_\lambda>)^2}{2\sigma^2_{\ln I}} \right). \qquad (8.3.10)$$

It follows from this formula that for the mean intensity, $<I_\lambda>$, and its dispersion, $\sigma^2_I = <[I_\lambda - <I_\lambda>]^2>$, we can write

$$<I_\lambda> = \exp\left(<\ln I_\lambda> + \tfrac{1}{2}\sigma^2_{\ln I}\right), \qquad \sigma^2_I = <I_\lambda>^2 \left[\exp(\sigma^2_{\ln I}) - 1\right], \qquad (8.3.11)$$

respectively. Using (8.3.11), the parameters $\sigma^2_{\ln I}$ and $<\ln I_\lambda>$ can be expressed in terms of $<I_\lambda>$ and the normalized dispersion of the intensity fluctuations, β^2, defined by the relation

$$\beta^2 = <[I_\lambda - <I_\lambda>]^2> / <I_\lambda>^2 = \frac{\sigma^2_{I_\lambda}}{<I_\lambda>^2} = \int_{-\infty}^{\infty} W_{I_\lambda}(\mathrm{f}) d\mathrm{f}\,. \qquad (8.3.12)$$

The quantity β^2 is the so-called *scintillation index* evaluated experimentally and $W_I(f)$ is the respective frequency spectrum (see (8.2.10)). Therefore, we have

$$\sigma^2_{\ln I} = \ln(1+\beta^2), \qquad <\ln I_\lambda> = \ln\left[\frac{<I_\lambda>}{\sqrt{1+\beta^2}}\right]. \qquad (8.3.13)$$

In turn, based on the hypothesis that $\ln I_\lambda$ is distributed normally, it is possible to relate the quantity

$$\sigma^2_{\ln I} = <[\ln I_\lambda - <\ln I_\lambda>]^2> \cong 4<[\ln A_\lambda / A_{\lambda 0}]^2> = 4\sigma^2_x = 4<X^2> \equiv \sigma^2_0 \qquad (8.3.14)$$

emerging from the theory, and the scintillation index, β^2, measured in the sensing experiment, as

$$4<X^2> = 4\sigma^2_x = \sigma^2_{\ln I} = \sigma^2_0 = \ln(1+\beta^2)\,. \qquad (8.3.15)$$

The combined formulas (8.3.15) and (8.3.5*) serve as a basis to determine the structure characteristic of the refractive index of air, C^2_n, from the known value of the scintillation index, β^2.

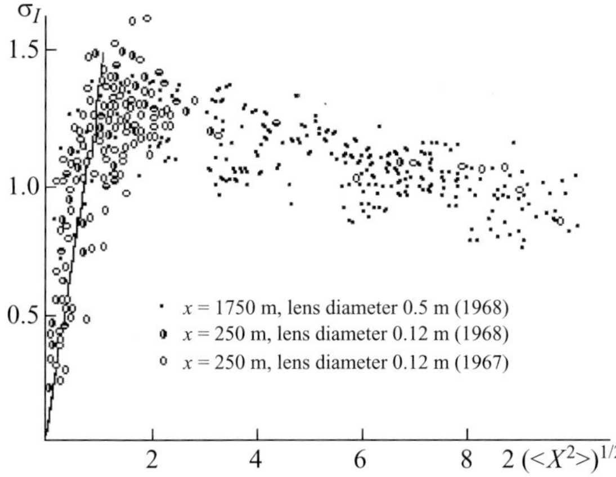

Fig. 8.3.1 Comparison of the measured dispersions of the light intensity fluctuations (open circles and filled squares) with the calculation results in the small and smooth perturbations (SSP) method (solid broken line). Evidently, the discrepancy between experimental and computed data for the amplitude fluctuations becomes significant for large values of the scintillation index, β ($< X^2 > \geq 0.7$), which essentially restricts the applicability domain of the SSP method. According to *Gurvich et al.,1976.*

Not going into mathematical details of the *SSP* method, we will only touch upon the point of its applicability limits. To evaluate the problem formally from the smallness requirement of the first discarded terms in the original wave equation, we have a very firm sufficient (though not simultaneously necessary) condition for the convergence of the *SSP* series for the amplitude variations,

$$\beta_0^2 \equiv 4 < X^2 > = 1.23 C_n^2 k^{7/6} S^{11/6} < 1. \tag{8.3.16}$$

This means that the dispersion, β_0^2, increases along the path length proportional to $S^{11/6}$. However, in reality the so-called saturation or even decrease of the intensity fluctuation dispersion occurs when the path length, *S*, grows unlimitedly. Figure 8.3.1 compares experimental data with the results of calculations in the *SSP* approximation. This figure testifies that the discrepancy of experimental and computed data for the amplitude fluctuations becomes significant for $< X^2 > \geq 0.7$. This threshold essentially restricts the applicability domain of the *SSP* method.

8.3.5. Strong fluctuations

Despite the constraints on using formula (8.3.15), the parameter $\beta_0^2 = 1.23 C_n^2 k^{7/6} S^{11/6}$ is commonly used in order to characterize the atmospheric turbulence effects on radiation propagation along a path S. As is known, either weak ($\beta_0^2 < 1$) or strong ($\beta_0^2 > 1$) intensity fluctuations governed by β_0^2 are distinguished. Under real atmospheric conditions,

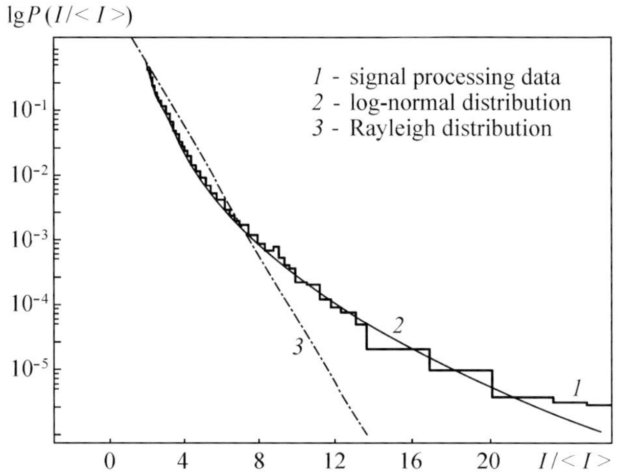

Fig. 8.3.2. Probability density, P, of the relative intensity fluctuations, $I/<I>$, in the case of strong intensity fluctuations of the universal parameter β_0^2=25. The solid curve (approximating the histogram obtained from the recorded data processing) depicts the probability density corresponding to the log-normal law with the mean value and the dispersion taken from the experiment; the dash-dotted line represents the exponential (Rayleigh) distribution of the amplitude intensity probabilities. It is clear that the log-normal distribution fits to the experimental data much better than the Rayleigh distribution.

the scintillation index, β^2, intricately depends on the path length and geometry of the beam propagation, atmospheric turbulence patterns, the wave number, $k = 2\pi/\lambda$, and other factors. The quantity β_0^2 is often regarded as the universal parameter to determine the scintillation index, $\beta^2 = \beta^2(\beta_0^2)$.

In the case of strong intensity fluctuations, the perturbation theory becomes inapplicable and no conclusions can be drawn concerning the intensity probability distribution. Based on some approximate calculations (*De Wolf, 1973*) or qualitative study of the scattered radiation field patterns (*Torrieri and Taylor, 1972*) it was argued that the Rayleigh mode of the probability distribution favors amplitude fluctuations in the saturation domain. The assertion that strong fluctuations of the radiation field follows a normal distribution proceeds from the perception that the total field at the arrival point of the beam results from the composition of fields scattered by a large number of independently re-radiating volumes.

However, such a concept is hardly applicable to the atmosphere whose large scale inhomogeneities exhibit small-angle forward scattering. In this case the interference produced by the strong correlation of phase differences of closely propagating rays appears to contribute mainly to the intensity fluctuations. Figure 8.3.2 shows an example of the histogram obtained for $\beta_0^2 = 25$. Here the solid curve depicts the probability density corresponding to the log-normal law with the mean value and the dispersion taken from the experiment, while the dash-dotted line represents the exponential distribution of the amplitude intensity probabilities,

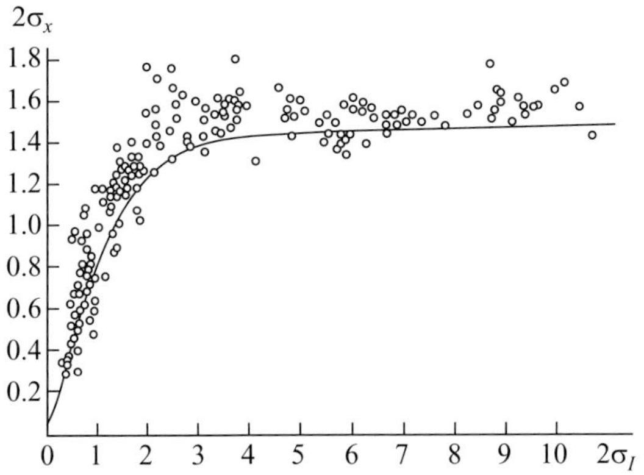

Fig. 8.3.3. The relationship between the mean square root fluctuations of the logarithm of the light intensity, $2\sigma_x$, (the experimental values, open dots) and those calculated in the first approximation of the method of smooth disturbances, $2\sigma_I$, (solid curve). It confirms that the curve obtained from the quantum field theory dependence $\sigma_I = f(\beta_0)$ in the domain of strong amplitude fluctuations fairly well complies with the experimental values. According to *Tatarskiy, 1967.*

$$P(I_\lambda) = (1/ <I_\lambda>) \exp[-I_\lambda / < I_\lambda >], \tag{8.3.17}$$

corresponding to the exponential (Rayleigh) distribution. It follows from this histogram that the log-normal distribution fits to the experimental data much better than (8.3.17).

For the sake of completeness let us mention that according to (*Ishimaru, 1981*), the intensity probability distributions are neither log-normal nor exponential in the range $0.3 < \beta_0^2 < 25$. At the same time, within the range $25 < \beta_0^2 < 100$ it seems to be closer to the log-normal law. Finally, it approaches the Rayleigh distribution for $\beta_0^2 \to \infty$.

It is also worthwhile to note that the advances in the intensity fluctuation theory were accomplished by applying the methods from the quantum field theory to determining the domain of strong amplitude fluctuations. The obtained dependence $\sigma_I = f(\beta_0)$ can be fairly well approximated by the formula

$$\sigma_I^2 = 1 - \left[\frac{1}{(1 + 1.5\beta_0^2)^{1/6}} \right], \tag{8.3.18}$$

which complies with the experimental values given in Figure 8.3.3. The following approximating formula for β^2 in the domain of strong fluctuations (*Prokhorov et al., 1974*) emerges from the analysis of all data available, which done with an accuracy up to the terms of the order of $\sigma_0^{-8/5}$:

$$\beta^2 = 1 + 0.87\sigma_0^{-4/3}, \quad \sigma_0^2 < 1, \quad (\text{or} \quad \sigma_0^2 = 0.706\,(\beta^2 - 1)^{-5/2}) . \tag{8.3.19}$$

8.4. STRUCTURE CHARACTERISTIC OF THE REFRACTIVE INDEX AND THE STELLAR SCINTILLATION INDEX

As it was said, stellar scintillations and the respective light flux fluctuations at a receiver are caused by inhomogeneities in the of refractive index of air, n, when the light beam propagates through the atmosphere. For the receiver mounted on an orbiting space vehicle, the recorded light rays, being subjected to unequal refraction, exhibit random brightness variations.

Visual observations of stellar scintillations from manned spacecraft, including the orbital stations *Salyut-7* and *Mir*, revealed some interesting phenomena in the structure of the temperature field of the terrestrial stratosphere (*Grechko et al., 1980; 1989; Gurvich et al., 1985*). These observations were done using photometers which allowed to record the properties of stellar scintillations within the tangent height range (the ray's "perigee") $z = 20$–45 km. The obtained *scintillation photocurrent* showed that scintillations occured everywhere and had a qualitatively similar pattern irrespective of place and time of the observations. Based on these data, the values of the scintillation index were calculated and the photocurrent spectral density and fluctuations at different heights were analyzed, including the estimates of their characteristic scales. Also, from the measured scintillation spectra in the height range of 20–40 km, the vertical spectra of temperature fluctuations were reconstructed.

These results are of primary methodical interest for the purpose we pursue because such measurements serve as a basis for calculating the structure characteristic, C_n^2, and thus set up the necessary prerequisite for modeling atmospheric turbulence. We first will address some particular problems relevant to the data processing involving the geometry of the observations, the specific patterns of the scintillation photocurrent, and the optics opening.

8.4.1. Geometry of the Observed Scintillations

The typical geometry of scintillations of setting stars observed from outer space is shown in Figure 8.4.1. The configuration roughly corresponds to the operational pa-

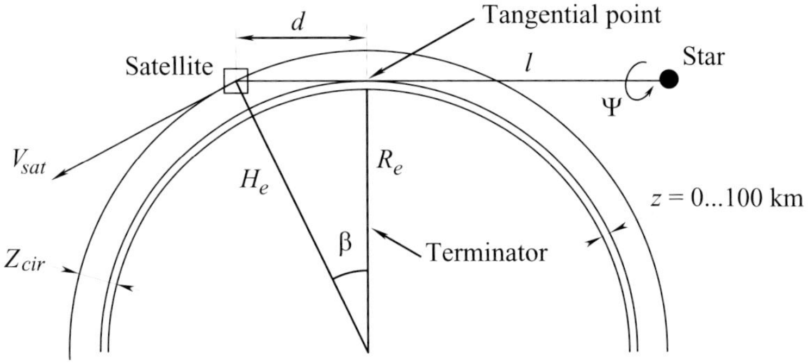

Fig. 8.4.1. Geometry of observations of scintillations generated by a light beam from an occulting star on board of an artificial satellite (a sketch). Position of the tangential point on the light of sight, l, is conditionally shown without refraction effects.

rameters of the *GOMOS* project for a satellite residing at a circular orbit at $Z_{cir} \approx 800$ km. In that case the effective thickness of the atmosphere along the light of sight, l, turns out to be several hundreds of kilometers. The distance from the "ray perigee" located above the terminator (for the given star) to the observer (a receiver on the satellite) makes up $d \approx 3000$ km. The refraction effect is not shown in the figure; it is evaluated in detail elsewhere (*Ioltukhovski and Marov, 1998*).

Assuming the terminator plane is orthogonal to the satellite orbital plane (in general, a correction factor for the azimuthal angle, ψ, is to be introduced), depending on the angle β between the radius-vectors of the satellite and tangential point, various setting trajectories in reference to the horizon in projection to the terminator plane are possible. In the two extreme cases, a star either sets vertically ($\beta = 0$), or there is no starset when a star touches the horizon ($\beta \geq \arcsin(R_e/H_e) \cong 71.4°$). Here R_e is the Earth's radius, and $H_e = R_e + Z_{cir}$ is the radial distance to the satellite. Thus, performing observations for different β, it is possible to probe the atmosphere within the height range $z \approx 0 - 100$ km above the surface and, generally, along the various azimuthal directions, determined by the selected stars and precession of the orbital plane.

Let us note that the problem of how to determine statistical features of the detected radiation can be simplified because the distance from the atmosphere to the observer considerably exceeds the atmosphere thickness. Then, for weak scintillations, $\sigma_I < 1$, the atmospheric impact on propagating radiation can be treated approximately as an equivalent phase screen located in the terminator plane.

8.4.2. The Scintillation Photocurrent

A photoelectric device placed in the focal plane of a telescope transforms the light flux into an electric voltage whose variations are used for statistical analysis.

As an example, Figure 8.4.2 (the upper part) shows the photocurrent recorded with observations of *Canopus* (*Alpha Carinae*) during its setting on November 25, 1987 (*Grechko et al., 1989*). The used color filter had a transmission half-width $\Delta\lambda = 0.05$ microns and was centered at 0.45 microns. The time, t, and the beam perigee height, z, are plotted along the horizontal axis while the photocurrent, I, (in arbitrary units) is plotted as ordinates. The middle and bottom parts of Figure 8.4.2 show several fragments of the same recording enhanced tenfold in time against the original record. Photocurrent fluctuations at heights $z \geq 70$–100 km, at which the atmosphere affects stellar images only weakly, are mainly caused by the shot noise of the photomultiplier. It corresponds to the mean signal level, $\langle I \rangle = I_0$, and is limited by the bandwidth of the receiver. As a starset occurs, $\langle I \rangle$ decreases and scintillations become progressively pronounced as the beam sinks into atmosphere. Beginning at about $z = 20$ km downward, chromatic dispersion due to refraction exerts a significant influence on scintillation patterns and simultaneously the low-frequency component of scintillations becomes more appreciable.

The pattern shown in Figure 8.4.2 is a typical record of light scintillations, with pronounced outbursts (random focusing) and severe fading, for which random inhomogeneities in the refractive index of air are responsible. Attenuation of the mean signal

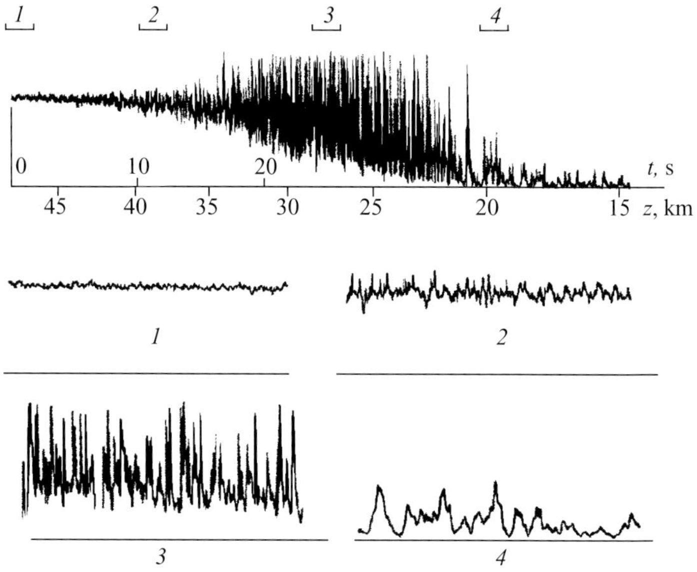

Fig. 8.4.2. An example of the photocurrent record caused by scintillations of *Canopus* (α *Carinae*) in the terrestrial atmosphere observed at starset on 25.11.1987. Fragments of the record 1 to 4 are spread in time tenfold. According to *Grechko et al., 1989.*

level is determined by the combined effects of molecular and aerosol scattering (these effects were treated in detail by Ioltukhovski and Marov (*1998*)), ozone absorption in the *Chappuis* band, and refractive extinction. The extinction due to scattering and absorption can be estimated assuming the refractive attenuation, $<I>$, varies with height exponentially depending on the atmospheric scale height, H_0. Note that as extinction is wavelength selective, the atmosphere serves as a color filter for the passing radiation. Processing the recorded scintillation patterns allows to find:

- The effective wavelength, λ, and the wavelength band, $\Delta\lambda$, forming the scintillation photocurrent, accounting for the stellar spectrum, the spectral sensitivity of the photomultiplier and the filter transmission;
- The spectra of the relative scintillation fluctuations, $W(\mathrm{f})$, using the Fourier fast transform software;
- The values of the mean current, $<I>$, in every part of the record used for normalization.

It was found that in order to achieve the correct data processing within the height intervals $\Delta z = v\Delta\tau \cong 3$–4 km, the duration of the record fragments, Δt, should not exceed 2 to 8 s. With Δt increasing further, the regular variations, rather than the random variations, of the photocurrent with characteristic scale of the order of H_0 have the largest effect on the scintillation spectra at the largest scale ranges.

8.4.3. Dependence of Scintillations on Instrument Optics

Fluctuations in the recorded light flux strongly depend on the telescope diameter (numerical aperture) and the diaphragm used, as well as on the zenith distance of the light source, its angular dimensions, and the meteorological conditions. The magnitude of fluctuations is described by the scintillation index,

$$\sigma_P^2 = <(P-<P>)^2>/<P>^2, \tag{8.4.1}$$

where P is the light flux through the lens of the telescope.

Obviously, $P = \iint_{\Sigma} I_\lambda(y,z)dydz$, where $I_\lambda(y,z)$ is the intensity of a monochromatic light wave at the lens surface with area Σ. According to what has been noticed earlier, the quantity $I_\lambda(y,z)$ follows a log-normal probability distribution. Similarly, if $\sqrt{\Sigma} \leq \sqrt{\lambda S}$, i.e. the lens diameter is less than the correlation radius of the intensity fluctuations and hence, light flux oscillations through different cells of the lens occur simultaneously, the quantity P also follows a log-normal law. If, however, $\sqrt{\Sigma} > \sqrt{\lambda S}$, i.e. the lens area encloses a large number of uncorrelated inhomogeneities in the light field, then the quantity P, by virtue of the central limit theorem, would be distributed normally. Note that because only 2–4 inhomogeneities in the field within Σ fit the dimensions of onboard telescopes, the distribution of P remains very close to log-normal. In this case the theoretically defined quantity $<(P/P_0)^2>$ is connected with the measured scintillation index of the light flux, σ_P^2, by the relation

$$<(P/P_0)^2> = \ln(1+\sigma_P^2). \tag{8.4.2}$$

On the other hand, for the dispersion, $<X^2>$, of weak amplitude fluctuations of a monochromatic light wave, the following theoretical relation for the quantity $<(P/P_0)^2>$ is valid:

$$<(P/P_0)^2> = 4<X^2>G(X,Y) = \beta_0^2 G(X,Y), \tag{8.4.3}$$

where

$$G = (1/\beta_0^2)\ln\left\{\frac{16}{\pi}\int_0^1 \exp[4B_A(Dx)](\arccos x - x\sqrt{1-x^2})xdx\right\}, \tag{8.4.4}$$

$X = D/(\lambda\Sigma)^{1/2}$; $\Psi = \beta_0^2$; D is the size of diaphragm; and $B_A(Dx)$ is the correlation function of wave amplitude fluctuations defined by (8.3.2). The function $G(X,Y)$ specifies the reduction of scintillations due to the lens averaging effect. Numerical integration of expression (8.4.4) for $\beta_0^2 = 4$ and $\beta_0^2 \to 0$ results in the plot shown in Figure 8.4.3. It is seen that the function G is much smaller than unity for $D >> (\lambda S)^{1/2}$. Thus, fluctuations of the total light flux passing through the telescope turn out to be attenuated substantially when diameter of the telescope diaphragm exceeds the correlation radius of fluctuations, $(\lambda S)^{1/2}$, considerably.

8.4.4. The Scintillation Index and the Structure Characteristic

As was shown earlier, the results of stellar scintillation measurements serve as a basis for determining the structure characteristic, C_n^2. Assuming that C_n^2 depends only on height, z, let us change the variable $x = z \approx \sec\theta$ in (8.3.5*), where θ is the zenith angle of the star. This yields the following formula for the mean square of the intensity fluctuations:

$$4 < X^2 > = \sigma_0^2 = <(I/I_0)^2> = 2.24\,k^{7/6}(\sec\theta)^{11/6}\int_0^\infty C_n^2(z) z^{5/6} dz\,. \quad (8.4.5)$$

In order to obtain the mean square of fluctuations of the light flux, P, it is necessary to multiply the right hand side of (8.4.5) by the function $G(D/\sqrt{\lambda S})$. As $S = H_0 \sec\theta$ (assuming noticeable fluctuations in the refractive index to occur within the scale height) the function $G(D/\sqrt{\lambda S})$ also depends on θ. Then

$$\ln(1+\sigma_P^2) = <(\ln P/P_0)^2> = \sigma_0^2 G(D/\sqrt{\lambda H_0 \sec\theta}) =$$
$$= 2.24 k^{7/6}(\sec\theta)^{11/6} G(D/\sqrt{\lambda H_0 \sec\theta})\int_0^\infty C_n^2(z) z^{5/6} dz \quad (8.4.6)$$

Now, in one way or another a height profile should be assigned to C_n^2 in order to evaluate this parameter from the data on scintillations of stellar images, i.e. from the known value of the dispersion, σ_p^2. For example, if $C_n^2(z)$ decreases with height as z^{-2}, it can be defined as $C_n^2(z) = C_{n0}^2[1+(z/H_0)^2]^{-1}$. Then the following relation emerges from (8.4.6):

$$<(\ln P/P_0)^2> = 13.6 k^{7/6}(\sec\theta)^{11/6} G(D/\sqrt{\lambda H_0 \sec\theta}) C_{n0}^2 H_0^{11/6}. \quad (8.4.7)$$

We may summarize that it is necessary to perform the following subsequent steps to estimate C_{n0}^2:

- to calculate $<(\ln P/P_0)^2> = \ln(1+\sigma_P^2)$ from the scintillation index of the light flux, σ_p^2, based on a log-normal distributed radiation intensity, I_λ, in the plane of the telescope lens;

- to transform $<(\ln P/P_0)^2>$, defining the measured scintillation of the light source with finite angular size, to a source with infinitesimal angular size using the curve depicted in Figure 8.4.3;

- finally, taking advantage of (8.4.7), to evaluate C_{n0}^2.

8.4.5. Spectra of Light Intensity Fluctuations

In section 8.2 we stressed the point that in order to estimate the relative dispersion of light flux fluctuations, σ_p^2, it is convenient to use the frequency spectra of the respective random fields. This is because exactly these spectra are measured while processing the experimental data, although from the theoretical viewpoint it is easier to determine the spatial structure of the random field of the structural parameter, f, under study. The "frozenness" hypothesis discussed in section 8.2.2, is used commonly when calculating the frequency spectral density of the wave amplitude fluctuations, $W_X(f)$.

Let us define the time correlation function of the wave amplitude fluctuations as $B_X(t) = <X(\boldsymbol{r}, t+\tau)X(\boldsymbol{r}, t)>$. Then, in view of the "frozenness" condition ($f(\boldsymbol{r}, t) = f(\boldsymbol{r} - \boldsymbol{w}t, 0)$, where $\boldsymbol{w}$ is the velocity of inhomogeneities travelling transversely to the direction of wave propagation), we have the following relation pertinent to the lens plane $x = const$:

$$B_X(t) = B_A(|\boldsymbol{w}|\tau)). \quad (8.4.8)$$

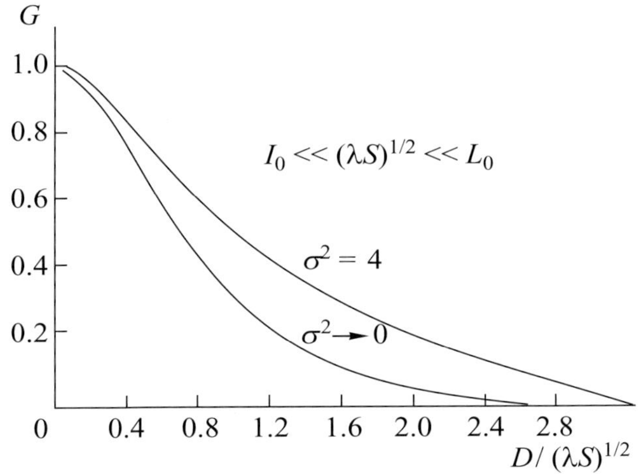

Fig. 8.4.3. Relative reduction of fluctuations of the recorded light flux depending on the lens' diameter (theoretical estimations). It is seen that the function $G << 1$ for $D >> (\lambda S)^{1/2}$. Therefore, fluctuations of the total light flux passing through the telescope is attenuated substantially when the diameter of the diaphragm, D, considerably exceeds the correlation radius of the fluctuations, $(\lambda S)^{1/2}$. According to *Tatarskiy, 1967.*

Here $B_A(r)$ is the spatial correlation function coupled with the two-dimensional spectral density, $F_A(\kappa,0)$, by expression (8.3.1). The frequency spectrum of the wave amplitude fluctuations, $W_X(\mathrm{f})$, and the corresponding correlation function, $B_X(t)$, are linked by the relation reciprocal to (8.3.2):

$$W_X(\mathrm{f}) = 4\int B_X(\tau)\cos(2\pi \mathrm{f}\,\tau)d\tau = 4\int_0^{\infty} B_A(|\boldsymbol{w}|\tau)\cos(2\pi \mathrm{f}\,\tau)d\tau . \qquad (8.4.9)$$

To evaluate $W_X(\mathrm{f})$ using (8.4.9), we first express $B_A(|\boldsymbol{w}|\tau)$ in terms of the two-dimensional spectral density of amplitude fluctuations, $F_A(\kappa,0)$; this yields

$$B_A(|\boldsymbol{w}|\tau) = 2\pi\int_0^{\infty} J_0(\kappa|\boldsymbol{w}|\tau)F_X(\kappa,S)\kappa d\kappa . \qquad (8.4.10)$$

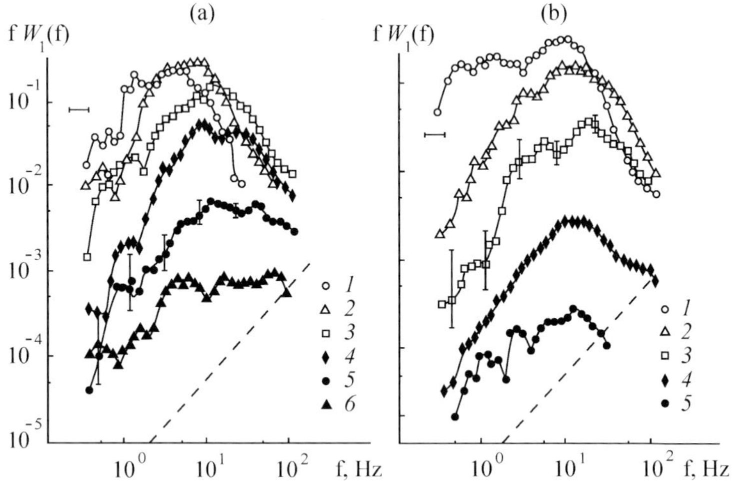

Fig. 8.4.4. Frequency scintillation spectra of *Arcturus* and *Canopus* taken during their occultations by *Alexandrov et al., 1990* from the manned spacecraft on 09.10.1987 **(a)** and 25.11.1987 **(b)**, respectively. Horizontal bars in the upper left – the width of averaging; tilted dashed lines below – estimation of the shot noise level. The measurements are assigned to the middle, z_0, of the covered height intervals with the respective mean square root of the light flux fluctuations, σ_p, as follows:

(a): 1. z_0 = 18.5 km, σ_p = 65%; 2. z_0 = 22.5 km, σ_p = 75%;
3. z_0 = 26.5 km, σ_p = 55%; 4. z_0 = 31.0 km, σ_p = 33%;
5. z_0 = 36.0 km, σ_p = 12%; 6. z_0 = 41.5 km, σ_p = 4.5%.

(b): 1. z_0 = 20.0 km, σ_p = 85%; 2. z_0 = 25.5 km, σ_p = 60%;
3. z_0 = 31.5 km, σ_p = 29%; 4. z_0 = 38.0 km, σ_p = 8%;
5. z_0 = 45.0 km, σ_p = 2.5%.

Substituting now this relation into (8.4.9) and using the *Weber* integral properties, we come to the following formula:

$$W_X(\mathrm{f}) = (8\pi/|\boldsymbol{w}|)\int_0^\infty F_A(\sqrt{\kappa^2 + (4\pi \mathrm{f}^2/|\boldsymbol{w}|^2}), S)\, d\kappa . \tag{8.4.11}$$

This formula allows us to calculate the frequency spectrum of wave amplitude fluctuations from the given two-dimensional spatial spectral density of amplitude fluctuations, $F_A(\kappa,0)$, the latter being defined theoretically. Both theoretical and experimental estimates of σ_p^2 should be fulfilled in this case taking into account the relation

$$\sigma_P^2 = \int_0^\infty W_X(\mathrm{f})\, d\mathrm{f} . \tag{8.4.12}$$

As an example, Figure 8.4.4 shows the frequency spectra obtained from the observations of two starsets at different heights, z_0, over the middle of the height intervals for which the spectra were computed (*Aleksandrov et al., 1990*). The scintillation frequency, f, and the dimensionless quantity $\mathrm{f}\, W_I(\mathrm{f})$ are plotted along the horizontal and vertical axes, respectively, $W_I(\mathrm{f})$ being the smoothed estimates of the spectral density of the relative fluctuation intensity of the light flux coming from a star to the receiver's lens.

With regard for the frozenness assumption, the scintillation frequency spectra can be recounted into one-dimensional spatial spectra. In this case the essential parameter is the velocity of the point where the light beam intersects the equivalent phase screen plane, $\boldsymbol{w}$. For the one-dimensional bilateral scintillation spectrum in the direction colinear to $\boldsymbol{w}$, it is possible to write

$$V_I(\kappa_s) = (|w|/4\pi) W_I(\kappa_s |w|/2\pi) , \tag{8.4.13}$$

where $\kappa_s = 2\pi \mathrm{f}/|\boldsymbol{w}|$ are the wave numbers.

Evidently, if inhomogeneities of the atmospheric refractive index are statistically isotropic, the scintillation spectra, $V_I(\kappa_s)$, should be independent of the starset angle. Otherwise, when inhomogeneities are strongly anisotropic and horizontally elongated, the frequency scintillation spectrum is controlled by the vertical velocity, w_v , and hence, the measured scintillation spectra are congruent only after their conversion to the vertical wave numbers, $\kappa_1 = 2\pi \mathrm{f}/v_\perp$, where $v_\perp$ is vertical component of the velocity $\boldsymbol{w}$. Accordingly, in this case the vertical scintillation spectra, κ_1, are defined as follows:

$$V_I(\kappa_1) = (v_\perp/4\pi)\, W_I(\kappa_1 v_\perp/2\pi) . \tag{8.4.14}$$

The spatial scintillation spectra of the stars *Achernar* (*Alpha Eridani*) and *Sirius* (*Alpha Canis Majoris*) for two heights recalculated from the frequency spectra using the relations (8.4.13) and (8.4.14), are shown in Figure 8.4.5 as dependence of the dimen-

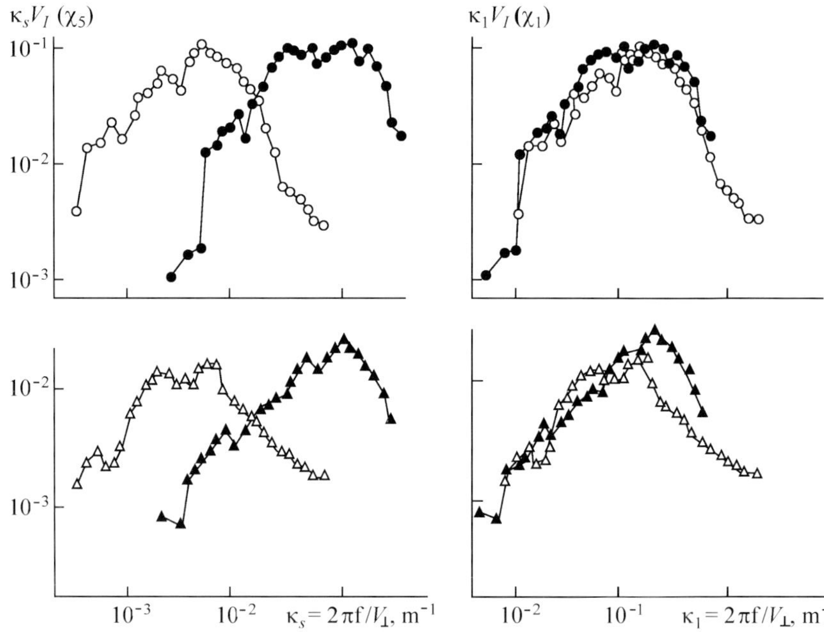

Fig. 8.4.5. Spatial scintillation spectra of *Achernar* (open quadrangles and triangles) and *Sirius* (black quadrangles and triangles) taken during their occultations from the manned spacecraft on 6.10.1987 and on 05.04.1988, respectively. Upper spectra – for the mean altitude $z_0 = 25$ km; lower spectra – for the mean altitude $z_0 = 33$ km. According to *Alexandrov et al., 1990*.

sionless quantities $\kappa_s V_I(\kappa_s)$ and $\kappa_1 V_I(\kappa_1)$ upon the wave numbers κ_s and κ_1. For the sake of clearness, two strongly different starset angles of $90^\circ - \alpha$ are chosen. It is seen that the scintillation spectra reduced to the wave numbers, κ_s, are mutually biased in x-axis by more than an order of magnitude. At the same time, the scintillation spectra, reduced to the vertical wave numbers, κ_1, coincide with each other quite satisfactorily. A similar picture is observed when analyzing the scintillation spectra for other starsets.

We will close by appreciating the described data treatment from observations of setting stars as a methodological background to extract the important information on atmospheric turbulence when dealing with middle/upper atmosphere space monitoring projects. This particular part of the overall program relies on utilizing virtually the same basic procedures of data processing involving real time acquisition of the turbulent exchange coefficients. Such an approach seems a powerful tool to refine the simultaneous reconstruction of the structure and the dynamic behavior of the middle atmosphere. The ultimate goal is to reveal the key features of minor species abundances, including their temporal-spatial variations and possible long-term trends.

Summary

- *Modeling the dynamic properties of the middle/upper atmosphere using light or radio occultation technique, is addressed as one of the important applications of the turbulence theory for multicomponent media. The measured spectra of occulting stars permitting to reveal simultaneously the minor atmospheric species contents and the statistical structure of the turbulent field, serve as a particular example of such an application.*

- *To ensure methodological support for space monitoring aimed to reveal patterns of atmospheric dynamics, a procedure was developed which involves calculation of the external turbulence scale,* L, *from the known structure characteristic of the refractive index of air,* C_n^2. *This parameter can also be evaluated from the results of remote sensing measurements on board spacecraft of light emitted by setting stars. The procedure to derive the scintillation index is based on the data processing of fluctuations of the recorded photocurrent produced by the stellar images occulted at different heights.*

- *The results from observations of starsets from manned spacecraft and the respective data analysis served as a methodological background to extract important information on atmospheric turbulence when dealing with the future middle/upper atmosphere space monitoring projects. Such projects are addressed as a powerful tool to refine the reconstruction of the atmospheric structure, composition, and dynamics.*

CHAPTER 9

THE PROCESSES OF HEAT AND MASS TRANSFER AND COAGULATION IN PROTOPLANETARY GAS-DUST NEBULA

We revert now to cosmogonic application of the developed theory. Some basic problems of the contemporary planetary cosmogony were addressed in Chapter 1. Here we will focus on the conditions for maintaining shear turbulence in the Keplerian accretion disk near a solar-type star during the evolution of the system, including forming a gas-dust subdisk in the vicinity of the disk's central plane, based on the model of a two-phase, differentially rotating medium containing particles with different amount and size distributions.

Special attention will be paid to an heuristic way of modeling the turbulent exchange coefficients for accretion disks. The formulas for these coefficients, taking into account the inverse effects of dust transfer and potential temperature on the maintenance of shear turbulence, are generalized to the protoplanetary gas-dust nebula including the respective expression for the turbulent viscosity coefficient in the so-called α-disks, where α is the dimensionless disk viscosity parameter (*Shakura and Syunyaev, 1973*). The derived defining relations for turbulent diffusion and heat flows are pertinent to the two-phase mixture rotating differentially with an angular velocity $\Omega(\boldsymbol{r}, z)$ and allow to describe dust and heat transfer in the direction perpendicular to its central plane.

The evolution problem of a protoplanetary nebula involves a limiting saturation by small dust particles in the "*cosmic fluid*" layer located slightly above or below the near-equatorial gas-dust subdisk. This incorporates the concurrent processes of mass transfer and coagulation in a two-phase medium of a differentially rotating disk, where shear turbulence occurs in the presence of polydisperse solid particles suspended in a carrying flow. The refined defining relations for diffuse fluxes of different particle sizes in the turbulent diffusion equations in a gravitational field are derived, which describe their coagulation growth in addition to convective transfer, turbulent mixing, and sedimentation of disperse dust grains onto the central plane of the disk.

A semi-empirical procedure developed for evaluating the turbulent viscosity and turbulent diffusion coefficients, takes into account the inverse effects of dust transfer on the disk's turbulence evolution, including the inertial difference between disperse solid particles. To accommodate a more rigorous solution of the problem of mutual influence of turbulent mixing and coagulation kinetics in forming the gas-dust subdisk, the plausible mechanisms (gravitational, turbulent and electric coagulation) are explored. The parametric method of moments for solving the *Smoluchowski integral-differential coagulation equation* for the different particle size distribution functions is discussed thoroughly.

9.1. TURBULENCE IN TWO-PHASE DISPERSE MEDIA. THE TURBULENT TRANSFER COEFFICIENTS IN ROTATING GAS-DUST ACCRETION DISKS

The gas-dust accretion disk is assumed to be a rotating, compressible, stratified, and, possibly, magnetized object. It is also recognized as a highly turbulized medium, which was emphasized in section 1.4.3. This is supported, first of all, by a very high value of the Reynolds number: $\boldsymbol{Re} = \Omega R^2/\nu \geq 10^{10}$. Besides, in addition to the energy from a new born star absorbed by the surrounding dust envelope and re-emitted at infrared wavelengths, dissipation of turbulent energy may also contribute to such radiation. Anyway, the energy recorded from some compact cosmic objects is incompatible with their laminar nature. Moreover, self-organization processes occurring on the background of turbulent motion of a substance could serve as an important mechanism to influence viscous gas-dust disk properties at different phases of its evolution and ultimately lead to accretion of planetary system bodies.

Thus the efficiency of competing mechanisms within a protoplanetary nebula significantly depends on the turbulization of the heterogeneous medium. Indeed, the dynamic processes of gas-dust interaction involve the inverse effect of dust particles on the disk turbulence and heat balance while turbulence influences the rates of phase transitions (evaporation, condensation) and particle accumulation (coagulation, fragmentation) in the substance flow. In turn, motion and heat exchange in the disperse phase are intimately related with the turbulent stresses, diffusion and heat fluxes caused by the participation of solid particles in the pulsation motions of the carrying phase.

Although detailed hydrodynamic aspects of the influence of solid particles on the flow, depending on their inertial properties and concentration, are not unambiguously defined, in general such effects are to be taken into account as particles play an important role in the mechanism of gradient production of turbulent energy from the averaged rotational energy of the disk matter; in the formation of a non-stationary structure of eddies in the flow separated from a large particle; in the diffusive turbulent transport of dust particles caused by the non-uniform distribution of the disperse phase in the disk and by the external mass forces; in the generation of disturbances due to colliding particles; etc.

Furthermore, turbulent mixing prevents the diffusive separation of dust and gas constituents in the disk's gravitational field (*gravitational sedimentation*), thereby impeding dust clustering. On the other hand, in a turbulent flow, the number of collisions between particles per a unit time considerably increases when compared to the laminar motion, as a result of the chaotic turbulent pulsations. Respectively, a non-gravitational accretion such as that related to the growth of nuclei by means of *turbulent coagulation*, becomes an efficient clustering mechanism. Thus, turbulence favors the effective sedimentation of dust particles on the disk’s central plane, the formation of a subdisk and the eventual accumulation of planetesimals. Unfortunately, until recently the most relevant coagulation mechanism has not been investigated in detail and needs more in-depth study.

9.1.1. Modeling Shear Turbulence in the Disk

Let us analyze quantitatively some modeling aspects of shear turbulence in a thin gas-dust disk having the half-thickness-to-radius ratio $h/R \ll 1$. An example of such a disk is the system of close binaries mentioned in Chapter 1, in which the substance streams from the star enlarged to the size of its Roche sphere to its more massive companion. The non-zero angular momentum of the overflowing substance is responsible for the formation of a rotating gas-dust disk around the second star. The origin of turbulence in such a disk is apparently caused by the substance velocity gradient due to the differential character of its angular rotation (see, e.g., *Gorkavyi and Friedman, 1994*). Obviously, because each layer with radius r in the differential rotation disk has a nearly Keplerian orbital velocity $V_\varphi = r\Omega_K(r) = \sqrt{GM_\odot / r}$, where G is the gravitational constant and $M_\odot$ is the mass of the central body, the angular rotation rate, $\Omega_K(r)$, is proportional to $r^{-3/2}$, which represents a typical example of a large-scale shear flow.

The turbulent viscosity coefficient for a gaseous disk was first estimated based on Kolmogorov's concept for dynamical turbulence, $\mu^T = \rho v_{turb} l_{turb}$ (*Shakura and Syunyaev, 1973*). The following relationship between the tangential Reynolds turbulent stresses, $R_{r\varphi}$, and gas pressure, p, was derived:

$$R_{r\varphi} = \mu^T r\partial_r(V_\varphi / r) = -\tfrac{3}{2}\rho v_{turb} l_{turb} \Omega_K(r) = -\alpha\, p\big|_{z=0}, \qquad \alpha \le 1. \tag{9.1.1}$$

Here v_{turb} is the root-mean-square velocity of turbulent pulsations, restricted by the speed of sound, $c_s\big|_{z=0}$, which is calculated for the disk's central plane ($v_{turb} \le c_s\big|_{z=0} \cong \sqrt{p/\rho}\big|_{z=0}$), and l_{turb} is the so-called Prandtl mixing length restricted by the disk thickness. The dimensionless free parameter α, which characterizes the degree of excitation of turbulent motions, can be calibrated empirically, using time-dependent spectra derived, in particular, from observations of bursts in binary systems with mass transfer, including white dwarfs. For these cases, values of α in the range from 0.01 to 1 were found (*Eardley et al., 1978*). Models constructed on the basis of relationship (9.1.1) pertain to the viscous α-disks. The value $\alpha \approx 10^{-3}$ was generally adopted based on the study of different physical mechanisms underlying the turbulence in a protoplanetary disk, such as heat convection and differential rotation (*Cabot et al., 1987; Dubrulle, 1993*), or the disk structure around *T-Tauri* stars (*Makalkin and Dorofeeva, 1991; 1995*).

The main advantage of an heuristic treatment of turbulence as we discussed in Chapters 1 and 6, is its relative simplicity: the effect of turbulence on regular gas motion is introduced through the turbulent viscosity coefficient instead of the molecular viscosity coefficient in the hydrodynamics equations. However, in this particular case, when deriving formula (9.1.1), the gas-dust composition of the disk was ignored, though inter-phase interactions seem to be important at the various stages of the gas-dust disk evolution. While small dust particles (making up only 2% of the protoplanetary nebula mass) appear to render a limited effect on the rotational motion of the carrier gas phase at the initial stage (when the central condensation is heated and primordial particles in the enclosed disk are mostly evaporated), the effect gradually grows. This is because the particle size increases due to *adhesion* (sticking) in collisions, sedimentation to the central plane of the accretion disk, and gas dissipation from the system.

The multiphase flow in the disk implies an inverse influence of particles on the flow, which results in the gradient production of turbulent energy from averaged shear motion; diffusive turbulent transport of particles due to the inhomogeneous spatial distribution of the disperse phase; generation of disturbances by interacting colliding particles, etc. Obviously, formula (9.1.1) must be modified by introducing the correction factor allowing for the decrease of turbulence energy in the disk.

The inverse effect emerges due to additional buoyancy forces for which the gradient of dust particle concentrations in the disk's interior regions (related, in particular, to the gravitational sedimentation of particles to the central plane of the nebula) is responsible. This impedes turbulence development. Moreover, in general, one should also correct the formula for ν^T to account for the inverse effect of the non-adiabatic temperature distribution in the disk on the turbulence intensity. The respective corrections in the turbulent exchange coefficients for two-phase disperse media can be found, for example, by using the *K*-theory for multicomponent turbulence that was developed and discussed in Chapter 4. Let us recall that this theory is based on the evolutionary transfer equations for the one-point second correlation moments of pulsating thermohydrodynamic mixture parameters, the equations being written in the local-equilibrium approximation, i.e. without convective and diffusive terms.

9.1.2. Turbulent Energy Balance in Two-Phase Disperse Media

Basically, the conditions for set up and maintenance of shear turbulence in a gas-dust accretion disk must be investigated in the framework of a model for a two-phase disperse compressible medium containing a different fraction of suspended dust particles. Their probable size distribution, allowing possible coagulation processes and, in particular, turbulent coagulation, should be introduced in the model because this function influences the dynamical and optical properties of the turbulent fluid. Indeed, the original conditions for heat and mass transfer and energy balance in the system, including disk transparency to thermal radiation, change in the process of particle accumulation and their sedimentation to the equatorial plane. This may result in decreasing convective instability and turbulence intensity because loading the gas flow by heavy particles weakens the energy of turbulent pulsations.

Unfortunately, when modeling the turbulent transport coefficients, it is difficult to exploit the particles size distribution function, jointly with the Smoluchowski kinetic equations describing the growth of polydisperse particles, as one of the accretion disk parameters. This is caused, in particular, by the mathematical complexity of the system of coagulation equations (especially those accounting for disintegrating colliding particles and inertial differences of different size particles), and, so far quite vague, ideas how to attack the problem of interplay between coagulation kinetics and turbulent mixing.

Hence we confine the nature of the disk medium assuming it only contains a suspension of small heavy particles with one fixed size. Further we assume that these particles are in the order of or less than the internal turbulence scale and form a continuous, thermodynamically active phase with its own parameters: density, temperature, viscosity, etc. This condition is satisfied by *Stokes particles* of any dispersity, from submicron to centimeter sizes. Considering a thermodynamically equilibrium, two-fluid mixture with equal temperatures and pressures of gas and solid particles ($T_g = T_d = T$, $p_g = p_d = p$), we permit the instantaneous velocity components in the direction of the flow to be coincident, while the transverse components may differ by a constant value, W_d, the latter being equal to the rate of gravitational sedimentation of solid particles in the fluid.

The overall mass density, ρ, and the hydrodynamic mean-mass velocity, v_j, of the heterogeneous mixture can be defined by the following relations:

$$\rho = \sum_\alpha \rho_\alpha s_\alpha = \rho_g(1-s) + \rho_d s,$$
$$v_j = \rho^{-1}\sum_\alpha \rho_\alpha s_\alpha v_{\alpha j} = \frac{\rho_g(1-s)}{\rho} v_{gj} + \frac{\rho_d s}{\rho} v_{dj}, \tag{9.1.2}$$

where ñ$_{\text{å}}$ is the proper mass density of α-phase (the subscript $\alpha = g$ refers to the carrier gas phase and the subscript $\alpha = d$, to the dust); $v_{\alpha j}$ is the hydrodynamic particle velocity; and s_α is the volume content of the α-phase, which is equal to the volume of α-particles contained in a unit volume of the whole medium ($\sum_\alpha s_\alpha = 1$). The subscript "d" in the quantity s_α is omitted.

The carrier gas phase is supposed to be a compressible, chemically homogeneous *baroclinic* medium that satisfies the equation of state for perfect gases, while the disperse phase (solid particles) is thought to be an incompressible medium. Then

$$p_g = \rho_g R_g T, \quad h_g = C_{Pg} T + h_g^0,$$
$$\rho_d = const, \quad h_d = C_d T, \tag{9.1.3}$$

where $R_g = \mathrm{R}/\mu_p$, R is the universal gas constant, μ_p is the mean atomic mass, h_α is the enthalpy per unit mass of the α-phase, h_g^0 is the gas enthalpy at zero temperature, and C_{Pg} and C_d are the heat capacities of gas (at constant pressure) and dust.

Assuming additivity for thermodynamic functions (internal energy, enthalpy, etc.) for the heterogeneous mixture with respect to the mass of phases in the mixture (due to the possibility of neglecting the influence of the near-surface *Knudsen layer* of a particle), we can write the equation of state and the specific enthalpy of the entire system in the form

$$p = \rho R T, \qquad R(Y_g, s) = R_g \rho_g / \rho = R_g Y_g /(1-s) \cong R_g Y_g,$$
$$h = \sum_\alpha Y_\alpha h_\alpha = \rho^{-1}[\rho_g (1-s)(C_{Pg} T + h_g^0) + \rho_d s C_d T] \cong C_p T + Y_g h_g^0, \tag{9.1.4}$$

where

$$C_P = \sum_{\alpha} Y_\alpha C_{P\alpha} \cong [\rho_g C_{Pg} + s\,\rho_d\, C_d\,]/\rho$$

is the specific heat capacity at constant pressure and $Y_\alpha = \rho_\alpha s_\alpha / \rho$ is the mass concentration of the α-phase ($\sum_{\alpha} Y_\alpha = 1$).

The approximate equality in formula (9.1.4) holds for gas suspensions with low volume content of the condensed phase (when $s_d << 1$ and $s_g \approx 1$). This condition is further on accepted in our consideration. Nevertheless, the dynamical effect of solid particles on the gas flow may be essential in view of the enormous influence of gravitational forces. Thus, such a gas-dust disk medium, as a whole, can be considered as a perfect gas with adiabatic exponent γ and sound speed c_s, which are determined by the relations

$$\gamma = \frac{C_P}{C_P - R} \cong \frac{\rho_g C_{Pg} + s\,\rho_d C_d}{\rho_g (C_{Pg} - R_g) + s\rho_d C_d}, \qquad 1 \le \gamma \le \gamma_g = \frac{C_{Pg}}{C_{Pg} - R_g},$$

$$c_s^2 = \gamma \frac{p}{\rho} = \gamma RT \cong c_{sg}^2 \frac{\gamma \rho_g}{\gamma_g \rho}, \qquad c_s < c_{sg} = (\gamma_g R_g T)^{1/2}, \tag{9.1.5}$$

where γ_g and c_{sg} are the adiabatic exponent and the sound speed in a pure gas, respectively. To simulate the averaged motion in such a mixture, one can take advantage of the apparatus for turbulized fluid hydrodynamics developed elsewhere (*Marov and Kolesnichenko, 1987*).

Let us now consider a simplified statistical scheme of turbulence based solely on the balance equation for turbulent energy for the shear flow that carries the gas suspension. For the above assumptions on the properties of the medium, this equation has the same appearance as the equation for multicomponent compressible mixtures (3.1.68). It can be written, therefore, as

$$\bar{\rho}\left(\partial_t < e > + V_j \partial_j < e >\right) = -\partial_j \left(\overline{\rho(e + p'/\rho) v_j'' - \pi_{ij} v_i''}\right) + R_{ij}\,\partial_j V_i +$$

$$+ \overline{p'\partial_j v_j''} - J^T_{(1/\rho)j} \partial_j \bar{p} + \sum_{\alpha} J^T_{\alpha j} F_{\alpha j} - \bar{\rho} < \varepsilon_e >. \tag{9.1.6}$$

Here $< e > = \frac{1}{2} < v_j'' v_j'' >$ is the mean kinetic energy (including the energy of the disperse phase) of turbulent pulsations (turbulent energy); $< \varepsilon_e >$ is the specific turbulent energy dissipation of the gas suspension under influence of viscosity; $F_{\alpha j}$ is the external force acting upon a unit mass of the α-phase (e.g., the electromagnetic *Lorentz force* for a conducting mixture); $R_{ij} \equiv -\bar{\rho} < v_i'' v_j'' >$ is the Reynolds turbulent stress tensor;

$J^T_{\alpha j} \equiv \overline{\rho} < (\rho_\alpha/\rho)'' v''_j >$ is the (α-phase) turbulent diffusion flow ($\sum_\alpha J^T_{\alpha j} = 0$); and $J^T_{(1/\rho)j} \equiv \overline{\rho} < (1/\rho)'' v''_j >$ is the turbulent specific volume flux.

Turbulent stresses in the multiphase continuum under consideration, as well as various turbulent flows, contain the terms caused by involving dust particles in pulsation motions of the carrier. Let us recall that in equation (9.1.6), similar to the previous chapters, two symbols are used to denote the mean values: the over-bar means the probability-theoretical averaging over an ensemble ($\overline{A} = \lim_{N\to\infty} \frac{1}{N} \sum_{p=1}^{N} A^{(p)}$), where summation is performed over the set of possible realizations, and the corresponding mean field is determined as a mathematical expectation of the parameter A for the ensemble of equivalent systems, whereas angle brackets denote Favre averaging ($< A > = \overline{\rho A}/\overline{\rho}$).

Like for a homogeneous fluid, the introduction of weighted averaged values is convenient in the case of modeling a turbulized multi-fluid (heterogeneous) continuum with variable density, ρ, as well. The first term in the right-hand side of equation (9.1.6) defines the transfer of kinetic energy of turbulent pulsations by turbulent diffusion. The quantity $R_{ij}\partial_j V_i$ describes the generation of turbulent energy caused by dissipation of the shear flow kinetic energy of the total averaged heterogeneous continuum. The third term defines the work done by pressure in the pulsation motion, while the fourth and fifth terms describe the turbulent energy production rate in a unit volume due to the buoyancy effects and non-gravitational forces. Finally, the sixth term indicates the turbulent kinetic energy dissipation rate to heat by molecular viscosity.

Let us examine in more detail the terms in equation (9.1.6) with application to the accretion disk. In the general, anisotropic case, the Reynolds stress tensor, R_{ij}, is related to the gradients of the weighted mean flow velocities by the following defining relation (see (3.3.19)):

$$R_{ij} = -\tfrac{2}{3}\overline{\rho} < e > \delta_{ij} + \overline{\rho}\nu^T_{ijsl}(\partial_l V_s + \partial_s V_l - \tfrac{2}{3}\delta_{sl}\partial_k V_k), \tag{9.1.7}$$

which represents the generalization of the classic Boussinesque and Prandtl models. Here ν^T_{ijsl} is the turbulent viscosity tensor (of rank four) for which, by *Monin's hypothesis*, the following formula is valid:

$$\nu^T_{ijsl} = \tfrac{1}{2}(\nu^T_{is}\delta_{jl} + \nu^T_{js}\delta_{il}); \qquad \nu^T_{ij} = \sqrt{< e >} L_{ij};$$

and L_{ij} is the symmetric tensor for different turbulence scales which give an indication for the mean distances the turbulent eddies can overcome along the coordinate axes (*Monin and Yaglom, 1971*). In the isotropic case one can take $L_{ij} = L\delta_{ij}$, and then

$$R_{ij} = -\tfrac{2}{3}\overline{\rho} < e > \delta_{ij} + \overline{\rho} L\sqrt{< e >}(\partial_i V_j + \partial_j V_i - \tfrac{2}{3}\delta_{ij}\partial_k V_k).$$

Assuming further on that the true dust density, ρ_d, is constant, while the true gas density, ρ_g, and the volume dust content (turbidity), s, oscillate as a result of turbulization, we find from expression (9.1.2) that

$$\rho'/\overline{\rho} \cong (1-\overline{s})\rho'_g/\overline{\rho}+\sigma s',$$

where $\sigma \equiv (\rho_d - \overline{\rho}_g)/\overline{\rho}$ is the relative excess of dust particle density over gas density. Then, using the formula $\overline{\rho}_g/\overline{\rho} = 1-\sigma\overline{s}$ we obtain the following expression for the turbulent pulsation of the total density, ρ':

$$\rho'/\overline{\rho} \cong (1-\overline{s})(1-\sigma\overline{s})\rho'_g/\overline{\rho}_g + \sigma s' = <Y_g> \rho'_g/\overline{\rho}_g + \sigma s'. \qquad (9.1.8)$$

Let us notice that the expressions for ρ' and $\overline{\rho}_g$ given above (as well as similar expressions derived below) hold on the assumption that the inequalities $\overline{A'B'}/\overline{A}\,\overline{B} << 1$ are valid for any pulsating thermodynamic parameters A and B (different from the instantaneous hydrodynamic flow velocity, v_j). This assumption is kept throughout this chapter.

Taking relation (9.1.8) into account, we obtain for the turbulent specific volume flux, $J^T_{(1/\rho)j} \equiv \overline{v''_j} = -\overline{\rho' v''_j}/\overline{\rho}$, entering the following expression in equation (9.1.6):

$$J^T_{(1/\rho)j} = -\sigma\overline{s'v''_j} - (1-\overline{s})(1-\sigma\overline{s})\frac{\overline{\rho'_g v''_j}}{\overline{\rho}_g} = -\sigma\overline{s'v''_j} - \left(1-\rho_d\frac{\overline{s}}{\overline{\rho}}\right)\frac{\overline{\rho'_g v''_j}}{\overline{\rho}_g}. \qquad (9.1.9)$$

The defining equation for the turbulent diffusion flux of the α-component of the mixture, $J^T_{\alpha j} \equiv \overline{Y''_\alpha v''_j}$, for the disperse phase of the disk takes the form

$$J^T_{dj} = -\overline{\rho}D^T\partial_j <Y_d> = -\overline{\rho}\,\rho_d D^T\partial_j\left(\frac{\overline{s}}{\overline{\rho}}\right) = -\frac{\rho_d\overline{\rho}_g}{\overline{\rho}}D^T[\partial_j\overline{s} - \overline{s}(1-\overline{s})\partial_j \ln\overline{\rho}_g], \qquad (9.1.10)$$

where D^T is the turbulent dust transfer coefficient. On the other hand, one can write the following expression for the turbulent dust flux:

$$J^T_{dj} \equiv \rho_d\overline{s\,v''_j} = \rho_d\overline{s}\;\overline{v''_j} + \rho_d\overline{s'v''_j} = \rho_d\,\overline{s}\,J^T_{(1/\rho)j} + \rho_d\overline{s'v''_j}.$$

From this relation and from (9.1.9) it follows that the turbulent specific volume flux of the heterogeneous mixture is determined by the formula

$$J^T_{(1/\rho)j} = -\sigma\frac{\overline{\rho}}{\rho_d\overline{\rho}_g}J^T_{dj} - (1-\overline{s})\frac{\overline{\rho'_g v''_j}}{\overline{\rho}_g} = \sigma D^T[\partial_j\overline{s} - \overline{s}(1-\overline{s})\partial_j\ln\overline{\rho}_g] - (1-\overline{s})\frac{\overline{\rho'_g v''_j}}{\overline{\rho}_g}. \qquad (9.1.11)$$

Now we can find the gradient representation for the correlation term in formula (9.1.11). Using the equations of state for pure gases, (9.1.3), and for heterogeneous mixtures, (9.1.4), it is easy to obtain the relationship

$$\rho'_g/\overline{\rho}_g = p'/\overline{p} - \rho_g T''/\overline{\rho}_g < T >,$$

which links turbulent pulsations of the true gas density, temperature, and pressure (here $\overline{p} \cong R_g \overline{\rho}_g < T >$). It is known that relative pressure pulsations in a gaseous flow with a small Mach number, $\boldsymbol{Ma}$, can be neglected in almost all cases as compared to the relative temperature pulsations. This hypothesis (*Morkovin, 1961*) is verified for up to $\boldsymbol{Ma} = 5$, and is probably also valid for turbulent motions in thin accretion disks where the matter moves along the r and z directions with subsonic velocities, while the rotation rate, V_φ, exceeds the speed of sound, c_s. In fact, the thin-disk approximation, $h << r$, coupled with the expression $h \approx c_s / \Omega$ for the disk thickness, suggests that $h/r \approx c_s / V_\varphi << 1$).

With this fact in mind, the correlation term in equation (9.1.11) containing turbulent pulsations of the true gas density can be rewritten in the form

$$-(1-\overline{s})\frac{\overline{\rho'_g v''_j}}{\overline{\rho}_g} \cong (1-\overline{s})\frac{\overline{\rho Y_g T'' v''_j}}{\overline{\rho}_g < T >} \cong (1-\overline{s})\frac{< T'' v''_j >}{< T >}.$$

For the final transformation of this expression, we will use the defining relation for the turbulent heat flux in a multi-liquid continuum (*Kolesnichenko, 1995*),

$$q_j^T \equiv \overline{\rho} < h'' v''_j > \cong \overline{\rho} < C_P >< T'' v''_j > + \sum_\alpha < h_\alpha > J_{\alpha j}^T, \tag{9.1.12}$$

which, in case of the considered heterogeneous medium, takes the form (compare with (3.3.15)):

$$J_{qj}^T \equiv q_j^T - \sum_\alpha < h_\alpha > J_{\alpha j}^T \cong -\lambda^T \left(\partial_j < T > - \frac{\partial_j \overline{p}}{\overline{\rho} < C_P >} \right), \tag{9.1.13}$$

where J_{qj}^T is the so-called reduced heat flux, λ^T is the turbulent heat conductivity coefficient of the gas-dust mixture, and $< C_P > = [\overline{\rho}_g (1-\overline{s}) C_{Pg} + \rho_d \overline{s} C_d] / \overline{\rho}$ is the averaged specific heat capacity (at constant pressure) for the total continuum.

Using formulas (9.1.11)–(9.1.13), we obtain the following final expression for the turbulent specific volume flux for the mixture:

$$J_{(1/\rho)j}^T = -\sigma \frac{\overline{\rho}}{\rho_d \overline{\rho}_g} J_{dj}^T + \frac{(1-\overline{s})}{\overline{\rho} < C_P >< T >} J_{qj}^T =$$

$$= \sigma D^T [\partial_j \bar{s} - \bar{s}(1-\bar{s})\partial_j \ln \bar{\rho}_g] - (1-\bar{s}) \frac{\chi^T}{<T>} \left(\partial_j <T> - \frac{\partial_j \bar{p}}{\bar{\rho} < C_P >} \right), \qquad (9.1.14)$$

where $\chi^T = \lambda^T / \bar{\rho} < C_P>$ is the turbulent thermal diffusivity coefficient.

9.1.3. Turbulent Energy Balance

We will focus here on the thin accretion disk's interior regions in order to deduce the viscosity law. Let us introduce the following assumptions:

- The disk configuration rotating around the fixed z-axis at an angular rate Ω is stationary in the inertial reference frame with the origin at the center of mass (a compact star), and the density of each mass element along its path remains constant ($d\rho / dt = 0$);
- equatorial symmetry plane exists for the baroclinic disk;
- the half-thickness-to-radius ratio of the disk is much smaller than unity, $h / R << 1$;
- the effect of self-gravitation of the disk substance is much smaller compared to the gravitational field of the central star and therefore, can be neglected;
- the chemical composition of the disk's gas phase is homogeneous;
- the radiation pressure in the disk is much smaller than the gas pressure, i.e. $p_R << p_g$;
- the ponderomotive Lorentz forces play an insignificant role;
- the developed turbulence in the disk is associated with the large-scale shear flow, i.e. with the differential rotation of substance around the central star.

Performing a nearly Keplerian rotation around the central star, each gas suspension element in the disk experiences retardation caused by viscous friction between the disk layers. As a result, the main mass of the disk substance slowly spirals inward towards the star along a gently sloping trajectory loosing angular momentum, while, due to the conservation law, the angular momentum of a smaller part of the disk mass is transferred outward, from the inner to the outer disk regions. The turbulent stresses resulting from the relative shift of elements of the gas suspension in the orbital motion, give rise to viscous heat dissipation. Nonetheless, the thinness of the disk implies that its temperature is relatively low and the pressure gradient is much smaller than two basic mechanical forces: gravitational and centrifugal (*Shapiro and Teukolsky, 1985*). Indeed, the temperature can be kept low if the viscous heat dissipated in the turbulized system is effectively radiated outward and does not accumulate in the disk. In a steady state, the major part of this heat would be emitted by the upper and lower surfaces of the disk.

We will use the inertial cylindrical coordinate system (r,φ,z) with the origin coincident with the center of gravity; the $z=0$ plane is taken to be coincident with the equatorial plane of the disk. Next, we assume that the motion of the "cosmic fluid" is realized on average in the azimuthal direction,

$$V_r = 0, \quad V_\varphi = \Omega\, r, \quad V_z = 0, \tag{9.1.15}$$

and the proper velocity of the gas-dust mixture flow pulsates in a random way about this mean value changing extremely irregularly in the meridional and azimuthal directions. It can be shown that, if the substance of the disk is in steady-state rotation in the inertial reference frame, it inevitably possesses axial symmetry:

$$\overline{s} = \overline{s}(r,z), \quad \overline{\rho}_g = \overline{\rho}_g(r,z), \quad \overline{\rho} = \overline{\rho}(r,z), \quad \overline{p} = \overline{p}(r,z),$$

$$< T > = < T(r,z) >, \quad \Omega = \Omega(r,z),$$

and so on (*Tassoul, 1982*). In this case, the coupled two-dimensional hydrodynamic equations for the axisymmetric flow split into separate equations for the averaged radial and vertical motions:

$$\frac{1}{\overline{\rho}}\partial_r \overline{p} = r\Omega^2(r,z) - \frac{GM_\odot r}{(r^2+z^2)^{3/2}} \equiv g_r \approx 0, \tag{9.1.16}$$

$$\frac{1}{\overline{\rho}}\partial_z \overline{p} = -\frac{GM_\odot z}{\left(r^2+z^2\right)^{3/2}} \equiv g_z, \qquad g_z = -\frac{GM_\odot z}{r^3}\left(1+\frac{z^2}{r^2}\right)^{-3/2} \cong -\Omega_K^2(r)\, z\,. \tag{9.1.17}$$

Here $\boldsymbol{g} = \{g_r, 0, g_z\}$ is the effective gravity force (per unit mass) corrected for centrifugal acceleration. It is worth noting that dissipation does not influence the r - and z - components of the equation of motion. Conversely, the φ - component, which can be reduced to the form

$$div(\mu^T r^2 grad\, \Omega) = c^{-2} div(\Omega\, r^2 \boldsymbol{F}_{\mathfrak{R}}) \cong 0, \tag{9.1.18}$$

where μ is the total shear viscosity coefficient and c is the speed of light in vacuum. This relation describes an irreversible change of angular momentum due to viscous friction and in general takes into account the transfer of the angular momentum by the total radiant flow, $\boldsymbol{F}_{\mathfrak{R}}$. Note that the approximate form of (9.1.18) is valid for a thin disk.

Neglecting diffusive transfer of turbulent pulsation energy and other small terms in (9.1.6), we can use formula (9.1.9) to rewrite this equation in cylindrical coordinates:

$$\partial_t b = -<v''_r v''_\varphi> r\partial_r (V_\varphi / r) - <v''_z v''_\varphi> \partial_z V_\varphi + \left[\sigma \overline{s' v''_z} + \left(1 - \rho_d \frac{\overline{s}}{\overline{\rho}}\right) \frac{\overline{\rho'_g v''_z}}{\overline{\rho}_g} \right] \frac{\partial_z \overline{p}}{\overline{\rho}} - \varepsilon \cong 0 \tag{9.1.19}$$

Hereafter, simpler notations are introduced for the mean kinetic energy of turbulent pulsations, $<e>$, and for the turbulent energy dissipation rate, $<\varepsilon_e>$, into heat due to molecular viscosity: $b \equiv <e>$, and $\varepsilon \equiv <\varepsilon_e>$.

Equation (9.1.19), takes into account that the gravity in the radial direction (perpendicular to the rotation axis) is balanced by the centrifugal force (see equation (9.1.16)). This means that the pressure gradient, $\partial_r \overline{p}$, for the suspension is very small and the rotation is virtually Keplerian, though this gradient eventually serves as a driving force for the radial drift of dust particles toward the disk's center. By contrast, the equilibrium in the z - direction (perpendicular to the central plane) is maintained by the pressure gradient, $\partial_z \overline{p} = \overline{\rho} g_z$, which specifies the main direction of the buoyancy force in the gravitational field of the central mass. This force promotes, in particular, additional generation of turbulent energy due to convective instability in the vertical direction.

In the case of isotropic viscosity, the defining relations (9.1.7) for the components of the Reynolds stress tensor take the form

$$R_{r\varphi} \equiv -\overline{\rho} <v''_r v''_\varphi> = \overline{\rho} \nu^T r \partial_r (V_\varphi / r) = \overline{\rho} \nu^T r \partial_r \Omega(r,z), \qquad (^1)$$

$$R_{rz} \equiv -\overline{\rho} <v''_r v''_z> = \overline{\rho} \nu^T \partial_z V_\varphi = \overline{\rho} \nu^T r \partial_z \Omega(r,z), \qquad (^2)$$

(9.1.20)

where ν^T is the kinematic turbulent viscosity coefficient. Let us note that since the rotation of the dusty matter defines the predominant direction in the disk system, its effect on the turbulent motions must be anisotropic, the radial direction being the main one. Thus, in general the turbulent viscosity coefficient, ν^T, is the tensor ν^T_{ik}, which has is diagonal in cylindrical coordinates (*Wasiutynski, 1946*):

$$\nu^T_{ik} = \nu^T \begin{pmatrix} 1 & 0 & 0 \\ 0 & k & 0 \\ 0 & 0 & k \end{pmatrix} = k\nu^T \begin{pmatrix} 1 & 0 & 0 \\ 0 & 1 & 0 \\ 0 & 0 & 1 \end{pmatrix} + (1-k)\nu^T \begin{pmatrix} 1 & 0 & 0 \\ 0 & 0 & 0 \\ 0 & 0 & 0 \end{pmatrix}. \tag{9.1.21}$$

Here $(1-k)$ is a free parameter, which serves as a measure of deviation from isotropy in turbulent motion. The term proportional to the factor $k\nu^T$ corresponds to the isotropic portion of turbulent viscosity, and the second term describes the superimposed radial viscosity that is responsible for the anisotropy of turbulent velocities.

The tangential component, $R_{r\varphi}$, of the Reynolds tensor can be represented as

$$R_{r\varphi} / \overline{\rho} \equiv -<v''_r v''_\varphi> = k \nu^T r \partial_r (<V_\varphi> / r) + (1-k)\nu^T (1/r) \partial_r (<V_\varphi> r) =$$

$$= k\nu^T r\partial_r\Omega + (1-k)\nu^T (1/r)\partial_r(\Omega r^2) = \nu^T (1/r)\partial_r(\Omega r^2) - 2k\nu^T\Omega. \qquad (9.1.22)$$

For isotropic momentum exchange, $k=1$ and formula (9.1.22) reduces to the ordinary hydrodynamic expression (9.1.20$^{(1)}$), while for purely radial exchange formula (9.1.22) takes Karman's form, $R_{r\varphi} \cong \bar{\rho}\nu^T(1/r)\ \partial_r(\Omega\ r^2)$ (*Karman, 1953*). In the case of Keplerian rotation, ($\Omega \sim r^{-3/2}$), these two limiting expressions coincide apart from a constant factor. At the same time, the relations (9.1.22) are not physically substantiated: expression (9.1.21) is actually not a tensor because it is applicable only to the specified coordinate system.

Evidently a certain intermediate relation appears to characterize the turbulent disk motion and hence the right value of $R_{r\varphi}$. Because of uncertainty in selecting the parameter k, we introduced a constant factor ξ^2 in the right-hand side of expression (9.1.20$^{(1)}$), taken as a basis. A similar approach was made to estimate the velocity dispersion in the system of gravitating bodies with Keplerian rotation; as a result, the value $\xi \approx 1/\sqrt{5}$ was found (*Safronov, 1972*).

For the steady-state regime, substituting the formulas (9.1.14), (9.1.17), (9.1.20$^{(1)}$), and (9.1.20$^{(2)}$) into equation (9.1.19) yields:

$$\nu^T\left[\xi^2\left(r\,\partial_r\Omega(r,z)\right)^2 + \left(r\,\partial_z\Omega(r,z)\right)^2\right] - \sigma D^T g_z[\partial_z\bar{s} - \bar{s}(1-\bar{s})\partial_z \ln\bar{\rho}_g] +$$

$$+(1-\bar{s})\chi^T \frac{g_z}{<T>}\left(\partial_z <T> - G_a\right) - \varepsilon = 0, \qquad (9.1.23)$$

where

$$G_a \equiv \frac{g_z}{<C_P>} = -\frac{1}{<C_P>}\frac{GM\ z}{r^3}\left(1+\frac{z^2}{r^2}\right)^{-3/2} \cong -\frac{\bar{\gamma}-1}{\bar{\gamma}}\frac{1}{<R>}\Omega_K^2\ z \qquad (9.1.24)$$

is the adiabatic temperature gradient in the disk, and

$$\bar{\gamma} = <C_P>/(<C_P> - <R>), \qquad <R> = R_g\bar{\rho}_g/\bar{\rho}$$

are the adiabatic exponent and "gas constant" for the averaged two-phase continuum, respectively.

We denote, as earlier, the turbulence scale by L (the mean eddy size that transport the main portion of the turbulent pulsation energy); the Schmidt turbulent number by $\boldsymbol{Sc}^T \equiv \nu^T/D^T$; and the Prandtl turbulent number by $\boldsymbol{Pr}^T \equiv \nu^T/\chi^T$. Then, assuming according to Kolmogorov's hypothesis that the kinematic turbulent viscosity coefficient, ν^T, and the turbulent energy to heat dissipation rate, ε, only depend on the turbulence energy, b, and on the turbulence scale, L, we obtain

$$\nu^T = L\sqrt{b}, \qquad D^T = \frac{1}{\boldsymbol{Sc}^T}L\sqrt{b}, \qquad \chi^T = \frac{1}{\boldsymbol{Pr}^T}L\sqrt{b}, \qquad \varepsilon = \beta^2\frac{b^{3/2}}{L}, \qquad (9.1.25)$$

where, in view of the uncertainty in L, the constant in the expression for ν^T can be set equal to unity, while the numerical factor β^2 is assumed, to first approximation, to be constant. Substituting these expressions into (9.1.23) and neglecting the second order terms in the volume content of the condensed phase, we obtain

$$L\sqrt{b}\left[\xi^2(r\,\partial_r\Omega)^2+(r\,\partial_z\Omega)^2\right]-\sigma\frac{1}{\boldsymbol{Sc}^T}L\sqrt{b}\,g_z[\partial_z\bar{s}-\bar{s}\partial_z\ln\bar{\rho}_g]+$$

$$+\frac{1}{\boldsymbol{Pr}^T}L\sqrt{b}\frac{g_z}{<T>}\left(\partial_z<T>-G_a\right)-\beta^2\frac{b^{3/2}}{L}=0,$$

or

$$\sqrt{b}\left\{L^2\left[\xi^2(r\partial_r\Omega)^2+(r\partial_z\Omega)^2\right](1-\boldsymbol{Rf}-\boldsymbol{Kf})-\beta^2 b\right\}=0. \qquad (9.1.26)$$

Here two dimensionless parameters are introduced: the Richardson dynamical_number (compare to formula (4.2.32)),

$$\boldsymbol{Rf}=\frac{\boldsymbol{Ri}}{\boldsymbol{Pr}^T}=\frac{1}{\boldsymbol{Pr}^T}\frac{\Omega_K^2\,z}{r^2}\frac{1}{<T>}\frac{(\partial_z<T>-G_a)}{(\xi^2\partial_r^2\Omega+\partial_z^2\Omega)}, \qquad (9.1.27)$$

($\boldsymbol{Ri}$ is the Richardson gradient number); and the Kolmogorov dynamical number (compare to the formula (4.2.34)),

$$\boldsymbol{Kf}\equiv\frac{\boldsymbol{Ko}}{\boldsymbol{Sc}^T}=-\frac{1}{\boldsymbol{Sc}^T}\sigma\frac{\Omega_K^2 z}{r^2}\frac{[\partial_z\bar{s}-\bar{s}\partial_z\bar{\rho}_g]}{(\xi^2\partial_r^2\Omega+\partial_z^2\Omega)}, \qquad (9.1.28)$$

($\boldsymbol{Ko}$ is the Kolmogorov gradient number). The former takes into account the effect of thermal stratification of the gas suspension on the evolution of the turbulent flux in the disk, while the latter accounts for the effect of the inhomogeneous distribution of the mixture's gas constituent on the disk thickness and the influence of turbidity stratification on turbulence development. For $\rho_g=const$, $\boldsymbol{Ko}$ determines the relative expenditure of the turbulent energy on the suspension of particles by the carrier flow.

The parameters $\boldsymbol{Kf}$ and $\boldsymbol{Rf}$ serve as criteria of dynamical activity in the baroclinic disk substance, in particular, the carrier gas compressibility, and the disk's thermal stratification, their impact on turbulence dynamics being defined by the inverse influence of stratification of suspended particles.

In what follows, we will assume that $\boldsymbol{Sc}^T=\boldsymbol{Pr}^T$ because the equality $\chi^T=D^T$ usually holds for turbulized mixtures. This is equivalent to assuming that the mixing lengths for both substance and heat within the disk are the same. Furthermore, in the full *dissipative function*, $\Phi_v\equiv R_{ij}\partial_j V_i$, we will neglect the vertical gradient of the azimuthal velocity, $\partial_z V_\varphi\equiv r\partial_z\Omega(r,z)$, which is small compared to the radial gradient, $r|\partial_r(V_\varphi/r)|\approx r\,|\partial_r\Omega_K(r)|=\frac{3}{2}\Omega_K$. Indeed, one may infer from equation (9.1.16) that

$$\frac{V_\varphi^2}{r} = \Omega_k^2 r(1-\frac{3}{2}\frac{z^2}{r^2}) + \frac{1}{\rho}\partial_r p ,$$

or

$$V_\varphi = \Omega_k r\left(1-\frac{3}{2}\frac{z^2}{r^2}+\frac{\partial_r p}{\rho\Omega_k^2 r}\right)^{1/2} \approx \Omega_k r\left(1-\frac{3}{4}\frac{z^2}{r^2}+\frac{\partial_r p}{2\rho\Omega_k^2 r}\right),$$

whence it follows that

$$\partial_z V_\varphi = -\frac{1}{2}\Omega_k^2\left(\frac{z\partial_r\rho}{\rho}+\frac{\partial_z\rho\partial_r p}{\rho^2\Omega_k^2}\right).$$

Obviously, plausible spatial distributions of density, pressure, temperature, opacity, and other disk parameters critically depend on the disk's nature, as well as on the distances from the central star and/or from the equatorial plane of the accretion disk.

Calculating the disk's velocity field (*Urpin, 1984*) showed that in the cold outer regions (at large radii, r, where $p_R \ll p_g$ and $\overline{\kappa}_{scatt} \ll \overline{\kappa}_{abs}$) the gradient of the azimuthal velocity, $\partial_z V_\varphi$, has the following asymptotic form:

$$\partial_z V_\varphi = 0.375\Omega_K\left(\frac{z}{r}\right)(-1-1.5\zeta+0.875\zeta^2+0.375\zeta^3-0.75\zeta^4+...)$$

This expression is valid for small values of the dimensionless parameter $\zeta \equiv z/2h$, where

$$2h = \sqrt{2}\left(\frac{\overline{p}}{\overline{\rho}}\right)_{z=0}^{1/2}\frac{1}{\Omega_K} = \sqrt{2}\left(R<T>|_{z=0}\right)^{1/2}\frac{1}{\Omega_K}$$

is the disk thickness, the latter being different in various regions, which can be found by integrating (9.1.17) over z from $z=0$ to $z=h$. The inequality $\overline{\kappa}_{scatt} \ll \overline{\kappa}_{abs}$ means that in the outer disk regions *Thomson scattering* of photons is much weaker than the opacity due to *free-free photon absorption* related to non-relativistic thermal bremsstrahlung.

One can infer from the expression given above that, for a thin gas-phase accretion disk, we have that

$$|\partial_z V_\varphi| \ll |r\partial_r(V_\varphi/r)| \cong |r\partial_r\Omega_K(r)| = \tfrac{3}{2}\Omega_K .$$

Similar asymptotic representations, although having different expansion coefficients defined by the selected disk model, can be derived as well for various accretion disk regions. Let us note that for protoplanetary disks the inequalities $p_R << p_g$ and $\overline{\kappa}_{scatt} << \overline{\kappa}_{abs}$ are also valid, while the opacity is caused by extinction by dust particles.

9.1.4. Turbulent Transfer Coefficients in the Gas-Dust Disk

Equation (9.1.26) decomposes into two equations: $b = 0$ corresponding to the laminar flow, and

$$L^2 \left[\xi^2 (r\partial_r \Omega)^2 + (r\partial_z \Omega)^2\right](1 - \boldsymbol{Rf} - \boldsymbol{Kf}) = \beta^2 b, \tag{9.1.29}$$

describing the steady-state disk turbulence. This equation has real solutions solely for $(1 - \boldsymbol{Rf} - \boldsymbol{Kf}) \geq 0$, whence it follows for the turbulent regime that $\boldsymbol{Ri} + \boldsymbol{Ko} \leq (\boldsymbol{Ri} + \boldsymbol{Ko})_{cr} \approx \boldsymbol{Pr}^T$. Note that for $\boldsymbol{Ri} + \boldsymbol{Ko} \geq \boldsymbol{Pr}^T$ the equation has a single real solution $b = 0$ pertaining to the laminar flow regime.

Let $\boldsymbol{Ri} + \boldsymbol{Ko} \leq \boldsymbol{Pr}^T$. Then

$$\sqrt{b} = \frac{1}{\beta} L\, r \sqrt{\xi^2 (\partial_r \Omega)^2 + (\partial_z \Omega)^2}\, \sqrt{1 - \frac{\boldsymbol{Ri} + \boldsymbol{Ko}}{(\boldsymbol{Ri} + \boldsymbol{Ko})_{cr}}} \cong \frac{\xi}{\beta} L\, r \left|\partial_r \Omega(r,z)\right| \sqrt{1 - \frac{\boldsymbol{Ri} + \boldsymbol{Ko}}{\boldsymbol{Pr}^T}}\,, \tag{9.1.30}$$

its approximate form being valid for nearly the complete disk except the regions near the star. From (9.1.30) we obtain the following expression for the turbulent viscosity coefficient:

$$\nu^T = \frac{\xi}{\beta} L^2\, r \left|\partial_r \Omega(r,z)\right| \sqrt{1 - \frac{\boldsymbol{Ri} + \boldsymbol{Ko}}{\boldsymbol{Pr}^T}} = \frac{\xi}{\beta} L^2 r \left|\partial_r \Omega(r,z)\right| \varphi(\boldsymbol{Ri}, \boldsymbol{Ko}), \tag{9.1.31}$$

Here

$$\varphi(\boldsymbol{Ri}, \boldsymbol{Ko}) \equiv \sqrt{1 - \frac{\boldsymbol{Ri} + \boldsymbol{Ko}}{\boldsymbol{Pr}^T}}$$

is the correction factor in the equation for turbulent momentum transport, which reflects the inverse effect of dust (heat) transfer on the efficiency of turbulent mixing.

Expression (9.1.30) shows that, in case of temperature-unstable stratification ($\boldsymbol{Ri} < 0$) of the solar type disk in the direction perpendicular to the equatorial plane, with a superadiabatic temperature gradient ($f_{r=10\,a.u.} \approx 0.2$ in $\partial_z \langle T \rangle - G_a = f\, \partial_z \langle T \rangle$), the turbulence energy promoted by convection increases. Conversely, in the adiabatic case ($\partial_z \langle T \rangle = G_a$; $\boldsymbol{Ri} = 0$), the temperature gradient in the disk does not influence the turbulent transport coefficients. Let us recall, however, that in the presence of suspended particles in the flow the turbulent energy decreases because $\boldsymbol{Ko} > 0$.

The inverse Prandtl turbulent number is accepted to be $1/\boldsymbol{Pr}^T = 1$ for shear stress turbulence caused by the disk's differential rotation, while it increases by a factor of

2–3 when the turbulence is due to thermal convection in the vertical direction (*Shakura and Syunyaev, 1973*; *Shakura et al., 1978*). It is easy to verify that expression (9.1.31) coincides with (9.1.1) for the gas-phase accretion disk (i.e., for $\boldsymbol{Ko}=0$ and $\boldsymbol{Ri}=0$), if one sets $\xi/\beta=4/9\,\alpha$ in (9.1.31). Indeed, substituting the angular velocity in the Keplerian rotation, $\Omega_K(r)=(GM_\odot)^{1/2}r^{-3/2}$, into (9.1.31) (in this case $r|\partial_r\Omega_K(r)|=3/2\,\Omega_K$), and taking the quantity

$$L=h=(\overline{p}/\overline{\rho})^{1/2}\big|_{z=0}(r^3/GM_\odot)^{1/2}=(\overline{p}/\overline{\rho})^{1/2}\big|_{z=0}\Omega_K^{-1},$$

as the integral turbulence scale (restricted by the disk thickness, $2h$), we obtain

$$\nu^T=3/2\,\beta^{-1}(\overline{p}/\overline{\rho})\big|_{z=0}\Omega_K$$

and, therefore

$$R_{r\varphi}=\overline{\rho}\nu^T r\partial_r\Omega_K=-9/4\,\xi\beta^{-1}\overline{p}\big|_{z=0}.$$

Thus

$$R_{r\varphi}=\overline{\rho}\,\nu^T r\,\partial_r\Omega(r,z),\qquad \nu^T=\alpha L^2\,r|\partial_r\Omega(r,z)|\sqrt{1-(\boldsymbol{Ri}+\boldsymbol{Ko})/\boldsymbol{Pr}^T}\,,\tag{9.1.32}$$

where

$$\boldsymbol{Ko}=\frac{\sigma}{\xi^2}\frac{GM_\odot z}{r^5}\frac{[\partial_z\overline{s}-\overline{s}\partial_z\overline{\rho}_g]}{\partial_r^2\Omega};\qquad \boldsymbol{Ri}=\frac{1}{\xi^2}\frac{GM_\odot z}{r^5}\frac{1}{<T>}\frac{(\partial_z<T>-G_a)}{\partial_r^2\Omega}.$$

Obviously, formula (9.1.32) generalizes the expression for tangential Reynolds stresses obtained for the gas-phase disk to the gas-dust accretion disk. Note that since the dimensionless viscosity parameter, α, cannot be estimated with sufficient accuracy, the factor $4/9$ in (9.1.32) has no fundamental meaning and can be omitted.

Now let us write down, using formulas (9.1.10), (9.1.13), (9.1.17), and (9.1.24), the rheological relations for diffusion and heat flows allowing to describe the transfer of dust and thermal energy in the direction perpendicular to the central plane of the disk in a turbulent flow of the gas-dust mixture:

$$J_{dz}^T=-\overline{\rho}\rho_d D^T\partial_z\left(\frac{\overline{s}}{\overline{\rho}}\right)=-\alpha\overline{\rho}\,\rho_d\frac{\varphi}{\boldsymbol{Sc}^T}L^2\,r|\partial_r\Omega(r,z)|\partial_z\left(\frac{\overline{s}}{\overline{\rho}}\right)=$$

$$=-\alpha\frac{\rho_d\overline{\rho}_g}{\overline{\rho}}\frac{\varphi}{\boldsymbol{Sc}^T}L^2r|\partial_r\Omega(r,z)|\left[\partial_z\overline{s}+\overline{s}\left(-\frac{1}{<H_g>}+\frac{\partial_z<T>}{<T>}\right)\right],\tag{9.1.33}$$

$$q_z^T \cong -\lambda^T\left(\partial_z < T > - \frac{\partial_z \overline{p}}{\overline{\rho} < C_P >}\right) + \sum_{\beta} < h_\beta > J_{\beta z}^T =$$

$$= -\alpha \frac{\varphi}{\boldsymbol{Pr}^T} \overline{\rho} < C_P > L^2 r |\partial_r \Omega(r,z)| (\partial_z < T > - G_a) + (< h_d > - < h_g >) J_{dz}^T =$$

$$= -\alpha \frac{\varphi}{\boldsymbol{Pr}^T} \overline{\rho} L^2 r |\partial_r \Omega(r,z)| (\partial_z < h > - g_z), \qquad (\boldsymbol{Pr}^T = \boldsymbol{Sc}^T), \tag{9.1.34}$$

where $< H_g > = \overline{p} / \overline{\rho}_g g_z$ is the averaged local scale height for a homogeneous medium. Because the formulas (9.1.32)–(9.1.34), in contrast to (9.1.1), bear a local character they can be used with higher efficiency for modeling the evolution of the protoplanetary gas-dust nebula.

For the numerical calculations of hydrodynamic motions in the interior regions of the two-phase disperse accretion disk, one should set the turbulence scale, L. As is known (see subsection 4.3.7), the scale L in the shear layers can be defined in two different ways: using either simple relationships or a modified Karman formula. In the first case, L is taken proportional to the thickness of a thin layer, $L = \kappa z$ or $L = \kappa z \Phi(\boldsymbol{Re}, \boldsymbol{Ri}, \boldsymbol{Ko})$, where Φ is a dimensionless function. In the second case, L is expressed as a partial derivative of the mean flow velocity, $L = \kappa |\partial_z V_\varphi / \partial_z^2 V_\varphi|$ for example, in the form of the following differential equation (*Laikhtman, 1970*):

$$L = -\kappa c^{1/4} \Psi / \partial_z \Psi, \qquad (\Psi = \sqrt{b} / L), \tag{9.1.35}$$

which enables one to calculate L in a layer with finite thickness in terms of local flow characteristics, taking into account the buoyancy forces (see formula (9.1.30)). Here and above, κ=0.4 is the Karman constant and c is an empirical constant.

9.1.5. The Rotating Gas-Dust Disk: A Model Example

As an example illustrating the capability of the developed approach, we will qualitatively consider a model for the height distribution of small, suspended dust particles in the thin disc layer located outside its equatorial plane. Such a subdisk is assumed to embody a steady-state flow and to be stratified neutrally in temperature ($\boldsymbol{Ri} = 0$). We will specially focus on the regime of limiting saturation by small dust particles.

Let us suppose that the disk's dust concentration is rather high, especially near the equatorial plane, $z = 0$, where particles form a flattened layer with increased density, though lower than the critical value to originate gravitational instability. Thus, the inverse effect of dust on the dynamics of the turbulent flow should be taken into account when describing the steady-state motion in such a medium. In view of the assumptions made above, the instantaneous gas and dust velocities are linked by $v_{dj} = v_{gj} - W_d \delta_{j3}$ (the index 3 in the Kronecker symbol refers to the vertical direction, i.e. the g_z component of the effective gravitational force, $\boldsymbol{g}$, corrected for centrifugal acceleration (see

(9.1.17)). Respectively, W_d is the gravitational sedimentation rate of a single dust particle.

For the assumed low turbidity ($s << 1$), the sedimentation rate, W_d, is thought to be independent of s. Then, the instantaneous (not averaged) velocity of the heterogeneous continuum can be written as

$$v_j = v_{gj} - \frac{\rho_d s}{\rho} W_d \, \delta_{j3}$$

and we obtain the following expression for the diffusive dust flow:

$$J_{dj} \equiv \rho_d s_d (v_{dj} - v_j) = -\rho_g \rho_d \frac{s(1-s)}{\rho} W_d \delta_{j3}, \qquad (J_{gj} + J_{dj} = 0) . \tag{9.1.36}$$

The averaged steady-state diffusion equation for dust particles in the z- direction, with allowance for expression (9.1.15) for the weighted mean velocity, $< v_j >$, takes the form

$$\partial_z (\bar{J}_{dz} + J^T_{dz}) = 0 ,$$

where $J^T_{dz} = -\frac{\rho_d \bar{\rho}_g}{\bar{\rho}} D^T \partial_z \bar{s}$ is the turbulent dust flux for $\rho_g = const$ (see (9.1.10)), and

$$\bar{J}_{dz} = -\rho_d \rho_g \overline{\frac{s(1-s)}{\rho}} W_d \cong -\frac{\rho_d \rho_g}{\bar{\rho}} W_d \bar{s}$$

is the averaged dust sedimentation flux. In the steady-state regime of the rotating disk the regular sedimentation of particles must be balanced by their upward turbulent transfer and hence, the total mass transfer in z-direction is, on average, absent. This yields

$$\bar{J}_{dz} + J^T_{dz} = const = 0 .$$

Assuming further that the equatorial plane ("a near-wall region" analog) only reflects and not absorbs the diffusing particles, we can write

$$D^T \partial_z \bar{s} = - W_d \bar{s} . \tag{9.1.37}$$

Evidently, up to the terms of order $|V|/c$, the φ-component of the averaged equation (9.1.18) for gas suspension motion reduces to the following equation:

$$\frac{1}{r} \partial_r (r \tau^{\Sigma}_{\varphi r}) + \partial_z \tau^{\Sigma}_{\varphi z} + \frac{\tau^{\Sigma}_{\varphi r}}{r} = 0 .$$

Here $\tau_{ij}^{\Sigma} \equiv R_{ij} + \overline{\tau}_{ij} \cong R_{ij}$ and $\overline{\tau}_{ij}$ are the averaged viscous stress tensor, which is important only near the flattened disk plane, $z = 0$. Since the dominant direction of motion is thought to be radial (*Safronov and Guseinov, 1990*), we can set for the "near-wall" equatorial region $\partial_z(\overline{\tau}_{\varphi z} + R_{\varphi z}) \cong 0$. Then

$$\nu^T r \partial_z \Omega(r,z) = V_*^2(r), \tag{9.1.38}$$

where

$$V_*(r) = \sqrt{\tau_0 / \rho} \quad \text{and} \quad \tau_0 \equiv \overline{\tau}_{\varphi z}\big|_{z=0} = \overline{\rho}\nu \partial_z V_\varphi\big|_{z=0}$$

are the dynamic velocity (friction rate) and the tangential friction stress on the disk's central plane at a distance r from the star, respectively.

Now, eliminating the quantity ν^T from (9.1.37) and (9.1.38), we obtain

$$\partial_z \ln \overline{s} = -\frac{\omega\kappa}{V_*} \partial_z \big(r\Omega(r,z)\big),$$

where $\omega(r) \equiv \dfrac{Sc^T W_d}{\kappa V_*(r)}$ is a dimensionless parameter. Integrating the last equation yields the dust's height distribution at the given distance r:

$$\overline{s}(r,z) = \overline{s}(r)\big|_{z=0} \exp\left\{-\frac{\omega\kappa}{V_*} r\big[\Omega(r,z) - \Omega(r,0)\big]\right\}. \tag{9.1.39}$$

Let us also determine the $z-$ distribution of the angular velocity of the turbulized flow that carries suspending particles for the considered steady-state regime of the flow. This velocity can be found using the equations (9.1.30), (9.1.37), and (9.1.38) and is written in the following form:

$$L\sqrt{b}\, \partial_z V_\varphi(r,z) = V_*^2, \qquad \frac{1}{\boldsymbol{Sc}^T} L\sqrt{b} \partial_z \overline{s} + W_d \overline{s} = 0,$$

$$b = \alpha V_*^2 \sqrt{1 - \boldsymbol{Kf}}, \quad \boldsymbol{Kf} = \frac{\sigma}{\xi^2} \frac{GM_\odot z}{r^3} \frac{W_d \overline{s}}{V_*^2 r \partial_r (V_\varphi / r)}, \quad L = \frac{\kappa}{\alpha^{1/2}} z \Phi(\boldsymbol{Kf}). \tag{9.1.40}$$

Here the turbulence scale L is expressed in terms of a certain universal function $\Phi(\boldsymbol{Kf})$ of the Kolmogorov number, $\boldsymbol{Kf}$, with the evident equality $\Phi(0) = 1$. Since the turbulence scale decreases under influence of dust, the function $\Phi(\boldsymbol{Kf})$ also declines as its argument increases.

The system (9.1.40) exhibits some specific properties distinctive for this type of gas suspension stream regimes. Note that peculiarities in the turbulent flow containing a heavy admixture was first studied for sediment motion (*Kolmogorov, 1954*) and atmospheric problems (*Barenblatt and Golitsyn, 1973*). First of all, the system only contains a velocity gradient, but not the velocity itself. This means that the situation known as *the limiting saturation regime* can develop, with an unlimited number of particles in the disk's vicinity and because of the inverse effect of particles on the flow dynamics. In other words, the turbulent gas suspension flow takes up the greatest possible amount of dust at the given dynamical velocity and other parameters. This regime (analogous to the onset of dust storms on Mars) must be described by the peculiar solution of the system (9.1.40), which is determined by the parameters entering only the differential equations themselves. This is why the limiting saturation regime does not require a specification of the boundary condition, $\bar{s}|_{z=0}$, for the dust concentration (see expression (9.1.39)).

A group analysis of the system (9.1.40) shows that it has the solution $\partial_z V_\varphi = C_1 / z$ and $\bar{s} = C_2 / z$, where C_1 and C_2 are constants (at a certain distance from the star) to be determined. Substituting these expressions into (9.1.40) and taking into account the relationship

$$\boldsymbol{Kf} \cong \frac{\sigma}{\xi^2} \frac{\Omega_K W_d C_2}{V_*^2} \equiv const\,,$$

we obtain

$$C_1 = \frac{V_*}{\Phi(\boldsymbol{Kf})(1-\boldsymbol{Kf})^{1/4}} = \frac{V_*}{\kappa\omega}, \qquad \omega \equiv \frac{\boldsymbol{Sc}^T W_d}{\kappa V_*}\,. \tag{9.1.41}$$

From this formula the functional equation $\Phi(\boldsymbol{Kf})(1-\boldsymbol{Kf})^{1/4} = \omega$ follows, which serves to determine the calculated Kolmogorov number, $\boldsymbol{Kf}$, specific for the motion regime under consideration at given r.

Since Φ is a non-increasing function, $\Phi(0) = 1$, and $\boldsymbol{Kf}$ is confined between zero and unity by its physical meaning, the functional equation has no roots for $\omega > 1$ which corresponds to a low flow velocity or to large particles. However, for $\omega < 1$, one, and only one, root exists (*Barenblatt and Golitsyn, 1973*). Taking (9.1.41) into account, we obtain from $\partial_z V_\varphi = C_1 / z$ ($\alpha\tau$ $\omega < 1$) that

$$V_\varphi(r,z) = r\Omega(r,z) = r\Omega(r,0) + \frac{V_*(r)}{\kappa\omega} \ln z\,. \tag{9.1.42}$$

In turn, from this relationship and formula (9.1.39) it is easy to obtain the following limiting steady-state height distribution for small dust particles in a thin near-equatorial disk layer:

$$\bar{s} = C_2 / z = \frac{\xi^2 V_*^2 \boldsymbol{Kf}}{\sigma \Omega_K W_d} \frac{1}{z}\,. \tag{9.1.43}$$

The flow tends to this limit in case of an unlimited reserve of dust on the "subjacent surface".

Formula (9.1.42) shows that the velocity distribution with height, z, in the ultimately loaded turbulent flow is logarithmic (as it would be in a turbulized fluid), but the presence of dust leads to the apparent decrease of the Karman constant, κ. This may be interpreted as if the gas flow is accelerated by dust particles as compared to a "pure" gas flow under the same external conditions, i.e. the same dynamical velocity, V_*. In other words, the velocity gradients near the equatorial plane of the disk increase, accounting for the *saltation* effect - the detachment and elevation of relatively small dust particles in the environment. The presence of such a dusty medium with increased concentration of suspended particles promote, in turn, more intensive *Brownian and turbulent coagulation*, resulting in increasing dust particle sizes and the their sinking velocity to the central disk region. Thus, one may assume that the limiting saturation regime of the rotating disk by small dust particles promotes subdisk formation from relatively larger and heavier particles, because they are not affected by turbulent friction stresses and hence, the turbulized flow fails to lift them.

9.2. COAGULATION AND MASS TRANSFER IN THE TURBULENT PROTOPLANETARY NEBULA

The phenomenological approach to modeling the turbulent transport coefficients developed in section 9.1 is based on the assumption that dust particles have the same size in the limiting regime in which they saturate the subdisk. However, the formation of a dust subdisk involves the concurrent processes of turbulent mass transfer and coagulation, with the account for dynamical distinctions in the behavior of solid particles with different sizes.

Indeed, following a routine scenario (see, e.g., *Safronov, 1987; Vityazev et al., 1990*), because formation of the subdisk is accompanied by the growth of sinking particles, its opacity decreases, convection in the disk ceases, and turbulence is damped out. It is further assumed that, while small particles remain still at higher *z*, the larger particles settle down and grow more rapidly increasing the dust density at the mid-plane up to a certain threshold value for which the central layer becomes gravitationally unstable and decomposes into a large number of dust clumps. In addition, the velocity gradients near the disk's central plane facilitate the saltation effect responsible for the separation and ascent of an appreciable number of small dust particles to the disk's periphery. This intensifies, in turn, various (turbulent, electric, gravitational, gravitational-turbulent, etc.) coagulation processes in the subdisk, which eventually lead to its saturation by particles with larger size composing a significant portion of the total mass. Thus, the particle size distribution becomes an important parameter of such a turbulized dispersed medium, and the system of kinetic equations describing the growth of polydisperse particles should be invoked.

Unfortunately, the relevant system of Smoluchowski type kinetic equations for coagulation (especially with the account for inertial distinctions between particles of different sizes) is complex and its solution encounters a considerable mathematical difficulties. The various attempts to attack the problem of mutual influence of the coagulation kinetics and turbulent mass-transfer processes which have been undertaken (*Mizuno et al., 1988; Schmitt et al., 1997; Dominik and Tielens, 1997*) were only partially successful. This particular problem will be addressed in this section. We will focus on the mathematical aspects of the theory for the combined mass transfer and coagulation processes in a two-phase medium in the presence of shear turbulence. The goal is to address in more detail the possible coagulation mechanisms in the accretion disk, the problem of mutual influence of turbulent mixing and coagulation kinetics, and the turbulent diffusion flow patterns taking into account the polydisperse nature of dust particles.

9.2.1. The Diffusion Equation for Dust Particle Coagulation

Let us analyze the initial phase of disk evolution. We confine ourselves by the coagulation and mass transfer processes in the gas-dust shear flow containing a suspension of small, heavy particles of different discrete sizes k ($k = 1,2...$) which is smaller than the internal turbulence scale, l_K. In this case the dust particles may be considered as those forming a thermodynamically active, multi-velocity continuum with its own thermodynamic and hydrodynamic parameters: velocity, density, temperature, viscosity, etc. Obviously, this condition is satisfied by *Stokes particles* with any dispersity, from submicron to centimeter sizes.

We will suppose hereafter that the disperse phase consists of solid particles with mass density, ρ_d, and we will designate the number density of particles with size k by n_k. In addition, we assume that particles of sort 1 constitute a group of particles with minimum size (primary particles); particles of sort 2, a group of particles with double size, etc., up to the maximum size. To simplify the analysis of the coagulation process occurring in the turbulized two-phase, polydisperse flow, we also assume that all solid particles are spherical or near-spherical in shape with *Feret* diameter d_k.

Since sizes of particles of the same chemical composition increases after adhesion as the cube-root of the number of primary particles, $d_k = d\sqrt[3]{k}$, the volume content of particles of sort k is determined by the relation $s_k = (\pi/6)d_k^3 n_k = U_1 k n_k$, where $U_1 = (\pi/6)d^3$ and $d \equiv d_1$ are the volume and diameter of dust grains with minimum size, respectively. Then, the volume concentration of dust (turbidity), s, the mass density, ρ, and the average-mass hydrodynamic velocity, $\boldsymbol{v}$, of the total multi-velocity continuum are determined by the relations

$$s \equiv \sum_{k=1} s_k = U_1 \sum_{k=1} k n_k \,, \tag{9.2.1}$$

$$\rho = \rho_g(1-s) + \rho_d s = \rho_g + U_1(\rho_d - \rho_g)\sum_{k=1} k n_k \,, \tag{9.2.2}$$

$$\boldsymbol{v} = \frac{\rho_g(1-s)}{\rho}\boldsymbol{v}_g + \sum_{k=1} \frac{\rho_d s_k}{\rho}\boldsymbol{v}_{dk} \,, \tag{9.2.3}$$

where ρ_g and $\boldsymbol{v}_g$ are the proper mass density and hydrodynamic velocity of the gas phase particles, and $\boldsymbol{v}_{dk}$ is the hydrodynamic velocity of dust particles of sort k, respectively.

Fragmentation (breakup) and coagulation (adhesion) are referred to as the main mechanisms governing the size distribution of relatively large particles in the gas-dust disk in the accumulation process. Because fragmentation of colliding solid bodies is quite well understood, it can be taken into account, if necessary. Consequently, in order to simplify the model, the particle fragmentation mechanism will not be specially addressed; its quite detailed analysis can be found elsewhere (see, e.g., *Vityazev et al., 1990*). We therefore assume that the change in number density, n_k, of particles of the group k is determined solely either by its decrease, due to unification with particles of other sizes, or by its increase, due to adhesion of smaller size particles.

The system of kinetic coagulation equations can then be written as follows (*Smoluchowski, 1936*):

$$\dot{n}_k = \frac{1}{2}\sum_{j=1}^{k-1} K_{j(k-j)} n_j n_{(k-j)} - \sum_{j=1}^{\infty} K_{kj} n_k n_j, \qquad (k = 1,2,.....), \tag{9.2.4}$$

Here $\dot{n}_k$ is the overall rate of change of the number of particles of sort k due to the coagulation process and K_{kj} is the coagulation coefficient of k- and j-particles, which is determined experimentally or by modeling the interaction of two equal-sized particles in the flow.

Using expression (9.2.4), the Reynolds-averaged diffusion equation that describes the transfer of dust particles of sort k in the turbulent flux takes the following form:

$$\overline{\rho}\frac{d}{dt}\left(\frac{\overline{n}_k}{\overline{\rho}}\right) + div(\boldsymbol{J}_k^{\Sigma} / m_k) = \dot{n}_k = \frac{1}{2}\sum_{j=1}^{k-1} K_{j(k-j)} \overline{n}_j \overline{n}_{(k-j)} - \sum_{j=1}^{\infty} K_{kj} \overline{n}_k \overline{n}_j . \tag{9.2.5}$$

Here $d(..)/dt = \partial(..)/\partial t + \boldsymbol{V}\cdot grad\,(..)$ is the substantive derivative of the mean-motion scale; $\boldsymbol{V} \equiv \langle \boldsymbol{v} \rangle$ is the weighted mean hydrodynamic velocity of the gas suspension; $\boldsymbol{J}_k^{\Sigma} = \overline{\boldsymbol{J}}_k + \boldsymbol{J}_k^{T}$ is the total diffusion flux of dust particles of sort k in the turbulized medium; $\overline{\boldsymbol{J}}_k$ $(\equiv \rho_d \overline{s_k(\boldsymbol{v}_k - \boldsymbol{v})})$ and $\boldsymbol{J}_k^{T}$ $(\equiv \overline{\rho} \langle Y_k'' \boldsymbol{v}'' \rangle)$ are, respectively, the averaged "true" diffusion flux and the turbulent diffusion flux for dust particles with size k; and $Y_k \equiv \rho_d s_k / \rho$ is the mass concentration of dust particles of sort k.

Assuming that only the volume concentration, s_k, of dust particles pulsate in a turbulized flow, while the true gas (ρ_g) and dust (ρ_d) densities remain constant, we obtain the following rheological relation for the turbulent diffusion flux,

$$\boldsymbol{J}_k^{T} \quad (= \rho_d \overline{s_k \boldsymbol{v}''} = k m_1 \overline{n_k \boldsymbol{v}''})$$

of particles with size k:

$$\boldsymbol{J}_k^T = -\bar{\rho} D_{dk}^T grad < Y_k > = -\bar{\rho}\rho_d D_{dk}^T grad\left(\frac{\bar{s}_k}{\bar{\rho}}\right) = -m_k \bar{\rho} D_{dk}^T grad\left(\frac{\bar{n}_k}{\bar{\rho}}\right) =$$

$$= -\rho_d D_{dk}^T \left[grad\,\bar{s}_k - \sigma\bar{s}_k\,grad\,\bar{s}\right] = -m_k D_{dk}^T \left[grad\,\bar{n}_k - \sigma\bar{n}_k\,grad\,\bar{s}\right]. \quad (9.2.6)$$

Here D_{dk}^T is the eddy diffusion coefficient for dust particles with size k; $\sigma \equiv (\rho_d - \bar{\rho}_g)/\bar{\rho}$; $m_k = km_1$; and $m_1 = \rho_d(\pi/6)d^3$ is the mass of the smallest dust particles. Recall that the more general case, in which both the true gas density, ρ_g, and the volume dust content pulsate as a result of the flow turbulization, was analyzed in section 9.1.

Taking formula (9.2.6) into account, we can write, in the case of developed turbulence, the equations (9.2.5) as

$$\frac{\partial \bar{n}_k}{\partial t} + div\left[\bar{n}_k \boldsymbol{V} - D_{dk}^T(grad\,\bar{n}_k - \sigma\bar{n}_k\,grad\,\bar{s})\right] =$$

$$= \dot{n}_k = \frac{1}{2}\sum_{j=1}^{k-1} K_{j(k-j)}\bar{n}_j\bar{n}_{(k-j)} - \sum_{j=1}^{\infty} K_{kj}\bar{n}_k\bar{n}_j. \quad (9.2.5^*)$$

Assuming equality of the diffusion coefficients ($D_{d1}^T = D_{d2}^T = ... = D_d^T$) and using the formula $\sum_k \bar{n}_k = N_d$ for the total number density of dust grains, summation of the left and right parts of (9.2.5*) over k yields

$$\bar{\rho}\frac{d}{dt}\left(\frac{N_d}{\bar{\rho}}\right) - div\left[\bar{\rho} D_d^T grad\left(\frac{N_d}{\bar{\rho}}\right)\right] = \sum_k \dot{n}_k = -\frac{1}{2}\sum_{k=1}^{\infty}\sum_{j=1}^{\infty} K_{kj}\bar{n}_k\bar{n}_j, \quad (9.2.7)$$

which enables us to find the quantity N_d. The right hand part of this equation is equal to half of the second term in the right part of (9.2.5*) because the overall number of dust particles in a unit volume during coagulation remains unchanged. In the spatially uniform case and when the coagulation coefficients are approximately the same, which is valid for equal diffusion (*Sinaiskii, 1997*), equation (9.2.7) can be written in a simpler form:

$$\partial N_d/\partial t = -(K/2)N_d^2, \qquad (K_{kj} = K).$$

The solution of this equation with the initial condition $N_d(0) = N_{d0}$ is $N_d = N_{d0}/(1+qt)$ with $q = KN_{d0}/2$. It allows the coagulation constant, K, to be determined from the slope of the straight line.

Let us multiply the left and right hand parts of the equations (9.2.5) by the volume $U_1 k$ of k-particles, sum them up over k, and use the conservation law

$\sum_k m_k \dot{n}_k = \rho_d U_1 \sum_k k \dot{n}_k = 0$ for the total mass of dust particles, the latter assuming to be modified in the course of coagulation. Then, admitting that $D_{d1}^T = D_{d2}^T = ... = D_d^T$, we obtain the following equation:

$$\bar{\rho}\frac{d}{dt}\left(\frac{\bar{s}}{\bar{\rho}}\right) + div(\boldsymbol{J}_d^T / \rho_d) = 0, \qquad (\boldsymbol{J}_d^T = \sum_k \boldsymbol{J}_k^T), \tag{9.2.8}$$

which enables us to calculate the total volume dust concentration, $\bar{s} = \sum_k \bar{s}_k$, in the two-phase polydisperse flow. Here

$$\boldsymbol{J}_d^T \equiv \rho_d \overline{s\boldsymbol{v}''} = -\rho_d D_d^T grad\left(\frac{\bar{s}}{\bar{\rho}}\right) = -\frac{\rho_d \rho_g}{\bar{\rho}} D_d^T grad\, \bar{s} = -\frac{\rho_d \rho_g}{\bar{\rho}} D_d^T grad\, \bar{s} \tag{9.2.9}$$

is the turbulent diffusion flux of the disperse phase in the gas-dust flow and D_d^T is the turbulent transport coefficient of all dust grains. It is worth to note that equation (9.2.8) is essentially the conservation law for the disperse phase mass and can therefore be obtained independently, without assuming equality of all eddy diffusion coefficients for dust particles with different sizes.

Upon determining the parameters N_d and $\bar{s}$, we can find the mean volume (size) of dust grains in the coagulating mixture in the turbulized two-phase flow using the formulas $\tilde{U}_d = \bar{s}/N_d$ and $\tilde{d}_d = \sqrt[3]{(6/\pi)(\bar{s}/N_d)}$.

9.2.2. Modeling the Eddy Diffusion Coefficient for the Gas-Dust Subdisk

Equations (9.2.5), (9.2.7), and (9.2.8) contain the unknown coefficients D_{dk}^T and K_{kj}. Let us first consider a phenomenological way for modeling the turbulent diffusion coefficients, D_{dk}^T, for dust particles with size k.

We will assume that the dust concentration in the medium is sufficiently high and therefore, the turbulized flow experiences inverse influence of dust outside the $z = 0$ plane where particles form a flattened dust layer. In the analysis it is convenient to take advantage of a simplified statistical turbulence scheme based entirely on the equation for turbulent energy balance in the suspension-carrying shear flow. With the above stated assumptions on the properties of the heterogeneous medium, this equation has the same form as for compressible multicomponent mixtures and can be written as follows (see (9.1.6)):

$$\begin{aligned}\bar{\rho}\frac{db}{dt} &= -div\left(\overline{\rho(e + p'/\rho)\boldsymbol{v}'' - \Pi \cdot \boldsymbol{v}''}\right) + \boldsymbol{R} : Grad\,\boldsymbol{V} + \\ &+ \overline{p' div\, \boldsymbol{v}''} - \boldsymbol{J}_{(1/\rho)}^T \cdot grad\, \bar{p} + \sum_\alpha \boldsymbol{J}_\alpha^T \cdot \boldsymbol{F}_\alpha - \bar{\rho}\varepsilon.\end{aligned} \tag{9.2.10}$$

Here $b = \langle e \rangle \equiv \frac{1}{2} \langle \boldsymbol{v}'' \cdot \boldsymbol{v}'' \rangle$ is the turbulent energy, i.e. the averaged kinetic energy of turbulent pulsations in the mixture, comprising the kinetic energy of the disperse phase; ε is the viscosity-related, turbulent energy dissipation of the gas suspension; $\boldsymbol{R} \equiv -\bar{\rho} \langle \boldsymbol{v}'' \boldsymbol{v}'' \rangle$ is the Reynolds turbulent stress tensor; $\boldsymbol{J}_\alpha^T \equiv \bar{\rho} \langle Y_\alpha'' \boldsymbol{v}'' \rangle$ is the turbulent diffusion flux of α-substance ($\sum_\alpha \boldsymbol{J}_\alpha^T = 0, \alpha = g, \underbrace{1, 2, \ldots}_{sort\ of\ dust}$) and $\boldsymbol{J}_{(1/\rho)}^T \equiv \bar{\rho} \langle (1/\rho)'' \boldsymbol{v}'' \rangle$ is the turbulent flux of the specific volume of the gas suspension. It should be emphasized once again that turbulent stresses in the considered gas-dust continuum, as well as various turbulent flows, include the terms related to the participation of dust particles in the pulsation motion of the carrying phase.

Let us examine the individual terms in equation (9.2.10). While the possible anisotropy of the coefficient ν^T in the rotating accretion disk has been analyzed in section 9.1, here we restrict ourselves to the isotropic case when the Reynolds stress tensor is related to the gradients of the weighted mean velocities of the gas-dust flow by the following defining relation:

$$\boldsymbol{R} = -\tfrac{2}{3} \bar{\rho} b \boldsymbol{I} + \bar{\rho} \nu^T \overset{\circ}{\boldsymbol{D}}, \qquad \overset{\circ}{\boldsymbol{D}} \equiv \boldsymbol{D} - \tfrac{1}{3} \boldsymbol{I}\, div\, \boldsymbol{V}), \tag{9.2.11}$$

where $\boldsymbol{I}$ is the unit tensor, $\boldsymbol{D} = \frac{1}{2}\left(Grad\, \boldsymbol{V} + (Grad\, \boldsymbol{V})^t\right)$ is the strain rate tensor, and ν^T is the kinematic turbulent viscosity coefficient of the gas suspension. Then the quantity $\boldsymbol{R} : Grad\, \boldsymbol{V}$ in the right part of equation (9.2.10), which describes generation of turbulent energy due to kinetic energy dissipation of the averaged shear flow of the whole continuum, takes the form

$$\boldsymbol{R} : Grad\, \boldsymbol{V} = \boldsymbol{R} : \boldsymbol{D} = \bar{\rho} \nu^T \overset{\circ}{\boldsymbol{D}} : \overset{\circ}{\boldsymbol{D}} - \tfrac{2}{3} \bar{\rho} b\, div \boldsymbol{V}. \tag{9.2.12}$$

Since, according to the above stated assumption, merely the volume dust content, s, pulsates due to turbulent mixing of the flow, we obtain from expression (9.2.2) that

$$\frac{\rho'}{\bar{\rho}} \cong \sigma s', \qquad \frac{\bar{\rho}_g}{\bar{\rho}} = 1 - \sigma \bar{s} = \frac{1}{1 + \sigma^* \bar{s}}, \tag{9.2.13}$$

where $\sigma^* \equiv (\rho_d - \bar{\rho}_g)/\rho_g$ is the relative extra density of dust particles over the gas density. Therefore, using the formulas (9.2.9) and (9.2.13), we can write down the following expression for the turbulent flux of the specific volume, $\boldsymbol{J}_{(1/\rho)}^T = -\overline{\rho' \boldsymbol{v}''} / \bar{\rho} = -\sigma \overline{s' \boldsymbol{v}''}$, appearing in (9.2.10):

$$\boldsymbol{J}_{(1/\rho)}^T = -\sigma \frac{\bar{\rho}}{\rho_d \bar{\rho}_g} \boldsymbol{J}_d^T = -\frac{\sigma^*}{\rho_d} \boldsymbol{J}_d^T = \sigma D_d^T\, grad\, \bar{s}. \tag{9.2.14}$$

Generally, dust particles with different sizes diffuse and coagulate in the two-phase polydisperse turbulized medium with different turbulent diffusion, D_{dk}^T, and turbulent coagulation (*Sinaiskii, 1997*) coefficients, thus affecting differently the turbulence evolution in the gas-dust medium. In order to take this circumstance into account, it is convenient to introduce the "individual" turbulent diffusion fluxes $\boldsymbol{J}_k^T = \rho_d \overline{s_k \boldsymbol{v}''}$ ($\boldsymbol{J}_d^T = \sum_k \boldsymbol{J}_k^T$) in expression (9.2.14), for which the governing relations are given by expression (9.2.6). Then, using the unequal turbulent Schmidt numbers for dust particles of different sizes k $(k = 1,2,...)$, and for the disperse phase as a whole,

$$\boldsymbol{Sc}_k^T \equiv \frac{\nu^T}{D_{dk}^T}, \text{ and } \boldsymbol{Sc}^T \equiv \frac{\nu^T}{D_d^T}, \tag{9.2.15}$$

respectively, we can rewrite expression (9.2.14) as

$$\boldsymbol{J}_{(1/\rho)}^T = -\sigma \frac{\overline{\rho}}{\rho_d \overline{\rho}_g} \sum_k \boldsymbol{J}_k^T = \sigma^* \sum_k D_{dk}^T \left[grad\, \overline{s}_k - \sigma \overline{s}_k grad\, \overline{s} \right] =$$

$$= \sigma^* \nu^T \sum_k \frac{1}{\boldsymbol{Sc}_k^T} grad\, \overline{s}_k - \sigma\sigma^* \nu^T \sum_j \frac{\overline{s}_j}{\boldsymbol{Sc}_j^T} grad\, \overline{s} =$$

$$= \sigma^* \nu^T \sum_k \frac{1}{\boldsymbol{Sc}_k^T} grad\, \overline{s}_k - \sigma^* \boldsymbol{J}_{(1/\rho)}^T \sum_j \frac{\boldsymbol{Sc}^T \overline{s}_j}{\boldsymbol{Sc}_j^T},$$

whence the final expression follows for the flux $\boldsymbol{J}_{(1/\rho)}^T$:

$$\boldsymbol{J}_{(1/\rho)}^T = \sigma D_d^T grad\, \overline{s} = \sigma^* \nu^T \sum_k \frac{1/\boldsymbol{Sc}_k^T}{1 + \sigma^* \boldsymbol{Sc}^T \sum_j \overline{s}_j / \boldsymbol{Sc}_j^T} grad\, \overline{s}_k . \tag{9.2.16}$$

Neglecting now the effect of diffusion transfer of turbulent pulsation energy and other small terms in equation (9.2.10), defining the turbulent energy of the gas-dust medium, we rewrite this equation, using formulas (9.2.12) and (9.2.16), as

$$\left(\frac{\partial b}{\partial t} + \boldsymbol{V} \cdot grad\, b \right) = \nu^T \overset{\circ}{\boldsymbol{D}} : \overset{\circ}{\boldsymbol{D}} - \tfrac{2}{3} b\, div \boldsymbol{V} - \sigma D_d^T \boldsymbol{g} \cdot grad\, \overline{s} - \varepsilon , \tag{9.2.17}$$

where $\boldsymbol{g}$ is the gravitational force per unit mass, $\boldsymbol{g} = grad\, \overline{p} / \overline{\rho}$.

In order to modify this equation into a form convenient for further transformations, we take advantage of Gibbs fundamental identity,

$$db = T_{turb}\, d\, S_{turb} - p_{turb}\, d\left(1/\overline{\rho}\right),$$

which defines the thermodynamic structure of the pulsation motion subsystem of the averaged gas-dust continuum (*Kolesnichenko, 1998*). Such a continuum involves two subsystems for the chosen phenomenological way of simulating turbulence: the subsystem of the averaged motion (averaged molecular chaos) and the subsystem of pulsation motion (averaged turbulent chaos). The Gibbs identity introduces the entropy, S_{turb}, pressure, $p_{turb} \equiv \frac{2}{3}\bar{\rho}b$, and turbulization temperature, T_{turb}, as it was shown in Chapter 5. Then equation (9.2.17) takes the form

$$T_{turb}\frac{dS_{turb}}{dt} = \frac{db}{dt} + p_{turb}\frac{d}{dt}\left(\frac{1}{\bar{\rho}}\right) = \nu^T \overset{\circ}{\boldsymbol{D}}:\overset{\circ}{\boldsymbol{D}} - \sigma D_d^T \boldsymbol{g}\cdot grad\,\bar{s} - \varepsilon\,. \tag{9.2.18}$$

Consider now the local steady state of the developed turbulent flow, when a certain internal equilibrium exists in the turbulence structure such that the turbulization entropy production is approximately equal to its sink. Likewise, for example, in the near-surface atmospheric layer, a similar situation is likely to exist in the subdisk for both forced and natural convection. In this case, $T_{turb}\, dS_{turb}/dt \approx 0$ and equation (9.1.18) takes the form

$$\nu^T(1-\boldsymbol{Kf})\,\overset{\circ}{\boldsymbol{D}}:\overset{\circ}{\boldsymbol{D}} - \varepsilon \approx 0\,. \tag{9.2.19}$$

Here by analogy with the Richardson dynamic number with account thermal stratification of the medium affecting turbulence evolution, the relationship

$$\boldsymbol{Kf} \equiv \sigma\frac{D_d^T}{\nu^T}\frac{\boldsymbol{g}\cdot grad\,\bar{s}}{\overset{\circ}{\boldsymbol{D}}:\overset{\circ}{\boldsymbol{D}}} = \sigma\frac{1}{\boldsymbol{Sc}^T}\frac{\boldsymbol{g}\cdot grad\,\bar{s}}{\overset{\circ}{\boldsymbol{D}}:\overset{\circ}{\boldsymbol{D}}} = \frac{\boldsymbol{Ko}}{\boldsymbol{Sc}^T} \tag{9.2.20}$$

defines the Kolmogorov dynamic number, $\boldsymbol{Kf}$, as a criterion of dynamical activity of dust particles in the gas-dust system of the turbulized shear flow. In turn, the Kolmogorov gradient number,

$$\boldsymbol{Ko} = \sigma\frac{\boldsymbol{g}\cdot grad\,\bar{s}}{\overset{\circ}{\boldsymbol{D}}:\overset{\circ}{\boldsymbol{D}}}\,,$$

accounts for the inverse effect of stratification of suspended dust particles on the turbulization dynamics in the flow.

Simulating the evolution of the gas-dust subdisk, when the polydisperse nature of the mixture dust component is essential, we should use the formulas

$$\boldsymbol{Kf} = \sum_j \frac{\boldsymbol{Ko}_j}{\boldsymbol{Sc}_j^T}\,, \tag{9.2.21}$$

$$\boldsymbol{Ko}_j = \frac{\sigma^*}{1+\sigma^*\sum_k \bar{s}_k \dfrac{\boldsymbol{Sc}^T}{\boldsymbol{Sc}_k^T}}\,, \tag{9.2.22}$$

which can easily be obtained from (9.2.16). Here the number $\boldsymbol{Ko}_j$ has the sense of the Kolmogorov gradient number for j-particles. Let us note that, in practical calculations, there is no need to take into account the entire spectrum of possible particle sizes in the formulas (9.2.21) and (9.2.22). Instead, it is sufficient to use a finite number of characteristic kinds of particles in the disperse medium (fine-grained, coarse-grained, etc.).

Assuming according to Kolmogorov's hypothesis that the kinematic turbulent viscosity coefficient, ν^T, and the dissipation rate of turbulent energy to heat, ε, depend only on two flow parameters – the turbulence energy, b, and the turbulence scale, L – we obtain $\nu^T = L\sqrt{b}$ and $\varepsilon = \frac{1}{\alpha^2}\frac{b^{3/2}}{L}$. In view of the uncertainty in L, it is reasonable, like in (9.1.25), to set the constant in the expression for ν^T equal to unity and to regard the numerical factor $1/\alpha^2$ in the expression for ε as a constant. Substituting these expressions into (9.2.19) gives

$$L\sqrt{b}(1-\boldsymbol{Kf})\,\overset{\circ}{\boldsymbol{D}}:\overset{\circ}{\boldsymbol{D}} - \frac{1}{\alpha^2}\frac{b^{3/2}}{L} \approx 0\,. \tag{9.2.23}$$

Similar to (9.1.26), equation (9.2.23) decomposes into two equations: $b = 0$, corresponding to the laminar flow regime, and

$$b = \alpha^2 L^2 (1-\boldsymbol{Kf})\,\overset{\circ}{\boldsymbol{D}}:\overset{\circ}{\boldsymbol{D}}\,, \tag{9.2.24}$$

which describes the stationary turbulent regime of motion in the system. Equation (9.2.24) has a real solution only for $(1-\boldsymbol{Kf}) \geq 0$, whence, for the turbulent regime, we obtain $\boldsymbol{Ko} \leq (\boldsymbol{Ko})_{\rm cr} \approx \boldsymbol{Sc}^T$. Note that the case $\boldsymbol{Ko} \geq \boldsymbol{Sc}^T$ corresponds to the single real solution, $b = 0$, pertaining to laminar motion. On the other hand, if $\boldsymbol{Ko} \leq \boldsymbol{Sc}^T$, then

$$\sqrt{b} = \alpha L\,\sqrt{\overset{\circ}{\boldsymbol{D}}:\overset{\circ}{\boldsymbol{D}}}\,\sqrt{1-\boldsymbol{Ko}/\boldsymbol{Sc}^T}\,, \tag{9.2.25}$$

from which it follows that the presence of suspended particles in the flow always leads to a decrease in turbulent energy, because $\boldsymbol{Ko} > 0$, and the turbulent viscosity coefficient is given by the

$$\nu^T = \alpha L^2 \sqrt{\overset{\circ}{\boldsymbol{D}}:\overset{\circ}{\boldsymbol{D}}}\,\sqrt{1-\sum_k \frac{\boldsymbol{Ko}_k}{\boldsymbol{Sc}_k^T}}\,. \tag{9.2.26}$$

A Rotating, Turbulent Gas-Dust Disk. Let us apply the derived general relations to the analysis of turbulence in a thin gas-dust disk. We choose the inertial cylindrical coordinate system (r,φ,z) with the origin coincident with the gravitational center; the $z = 0$ plane is considered to be coincident with the equatorial plane of the disk. Further, we assume that circular motion of the "cosmic fluid" is implemented, on average, only in the azimuthal direction, such that $V_r = 0$, $V_\varphi = \Omega\, r$, $V_z = 0$, and the true velocity of the gas-dust flow pulsates about this mean value, changing extremely irregularly in the meridional and azimuthal directions. It can be shown that if the disk matter is in steady-state rotation in the inertial frame, it necessarily possesses axial symmetry:

$$\bar{s}=\bar{s}(r,z),\quad \bar{\rho}=\bar{\rho}(r,z),\quad \bar{p}=\bar{p}(r,z),\quad \Omega=\Omega(r,z)\ ,$$

etc (see, e.g., *Tassoul, 1979*). For the model under consideration, the coupled two-dimensional hydrodynamic equations for axisymmetric flows decompose into two separate equations for the averaged radial (9.1.16) and vertical (9.1.17) motions. Equation (9.1.16) assumes that the gravitational force in the radial direction perpendicular to the rotation axis is almost counterbalanced by the centrifugal force. This means that the gas-suspension pressure gradient is very small and the rotation is almost Keplerian. Conversely, the equilibrium in the z-direction perpendicular to the central plane is maintained by the pressure gradient, $\partial \bar{p}/\partial z=\bar{\rho} g_z$, which determines the main direction of the buoyancy force in the gravitational field of the central mass, which promotes turbulence damping.

In case of isotropic viscosity, the non-zero components of the Reynolds stress tensor are determined, according to (9.2.11), by the formulas (9.1.20$^{(1)}$) and (9.1.20$^{(2)}$). Therefore, formula (9.2.25) for the turbulent viscosity coefficient, using (9.1.16), (9.1.17) and (9.1.20$^{(2)}$), takes the form

$$\nu^T=\alpha L^2 r\sqrt{(\partial_z\Omega)^2+(\partial_r\Omega)^2}\,\sqrt{1-\frac{\boldsymbol{Ko}}{\boldsymbol{Sc}^T}}=\alpha L^2 r\sqrt{(\partial_z\Omega)^2+(\partial_r\Omega)^2}\,\sqrt{1-\sum_k\frac{\boldsymbol{Ko}_k}{\boldsymbol{Sc}_k^T}}\,, \tag{9.2.27}$$

where

$$\boldsymbol{Ko}=\sigma\frac{GM_\odot z}{r^5}\frac{\partial\bar{s}/\partial z}{(\partial\Omega/\partial r)^2+(\partial\Omega/\partial z)^2}\,, \tag{9.2.28}$$

and

$$\boldsymbol{Ko}_j=\frac{\sigma^*}{1+\sigma^*\sum_k \bar{s}_k\dfrac{\boldsymbol{Sc}^T}{\boldsymbol{Sc}_k^T}}\,\frac{GM_\odot z}{r^5}\,\frac{\partial\bar{s}_j/\partial z}{(\partial\Omega/\partial r)^2+(\partial\Omega/\partial z)^2} \tag{9.2.29}$$

are the Kolmogorov gradient numbers for the disperse phase as a whole and for particular kinds (sizes) of dust grains.

This important formula extends the expression for the gas-phase α-disks to gas-dust polydisperse accretion disks. To provide numerical modeling of hydrodynamic motions in the two-phase disk interior, one should specify the turbulence scale, L . The known procedure can be used (see, e.g., *Monin and Yaglom, 1971*) to assign this parameter adequately to the layers susceptible to shear motion.

9.2.3. Coagulation Coefficients in the Turbulent Gas-Dust Disk

We will now proceed to studying the coagulation coefficients, K_{kj}, appearing in the diffusion equations (9.2.5), (9.2.7) and (9.2.8). Several coagulation mechanisms in the accretion disk, responsible for the growth of dust particles and evolution of their size distribution, are possible (*Morfill et al., 1993*) and must be considered first.

As is known, the coagulation process is realized exclusively through random motions of thermal or turbulent origin and subsequent particle collisions (*Voloschuk, 1984; Reist, 1984*). Generally, the efficiency of accumulating dust particles in a laminar or turbulent flow is affected by external forces, such as hydrodynamical, gravitational, and electric forces. Neglecting the repulsive forces, we will assume so-called fast coagulation, i.e. associating particles in every collision, though the adhesion mechanism will not be discussed specially. The coefficient K_{kj} is regarded as one of the main parameters determining the coagulation efficiency. It is defined as the mean number of collisions of particles with size d_k and d_j per unit volume, per unit time, and per unit particle of both sizes concentration. We will briefly discuss the methods to evaluate the coefficients $K_{kj}(d_k, d_j)$ based on the results of elaborative studies (see, e.g., *Voloschuk, 1984; Mazin, 1971; Sinaiskii, 1997*) of different coagulation mechanisms.

Brownian (Perkinetic) Coagulation. Basically, coming together of highly dispersed dust particles suspended in the gas flow occurs through a random thermal motion of gas molecules. In that case, the collision probability (the number of collisions per unit volume and per unit time) is given by the expression (*Fuks, 1955*)

$$\beta_{kj} = 2\pi \mathcal{D}_{kj}(d_k + d_j) n_k n_j, \tag{9.2.30}$$

where $\mathcal{D}_{kj} = (\mathcal{D}_k + \mathcal{D}_j)$; $\mathcal{D}_k$ is the translational diffusion coefficient defined by the *Stokes-Einstein* formula (see *Smoluchowski, 1936*),

$$\mathcal{D}_k = \frac{\mathrm{k}T}{3\pi\mu_g d_k}; \tag{9.2.31}$$

and $\mu_g = \rho_g \nu_g$, where ν_g is the kinematic gas viscosity coefficient. This yields

$$K_{kj}^{brown} = K_{jk}^{brown} = 2\pi \mathcal{D}_{kj}(d_k + d_j) = \frac{2}{3}\frac{kT}{\mu_g}\left(\frac{1}{d_k} + \frac{1}{d_j}\right)(d_k + d_j). \tag{9.2.32}$$

For the monodisperse dust phase ($d_k = d_j$), formula (9.2.32) takes the simplified form $K_{kk}^{brown} = \frac{4}{3}\frac{kT}{\mu_g}$. Note that the expression resulting from (9.2.32) must be divided by 2 because collisions are taken into account twice – firstly, the particle is considered as trial and, secondly, as diffusing. For the monodisperse suspension, the rate of change of the dust particle concentration is given by the relation $\dot{n}_j = -K_{jj}^{brown} n_j^2$, which corresponds to a second-order chemical reaction.

Gravitational Coagulation. This particular type of coagulation is related to the collisions of solid particles caused by the difference between their infall velocities in the gravitational field. Let a spherical particle with size d_k fall in the direction perpendicular to the disk's central plane at a higher speed compared to a smaller particle with size d_j. Then the collision probability when the larger particle overtakes the smaller one is determined by the relationship

$$\beta_{kj}^{gravit} = \tfrac{1}{4}\pi(d_k+d_j)^2 \left|W_{dk} - W_{dj}\right| E_g(d_k,d_j) n_k n_j =$$
$$= \tfrac{1}{4}\pi(d_k+d_j)^2 \left|g_z(t_{relax}^k - t_{relax}^j)\right| E_g(d_k,d_j) n_k n_j, \tag{9.2.33}$$

where W_{dk} is the gravitational sedimentation rate (along the z- axis) for a single dust particle of sort k, which can be defined as $W_{dk} = g_z\, t_{relax}^k$. Here t_{relax}^k is the relaxation time for particles, and $E_g(d_k,d_j)$ is the so-called trapping coefficient which is defined as the ratio between the volume from which all small particles are captured by a large particle in a time Δt, and the volume cut out by the maximum collisional cross section over the same time. For small spherical particles having diameters less than the free path length for gas molecules, this time is determined by the expression $t_{relax}^k = \dfrac{d_k^2\, \rho_d}{18\, \mu_g}$.
The coefficient E_g was evaluated quite comprehensively by (*Davis and Sartor, 1967*).

Based on this premise, the gravitational coagulation coefficient in the accretion disk is given by

$$K_{kj}^{gravit} = K_{jk}^{gravit} = \pi \frac{\rho_d}{72\mu_g} \Omega_K\, z(d_k+d_j)^3 \left|d_k - d_j\right| E_g(d_k,d_j). \tag{9.2.34}$$

For coarse-grained spherical particles with diameters in excess of the free path length for gaseous molecules, this formula is somewhat modified (*Mednikov, 1981*), because in this case the relaxation time is determined by the expression

$$t_{relax}^k = \frac{\pi d_k^3 \rho_d}{6} \frac{u_{gd}}{F_k} = \frac{4d_k}{3} \frac{\rho_d}{\rho_g\, \varsigma \left|u_{gd}\right|},$$

where $F_k = \tfrac{1}{8}\varsigma\rho_g \left|u_{gd}\right| u_{gd} \pi\, d_k^2$ is the drag force rendered by the medium on a *k*-particle; $u_{gd} = v_g - v_d$ is the gas velocity relative to the particle; $\varsigma = A/(\boldsymbol{Re}_d)^n$ is the particle drag coefficient; $\boldsymbol{Re}_d = \left|u_{gd}\right| d/\nu_g$ is the Reynolds number for dust; and A and n are coefficients depending on the particle's dispersity. It was found $\varsigma = 24/\boldsymbol{Re}_d$ for $\boldsymbol{Re} \le 1$

(Stokes' law); $\varsigma = 24 / \boldsymbol{Re}_d^{3/5}$ for $1 \le \boldsymbol{Re} \le 10^3$; and $\varsigma = 0{,}44$ for $\boldsymbol{Re} \ge 10^3$ (*Vityazev et al., 1990*).

Electric Coagulation. Adhesion and merging of particles due to electrostatic attraction is probably one of the most important initial accretion processes in the jet's gas-dust stream (*Alfven and Arrhenius, 1976*). The main physical mechanisms affecting the charge of a dust grain are photo-electron emission caused by intense irradiation, and collisions with electrons and positive ions. We restrict the analysis to pair interactions between charged particles and ignore the screening effects that seem to play a minor role in the rarefied gas-dust media (*Spitzer, 1979*).

Taking into consideration only the Coulomb forces, $F = q_k q_j / r^2$, (where $r = (d_k + d_j)/2$ and q_k is the mean charge of a k-particle proportional to its radius), we obtain the following expression for the coagulation coefficient:

$$K_{kj}^{elect} = K_{jk}^{elect} = \frac{2|q_k q_j|}{3\mu_g} \frac{d_k + d_j}{d_k d_j}. \qquad (9.2.35)$$

Obviously, the occurrence of an electric field results in the displacement of charged particles, the induction of additional dipole moments on particles, and the interaction between particles. However, these effects are usually little important when stimulating particle collisions (see, e.g., *Sinaiskii, 1997*).

Turbulent Coagulation. Due to turbulent pulsations the number of solid particles collisions per unit time in the turbulent gas-suspension flow increases considerably as compared to the laminar flow, i.e., turbulence favors the rise of the coagulation rate.

The collision probability for a particle d_k with a particle d_j is given by the relation

$$\beta_{kj} = \tfrac{1}{12}\sqrt{2\pi}\,(D_d^T / L^2)(d_k + d_j)^3 \bar{n}_k \bar{n}_j,$$

where L is the turbulence scale. The turbulent diffusion coefficient for dust particles, D_d^T, depends only on the turbulent pulsation scale in the fluid and can be evaluated as (*Levich, 1959*)

$$D_d^T \sim D_g^T \sim \sqrt{b}L = \begin{cases} (\varepsilon L)^{1/3} L, & for \quad L > l_k \\ \sqrt{\varepsilon / \nu_g}\, L^2, & for \quad L < l_k \end{cases}, \qquad (9.2.36)$$

where the following notations were used: $D_g^T \; (= \nu_g^T)$ is the turbulent viscosity coefficient for the carrying phase; $\varepsilon \cong b^{3/2} / L \cong \nu_g^3 / l_K^4$ is the turbulent energy dissipation rate in the gas medium; and $l_K \cong (\nu_g^3 / \varepsilon)^{1/4}$ is the Kolmogorov (internal) turbulence scale.

The turbulent diffusion mechanism is based on assuming complete entrapment of particles by turbulent pulsations with scales that are most important in the particle approach mechanism. The main diffusion resistance to the motion of dust particles corresponds to the scale range $r < l_K$. Therefore, the upper limit of frequency inherent in small-scale pulsations (with the scale l_K) is determined by the relation $D_d^T / L^2 \cong (\varepsilon / \nu_g)^{1/2}$. Hence, we have the following expression for the turbulent coagulation coefficient (*Levich, 1959*):

$$K_{kj}^{turb} = K_{jk}^{turb} = \tfrac{1}{12}\sqrt{2\pi}\,(\varepsilon/\nu_g)^{1/2}(d_k + d_j)^3 =$$
$$= \tfrac{1}{12}\sqrt{2\pi}\,(\sqrt{b})^{3/2}/(L\nu_g)^{1/2}(d_k + d_j)^3 = \tfrac{1}{12}\sqrt{2\pi}\,\nu_g \boldsymbol{Re}^{3/2} L^{-2}(d_k + d_j)^3, \qquad (9.2.37)$$

where $\boldsymbol{Re} = \sqrt{b}L/\nu_g$ is the Reynolds characteristic number for the loaded gas flow. Using (9.2.37), it is easy to obtain the expression for the rate of change in the number density of monodisperse dust ($d_k = d_j$):

$$\dot{n}_j = -\frac{\sqrt{2\pi}}{3}\nu_g \frac{\boldsymbol{Re}^{3/2}}{L^2} d_j^3 \bar{n}_j^2 .$$

This means that the coagulation rate in the turbulent flow corresponds to a second-order reaction.

Comparing Brownian and turbulent coagulation rates,

$$\frac{\beta_{kj}^{brown}}{\beta_{kj}^{turb}} \sim \left(\frac{\nu_g}{\varepsilon}\right)^{1/2} \frac{\mathcal{D}_{kj}}{(d_k + d_j)^2},$$

shows that the inequality $\beta_{kj}^{brown} < \beta_{kj}^{turb}$ holds for particles larger than 0.1 μ m, because the turbulent diffusion coefficient may be seven orders of magnitude larger than the thermal diffusion coefficient (*Sinaiskii, 1997*).

Let us notice that all the expressions for the coagulation coefficients given above were obtained without regard for the external interaction of particles. In general, the turbulent diffusion coefficient for the disk must include the hydrodynamic drag to particle motion, taking into account the distortion of the velocity field in the carrying flow. If particles are small and, therefore, their motion does not virtually differ from the motion of the carrying gas moles, the turbulent diffusion coefficients for particles and the medium are equal, $D_d^T = D_g^T$, which is assumed by formula (9.2.37) defining the coagulation coefficient in the case of the turbulent diffusion mechanism. However, a large number of experiments (*Mednikov, 1981*) confirm the identity $D_d^T = D_g^T$ for very small particles only or, more precisely, for the inertness index $\omega_E t_{relax} \ll 1$ which characterizes the degree of particle entrapment by turbulent pulsations and, therefore, the particle inertness in the turbulent field. Here $\omega_E = 4\omega_0 / \boldsymbol{Re}^{1/8}$ is the pulsation frequency (related to the variation of amplitude of hydrodynamical velocity per second), which pertains to large-scale eddies, i.e., it is characteristic for low-frequency pulsations. $\omega_0 = V/L$ is the lower frequency limit for turbulent pulsations, corresponding to the largest eddies with scale L.

If the particles are not too small (and, therefore, the gas carrying moles entrain them), their relative velocities, gained by turbulent pulsations, depend on mass substantially. The difference between pulsation velocities of particles with different sizes is responsible for their mutual approach and increases the collision probability. This determines the so-called *inertial mechanism* of particle coagulation in the turbulent flow. In this case the following formula is valid:

$$D_d^T = \frac{1}{1+\omega_E \tau} D^T,$$

or

$$\boldsymbol{Sc}^T = \frac{\nu^T}{D_d^T} = (1+\omega_E \tau) \cong (1+\boldsymbol{Stc}), \qquad (9.2.38)$$

which is reliably supported by experimental studies. Here $\boldsymbol{Sc}^T$ and $\boldsymbol{Stc} = V\tau / L$ are the Schmidt turbulent number and the Stokes number, respectively.

We may conclude that the particle coagulation processes in the gas-dust flow might be caused by the simultaneous influence of different collisional mechanisms in various combinations. Brownian coagulation of charged particles, turbulent Brownian coagulation of charged and neutral particles, turbulent Brownian coagulation of charged particles in the gravitational field are addressed as the most important concurrent mechanisms when modeling gas-dust accretion disks. Unfortunately, this problem is still far from the proper solution and it faces a real challenge for future studies.

9.2.4. The Coagulation Kinetics Equation: Solution by the Moments Method

The number of non-linear differential equations (9.2.4) required to describe the dust particle size distributions and space-temporal variations throughout the disk is generally unlimited. Therefore, when a confined set of equations (9.2.4) for numerical modeling the coagulation processes, is utilized, "loss" of material is possible because the sizes of some coagulating particles can exceed the maximum size adopted in the model. This is why another, integral representation of the system (9.2.4) seems more appropriate.

To obtain this integral form let us assume that the number of particles in the volume range from U to $U+dU$ near a point $\boldsymbol{r}$ at a time t is $f(U,\boldsymbol{r},t)\,dU$. The function $f(U,\boldsymbol{r},t)$, describing the spectrum of particle sizes, satisfies, by definition, the following normalizing relation:

$$N_d(\boldsymbol{r},t) = \int_0^\infty f(U,\boldsymbol{r},t)dU. \qquad (9.2.39)$$

Obviously, the formula

$$s(\boldsymbol{r},t)=\int_0^{\infty} U\, f(\boldsymbol{r},t,U)dU \tag{9.2.40}$$

determines the overall bulk concentration of dust particles.

Since the volume filled by particles with size k is kU_1, the number density, n_k, can be expressed (in terms of the function $f(U,\boldsymbol{r},t)$) as follows:

$$n_k = f(kU_1,t)U_1. \tag{9.2.41}$$

Then, using this relationship we can obtain the following kinetic equation for coagulation from (9.2.5*), after the transition operation $U_1 \to dU$:

$$\frac{\partial f(U,\boldsymbol{r},t)}{\partial t}+div\left[f(U,\boldsymbol{r},t)\boldsymbol{V}-D_d^T(U)\left(grad\, f(U,\boldsymbol{r},t)-\sigma f(U,\boldsymbol{r},t)grad s\right)\right]=$$

$$=\frac{1}{2}\int_0^U f(W,\boldsymbol{r},t)f(U-W,\boldsymbol{r},t)K(W,U-W)\,dW - f(U,\boldsymbol{r},t)\int_0^{\infty} f(W,\boldsymbol{r},t)\,K(W,U)\,dW \tag{9.2.42}$$

Here $K(W,U)$ is the coagulation kernel symmetric in its arguments that determines the time behavior of the disperse medium. Note that this equation extends the known *Muller* equation for coagulating disperse systems to spatially heterogeneous turbulent motions.

The kinetic equation (9.2.42) is a non-linear integro-differential equation. Its solution must satisfy the conditions $f(U,\boldsymbol{r},t)\to 0$ for $U\to 0$ and $U\to\infty$, and the initial $f(U,\boldsymbol{r},0)=f_0(U,\boldsymbol{r})$ and boundary conditions must be specified. However, in general the solution can only be obtained by numerical methods, because the terms describing convection and turbulent diffusion extremely complicate the standard coagulation equation (see, e.g., *Lissauer and Stewart, 1993*).

A series of exact analytic solutions of the unsteady, spatially uniform analog of (9.2.42) was derived for some coagulation kernels based on the *Laplace* integral transformation, which is simple in structure and linear with respect to each of its arguments (*Safronov, 1972; Voloschuk, 1984*). Moreover, to date the most advanced theoretical studies of coagulation processes are carried out for the kernels $K(W,U)=\Lambda_0$, which are independent of the coagulating particle volumes. As far as the coagulation equation with the kernel $K(W,U)=\Lambda_1 WU$ is concerned, its solution can hardly be physically realized because it is not continuous in time; indeed, starting at a certain time, the number of particles in the system becomes a negative value. Note that in application to the evolution of a primordial gas-dust nebula, the kinetic equation with the kernel $K(W,U)=\Lambda_2(W+U)$, which is proportional to the sum of the coagulating particle volumes, has been solved analytically (*Safronov, 1972*). However, no disperse system has been found so far for which the microphysics of the coagulation process would have been associated with such a kernel.

At the same time, as was shown in subsection 9.2.2, for hydrodynamic simulations of disperse mixtures, when the inverse influence of solid particles transport on turbulization of the carrying flow is important, it is often sufficient to know only several first moments of the particle size distribution function and their temporal behavior, rather than the function itself. Then one of the possible approximate methods for solving the kinetic equation for coagulation – the moments method – can be used. Let us discuss the possibilities offered by this method using as an example an equation analogue to equation (9.2.42) describing the unsteady spatially uniform coagulation process in the disk.

As is known, the essence of the moments method is reducing the basic equation to a system of ordinary differential equations relative to the moments of the distribution function $f(U,\mathbf{r},t)$,

$$M_p = \int_0^\infty U^p f(U,t)dU \; (p = 0,1,2,\dots).$$

To obtain such a system, we multiply both parts of the simplified equation (9.2.42) by U^p and integrate the result over U from zero to infinity. This yields

$$\frac{\partial M_0}{\partial t} = \frac{\partial N_d}{\partial t} = -\frac{1}{2}\int_0^\infty\int_0^\infty K(U,W)\, f(U,t) f(W,t) dW dU\,,$$

$$\frac{\partial M_1}{\partial t} = \frac{\partial s}{\partial t} = \frac{1}{2}\int_0^\infty\int_0^\infty (W-U)K(U,W)\, f(U,t) f(W,t) dW dU = 0\,,$$

$$\frac{\partial M_p}{\partial t} = \int_0^\infty\int_0^\infty K(U,W)\left[\tfrac{1}{2}(U+W)^p - U^p\right] f(U,t) f(W,t)\, dW\, dU\,, \quad (p = 2,3\ \dots). \tag{9.2.43}$$

In order to express the right-hand sides of these equations in terms of moments, one should specify the coagulation kernel and make an assumption about the initial form of the particle distribution function.

Let us consider the following class of kernels:

$$K(U,W) = \Lambda \sum_{j=0}^{K} \beta_j (U^{\alpha-\alpha_j} W^{\alpha_j} + U^{\alpha_j} W^{\alpha-\alpha_j})\,, \tag{9.2.44}$$

where Λ is a factor defined by the external conditions under which coagulation occurs. This class includes the kernels for which the kinetic equation has exact solutions. Moreover, formula (9.2.44) can approximate the kernels corresponding to the different coagulation mechanisms. In particular, the kernels discussed in subsection 9.2.4, are represented by homogeneous functions $K(U,W) = U^{\alpha} K(1,x)$, α being the degree of homogeneity and $x \equiv W/U$.

Such functions can be approximated by the polynomial

$$K(U,W) = \Lambda U^{\alpha} \sum_{j=0}^{K} \beta_j (x^{\alpha_j} + x^{\alpha - \alpha_j}). \qquad (9.2.44^*)$$

To determine the unknown coefficients β_j, it is necessary to choose s interpolation points x_i, s being the number of unknown coefficients, and take

$$K(1,x_i) = \Lambda \sum_{j=0}^{K} \beta_j (x_i^{\alpha_j} + x_i^{\alpha - \alpha_j}) \qquad (i = 1,2,..., s).$$

For the moments of the order exceeding the degree of homogeneity of the kernel to appear in the subsequent integration of expansion (9.2.44*), the quantity α_j must be defined by the equality $\alpha_j = \alpha j/(K+1)$. Note that for various K values different interpolation polynomials can be obtained (see *Sinaiskii, 1990* for further discussion).

Substituting (9.2.44) into (9.2.43) gives the following infinite system of equations for the desired moments:

$$\frac{\partial M_0}{\partial t} = \frac{\partial N_d}{\partial t} = -\Lambda \sum_{j=0}^{K} \beta_j M_{\alpha - \alpha_j} M_{\alpha_j}, \quad \frac{\partial M_1}{\partial t} = \frac{\partial s}{\partial t} = 0$$

$$\frac{\partial M_p}{\partial t} = \frac{\Lambda}{2} \sum_{j=0}^{K} \beta_j \sum_{k=1}^{p-1} C_p^k (M_{\alpha - \alpha_j + p - k} M_{\alpha_j + k} + M_{\alpha - \alpha_j + k} M_{\alpha_j + p - k}), \quad (p = 2,3\ \ldots). \qquad (9.2.45)$$

To solve this system (the right hand parts of which in general contain fractional moments), it is necessary to supplement it by the equations connecting the fractional and integer moments. Different approaches can be utilized: either the procedure based on the Lagrange polynomial representation of the fractional moments through the integer moments (*Loginov, 1979*), or the parametric method.

We restrict our analysis to the parametric method which allows to assign the desired distribution function $f(U,t)$ to a definite parametric class of distributions. For simplicity, we will assume that, as the result of the coagulation process, the distribution mode, $f(U,t)$, remains in the same class of distributions as the initial one, and that only statistical parameters (mean value, dispersion, etc.) are time dependent. As an initial particle size distribution in the gas-dust accretion disk, the two-parameter log-normal distribution is accepted, by the analogy with atmospheric aerosols.

The probability density for the log-normal law depends on the mean value $< \ln d >$, where d is the dust particle diameter, and on the scattering (dispersion) index for the logarithm of the diameter, $\sigma_L^2 \equiv < (\ln d - < \ln d >)^2 >$:

$$p(d;\mu^*,\sigma_L)=\frac{N_d}{\sigma_L d\sqrt{2\pi}}\exp\left[-\frac{(\ln d-<\ln d>)^2}{2\sigma_L^2}\right]=\frac{N_d}{\sigma_L d\sqrt{2\pi}}\exp\left[-\frac{\ln^2(d/\mu^*)}{2\sigma_L^2}\right]. \quad (9.2.46)$$

The median of the distribution is determined from (9.2.46) by the relation $\mu^*=\exp(<\ln d>)$, while the mean diameter and its dispersion values are defined, respectively, by the formulas

$$<d>=\exp\left(\frac{1}{2}\sigma_L^2+\ln\mu^*\right), \qquad (^1)$$

$$\sigma^2\equiv<(d-<d>)^2>=<d>^2\left[\exp\sigma_L^2-1\right]. \qquad (^2) \quad (9.2.47)$$

Using these relations, the following formulas are obtained for the statistical parameters σ_L^2 and μ^* depending only on $<d>$ and relative dispersion, $\beta^2\equiv\frac{<(d-<d>)^2>}{<d>^2}$:

$$\sigma_L^2=\ln(1+\beta^2), \qquad \mu^*=<d>/\sqrt{1+\beta^2}. \quad (9.2.48)$$

Let us use now the transition formula $f(U)=p[d(U)]\left|\frac{dd}{dU}\right|$, which is valid for a rigorously increasing function $U=U(d)$ of the random quantity d (*Hahn and Shapiro, 1967*), and the distribution (9.2.46) for determining the density of the initial particle volume distribution, $U=(\pi/6)d^3$:

$$f(U;\sigma_L,\mu)=\frac{N_d}{3\sqrt{2\pi}\sigma_L U}\exp\left[-\frac{\ln^2(U/\mu)}{18\sigma_L^2}\right], \qquad (\mu=(\pi/6)\mu^{*3}). \quad (9.2.49)$$

We assume that the coagulation process in the disk does not modify this distribution and only the parameters $\mu(t)$ and $\sigma_L^2(t)$ change with time. Thereupon we introduce the moments of the log-normal distribution:

$$M_p(t)=\frac{N_d}{3\sqrt{2\pi}\sigma_L(t)}\int_0^\infty U^{p-1}\exp\left[-\frac{\ln^2[U/\mu(t)]}{18\sigma_L^2(t)}\right]dU. \quad (9.2.50)$$

The p-th order moment can be represented as (*Lee, 1983*)

$$M_p=M_1\mu^{p-1}\exp\left[\frac{3}{2}(p^2-1)\sigma_L^2\right], \qquad M_1=s=const, \quad (9.2.51)$$

which enables the fractional moments in (9.2.45) to be expressed in terms of M_1,μ, and

σ_L^2. As a result, we obtain the following parametric system of two ordinary differential equations (their number should be equal to the number of unknown coefficients) for determining the parameters $\mu(t)$ and $\sigma_L^2(t)$ having the initial values $\mu(0)$ and $\sigma_L^2(0)$:

$$N_d \equiv M_0 = s\mu^{-1} exp\,(-\tfrac{3}{2}\sigma_L^2), \quad s \equiv M_1 = const\,, \quad M_2 = s\mu \exp(\tfrac{9}{2}\sigma_L^2)\,; \tag{9.2.52}$$

$$\frac{\partial M_0}{\partial t} = -\Lambda \sum_{j=0}^{K} \beta_j M_{\alpha-\alpha_j} M_{\alpha_j} = -\Lambda M_1^2 \mu^{\alpha-2} \sum_{j=0}^{K} \beta_j \exp\left[\frac{3}{2}[\alpha_j^2 + (\alpha-\alpha_j)^2 - 2]\sigma_L^2\right] =$$

$$= -\Lambda \mu^{\alpha} M_0^2 \sum_{j=0}^{K} \beta_j \exp\left[\frac{3}{2}[\alpha_j^2 + (\alpha-\alpha_j)^2]\sigma_L^2\right], \tag{9.2.53}$$

$$\frac{\partial M_2}{\partial t} = 2\Lambda \sum_{j=0}^{K} \beta_j M_{\alpha-\alpha_j+1} M_{\alpha_j+1} =$$

$$= 2\Lambda M_1^2 \mu^{\alpha} \sum_{j=0}^{K} \beta_j \exp\left[\frac{3}{2}[(\alpha_j+1)^2 + (\alpha-\alpha_j+1)^2 - 2]\sigma_L^2\right],$$

$$= 2\Lambda \mu^{\alpha+2} \sum_{j=0}^{K} \beta_j \exp\left[\frac{3}{2}[\alpha_j^2 + (\alpha-\alpha_j)^2]\sigma_L^2\right]. \tag{9.2.54}$$

This set of non-linear parametric equations which can be solved only numerically, serves as a valuable tool to simulate dust particle sedimentation to the central plane of the disk. However, the change of the mean particle number with time, $N_d(t)$, can be evaluated just by assuming that the dispersion, σ_L^2, remains constant. In this case, keeping only two first moments, we obtain from (9.2.53)

$$\frac{\partial N_d}{\partial t} = -\Lambda \mu^{\alpha} N_d^2 \sum_{j=0}^{K} \beta_j \exp\left[\frac{3}{2}[\alpha_j^2 + (\alpha-\alpha_j)^2]\sigma_L^2\right]. \tag{9.2.55}$$

The solution to this equation, obtained with the initial condition $N_d(0) \equiv N_{d0} = s/\widetilde{U}(0)$, has the form

$$N_d(t) = \frac{s}{\widetilde{U}(0)} \frac{1}{1+qt}, \qquad (^1)$$

$$q = \Lambda \mu^{\alpha} \frac{s}{\widetilde{U}(0)} \sum_{j=0}^{K} \beta_j \exp\left[\frac{3}{2}[\alpha_j^2 + (\alpha-\alpha_j)^2]\sigma_L^2\right] \qquad (^2), \tag{9.2.56}$$

where

$$\tilde{U}(0) = (\pi/6) < d >^3 = \mu(0)\exp(3/2\,\sigma_L^2)$$

is the initial value of the mean volume (this formula follows from (9.2.48)). Hence, using the relation $\tilde{U}(t) = s/N_d$, one may find how the mean particle volume changes with time. For relatively large time intervals, when $qt >> 1$, it follows from (9.2.56) that

$$N_d(t) = 1/\Lambda\mu^{\alpha}t\sum_{j=0}^{K}\beta_j\exp\left[\frac{3}{2}[\alpha_j^2 + (\alpha - \alpha_j)^2]\sigma_L^2\right]. \qquad (9.2.57)$$

This expression states that, for a sufficiently large coagulation time, the mean particle number in the system becomes independent of their initial size distribution. In other words, the current mean number "forgets its past" and can therefore be described by a universal function whose appearance is determined only by the coagulation kernel.

A similar approach can be applied to some other particle size distribution laws in the turbulent flow, for example the gamma-distribution. Moreover, with minor changes, the outlined treatment can also be used when the particle size distribution depends on a single spatial coordinate z. In our case, this might correspond to the steady-state regime in the dust layer, when particles settle down due to gravity. Then, if one poses no constraints on the form of the distribution function, it is possible to use, as was mentioned above, the moments method by representing the fractional moments, $M_{p+\alpha}$, in terms of the integer moments, M_q, on the time interval $[t, t+\Delta t]$ with the Lagrange interpolation polynomial:

$$M'_{p+\alpha}(t) = \prod_{j=k}^{k+n}\left[M'_q(t)\right]^{L_q^n}, \qquad k < p < k+n, \qquad M'_l(t) = M_l(t)/M_l(0),$$

$$L_q^n(x) = \frac{1}{n!}\prod\nolimits_{n+1}\frac{(-1)^{n-j}C_n^j}{x-j}, \qquad x = p+\alpha-j, \qquad \prod\nolimits_{n+1}(x) = x(x-1)...(x-n). \qquad (9.2.58)$$

The relative error of such an interpolation decreases monotonically with n, with an increasing dispersion in the distribution under study, and with the order of the interpolated moment. In particular, for a log-normal distribution, expression (9.2.58) becomes accurate for $n \geq 2$. Upon determining the distribution moments from the respective system of differential equations, with chosen initial distribution, the first five moments can be used to evaluate the form of the desired particle size distribution with the help of Pearson's diagram (*Hant and Shapiro, 1967; Loginov, 1979*).

Summary

- *A phenomenological approach to modeling the turbulent transport coefficients for the gas-dust accretion disks around young stars was developed. The derived formulas for these coefficients, including the inverse effects of dust transfer and potential temperature on turbulence damping or maintenance, generalize the ex-*

pression derived for the coefficient of turbulent viscosity in gas-phase α *-disks to gas-dust disks. The rheological relations are obtained for turbulent diffusion and thermal fluxes, which describe, in case of differential rotation of a two-phase mixture with angular velocity* $\Omega(r,z)$, *the dust and thermal energy transfer in the direction perpendicular to the disk's equatorial plane.*

- *The limiting saturation regime of the model for gaseous flows by small dust particles in a thin gas-dust layer adjacent to the dust subdisk was examined. It is shown that in such a turbulized disk layer saturated by tiny dust particles, they can efficiently grow by the turbulent coagulation mechanism.*

- *The combined mass transfer and coagulation processes were studied in a two-phase medium of differentially rotating primordial protoplanetary nebula, containing polydisperse dust particles suspended in the carrying flow in the presence of shear turbulence. Discussed coagulation mechanisms in the accretion disk (gravitational, turbulent and electric coagulation) aimed to treat rigorously the problem of mutual influence of turbulent mixing and coagulation kinetics on the formation of the gas-dust subdisk.*

- *The parametric moments method was applied, as an example, to the solution of the Smoluchowski integro-differential coagulation kinetics equation, when only statistical parameters (the mean value, dispersion, etc.) of a definite class of particle size distribution functions change with time. With minor modifications this approach can be used to evaluate steady-state motion in the turbulized saturated disk layer, where particles effectively grow via the turbulent coagulation mechanism.*

CONCLUSION

This book is addressed to multicomponent natural media, specifically those exhibiting turbulent dynamics. Interest to this field, occupying a position between mechanics and astrophysics, continuously grows, being stimulated by the dramatic expansion of our knowledge on outer space, the Earth as a solar system planet and its closest neighbourhood, and the diverse objects and processes observed in the universe. Methods of mathematical modeling allows us to place important constraints on the key physical mechanisms underlying the natural phenomena and invoked to explain the gained experimental information. The model improvements and verification depend, however, on further development and breakthroughs in the theories related to the study. This is especially important when dealing with fully inaccessible objects or poorly known mechanisms, stellar and planetary interiors and celestial body formation and evolution being brilliant examples.

An overwhelming part of the hydrodynamic processes and heat and mass transfer in natural objects, scaling from propagation of small impurities in a regional volume of the planetary atmosphere to the formation of gigantic gas-dust nebulae, stellar associations and galactic condensations, are turbulent in their nature. Turbulence bears some specific features in multicomponent gas mixtures which is usually the case for natural environments. These effects are most pronounced in mixed gases with low densities, which is characteristic, for example, for the rarefied gaseous envelopes of planets (their middle/upper atmospheres) where turbulence is concurrent with numerous elementary processes originating from solar EUV and X-ray radiation. This is a particular case of turbulence in multicomponent reactive gas mixtures, which is one of the cornerstones of planetary aeronomy. Another example is galaxy and star formation and evolution directly related to the problem of the origin of the solar system and other planetary systems around stars, the latter stimulating by observations of primordial disks and the discoveries of extrasolar planets. Shear turbulence in the gas-dust media of Keplerian accretion disks is of key importance to advance the theory of planet formation. Because studying and modeling aeronomic and cosmogonic problems bears a rather general character, this allows us to expand the basic concepts over a broad class of natural turbulent media. This is the goal of this book which combines the theory of developed turbulence of reacting gases with applications to some particular examples, to get insight into macroscopic models for turbulent continua with complicated properties.

Let us summarize the main issues of the book.

The system of closed relations was obtained, which includes the hydrodynamic equations for a scale of mean motion, the equations of chemical kinetics for turbulent mixtures, as well as the defining (rheological) relation for the turbulent flows of matter, momentum and energy, taking into account variability of thermophysical properties and compressibility of multicomponent gas media, diffusive heat and mass transfer, chemical reactions, and the effects of a gravitational field. Besides the conventional weightless averaging of the basic hydrodynamic equations for mixtures, the weighted Favre averaging was utilized systematically, which permitted to simplify essentially the analysis of the mean equations for mixtures with variable density.

Based on the suggested averaging procedure of the reaction rates in a chemically active medium, the respective relations for the chemical transformation rates in a turbulent flow of reacting substances were derived. Respectively, the presence of strongly non-linear source terms of substance production in the set of chemical reactions (having an exponential nature) and the appearance of a large number of additional correlations (related to pulsating temperature and composition in the flow) were taken into account in the averaged hydrodynamic equations. The rheological relations for the turbulent diffusion and heat flows and the Reynolds stress tensor were deduced using, in particular, the conventional concept of a mixing path. This allowed us to generalize the results obtained earlier for homogeneous incompressible fluids to shear flows of multicomponent mixtures stratified in a gravitational field. In addition, for turbulent flows with small Mach numbers, once it is possible to neglect the relative density changes caused by pressure fluctuations as compared to the changes related to temperature and composition fluctuations, the algebraic equations for the defining correlations with mass density fluctuations were also obtained.

Parallel with this simplified approach, the complicated mathematical model for geophysical turbulence was developed for which, along with the basic hydrodynamic equations for mean motion, the evolutionary transfer equations for the single-point moments of second order for pulsating thermohydrodynamical parameters of multicomponent reacting gas mixtures were derived. The model includes the evolutionary transfer equations for the components of the Reynolds turbulent stress tensor and the vectors of heat and eddy diffusion turbulent flows; the transfer equation for turbulent energy and dispersion of enthalpy fluctuations in the medium; and the transfer equation for the pair correlations of enthalpy and composition fluctuations in the mixture, as well as for the mixed pair correlations of number density fluctuations of mixture species. Such an approach gives the opportunity to calculate the complex flow patterns of multicomponent reacting gases with variable density, when the diffusive turbulence transfer, the convective terms, and the pre-history of the flow are essential, and hence simpler models based on the idea of isotropic turbulent exchange coefficients, turn out inappropriate.

Special focus was given to the closing problem which is of key importance in the turbulence theory. The thermodynamic approach to closing the hydrodynamic equations for mean motion at the level of second order models allowed to derive more general rheological (gradient) relations for the turbulent flows of matter, momentum and energy in the multicomponent continuum, as compared to what was obtained when employing the mixing path concept. A turbulized continuum was represented as a thermodynamic complex consisting of two subsystems: the subsystem of mean motion (averaged molecular and turbulent chaos) and the subsystem of pulsating motion (turbulent superstructure). This idea served as a basis for the thermodynamic analysis.

The Onsager formalism of non-equilibrium thermodynamics resulted in the defining relations between the thermodynamic flows and forces for the three main turbulent flow domains: the laminar sublayer, the buffer zone (an intermediate region where the effects of molecular and turbulent transfer are comparable), and the developed turbulent flow. For the particular case of a locally stationary state of a developed turbulent field (when production and sink of the turbulization entropy are nearly balanced) the generalized Stefan-Maxwell relations for multicomponent eddy diffusion and the respective expression for the turbulent heat flow, as well as the algebraic equations for determining the multicomponent eddy diffusion coefficients through the binary eddy diffusion coefficients, were derived.

The deduced relations are addressed as the most complete form of describing the heat-and-mass transfer in turbulent multicomponent media, which generalizes the results earlier obtained for the description of regular (molecular) transfer processes over the case of turbulent flows. In such a case, the defining relations are resolved with respect to the diffusive thermodynamic forces through the flows. In other words, they are written in terms of Stefan-Maxwell relations using the diffusion coefficients for binary gas mixtures instead of the multicomponent diffusion coefficients.

The developed theoretical approach to the description of multicomponent turbulent media was applied to modeling some planetary aeronomy and planetary cosmogony problems, intended to promote their further advancement.

Diffusive transfer in the terrestrial upper atmosphere was studied in detail based on the generalized Stefan-Maxwell relations. In the framework of diffusive-photochemical composition and temperature models, the value of the time-averaging eddy diffusion coefficient was evaluated. The procedure was developed to simulate semi-empirically the isotropic turbulent exchange coefficients of multicomponent gas flows stratified in a gravitational field and experienced a transverse hydrodynamic velocity shear. For such flows the possibility was shown to take into account the non-equilibrium character of turbulence with respect to the mean velocity and temperature fields. Simultaneously, the universal algebraic expressions were derived for defining the turbulent viscosity and thermal diffusivity coefficients in the vertical direction, depending on the local values of the kinetic energy of turbulent fluctuations, the dynamic Richardson and Kolmogorov numbers and the turbulent Prandtl number, as well as the external turbulence scale. On the basis of the developed technique, numerical modeling the turbulent viscosity and thermal diffusivity coefficients in the terrestrial homopause, rendering strong influence on the structure and thermal regime of the near-Earth space, was performed.

Attention was paid to evaluating the transfer coefficients in the middle atmosphere based on the data on the height distribution of the structural parameter of the refractive index of air. The structural parameter is considered as the major characteristic of microstructure in the atmospheric turbulent field. Proceeding from the model transfer equations for the components of the Reynolds stress tensor and the turbulent heat flow, as well as the equations for the turbulent energy and the mean square enthalpy fluctuations, the technique was developed to calculate self-consistently the turbulent transfer factors, depending on the structural characteristic of refractive index fluctuations. In the general anisotropic case, the procedure can account for the different intensities of turbulent fluctuations of the structure, velocity and temperature along the different coordinate axes. The developed approach is grounded eventually on the capability to define an external turbulence scale from both the gradients of the mean thermohydrodynamical parameters, and the experimentally defined structural characteristic of the refractive index of air, accounting for its height distribution. The designed technique can be applied in the project of continuously space monitoring the terrestrial middle atmosphere, to inspect specifically ozone depletion, using the light emitted by a reference star for probing the atmosphere. As the basic statistical parameter of a sounding light wave, it is convenient to utilize for the purpose, for example, the dispersion of amplitude fluctuations, being calculated from the stellar scintillations index measured by a light receiver aboard a satellite.

As a cosmogonic application of the developed theory, the focus was placed on the conditions for maintaining shear turbulence in the Keplerian accretion disk near a solar-type

star during the system's evolution, including the formation of a gas-dust subdisk in the vicinity of the disk's central plane, based on the model of a two-phase differentially rotating medium containing particles with different size distributions in different amounts. Special attention was paid to an heuristic way of modeling the turbulent exchange coefficients for accretion disks. The derived formulas for these coefficients, taking into account the inverse effects of dust transfer and potential temperature on the maintenance of shear turbulence, generalize the respective expression for the turbulent viscosity coefficient in gaseous α-disks to protoplanetary gas-dust disks.
Of special interest and importance are the concurrent processes of mass transfer and coagulation in the two-phase medium of a differentially rotating disk, where shear turbulence occurs in the presence of polydisperse solid particles suspended in a carrying flow. The mathematical aspects of the theory for the combined processes of mass transfer and coagulation, the mutual influence of turbulent mixing and coagulation kinetics, and the turbulent diffusion flow patterns accounting for the polydisperse nature of dust particles were addressed and thoroughly analyzed. The refined defining relations for diffuse fluxes of different particle sizes in the equations for turbulent diffusion in a gravitational field were derived, which describe their coagulation growth in addition to convective transfer, turbulent mixing, and sedimentation of disperse dust grains onto the central plane of the disk. To accommodate a more rigorous solution of the problem of the mutual influence of turbulent mixing and coagulation kinetics in forming the gas-dust subdisk, the plausible mechanisms of gravitational, turbulent and electric coagulation were explored. The parametric moments method for solving the Smoluchowski integro-differential coagulation equation for the different particle size distribution functions was developed and applied to the model.

Although the reduced examples by far not exhaust the results of simulations, they testify to the effectiveness of the developed approach to numerous applications in the study of various natural phenomena related to turbulent mixtures. Advancement of the theory of multicomponent reacting turbulent media opens perspectives for further extending models, quantitative evaluations of the processes involved, and thus improving our understanding of planetary aeronomy, planetary cosmogony, and space environment.

REFERENCES

Acuna, M.H., Connerney, J.E.P., Ness, N.F., Lin, R.P., Mitchel, D., Carlson, C.W., McFadden, J., Anderson, K.A., Reme, H., Mazellle, C., Vignes, D., Wasilewski, P., and Cloutier, P. (1999) Global distribution of crustal magnetization discovered by the Mars Global Surveyor MAG/ER experiment, *Science*, **284**, 794.

Akasofu, S.-I., and Chapman, S. (1972) *Solar-Terrestrial Physics*, Oxford University Press, London.

Akmaev, R.A., and Shved, G.M. (1978) Simulation of O and O_2 distribution in the lower thermosphere with the account for of turbulent intermixing and vertical motions, *Geomagnetism and Aeronomy*, **18**, 487.

Aleksandrov, A.P., Grechko, G.M., Gyrvich, A.S., Kan V., Manarov, M.Ch., Pachomov, A.I., Romanenko, Yu.V., Savchenko, S.A., Serova, S. I., and Titov, V.N. (1990) Spectra of temperature variations in the stratosphere based on the observations of stellar scintillations from space, *Reports of USSR Acad. Sci., Physics of Atmosphere and Ocean*, **26**, 5 (in russian).

Alfven, H. (1978) From dark interstellar clouds to planets and satellites, in: *Protostars and Planets* (ed. T.Gehrels), The University of Arizona Press, Tucson, Arizona, 533.

Alfven, H., and Arrhenius, G. (1976) *Evolution of the solar system*, Sci & Techn. Inf. Office, NASA, Washington, D.C.

Allison, M., and Lumetta, J.T. (1990) A simple inertial model for the Neptune zonal circulation, *Geophys. Res. Lett.*, **17,** 2269.

Andre, J.C., De Moor, G., Lacarrere, P., Therry, G., and Vacht, R. (1978) Modeling the 24 hour evolution of the mean and turbulent structures of the planetary boundary layer, *J. Atmos. Sci.*, **35**, 1861.

Andre J.C., De Moor, G., and Vachat, R. (1976) Turbulence approximation for inhomogeneous flow: Part II. The numerical simulation of a penetrative convection experiment, *J. Atmos. Sci.*, **33**, 482.

Andrews, D.G., Holton, J.R., and Leovy, C.B (1987) *Middle Atmosphere Dynamics*, Academic Press, Orlando.

Anfimov, N.A. (1962) Laminar boundary layer in a multicomponent mixture of gases, *Reports of USSR Acad. Sci., Mechanics and Engineering*, **1**, 25 (in russian).

Armitage, P.J., and Hansen, B.M.S.(1999) Early planet formation as a trigger for further planet formation, *Nature*, **402**, 633.

Artymowicz, P. (2000) Beta Pictoris and other solar systems, *Space Sci. Reviews*, **92**, 69.

Atkinson, D.H., Pollack, J.B., and Seiff, A. (1998) The Galileo probe Doppler wind experiment: Measurement of the deep zonal winds on Jupiter, *J. Geophys. Res.*, **103**, (E10), 22,911.

Atmosphere Handbook (1991) Hydrometeoizdat, Leningrad (in russian).

Atreya, S.K., Donahue, T.M., and Festou, M.C. (1981) Jupiter: Structure and composition of the upper atmosphere, *Astrophys J.*, **247**, L43.

Atreya, S.K., Festou, M.C., Donahue, T.M., Kerr, R.B., Barker, E.S., Cochran, W.D., Bertaux, J.L., and Upson, W.L. II. (1982) Copernicus measurement of Jovian Lyman–alpha emission and its aeronomical significance, *Astrophys J.*, **262**, 377.

Atreya, S.K., Waite, J.H., Jr., Donahue, T.M., Nagy, A.F., and McConnell, J.C. (1984)
Theory, measurements, and models of the upper atmosphere and ionosphere of Saturn, in: *Saturn* (eds. T. Gehrels and M.S. Matthews), The University of Arizona Press, Tucson, Arizona, 239.

Atreya, S.K., Sandel, B.R., and Romani, P.N. (1991) Photochemistry and vertical mixing in Uranus, in: *Uranus* (eds. J. Bergstrlah et al.), The University of Arizona Press, Tucson, Arizona, 110.

Avduevsky, V.S. (1962) A method of calculation of spatial turbulent boundary layer in the compressed gas, *Reports of USSR Acad. Sci., Mechanics and Mashine Building,* **4,** 3 (in russian).

Avduevsky, V.S., and Medvedev, K.I. (1968) Physical peculiarities of the flow in three-dimensional tear-off zones, in: *Heat and Mass transfer*, Part 1, Energia, Moskow, 140 (in russian).

Balbus, S.A., and Hawley, J.F. (1991) A powerful local shear instability in weakly magnetised disks: 1. Linear analysis, *Astrophys J.*, **376**, 214.

Balbus, S.A., and Hawley, J.F. (2000) Solar nebula magnetohydrodynamics, *Space Sci. Rev.*, **92**, 39.

Banks, P.M.. (1966) Charged particle temperatures and electron thermal conductivity in the upper atmosphere, *Ann.Geophys*, **22**, 577.

Banks, P.M. (1966) Collision frequencies and energy transfer, *Planet. Space Sci.*, **14**, 1085.

Banks, P.M., and Holzer, T.E. (1969) Features of plasma transport in the upper atmosphere, *J. Geophys. Res.* **74**, 6304.

Banks, P.M., and Kockarts, G. (1973) *Aeronomy*, Parts A and B, Academic press, New York-London.

Baranov, V.B., and Krasnobaev, K.V. (1977) *The Hydrodynamic Theory of Space Plasma*, Nauka, Moskow (in russian).

Barenblatt, G.I. (1955) On the motion of suspended particles in turbulent flow occupying a half-space or a flat open channel of finite depth, *Applied Mathematics and Mechanics*, **19**, 61 (in russian).

Barenblatt, G.I. (1978) *Similarity, Self-Similarity, Intermediate Asymptotics (The Theory and Applications to Geophysical Hydrodynamics)*, Hidrometeoizdat, Leningrad (in russian).

Barenblatt, G.I., and Golitsyn, G.S. (1973) *A Local Structure of the Developed Dust Storms*, Moscow State Univ., Moscow (in russian).

Barth, C.A., Stewart, A.I.F., Bougher, S.W., Hunten, D.M., Bauer, S.J., and Nagy, A.F. (1992) Aeronomy of the current Martian atmosphere, in: *Mars* (eds. H.H. Kieffer et al.), The University of Arizona Press, Tucson and London, 1054.

Batchelor, G.K. (1953) *The Theory of Homogeneous Turbulence*, Cambridge University Press, Cambridge.

Batchelor, G.K. (1970) *An Introduction to Fluid Dynamics*, Cambridge Univ. Press, Cambridge.

Bauer S.J. (1973) *Physics of Planetary Ionospheres*, Springer-Verlag, Berlin-Heidelberg-New York.

Bazhinov, I.K., Butsko, P.A., and Volkov, I.I. (1985) GOST–84. Upper atmosphere of the Earth. Model of density for ballistic flights maintenance of artificial Earth satellites, *GOST 25645. 115–84.* Sandards, Moskow (in russian).

Belotserkovskii, O.M. (1985) Direct numerical modeling of free developed turbulence, *J. Comp. Mathem. and Math. Phys*., **25**, 1856 (in russian).

Belotserkovskii, O.M. (1994) *Numerical Modeling in the Continuum Mechanics*, 2nd ed., Nauka, Moscow (in russian).

Belotserkovskii, O.M (1997) *Numerical Experiment in Turbulence: From Order to Chaos*, Nauka, Moscow (in russian).

Bertaux, J.L., Chassefiere, E., Dalaudier, F., Godin, S., Goutail, F., Hauchecorne, A., Le Texier, H., Megie, G., Pommereau, G., Simon, P., Brasseur, G., Pellinen, R., Kyrola, E.,Tuomi, T., Korpela, S., Leppelmeir, G., Visconti, G., Fabian, P., Isaksen, S.A., Larsen, S.H., Stordahl, F., Carriolle, D., Lenoble, J., Naudet, J.P.,and Scott, N.. (1988) GOMOS – Global Ozone Monitoring by the Occultation of Stars. *Proposal in Response to ESA EPOP 1. A.O. Service d'Aeronomie du CNRS*, Verrieres le Buisson.

Bethe, H.A. (1982) The theory of supernovae, in: *Essays in Nuclear Astrophysics* (eds. C.A. Barnes et al.), Cambridge University Press, Cambridge-London-NewYork.

Bikova, L.P. (1973) An experience of calculation of the atmospheric boundary layer characteristics from the given parameters of roughness sublayer, *The Main Geophys. Obs. Proceedings*, **297**, 12 (in russian).

Bishop, J., Atreya, S.K., Romani, P.N., Orton, G.S., Sandel, B.R., Yelle, R.V. (1995) The middle and upper atmosphere of Neptune, in: *Neptune and Triton* (ed. D. Cruikshank), The University of Arizona Press, Tucson, Arizona, 427.

Bisikalo, D.V, Boyarchuk, A.A, Kuznetsov, O.V., and Chechetkin, V.M. (1997) Mass transfer in binary stars, *J. of Journals (Review of Global Scientific Achievements)*, **1**, 12.

Blackadar, A.K. (1955) Extension of the laws of thermodynamics to turbulent system, *J. Meteorology*, **12**, 165.

Bobuleva, I.M., Zilitinkevich, S.S., and Laychtman, D.L. (1967) Turbulent mode in a stratified boundary layer of the atmosphere, in: *Atmospheric Turbulence and Radio Propagation*, Nauka, Moskow, 179 (in russian).

Bobuleva, I.M. (1970) Calculation of the turbulence characteristics in the planetary boundary layer of the atmosphere, *LSMI Proceedings*, **40**, 3, Leningrad (in russian).

Borgi, R. (1980) Models for numerical calculations of turbulent combustion, in: *Prediction Methods for Turbulent flows*, Hemisphere Publ. Corp., 329

Boss, A.P., and Vanhala, A.T. (2000) Triggering protostellar collapse, injection, and disk formation, *Space Sci. Rev.* **92**, 13.

Bougher, S.W., Dickinson, R.E., Ridley, E.G., and Roble, R.G. (1988) Venus mesosphere and thermosphere. III. Three dimensional general circulation with coupled dynamics and composition, *Icarus*, **73**, 545.

Bougher, S.W., Alexander, M.J., and Mayr, H.G. (1997) Upper atmosphere dynamics: Global circulation and gravity waves, in: *Venus II, Geology, Geophysics, Atmosphere, and Solar Wind Envinroment* (eds. S.W. Bougher et al.), The University of Arizona Press, Tucson, 259.

Boussinesque J. (1977) *Essai sur la Theoriedes Eaux Courantes. Memoires Presentees par Diverses Savants a l'Acad. d. Sci.*, **23**, 46, Paris.

Brace, L.H., Mayr, H.G., and Carignan, G.R. (1969) Measurements of electron cooling rates in the midlatitude and auroral zone thermosphere, *J. Geophys. Res.*, **74**, 257.

Brace, L.H., Mayr, H.G., and Findlay, J.A. (1969) Electron measurements bearing on the energy and particle balance of the upper F region, *J. Geophys. Res.*, **74**, 2952.

Bradbury, L.J.S. (1965) The structure of a self-preserving turbulent plane jet, *J. Fluid Mech.*, **23**, Pt.1, 31

Bradshaw, P. (1969) The analogy between streamline curvature and buoancy in turbulent shear flow, *J. Fluid Mech.*, **36**, 117.

Bradshaw, P. (1971) *An Introduction to Turbulence and Its Measurement*, Pergamon Press, New York.

Brasseur, G.B., and Solomon, S. (1984) *Aeronomy of the Middle Atmosphere (Chemistry and Physics of the Stratosphere and Mesosphere)*, D. Reidel Publishing Company, Dordrecht.

Briggs, G.A., and Leovy, C.B. (1974) Mariner 9 observations of the Mars North polar hood, *Bull. Amer. Meteorol. Soc.*, **55**, 278.

Brinkmann, W, Fabian, A.C., and Giovanelli, eds. (1990) *Physical Processes in Hot Cosmic Plasma*, Kluwer Academic Publishers, Dordrecht.

Bruyatsky, E.V. (1986) *The Turbulent Stratified Stream Flows*. Naukova Dumka, Kiev (in russian).

Bryunelli, B.E., and Namgaladze, A.A. (1988) *Physics of the Ionosphere*, Nauka, Moskow.

Busse, F.H. (1970) Differential rotation in stellar convection zones, *Astrophys. J.*, **159**, 629.

Busse, F.H. (1976) A simple model of convection in the Jovian atmosphere, *Icarus*, **29**, 255.

Butler, R. P., and Marcy, G.W. (1997) The Lick Observatory planet search, in: *Astronomical and Biochemical Origins and the Search for Life in the Universe* (eds. C.B. Cosmovici et al.), Editrice Compositori, Bologna, 331.

Cabot, W., Canuto, V.M., Hubickyi, O., and Pollack, J.B. (1987) The role of turbulent convection in the primitive solar nebula, *Icarus*, **69**, 387.

Cantwell, B. J. Organized motion in turbulent flow (1981) *Ann. Rev. Fluid Mech.*, 13, 457.

Carlson, R.W., Baines, K.H., Encrenaz, T., Drossart, P., Roos–Serote, M., Taylor, F.M., lrvin, P., Weir, A., Smith, P., and Calcutt, S.. (1998) Near–IR spectroscopy of the atmosphere of Jupiter, in: *Highlights of Astronomy,* v.11B *(*ed. J.Andersen*)*; *SPSI: The Galileo Mission to the Jupiter System* (eds. M.Ya. Marov and R.W. Carlson), Kluwer Acad. Publ., Dordrecht, 1050.

Cassen, P. (1994) Utilitarian models of the solar nebula, *Icarus*, 112, 405.

Cassen, P., and Woolum, D.S. (1999) The origin of the solar system, in: *Encyclopedia of the Solar System* (eds. P.R. Weissman, L.-A. McFadden, and T.V. Johnson), Academic Press, San Diego-London-Boston-New York-Sydney-Tokyo-Toronto.

Chandra, S., and Sinha, A.K. (1974) The role of eddy turbulence in the development of self–consistent models of the lower and upper thermospheres, *J. Geoph. Res.*, **79**, 1916.

Chamberlain, J.W., and Hunten, D.M. (1987) *Theory of Planetary Atmospheres. An Introduction to Their Physics and Chemistry,* 2nd ed., Academic Press, London.

Chapman, S., and Cowling, T.G. (1952) *The Mathematical Theory of Non-uniform Gases*, Cambridge University Press, Cambridge.

Chapman, D.P. (1980) Computing aerodynamics and the perspectives of its development: The Driden lecture, *Rocketry and Astronautics*, **18**, 3.

Charney, J.G., and Stern, M.E. (1962) On the stability of internal baroclinic jets in a rotating atmosphere, *J. Atmos. Sci.,* **19**, 159.

Chassefiere, E., Roseqvist, J., and Theodore, B. (1994) Ozone as a tracer of turbulence induced by breaking gravity waves on Mars, *Planet. Space Sci.*, **42,** 825.

Chemical Kinetics and Photochemical Data for Use in Stratospheric Modeling (1987) Jet Propulsion Laboratory Publ. No. 87–41, JPL., Pasadena, California.

Chorin, A.J. (1994) *Vorticity and Turbulence*, Springer, Berlin.

Clayton, D.D., Cox, D.P., and Michel, F.C. (1986) A local recent supernova: Evidence from X rays, ^{26}AI radioactivity and cosmic rays, in: *The Galaxy and the Solar System* (eds. R. Smoluchowski et al.), The University of Arizona Press, Tucson, Arizona, 129.

Collegrove, F.D., Hanson, W.B., and Johnson, F.S. (1965) Eddy diffusion and oxygen transport in the lower thermosphere, *J.Geophys.Res.*, **70**, 4931.

Collegrove, F.D., Johnson, F.S., and Hanson, W.B. (1966) Atmospheric composition in the lower thermosphere, *J.Geophys. Res.*, **71**, 2227.

COSPAR International Reference Atmosphere, CIRA–72 (1972), Acad. Verlag, Berlin.

Curtiss, C.F. (1968) Symmetric gaseous diffusion coefficients, *J. Chem. Phys.*, **49**, 2917.

Cuzzi, J.N., Dobrovolskis, A.R., and Champney, J.M. (1993) Particle–gas dynamics in the midplane of a protoplanetary nebula, *Icarus*, **106**, 102.

Danilov, A.D., and Kalgin, Yu.A. (1992) Seasonal – latitudinal variations of the factor of eddy diffusion in the lower thermosphere and mesosphere, *Geomagnetism and Aeronomy*, **32**, 69 (in russian).

Davis, M.H., and Sartor, T.D. (1967) Theoretical collision efficiencies for small cloud droplets in Stokes flow, *Nature*, **215**, 1371.

Davydov, B.I. (1959) On statistical dynamics of incompressible turbulent fluid, *USSR Acad. Sci. Dokladi*, **127**, 768 (in russian).

Davydov, B.I. (1959) On statistical turbulence, *Reports of USSR Acad. Sci.*, **127**, 980 (in russian).

Davydov, B.I. (1961) On statistical dynamics of incompressible turbulent fluid, *USSR Acad. Sci. Dokladi*, **136**, 47 (in russian).

Deardorff, J.W. (1973) The use of subgrid transport equations in a three–dimensional model of atmospheric turbulence, *J. Fluids Eng.*, **95**, 429.

De Groot S.R., Mazur P. (1962) *Non-equilibrium Thermodynamics*. North-Holland Publishing Company, Amsterdam.

Devoto, R.S. (1966) Transport properties of ionized monotomik gases, *Phys. Fluids*, **9**, 1230.

De Wolf, D. A. (1973) Strong rradiance fluctuations in turbulent air: plane waves, *J. Opt. Soc. Amer.*, **63**, 171.

Dickinson, R.E., Ridley, E.C., and Roble, R.G. (1984) Thermospheric general circulation with coupled dynamics and composition, *J. Atmos. Sci.*, **43**, 205.

Dominik, C., and Tielens, A.G.M. (1997) The physics of dust coagulation and the structure of dust aggregates in space, *Astrophys. J.*, **480**, 647.

Donaldson, S.R (1972) The calculation of turbulent flows in an atmosphere and isolated vortex, *Rocketry and Cosmonautics*, **10**, 4 (in russian).

Doroshkevich, A.G., Zeldovich, Ya.B., and Sunyaev, R.A. (1976) The adiabatic theory of galaxies formation, in: *Genesis and Evolution of Galaxies and Stars* (ed. S.B. Pickelner), Nauka, Moscow, 65 (in russian).

Dyarmati, I. (1970) *Non-Equilibrium Thermodynamics. Field Theory and Variational Principles*, Springer-Verlag, Berlin-Heidelberg-New York.

Dulbrulle, B (1993) Differential rotation as a source of angular momentum transfer in the solar nebula, *Icarus*, **106**, 59.

Eardley, D.M., Lightman, A.P., Payne, D.G., and Shapiro, S.L. (1978) Accretion disks around massive black holes: Persistent emission spectra, *Astrophys. J.*, **234**, 187.

Eneev, T.M., and Kozlov, N.N. (1981a) Model of the accumalation process for forming planetary systems, 1: Numerical experiments, *Solar System Research*, **15**, 80 (in russian).

Eneev, T.M., and Kozlov, N.N. (1981b) Model of the accumalation process for forming planetary systems, 2: Rotation of planets and relationship of the model with the theory of gravitational instability, *Solar System Research*, **15**, 131 (in russian).

Encrenaz, Th., Feuchgruber, H., Atreya, S.K., Bezard, B., Lellouch, E., Bishop, J., Edgington, S., de Graauw, Th., Griffin, M., and Kessler, F. (1998) ISO Observations of Uranus: The stratospheric distribution of C_2H_2 and the eddy diffusion coefficient, *Astron. Astrophys.*, **333**, L43.

Esposito, L.W., Knollenberg, R.G., Marov, M.Ya., Toon, O.B., and Turco, R.P. (1983) The clouds and hazes of Venus, in: *Venus* (ed. D.M. Hunten et al.), The University of Arizona Press, Tucson, Arizona, 484.

Favre, A. (1969) Statistical equations of turbulent gases, in: *Problems of Hydrodynamics and Continuum Mechanics*, SIAM, Philadelphia, 231.

Fels, S.B., and Lindzen, R.S. (1974) The interaction of thermally excited gravity waves with mean flows, *Geophys. Fluid Dyn.*, **6**, 149.

Ferziger, J.H., and Kaper, H.G. (1972) *Mathematical Theory of Transport Processes in Gases*. North-Holland Publishng Company, Amsterdam-London.

Fesenkov, V.G. (1973) On probable genesis of carbonaceous chondrites, *Meteoritics*, **32**, 3 (in russian).

Frisch, U. (1995). *Turbulence. The Legacy of A.N. Kolmogorov*. Cambridge Univ. Press, Cambridge.

Fuks, N.A. (1955) *Mechanics of Aerosols*, USSR Acad. Sci., Moscow.

Gavrilov, N.M. (1974) Thermal effect of internal gravity waves in the upper atmosphere, *Reports of USSR Acad. Sci., Physics of Atmosphere and Ocean*, **10** , 83 (in russian).

Gavrilov, N.M., and Shved, G.M. (1975) On the closure of equation system for turbulized layer of the upper atmosphere, *Ann. Geophys*., **31**, 375.

Geophysical Institute University of Alaska Biennial Report 1983/1984, Publ. No. **99775–0800**, Fairbanks, Alaska.

Gibson C.H.., Launder D.E.. (1976) On the calculation of free horizontal turbulent shear flows under influence of free convection, *Heat Transfer*, series C, **98**, 86 (in russian).

Gilman, P.A. (1977) Nonlinear dynamics of Boussinesque convection in a deep rotation spherical shell, *Geophys. Astrophys. Fluid Dyn*., **8**, 93.

Gilman, P.A. (1979) Model calculations concerning rotation at high solar latitudes and the depth of solar convection zone, *Astrophys. J*., **231**, 284.

Ginzburg, E.I., and Kuzin, G.I. (1981) Effect of swelling of short internal gravity waves as a possible reason of mesosphere and lower thermosphere turbulization. *Geomagnetism and Aeronomy*, 21, 489 (in russian).

Gledzer, E.B., Dolzhansky, F.V., and Obuhov, A.M. (1981) *Systems of Hydrodynamic Type and Their Application*, Nauka, Moskow (in russian).

Golitsyn, G. S. (1970) A similarity approach to the general circulation of planetary atmospheres, *Icarus*, **13**, 1.

Golitsyn, G. S. (1973) *Introduction to the Dynamics of the Planetary Atmospheres*, Hydrometeoizdat, Leningrad (in russian).

Golitsyn, G.S. (1997) Principle of the fastest response in hydrodynamics, geophysics, and astrophysics, *Russian Acad. Sci.Dokladi*, **356**, 321, (in russian).

Gordiets, B.V., Kulikov, Yu.N., Markov, M.N., and Marov, M.Ya. (1979) Idealized model of the Earth's thermosphere accounted for infrared emission, *P.N. Lebedev Physical Institute preprint № 112*, Moscow (in russian).

Gordiets, B.V., and Kulikov, Yu.N. (1981) Influence of turbulence and IR-radiation on thermal regime of the Earth's thermosphere, *Space Research*, **19**, 539 (in russian).

Gordiets, B.V., Kulikov, Yu.N., Markov, M.N., and Marov, M.Ya. (1982a) Numerical modeling of gas heating and cooling in the near-Earth space, in: Infrared Spectroscopy of Space Matter and Properties of Space Medium, *P.N. Lebedev Physical Institute Proceedings*, **130**, 3, Moscow (in russian).

Gordiets, B.V., and Kulikov, Yu.N. (1982) On the role of turbulence and infrared radiation in heat balance of the lower thermosphere, *Ibid*, 29 (in russian).

Gordiets, B.V., Kulikov, Yu.N., Markov, M.N., and Marov, M.Ya. (1982b), Numerical modeling of the thermospheric heat budget, *J. Geophys. Res*., **87** (A6), 4504.

Gor'kavyi, N.N., and Fridman, A.M. (1994) *Physics of the Planetary Rings*, Nauka, Moscow (in russian).

Grad, H. (1949) On the kinetic theory of rarefied gases, *Comm. Pure Appl. Math*., **2**, 331.

Grad, H. (1963) Asymptotic theory of the Boltzmann equation, *Phys. Fluids*, **6**, 147.

Grafov, B.M., Martem'yanov, S.A., and Nekrasov, L.N. (1990) *Turbulent Diffusive Layer in the Electrochemical Systems*, Nauka, Moscow (in russian).

Grechko, G.M., Gurvich, A.S., and Romanenko, Yu.V. (1980) Structure of density inhomogeneities in the atmosphere based on observations from the orbital station Salyut 6, *Reports of USSR Acad. Sci., Physics of Atmosphere and Ocean*, **16**, 339 (in russian).

Grechko, G.M., Gurvich, A.S., and Romanenko, Yu.V. (1981) Layered structure of temperature field in the atmosphere based on the refraction measurements from the orbital station Salyut 6. *Reports of USSR Acad. Sci., Physics of Atmosphere and Ocean*, **17**, 115 (in russian).

Grechko, G.M., Gurvich, A.S., and Dzhanibekov, B.A. (1989) Investigation of density and temperature variations in the stratosphere based on observations of stellar scintillations from space, *Research of the Earth from Space*, **4**, 22 (in russian).

Guillot, T. (1999) Interiors of giant planets inside and outside the solar system, *Science* **286**, 72.

Gurvich, A.S. (1968) Definition of the turbulence characteristics from experiments on light propagation, *Reports of USSR Acad. Sci., Physics of Atmosphere and Ocean*, **4**, 160 (in russian).

Gurvich, A.S., Kon, A.I, Mironov, V.L., and Chmelevtsov, S.S. (1976) *Laser Radiation in the Turbulent Atmosphere,* Nauka, Moscow (in russian).

Gurvich, A.S., Zacharov, I. A., Kan, V., Lebedev, V.V., Nesterenko A.A., Neuzhil L., Paxomov A.I., Savchenko S.A. (1985) Stellar scintillations based on observations from the orbital station Salyut 7, *Reports of USSR Acad. Sci., Physics of Atmosphere and Ocean*, **21**, 1235 (in russian).

Hahn, G. and Shapiro, S. (1967). *Statistical Models in Engineering*, John-Wiley & Son, New York.

Haken, H. (1978) *Synergetics*, Springer-Verlag, Berlin-Heidelberg-New York.

Handbook of Turbulence (eds. W. Frost and T.H. Moulden) (1977), Plenum Press, New York and London.

Harris, I., and Priester, W. (1962) Time dependent structure of the upper atmosphere, *J. Atmos. Sci.*, **19**, 286.

Hines, C.O. (1965) Dynamical heating of the upper atmosphere, *J. Geophys. Res.*, **70**, 177.

Hinze, J.O. (1959) *Turbulence*, McGraw-Hill, New York.

Hirschfelder, J.O., Curtiss, Ch. F., and Bird, R.B. (1954) *Molecular Theory of Gases and Liquids*, John Wiley and Sons, New York and London.

Hedin, A.E. (1987) MSIS-86 thermospheric model, *J. Geophys. Res.*, **92**, 46.

Hedin, A.E. (1988) CIRA 1986, COSPAR International Refrence Atmosphere. Part I, Chapter 1: Thermosphere Models, in: *Adv. Space Res.* (ed. D. Ress), **8**, No. 5.

Hedin, A.E. (1991) Extension of the MSIS thermosphere model into the middle and lower atmosphere, *J. Geophys. Res.*, **96** (A2), 1159.

Hocking, W.K. (1990) CIRA 1986. Part II, Chapter 5: Turbulence in the region 80–120 km, in: *Adv. Space Res.* (eds. D. Rees et al.), **10**, 153.

Holton, J.R. (1982) The role of gravity wave induced drag and diffusion in the momentum budget of the mesosphere, *J. Atmos. Sci.*, **40**, 2497.

Hunten, D.M. (1974) Energetics of thermospheric eddy transport, *J. Geophys. Res.*, 79, 2533.

Hunten, D.M. (1975) Vertical transport in atmospheres, in: *Atmospheres of Earth and Planets* (ed. B.M. McCormac), D. Reidel, Dordrecht, 59.

Hunten, D.M. (1982) Thermal and non–thermal escape mechanisms for terrestrial bodies, *Planet. Space Sci.*, **30**, 773.

Hunten, D.M., Pepin, R.O., and Owen, T.C. (1988) Planetary atmospheres, in: *Meteorites and the Early Solar System* (eds. J.F. Kerridge and M.S. Matthews), The University of Arizona Press, Tucson, Arizona, 565.

Hunten, D.M., Tomasko, M.G., Flasar, F.M., Samuelson, R.E., Strobel, D.F., and Stevenson, D.J. (1984) Titan, in: *Saturn* (eds. T. Gehrels and M.S. Matthews), The University of Arizona Press, Tucson, Arizona, 671.

Ievlev, V.M. (1971) Methods of calculation of the turbulent boundary layer for high–temperature gaseous flow, *9th Aerospace Sci. Meeting, AIAA Paper No. 71–165*, New York.

Ievlev V.M. (1975) *The Turbulent Motion of High-Temperature Continuum*, Nauka, Moscow (in russian).

Ievlev V.M. (1990) *Numerical Modeling of Turbulent Flows*, Nauka, Moscow (in russian).

Ingersoll, A.P. (1981) Jupiter and Saturn, in: *The New Solar System* (eds. J.K. Beatty et al.), Cambridge Univ. Press, Cambridge, 117.

Ingersoll, A.P., and Pollard, D. (1982) Motion in the interiors and atmospheres of Jupiter and Saturn: Scale analysis, an elastic equations, barotropic stability criterion, *Icarus*, **52**, 62.

Ingersoll, A.P. (1990) Atmospheric dynamics of the outer planets, *Science*, **248**, 308.

Ingersoll, A.P., Beebe, R.F., Conrath, B.J., and Hunt, G.E. (1984) Structure and dynamics of Saturn's atmosphere, in: *Saturn*, (eds. T. Gehrels and M.S. Matthews), The University of Arizona Press, Tucson, Arizona, 195.

Ingersoll, A.P., Barnet, C.D., Beebe, R.F., Flasar, P.M., Hinson, D.P., Limaye, S.S., Sromovsky, LA., and Suomi, V.E. (1995) Dynamic meteorology of Neptune, in: *Neptune and Triton* (ed. D. Cruikshank), The University of Arizona Press, Tucson, Arizona, 613.

Ingersoll, A.P., Vasavada, A.R. and The Galileo Imaging Team (1998) Dynamics of Jupiter's atmosphere, in: *Highlights of Astronomy,* v.11B *(*ed. J.Andersen*)*; *SPSI: The Galileo Mission to the Jupiter System* (eds. M.Ya. Marov and R.W. Carlson), Kluwer Acad. Publ., Dordrecht, 1042.

Ishimaru, A. (1978) *Wave Propagation and Scattering in Random Media. Vol. 2. Multiple Scattering, Turbulence, Rough Surfaces and Remote Sensing,* Academic Press, New York-London.

Ivanov, M.N. (1967) On the structure of the neutral upper atmosphere, *Space Sci. Rev.*, **7**, 579 (in russian).

Ivanovsky, A.I., Repnev, A.I., and Shvidkovsky, E.T. (1967) *The Kinetic Theory of the Upper Atmosphere,* Hydrometeoizdat, Leningrad (in russian).

Izakov, M.N., Morozov, S.K., and Shnol, E.E. (1972) Idealized model of diurnal variations of temperature, density and winds in the Earth's equatorial thermosphere during the equinox, *Space Research Inst. Preprint*, Moscow (in russian).

Izakov, M.N. (1977) Estimation of turbulent mixing factor and hydropause altitudes on Venus, Mars and Jupiter, *Space Research*, **15,** 248 (in russian).

Izakov, M.N. (1978) Influence of turbulence on thermal regime of planetary thermospheres, *Space Research*, **16**, 403 (in russian).

Izakov, M.N., and Marov, M.Ya., (1989) The upper atmosphere of Venus, in: *The Planet Venus* (ed. V.L. Barsukov), Nauka, Moscow, 114 (in russian).

Izrael, Yu. A. (1979) *Ecology and Control of the Environment Condition*, Hydrometeoizdat, Leningrad (in russian).

Jeans, J.H. (1969) *Astronomy and Cosmology*, Cambridge Univ. Press, London and New York.

Johnson, F.S., and Gotlieb, B. (1970) Eddy mixing and circulation at ionospheric levels, *Planet. Space Sci.*, **18**, 1707.

Johnson, F.S. (1975) Transport processes in the upper atmosphere, *J.Atmos. Sci.* **32**, 1658.

Justus, C.G. (1967) The eddy diffusivity, energy balance parameters and heating rate of upper atmospheric turbulence, *J. Geoph. Res.*, **72**, 1035.

Justus, C.G., and Roper, R.G. (1968) Some observations of turbulence in the 80 to 110 km region of the upper atmosphere, *Meteorol. Monographs*, **9**, 122.

Justus, C.G. (1969) Dissipation and diffusion by turbulence and irregular winds near 100 km, *J. Atmos. Sci.*, **26**, 1137.

Kalledrode, M.-B. (1998) *An Introduction to Plasmas and Particles in the Heliosphere and the Magnetosphere*, Springer-Verlag, Berlin.

Kampe de Feriet, J. (1959) Statistical mechanics and theoretical models of diffusion processes, in: *Atmospheric Diffusion and Air Pollution* (eds. F.N. Frenkiel and P.A. Sheppard), Academic Press, New York and London, 134.

Karman, T. (1953) Introductory notes on the turbulence problem, in: *Space Aerodynamics Problems,* Inostr. Lit., Moscow.

Kasprzak, W.T., Keating, G.M., Hsu, N.C., Stewart, A.I.F., Colwell, W.B., and Bougher, S.W. (1997). Solar activity behavior of the thermosphere. In: *Venus II* (eds. S.W. Bougher, D.M. Hunten, and R.J. Phillips), The University of Arizona Press, Arizona, Tucson, p. 225.

Keller, L.V., and Friedman, A.A. (1924) Differentialgleichungen fur die turbulente Bewegung einer kompressiblen Flussigkeit, in: *Proc. I Intern. Congress Appl. Mech.*, Delft., 395.

Kerzhanovich, V.V., and Marov, M.Ya. (1983) The atmospheric dynamics of Venus according to Doppler measurements by the Venera entry probes, in: *Venus* (eds. D.M. Hunten et al.), The University of Arizona Press, Tucson, Arizona, 766.

Klimontovich, Yu. L. (1986) *Statistical Physics*, Harwood Academic Publisher, New York.

Klimontovich Yu. L. (1991) *Turbulent Motion and Structure of Chaos*, Kluwer Academic Publisher, Dordrecht.

Kolesnichenko, A.V. (1978) Kinetic energy of diffusion in thermodynamic simulation of multicomponent continuum, *M.V. Keldysh Inst. Applied Math. Preprint No. 84*, Moscow (in russian).

Kolesnichenko, A.V. (1978) Basic hydrothermodynamics equations for multiphase multicomponent continuum with chemical reactions: Phenomenological approach, *M.V. Keldysh Inst. Applied Math. Preprint No. 125*, Moscow (in russian).

Kolesnichenko, AV. (1979) Stefan-Maxwell relations and heat flow in the highest approximations of the transfer coefficients for partially ionized gas mixtures, *M.V. Keldysh Inst. Applied Math. Preprint No. 66*, Moscow (in russian).

Kolesnichenko, AV. (1980) Methods of a non-equilibrium thermodynamics for description of the turbulent multicomponent hydrothermodynamic systems with chemical reactions, *M.V. Keldysh Inst. Applied Math. Preprint No. 66,* Moscow (in russian).

Kolesnichenko, A.V. (1981) Methods of continuum mechanics for description of the turbulent multicomponent mixtures with chemical reactions and the processes of heat-and-mass transfer, *Proceedings of the 5th USSR Congresses on Theoretical and Applied Mechanics*, Alma-Ata, 123 (in russian).

Kolesnichenko, A.V. (1982) Stefan-Maxwell relations and heat flow in the highest approximations of transfer coefficients for the multicomponent ionized gas mixtures in the magnetic field, *M.V. Keldysh Inst. Applied Math. Preprint No. 14,* Moscow (in russian).

Kolesnichenko, A.V. (1994) On the macroscopic theory of the processes of diffusion transfer in gases, *M.V. Keldysh Inst. Applied Math. Preprint No. 42,* Moscow (in russian).
Kolesnichenko, A.V. (1995) On the theory of turbulence in the planetary atmospheres: Numerical modeling of structural parameters, *Solar System Research*, **29**, 114.
Kolesnichenko, A.V. (1998) Stefan-Maxwell relationships and heat flux for turbulent multicomponent continuous media, in: *Problems of Contemporary Mechanics,* Moscow State Univ., Moscow, 52.
Kolesnichenko, A.V. (2000) Modeling turbulent transfer coefficients for gas-dust accretion disks, *Solar System Research*, **34**, 469.
Kolesnichenko, A.V., and Krasitsky, O.P. (1994) Diffusion in the thermosphere, *Solar System Research*, **28**, 125 .
Kolesnichenko, A.V., and Marov, M.Ya. (1979) Turbulence impact on the structure and energetics of the lower thermosphere of a planet, *M.V. Keldysh Inst. Applied Math. Preprint No. 175,* Moscow (in russian).
Kolesnichenko, A.V., and Marov, M.Ya. (1980) Simulation of turbulent transfer coefficients with the aeronomy applications, *M.V. Keldysh Inst. Applied Math. Preprint No. 20,* Moscow (in russian).
Kolesnichenko, A.V., and Marov, M.Ya. (1982) On the highest approximations to the diffusion coefficients of neutral and charged particles mixture in the magnetic field, *M.V. Keldysh Inst. Applied Math. Preprint No. 142,* Moscow (in russian).
Kolesnichenko, A.V., and Marov, M.Ya. (1984) Simulation of the second order of the factors of turbulent exchange for shear flows of the multicomponent compressed media, *M.V. Keldysh Inst. Applied Math. Preprint No. 31,* Moscow (in russian).
Kolesnichenko, A.V., and Marov, M.Ya. (1984) To the problem of closing in the theory of turbulent shift flows of multicomponent mixtures reactive of gases. *M.V. Keldysh Inst. Applied Math. Preprint No. 31,* Moscow (in russian).
Kolesnichenko, A.V., and Marov, M.Ya. (1985) Methods of non–equilibrium thermodynamics for description of the multicomponent turbulent gas mixtures, *Arch. Mech.*, **37**, No. 1/2, 3 , Warsaw.
Kolesnichenko, A.V., and Marov, M.Ya. (1994) On simulation of the parameters of multicomponent turbulence in the free atmosphere based on the optical measurements of fluctuations of the index of air refraction, *M.V. Keldysh Inst. Applied Math. Preprint No. 41,* Moscow (in russian).
Kolesnichenko, A.V., and Marov, M.Ya. (1996) On simulation of the parameters of multicomponent turbulence in the free atmosphere based on the optical measurements of fluctuations of the index of air refraction, *Space Research)*, **34**, 36 (in russian).
Kolesnichenko, A.V., and Tirsky, G.A. (1976) Stefan-Maxwell relations and heat flow for imperfect multicomponent continuum, *The Numerical Methods of Continuum Mechanics*, **7**, 106 (in russian).
Kolesnichenko, A.V., and Tirsky, G.A (1979) Thermodynamic analysis of partially ionized mixtures of imperfect gases flows under incomplete chemical equilibrium, *M.V. Keldysh Inst. Applied Math. Preprint No. 77,* Moscow (in russian).
Kolesnichenko, A.V., and Vasin, V.G. (1984) Simulation of the transfer coefficients for the turbulent flows of multicomponent gas mixtures in the Earth's upper atmosphere, *Reports of USSR Acad. Sci., Physics of Atmosphere and Ocean*, **20**, 683 (in russian).
Kolmogorov, A.N. (1941) Local structure of turbulence in the incompressible fluid at very large Reynolds numbers, *USSR Acad. Sci.Dokladi*, **30**, 299 (in russian).
Kolmogorov, A.N. (1942) Equations of turbulent motion of the incompressible fluid, *Reports of USSR Acad. Sci., Physics*, **6**, 56 (in russian).
Kolmogorov, A.N. (1954) On the new version of the gravitational theory of suspended sediments, *Moscow State Univ. Herald*, **3**, 41 (in russian).
Komponiets, V.Z., Polak, L.S, and Epshtein, I.L. (1977) Methods of mathematical modeling of turbulent flows, in: *Plasma Chemical Reactions and Processes*, Nauka, Moscow, 135 (in russian).
Komponiets, V.Z., Ovsyannikov, A.A., and Polak, L.S, (1979) *Chemical Reactions in Gas and Plasma Turbulent Flows*, Nauka, Moscow (in russian).
Koshelev, V.V. (1976) Diurnal and seasonal variations of oxygen, hydrogen and nitrogen components at heights of mesosphere and lower thermosphere, *J. Atmos. Terr. Phys.*, **38**, 991.
Koshelev, V.V., Klimov, N.N., and Sutirin, N.A. (1983) *Aeronomy of Mesosphere and Lower Thermosphere*, Nauka, Moscow (in russian).
Krasnopolsky, V.A. (1987). *Physics of Luminiscence of the Atmospheres of Planets and Comets,* Nauka, Moscow (in russian).

Krasnopolsky, V.A., and Parshev, V.A. (1983) Photochemistry of the Venus atmosphere, in: *Venus* (eds. D.M. Hunten et al.), The University of Arizona Press, Tucson, Arizona, 431.

Krymskii, A.M., Breus, T.K., Ness, N.F., and Acuna, M.H. (2000) The IMF pile up regions near the Earth and Venus: Lessons for the solar wind-Mars interaction, *Space Sci. Rev.*, **92**, 535.

Kulikov, Yu.N., and Pavlyukov, Yu.B. (1987) Simulation of diurnal variations of odd nitrogen in mid-latitude thermosphere of the Earth, in: *Mathematical Problems of the Applied Aeronomy*, (ed. M.Ya. Marov), Proceedings of M.V. Keldysh Inst. Applied Math., Moscow, 77 (in russian).

Kumantsev, M.A., Ovsyannikov, A.A., and Polak, L.S. (1974) Investigation of mixing of plasma jet with a cold gas at the presence of chemical reaction, in: *Fundamental Theory of Chemical Technology,* **8**, 872 (in russian).

Kuznetsov, V.P. (1969) Influence of temperature and density fluctuations on a chemical reaction mean rate in the turbulent flow, *Proceedings of the Second USSR Symposium on Combustion and Detonation,* USSR Acad. Sci. Chem.-Phys. Inst., Moscow, 99 (in russian).

Kuzmin, A.D., and Marov, M.Ya. (1974) *Physics of the Planet Venus*, Nauka, Moscow; NASA Transl. TT-F-16226 (1975), NASA, Washington, D.C.

Laikhtman, D.L. (1970) *Physics of the Atmospheric Boundary Layer*, Gidrometeoizdat, Leningrad (in russian).

Landau, L.D., and Lifshits, E.M. (1954) *Mechanics of Continuum*, Gostechteorizdat, Moscow-Leningrad (in russian).

Landau, L.D., and Lifshits, E.M. (1988) *Hydrodynamics*, Nauka, Moscow (in russian).

Lapin, Yu.V., Strelets, N.Ch., and Schur, N.L. (1984) Calculation of heat-and-mass transfer in the mixtures of viscous non-equilibrium gas flows with the account for effects of multicomponent diffusion, in: *Heat and Mass Transfer VII*, **3**, Inst. of Heat and Mass Transfer, Minsk, 121 (in russian).

Lapin, Yu.V., and Strelets, N.Ch. (1989) *Internal Flows of Mixed Gases*, Nauka, Moscow (in russian).

Launder, B.E., and Spalding, D.B. (1972) *Mathematical Model of Turbulence*, Acad. Press, London.

Launder, B.E. (1975) On the effects of gravitational field on the turbulent transport of heat and momentum, *J. Fluid Mech.,* **67**, 569.

Launder, B.E., and Morse, A. (1979) Numerical calculation of axisymmetrical free shear flows with the use of short closings for stresses, in: *Turbulent Shear Flows I* (eds. F. Durst et al.), Springer-Verlag, Berlin-Heidelberg-New York.

Lee, K.W. (1983) Change of particle size distribution during Brownian coagulation*, J. Colloid and Interface Sci.*, **92**, 307.

Lee, T., Papanastassiou, D.A., and Wasserburg, G.J. (1976) Demonstration of ^{26}Mg excess in Allende and evidence for ^{26}Al, *Geophys. Res. Lett.,* **3**, 109.

Leovy, C.B. (1982) Martian meteorological variability, *Adv. Space Res*., **2**, 19.

Lesieur, M. (1990) *Turbulence in Fluids*, 2nd ed., Kluwer Academic Publishers, Dordrecht, Holland.

Lettau, H. (1951) Diffusion in the upper atmosphere, in: *Compendium of Meteorology*, (ed. T.F Malone) Amer. Meteorol. Soc., New York, 320.

Levich, V.G. (1959) *Physicochemical Hydrodynamics*, Fizmatgiz, Moscow (in russian).

Lewellen, W.S. (1977) Method of invariant simulation, in: *Handbook of Turbulence* (eds. W. Frost and T.H. Moulden), Plenum Press, New York and London, 262.

Libby, P.A. (1972) On turbulent flow with fast chemical reactions. Pt. I: The closure problem, *Combus. Sci. and Technol*., **6**, 23.

Lindzen, R.S. (1981) Turbulence and stress due to gravity wave and tidal breakdown, *J.Geophys.Res.*, **86**, 9707.

Lissauer, J.J. (1995) Urey Prize Lecture: On the diversity of plausible planetary systems, *Icarus*, **114**, 217.

Lissauer, J.J., and Stewart, G.R. (1993) Growth of planets from planetesimals, in: *Protostars and Planets III* (eds. E.H. Levy and I.J. Lunine), Arizona Univ. Press, Tucson, 1061.

Lloyd, K.N., Low, C.H., McAvaney, Rees, D., and Roper, R.G. (1972) Thermospheric observations combining chemical seeding and ground based techniques. 1.Winds, turbulence and parameters of the neutral atmosphere, *Planet. Space Sci*., **20**, 761.

Lloyd, K.N. (1974) Investigations into the nature of the turbopause, *Proc. Int. Conf. on Structure, Composition and General Circulation of Upper and Lower Atmospheres and Possible Antropogenic Perturbation*, **2,** Melburne, 653.

Loginov, V.I. (1979) *Dehydration and Desalinization of Oil*, Khimiya, Moscow (in russian).

Lumley, J.L., and Panofsky, H.A. (1964) *The Structure of Atmospheric Turbulence*, John Wiley and Sons, New York-London-Sydney.
Luschik, V.G., Pavel'ev, A.A., and Yakubenko, A.E. (1978) Three-parametric model of shear turbulence, *Reports of USSR Acad. Sci., Mechanics of Gas and Fluids*, **3**, 13 (in russian).
Luschik, V.G., Pavel'ev, A.A., and Yakubenko, A.E. (1986) Three-parametric model of shear turbulence: Calculation of heat exchange, *Reports of USSR Acad. Sci., Mechanics of Gas and Fluids*, **2**, 40 (in russian).
Lykosov, V.N. (1991) On the antigradient moment transfer in the low level stream flow, *Reports of USSR Acad. Sci., Physics of Atmosphere and Ocean*, **27**, 325 (in russian).
Mahaffy, P.R., Atreya, S.K., Niemann, H.B., and Owen, T.C. (1998) In–situ chemical and isotopic measurements of the atmosphere of Jupiter, in: *Highlights of Astronomy* , v.11B (ed. J.Andersen); *SPSI: The Galileo Mission to the Jupiter System* (eds. M.Ya. Marov and R.W. Carlson), Kluwer Acad. Publ., Dordrecht, 1057.
Major, M., and Queloz, D. (1995) A Jupiter-mass companion to a solar-type star, *Nature* **378**, 355.
Major, M., Queloz, D., and Udry, S. (1997) From brown dwarfs to planets, In: *Astronomical and Biochemical Origins and the Search for Life in the Universe* (eds. C.B. Cosmovici et al.), Editrice Compositori, Bologna, 313.
Makalkin, A.B., and Dorofeeva, V.A. (1991) Temperatures in the protoplanetary disk: Models, constraints, and consequences for planets, *Izvestiya USSR Acad. Sci, Physics of the Earth*, **8**, 34.
Makalkin, A.B., and Dorofeeva, V.A. (1995) The structure of the protoplanetary accretion disk around the Sun at the T Tauri phase. I. Initial data, equations, and methods of modeling, *Solar System Research*, **29**, 85.
Makalkin, A.B., and Dorofeeva, V.A. (1996) Structure of protoplanetary accretion disk around the Sun at the T-Tauri stage: II. Results of the model calculations, *Solar System Research*, **30**, 441 (in russian).
Malhotra, R. (1999) Chaotic planet formation, *Nature*, **402**, 599.
Marchuk, G.I. (1982) *Mathematical Modeling in the Problem of Environment*, Nauka, Moscow (in russian).
Marcy, G.W., and Butler, R.P. (1998) Detection of extrasolar giant planets, *Ann. Rev. Earth Planet. Sci.* **36**, 57.
Marov, M.Ya. (1986) *Planets of the Solar System*, 2nd Edition, Nauka, Moscow.
Marov, M.Ya. (1992) Mars, Atmosphere, in: *The Astronomy and Astrophysics* (ed. S.P. Maran), Cambridge University Press, 407.
Marov M.Ya. (1994) The inner planets of the Solar System, in: *Space Biology and Medicine. I. Outer Space and its Utilization* (ed. V.A. Kotelnikov and D.D. Rummel), Nauka/AIAA, Moscow-Washington, 168.
Marov, M.Ya (1997) Methods of mechanics in the aeronomy problems, *M.V. Keldysh Inst. Applied Math. Preprint No. 18,* Moscow (in russian).
Marov, M.Ya.., and Grinspoon, D.H. (1998) *The Planet Venus*, Yale University Press, New Haven and London.
Marov, M.Ya, and Kolesnichenko, A.V. (1971) On simulation of the planetary heterosphere, *Geomagnetism and Aeronomy*, **1**, 40.
Marov, M.Ya, and Kolesnichenko, A.V. (1981) On some problems of theoretical simulation of the upper atmosphere, in: *Physics of the Upper Atmosphere, Proc. of Intercosmos Workshop*, **6**, Baya-Moscow, 87 (in russian).
Marov, M. Ya., and Kolesnichenko, A.V. (1983) On the upper atmosphere theoretical modeling, in: *Observation of Artificial Earth's Satellites*, **20,** Budapest-Moscow, 29 (in russian).
Marov, M,Ya., and Kolesnichenko, A.V. (1983) Methods of non–equilibrium thermodynamics for discribing turbulent multicomponent gas mixtures with chemical reactions and heat–mass transfer processes, in: *XVI Symposium on Advanced Problems and Methods in Fluid Mechanics*, Spala, Poland, 22.
Marov, M,Ya., and Kolesnichenko, A.V. (1987) *Introduction to the Planetary Aeronomy*, Nauka, Moscow (in russian).
Marov, M,Ya., Kolesnichenko, A.V., Krasitsky, O.P., Vasin, V.G., and Pavlyukov, Yu.B. (1989) The comparative analysis and graphic maintenance of empirical models of the Earth's thermosphere, *M.V. Keldysh Inst. Applied Math. Preprint No. 45,* Moscow (in russian).
Marov, M.Ya., and Krasitsky, O.P. (1990) Some comments of the CIRA–86 model, *Adv. Space Res.*, **10**, 117.

Marov, M. Ya.., Volkov V.P., Surkov, Yu.A., and Ryvkin, M.L. (1989) The lower atmosphere, in: *The Planet Venus* (ed. V.L. Barsukov), Nauka, Moscow, 25 (in russian).
Marov, M.Ya., Shematovich, V.I., and Bisikalo, D.V. (1996) Nonequilibrium Aeronomic Processes. A Kinetic Approach to the Mathematical Modeling, *Space Sci. Rev*., 76, 1.
Marov, M.Ya., Shematovich, V.I., Bisikalo, D.V., and Gerard, J.C. (1997) *Nonequilibrium Processes in the Planetary and Cometary Atmospheres: Theory and Applications*, Kluwer Academic Publishers, Dordrecht-Boston-London.
Maslov, L.A., and Petrovskaya, T.S. (1975) Calculation of turbulent boundary layer for the revolution bodies under angle of attack, *Proc. of Central Aerogydrodynamic Inst.* **1661**, 3.
Mason, E.A. (1954) Transport properties of gases obeying a modified Buckingham ($\exp^{-6}$) potential, *J.Chem. Phys.* **22**, 169.
Mason, E.A., and Saxena, S.C. (1958) Approximate formula for the conductivity of gas mixtures, *Phys. Fluids*, **1**, 361.
Mason, E.A. (1974) The Onsager reciprocal relations. Experimental evidence, in: *Foundations of Continuum Thermodynamics*, Mac-Millan, London and Basingstoke.
Maxwell, J.C. (1890) *The Scientific Papers of James Clerk Maxwell* (ed. W.D. Niven), **1**, 377, Cambridge University Press, Cambridge.
Mazin, I.P., (1971) Theoretical evaluation of the coefficient of coagulation of droplets in clouds, *TsAO Proceedings*, **95**, 12.
McConnell, J.C., Hoiberg, J.B., Smith, G.R., Sandel, B.R., Shemansky, D.E., and Broadfoot, A.L. (1982) A new look at the ionosphere of Jupiter in light of the UVS occultation results, *Planet. Space Sci*., **30**, 151.
Mednikov, E.P. (1981) *Turbulent Transport and Sedimentation of Aerosols*, Nauka, Moscow (in russian).
Meixner, J. (1969) Processes in simple thermodynamic materials, *Arch. Ration. Mech. Anal*., **33**, 33.
Mellor, G.L. (1973) Analitic prediction of the properties of stratified planetary surface layers, *J.Atmos.Sci*., 30, 1061.
Mellor, G.L., and Herring, H.J. (1973) Review of the models for closing of the equations of averaged turbulent flow, *Rocketry and Astronautics*, **11**, 17.
Mellor, G.L., and Yamada, T. (1974) A hierarhy of turbulence closure models for planetary boundary layers, *J. Atmos. Sci*., **31**, 1791.
Mellor, G.L., and Yamada, T. (1982) Development of a turbulence closure model for geophysical fluid problems, *Rev. Geoph. and Space Phys*., **20**, 851.
Meteorology and Atomic energy (1959; 1971) Inostr. Liter., Moscow (in russian).
Middle Atmosphere Program, Handbook for MAP (1985) (eds. K. Labitzke et al.), SCOSTEP Publication, 16, University of Illinois, Urbana, Illinois.
Millionschikov, M.D. (1941) On the theory of a homogeneous isotropic turbulence, *USSR Acad. Sci.Dokladi*, **32**, 611 (in russian).
Monin, A.S. (1950) Dynamic turbulence in the atmosphere, *Reports of USSR Acad. Sci., Geography and Geophysics,* **14**, 232 (in russian).
Monin, A.S. (1962) On the spectrum of turbulence in a temperature–inhomogeneous medium, *Reports of USSR Acad. Sci., Geophysics,* **3**, 397 (in russian).
Monin, A.S. (1962) General review of the atmospheric diffusion, in: *Atmospheric Diffusion and Air Pollution* (eds. F.N. Frenkiel and P.A. Sheppard), Academic Press, New York and London, 29.
Monin, A.S. (1965) On the properties of symmetry of turbulence in the air near-ground layer, *Atmospheric Physics and Ocean*, **1**, 45 (in russian).
Monin, A.S. (1969) *The Weather Forecast as the Physics Problem*, Nauka, Moscow (in russian).
Monin, A.S. (1988) *Theoretical Fundamentals of Geophysical Hydrodynamics*, Hydrometeoizdat, Leningrad (in russian).
Monin, A.S., and Yaglom, A.M. (1971) *Statistical Fluid Mechanics*, v.1, MIT Press (J. Lumley ed.), Cambridge, Mass.
Monin, A.S. and Yaglom, A.M. (1975) *Statistical Fluid Mechanics*, v.2, MIT Press (J. Lumley ed.), Cambridge, Mass.
Monin, A.S. Polubarinova-Kochina, P.Ya., and Chlebnikov, V.I. (1989) *Cosmology, Hydrodynamics, and Turbulence*, Nauka, Moscow (in russian).
Morfill, G., Spruit, H., and Levy, E.H. (1993) Physical Processes and Conditions Associated with the Formation of Protoplanetary Disks, in: *Protostars and Planets III* (eds. E.H. Levy and I.J. Lunine), Univ. Arizona Press, Tucson, 939.

Morkovin, M.V. (1961) Effects of Compressibility on Turbulent Flow, in: *Mechanics of Turbulence*, Gordon and Breach, New York, 367.

Morozov, S.K., and Krasitsky, O.P. (1978) The numerical method of solution of the systems of non-stationary, non-linear one-dimensional differential equations, *Space Research Inst. Preprint No. 396*, Moscow (in russian).

Muckenfuss, C., and Curtiss, C.F. (1958) Thermal Conductivity of Multicomponent Gas Mixtures, *J. Chem. Phys.*, **29**, 1273.

Muckenfuss, C. (1973) Stefan–Maxwell relations for multicomponent diffusion and the Chapman–Enskog solution of the Boltzmann eguations, *J. Chem. Phys.* **59**, 1747.

Munk, W.H., and Anderson E.R. (1948) Notes on the theory of the thermocline, *J. Marine Research*, **1**, 276.

Munn, R.J., Smith, F.J., and Mason, E.A. (1965) Transport collision integrals for quantum gases obeying a (12–6) potential, *J. Chem. Phys.*, 42, 537.

Murtagh, D.P., Witt, G., Stegman, J. et. al. (1990) An assessment of proposed $O(^1S)$ and $O_2(b^1\Sigma_g^+)$ nightglow excitation parameters, *Planet. Space Sci.*, 38, 43.

Myasnikov, V.P., and Titorenko, V.I. (1989a) Evolution of self-gravitating clots of gaseous-dusty nebula involved in the accumulation of planetary bodies, *Solar System Research*, **23**, 14.

Myasnikov, V.P., and Titorenko, V.I. (1989b) Evolution of self-gravitating gaseous-dusty clots with the account for radiation transfer in the diffusive approximation, *Solar System Research*, **23**, 207.

Nagy, A.F., Cravens, T.E., and Gombosi, T.I. 1983). Basic theory and model calculations of the Venus ionosphere. In: *Venus* (D.M. Hunten et al. eds.), The University of Arizona Press, Arizona, Tucson, 841.

Nakamura, R., Kitada, Y., and Mukai, T. (1994) Gas drag forces on fractal aggregates, *Planet. Space Sci.*, **42**, 721.

Namgaladze, A.A., Korenkov, Yu.N., Klimenko, V.V., Karpov, I.V., Bessarab, F.S., Surotkin, V.A., Gluschenko, T.A., and Naumova, N.M. (1990) Global numerical model of thermosphere, ionosphere and protonosphere of the Earth, *Geomagnetism and Aeronomy*, **30**, 612 (in russian).

Nevzglyadov, V.G. (1945) On the phenomenological theory of turbulence, *USSR Acad. Sci.Dokladi*, **47**, 169 (in russian).

Nevzglyadov, V.G. (1945) On the statistical theory of turbulence. *USSR Acad. Sci.Dokladi*, **47**, 482 (in russian).

Nigmatulin, R.I. (1987) *Dynamics of Multiphase Media. Part. I,* Nauka, Moscow (in russian).

Nicolet, M. (1984) Photodissociation of molecular oxygen in the terrestrial atmosphere: Simplified numerical relations for the spectral range of the Shuman–Runge bands, *J. Geophys. Res.*, **89** (D2), 2573.

Nicolis, G., and Prigogine, I. (1977) Self-Organization in Non-Equilibrium System, John Wiley & Sons, New York.

Nikolaevsky, V.N., (1984) Spatial average and the theory of turbulence, in: *Eddies and Waves*, Mir, Moscow, 266 (in russian).

Nicolis, G., and Prigogine, I. (1977) *Self-Organization in Nonequilibrium Systems*, John Wiley & Sons, New York.

Niemann, H.B., Kasprzak, W.T., Hedin, A.E., Hunten, D.M. and Spencer, N.W. (1980) Mass spectrometric measurements of the neutral gas composition of the thermosphere and exosphere of Venus, *J. Geophys. Res.,* **85**, 7817.

Nier, A.O, and McElroy, M.B. (1977) Composition and structure of Mars upper atmosphere: Results from the neutral mass spectrometers on Viking I and 2, *J. Geophys. Res.,* **86**, 9945.

Nikuradze, I. (1936) Regularities of turbulent motion in smooth tubes, in: *Problems of Turbulence*, ONTI, Moscow, 75.

O'Brien, E.E. (1980) Method of probability density functions in the theory of turbulent flows with chemical reactions, in: *Turbulent Reacting Flow* (ed. P.A. Libby and F.A.Williams), Springer-Verlag, Berlin-Heidelberg-New York.

Obuchov, A.M. (1941) On the energy distribution in the spectrum of turbulent flow, *Reports of USSR Acad. Sci., Geography and Geophysics,* **5**, 453 (in russian).

Obuchov, A.M. (1949) Structure of a temperature field in a turbulent flow, *Reports of USSR Acad. Sci., Geography and Geophysics,* **13**, 58 (in russian).

Obuchov, A.M. (1953) On influence of weak atmospheric inhomogeneities on the acoustic and light propagation, *Reports of USSR Acad. Sci., Geophysics,* **2,** 155 (in russian).

Obuchov, A.M. (1959) On influence of buoyancy forces on the structure of temperature field in a turbulent flow, *USSR Acad. Sci.Dokladi*, **125**, 1246 (in russian).

Obuchov, A.M. (1962) Description of turbulence in the Lagrangian coordinates, in: *Atmospheric Diffusion and Air Pollution* (eds. F.N. Frenkiel and P.A. Sheppard), Academic Press, New York and London, 112.

Oran, E.S., and Boris, J. (1987) *Numerical Simulation of Reactive Gases*, Elsivier Sci. Publ., New York-Amsterdam-London.

Owen, T. (1994) Outer planets of the Solar System, in: *Space Biology and Medicine I. Outer Space and its Utilization* (ed. V.A. Kotelnikov and D.D. Rummel), Nauka/AIAA, Moscow-Washington, 247.

Ozernoy, L.M., and Chibisov, G.V. (1970) Galaxies dynamic parameters as a consequence of the cosmological turbulence, *Sov. Astron. J.*, **47**, 769 (in russian).

Ozernoy, L.M. (1976) Vortex theory of the origin of galaxies and their systems, in: *Origin and Evolution of Galaxies and Stars* (ed. S.B. Pikelner), Nauka, Moscow, 105 (in russian).

Petrovay, K. (2001) Turbulence in the solar photosphere, *Space Sci. Rev.*, **95**, 9.

Petuchov, B.S., Dvortsov, V.N., Charin, B.E., and Shikov, V.K. (1984) Quasi-three-dimensional model and method of calculation of flow and heat exchange at the initial part of a channel of rectangular cross-section, *High Temperature Thermal Physics*, **22**, 74 (in russian).

Pikelner, S.B., and Kaplan, S.A. (1976) Fundamentals of the theory of star formation. An origin of stars of the first generation, in: *Origin and Evolution of Galaxies and Stars* (ed. S.B. Pikelner), Nauka, Moscow, 190 (in russian).

Pikelner, S.B., Kaplan, S.A., and Zasov, A.V. (1976) Large-scale dynamics of interstellar medium and the formation of stars of the flat subsystem, in: *Origin and Evolution of Galaxies and Stars* (ed. S.B. Pikelner), Nauka, Moscow, 235 (in russian).

Polack, L.S., Ovsyannikov, A.A. Slovetsky, D.I., and Vurzel, F.B. (1975) *Theoretical and Applied Plasmochemistry,* Nauka, Moscow (in russian).

Prandtl, L. (1925) Bericht uber untersuchungen zur ausgebildeten turbulenz, *Z. Angew. Math. Mech.*, **5**, 136.

Prandtl, L. (1942) Bemerkungen zur theorie der freien turbulenz, *Z. Angew. Math. Mech.*, **22**, 241.

Prigogine, I., and Defay, R. (1954) *Chemical Thermodynamics*, Longmans Green and Co., London-New York-Toronto.

Prigogine, I., and Stengers, I. (1984) *Order out of Chaos*, Heinemann, London.

Prokhorov, A.M. (1974) Distribution of a laser radiation in randomly inhomogeneous media, *Progress in Physical Sciences*, **114**, 3 (in russian).

Pudovkin, M.I., and Semenov, V.S. (1985) Magnetic field reconnection theory and the solar wind - magnetosphere interaction: A review, *Space Sci. Reviews*, **41**, 1.

Rees, D., Roper, R.G., Lloye, K.H., and Low, C.H. (1972) Determination of the structure of the atmosphere between 90 and 250 km by means of contaminant releases at Woomera, May, 1968, *Phys. Trans. Roy. Soc. London*, **A271**, 631.

Reist, P. (1984) *Introduction to Aerosol Science*, Macmillan, New York.

Richardson, L.F. (1926) Atmospheric diffusion shown on a distance–neighbour graph, *Proc. Roy. Soc.*, **A110**, 709.

Rishbeth, H., and Garriott, O.K. (1969) *Introduction to Ionospheric Physics*, Academic Press, New York.

Roper, R.G. (1966) Dissipation of wind energy in the height range 80 to 140km, *J. Geophys. Res.*, **71**, 4427.

Roper, R.G. (1974) The dynamics of the turbopause, *Proc. Int. Conf. on Structure, Composition and General Circulation of the Upper and Lower Atmospheres and Possible Anthropogemic Perturbation*, **2**, 642, Melburn.

Rosenberg, N. (1968) Statistical analysis of ionospheric winds – II, *J. Atmos. Terr. Phys.*, **30**, 907.

Rotta, J. (1951) Statistische theorie nichthomogener turbulenz. Teil 1, *Physik*, **129**, 547.

Rotta, J. (1951) Statistische theorie nichthomogener turbulenz. Teil 2, *Physik*, **131**, 51.

Rottger, L., Czechowsky, P., and Schmidt, G. (1981) First low-power VHF radar observations of tropospheric, stratospheric and mesospheric winds and turbulence at the Arecibo Observatory, *J. Atmos. Terr. Phys.* **43**, 789.

Ruden, S.P., and Pollack, J.B. (1991) The dynamical evolution of the protosolar nebula, *Astrophys. J.*, **375**, 740.

Russell, C.T. , ed. (1991) *Venus Aeronomy*, Kluwer Academic Publishers, Dordrecht.

Ruzmaikina, T.V. and Makalkin, A.B. (1991) Formation and Evolution of the Protoplanetary Disk, in: *Planetary Sciences. American and Soviet Research, Proceed from US-USSR Workshop* (T. Donahue ed.), National Academy Press, Washington, D.C., 44.

Rytov, S.M. (1937) Diffraction of light on ultrasonic waves, *Reports of USSR Acad. Sci., Physics,* **2**, 223 (in russian).

Rytov, S.M., Kravtsov, Yu.A., and Tatarsky, V.I. (1978) *Introduction to Statistical Radio Physics. Part 2*, Nauka, Moscow (in russian).

Sadovnichy, V.A., ed. (1998) *Space Physical Geography,* Moscow State University, Moscow (in russian).

Safronov, V.S. (1969) *Evolution of the Protoplanetary Cloud and Formation of the Earth and the Planets*, Nauka, Moscow (in russian). Trans. NASA TTF-677, 1972.

Safronov, V.S. (1982) Current state of the theory of the Earth's origin, *Reports of USSR Acad. Sci, Physics of the Earth*, **6**, 5.

Safronov, V.S. (1987) *The Origin of the Earth*, Znanie, Moscow (in russian).

Safronov V.S. (1991) Kuiper Prize Lecture: Some problems in the formation of planets, *Icarus*, **94**, 260.

Safronov V.S. (1996) Accumulation of small bodies at the outer boundary of the planetary system, *Solar System Research*, **30**, 251.

Safronov, V.S., and Guseinov, K.M. (1990) On the possibility of in situ comet origin, *Solar System Research*, **24**, 159.

Safronov, V.S., and Ruzmaikina, T.V. (1985) Formation of the Solar Nebula and Planets, in: *Protostars and Planets II* (eds. D.C. Black and M.S Matthews), The Univ. of Arizona Press, Tucson, 958.

Samarsky, A.A. (1977) *Theory of Difference Schemes*, Nauka, Moscow (in russian).

Samarsky, A.A., and Nikolaev, E.S. (1978) *Methods of Solution of Network Equations*, Nauka, Moscow (in russian).

Sandel, B.R., McConnell, J.C. and Strobel, D.F. (1982) Eddy diffusion at Saturn's homopause, *Geophys. Res. Letters*, **9**, 1077.

Sassi, F., and Visconti, G. (1990) Validation of parameterization scheme for eddy diffusion from satellite data, *J. Atmos. Sci.*, **47**, 2505.

Schetinkov, E.S. (1965) *Physics of Gas Combustion*, Nauka, Moscow (in russian).

Sedov, L.I. (1965) *Similarity and Dimension Methods in Mechanics*, Nauka, Moscow, (in russian).

Sedov, L.I. (1973) *Thoughts about Scientists and Science of Past and Present*, Nauka, Moscow, (in russian).

Sedov, L.I. (1980) On the perspective directions and problems in the mechanics of continuum, in: *Speculations about Science and Scientists*, Nauka, Moscow, 173 (in russian).

Sedov, L.I. (1984) *Mechanics of Continuum. Part. 2*, Nauka, Moscow (in russian).

Seiff, A. (1983) Thermal structure of the atmosphere of Venus, in: *Venus* (D.M. Hunten et al. eds.), The University of Arizona Press, Tucson, Arizona, 215.

Shakura, N.I., and Sunyaev, R.A. (1973) Black holes in binary systems: Observational appearance, *Astron. Astrophys.*, **24**, 337.

Shakura, N.I., Sunyaev, R.A., and Zilitinkevich, S.S. (1978) On the Turbulent Energy Transport in Accretion Disk, *Astron. Astrophys.*, **62**, 179.

Shapiro, S., and Teukolsky, S. (1983) *Black Holes, White Dwarfs and Neutron Stars*, John Wiley & Sons, New York.

Shapiro, V.D., and Shevchenko, V.I. (1987) Plasma turbulence in space, in: *The Results of Science and Engineering. Astronomy (*ed. R.A. Sunyaev), **32**, VINITI, 235.

Schmidt, O.Yu. (1957) *Four Lectures on the Earth's Origin*, 3rd edition, USSR Acad. Sci., Moscow

Schmitt, W., Henning, T., and Mucha, R. (1997) Dust Evolution in Protoplanetary Accretion Disks, *Astron. Astrophys.*, **325**, 569.

Schneider, J. (1999) Extrasolar planet encyclopedia site, http://www.obspm.fr/planets

Shimazaki, T. (1971) Effective eddy diffusion coefficient and atmosphere composition in the lower thermosphere, *J. Atmos. Terr. Phys.*, **33**, 1383.

Shimazaki, T. (1972) On boundary conditions in theoretical model calculations of the distribution of minor neutral constituents in the upper atmosphere, *Radio Sci.*, **7**, 695.

Shu, F.H., Adams, F.c., and Lizano, S. (1987) Star formation in molecular clouds: Observation and theory. *Ann. Rev. Astron. Astrophys.,* **25**, 23.

Sinaiskii, E.G. (1997) *Hydrodynamics of Physicochemical Processes*, Nedra, Moscow.

Smoluchowski, M. (1936) Three lectures on diffusion, Brownian molecular motion and coagulation of colloid particles, in: *Colloid Coagulation*, ONTI, Moscow.

Smith, E.J., Strobel, D.F., Broadfoot, A.L, Sandel, B.R., Shemansky, D.E., and Holberg, J.B. (1982) Titan's upper atmosphere: Composition and temperature from the EUV solar occultation results, *J. Geophys. Res.,* **87**, 1351.

Solar Terrestrial Research for 1980s (1981) US National Academy, Washington, D.C.

Spitzer, L., Jr. (1978) *Physical Processes in Interstellar Medium*, Wiley, New York.

Srinivasan, R., Giddens, D.P., Bangret, L.H., and Wu, J.C. (1977) Turbulent plane Couette flow using probability distridution functions, *Phys. Fluids*, **20**, 557.

Stefan, J. (1871) *Wien Sitzungsber*, **63**, 63.

Sterzik, M.F., and Morfill, G.E. (1994) Evolution of protoplanetary disks with condensation and coagulation, *Icarus*, **111**, 536.

Sunyaev, R.A., and Zeldovich, Ya.B. (1972) Formation of clusters of galaxies: Protocluster fragmentation and intergalactic gas heating, *Astron. Astrophys*., **20**, 189.

Tassoul, J.-L. (1979) *Theory of Rotating Stars*, Princeton Univ., Princeton.

Tatarsky, V.I. (1967) *Wave Propagation in a Turbulent Atmosphere*, Nauka, Moscow (in russian).

Terquem, C., Papaloizou, J.C.B., and Nelson, R.P. (2000) Disks, extrasolar planets and migration, *Space Sci. Reviews*, **92**, 323.

The Hubble Atlas of Galaxies (1961) Carnegie Institute of Washington, **618**.

Tichonov, A.N., and Arsenin, V.Ya. (1986) *Methods of Solution of the Ill-Posed Problems*, Nauka, Moscow (in russian).

Tirsky, G.A. Definition of the effective diffusion coefficients in the laminar boundary layer, *USSR Acad. Sci.Dokladi*, **155**, 1278 (in russian).

Tobiska, W.K. (1991) Revised solar extreme ultraviolet flux model, *J. Atmos. Terr. Phys.,* **53**, 1005.

Tobiska, W.K. (1993) Recent solar extreme ultraviolet irradiance observations and modeling: A review, *J. Geophys. Res.,* **98** (A11), 18,879.

Tobiska, W.K., and Barth, C.A. (1990) A solar EUV model, *J. Geophys. Res.*, **95** (A6), 8243.

Torr, M.R., Torr, D.G., Ong, R.A., and Hinteregger, H.E. (1979) Ionization frequencies for major thermospheric constituents as a function of solar cycle 21, *J. Geophys. Res. Lett*., **6**, 771.

Torr, M.R., and Torr, D.G. (1985) Ionization frequencies for solar cycle 21: Revised, *J. Geophys. Res*., **90** (A7), 6675.

Torrieri, D.J., and Taylor, L.L. (1972) Irradiance fluctuations in optical transmission through the atmosphere, *J. Opt. Soc. Amer*., **62**, 145.

Townsend, A.A. (1956) *The Structure of Turbulent Shear Flow*, Cambridge Univ. Press, Cambridge.

Truesdell, C. (1962) Mechanical basis of diffusion, *J.Chem. Phys*., **37**, 2236.

Turbulent Shear Flows – I (1979) (eds. F. Durst et al.) Springer-Verlag, Berlin-Heidelberg-New York.

Turbulent reacting Flow (1980) (eds. P.A. Libby and F.A. Williams), Springer-Verlag, Berlin-Heidelberg-New York.

Upps, F.B. (1971) Two–dimensional numerical experiments in thermal convection with vertical shear, *J. Atmos. Sci*., **28**, 3.

Urpin, V.A. (1984) Hydrodynamic motions in accretion disks, *Astron. J.,* **61**, 84.

Vager, B.G., and Nadezhina, E.D. (1979) *Boundary Layer of an Atmosphere in the Conditions of Horizontal Non-Uniformity*, Hydrometeoizdat, Leningrad (in russian).

Van de Ree, J. (1967) On the definition of the diffusion coefficients in reacting gases, *Physica,* **36**, 118.

Van Driest, E.R. (1952) Turbulent boundary layer in compressed medium, in: *Mechanics*, **1**, 27.

Van Mieghem, J. (1973) *Atmospheric Energetics*, Clarendon Press, Oxford.

Vityazev, A.V., Pechernikova, G.V., and Safronov, V.S. (1990) *Terrestrial Planets: Origin and Early Evolution*, Nauka, Moscow (in russian).

Vlasov, M. N., and Davydov, V. E. (1982) Theoretical description of the main neutral constituents in the Earth's upper atmosphere, *J. Atmos. Terr. Phys*., **44**, 641.

Volkov, V.P., Marov, M.Ya., Lebedev, V.N., Sidorov, Yu.I., and Schari, V.P. (1989) Clouds, in: *Planet Venus* (ed. V.L. Barsukov), Nauka, Moskow, 68 (in russian).

Voloshchuk, V.M. (1984) *Kinetic Theory of Coagulation*, Gidrometeoizdat, Moscow (in russian).

Von Zahn, V. (1967) Mass spectrometric measurement of atomic oxygen in the upper atmosphere: A critical review, *J.Geophys. Res.,* **72**, 5933.

Von Zahn, U., Fricke, K.H., Hunten, D.M., Krankowsky, D., Mauersberger, K., and Nier, A.O. (1980) The upper atmosphere of Venus during morning conditions, *J. Geophys. Res*., **85**, 7829.

Wasiutynski, J. (1946) Studies in hydrodynamics and structure of stars and planets, *Astrophys. Norvegica*, **4**, 86.

Wasserburg, G. J., and Papanastassiou, D.A. (1982) Some short-lived nuclides in the early solar system - connection with the original interstellar medium, in: *Essays in Nuclear Astrophysics* (eds. C.A. Barnes et al.), Cambridge University Press, Cambridge-London-NewYork.

Weidenschilling, S.J. (1980) Dust to planetesimals: Settling and coagulation in the solar nebula, *Icarus*, **44**, 172.

Weidenschilling, S.J., and Cuzzi, J.N. (1993) Formation of planetesimals in the solar nebula, in: *Protostars and Planets III* (eds. E.Levy and J. Lunine), Univ. Arizona Press, Tucson, 1031.

Weidenschilling, S.J. (2000) Formation of planetesimals and accretion of the terrestrial planets, *Space Sci. Rev.*, **92**, 295.

Williams F.A. (1965) *Combustion Theory*, Addison-Wesley, Palo Alto-London.

Williams, G.P. (1978) Planetary circulations. 1. Barotropic representations of Jovian and terrestrial turbulence, *J. Atmos. Sci.*, **35**, 1399.

Williams, G.P. (1979) Planetary circulations. 2. The Jovian quasi–geostrophic, *J. Atmos. Sci.*, **36**, 932.

Woo, R., Armstrong, J.W., and Kliore, A.J. (1982) Small scale turbulence in the atmosphere of Venus, *Icarus*, **52**, 335.

Yelle, R.V., Herbert, F., Sandel, B.R., Vervack, R.J., Jr., and Wentzel, T.M. (1993) The distribution of hydrocarbons in Neptune's upper atmosphere, *Icarus*, **104**, 38.

Zeldovich, Ya. B. (1981) On the friction fluids between rotating cylinders, *Proc. Roy. Soc.*, **A374**, 299.

Zeldovich, Ya.B., and Reiser, Yu.P. (1966) *Physics of Shock Waves and High-Temperature Hydrodynamic Phenomena,* Acad. Press, New York.

Zeldovich, Ya. B., and Novikov, I.D. (1975) *Structure and Evolution of the Universe*, Nauka, Moscow (in russian).

Zeldovich, Ya. B. (1981) On the friction fluids between rotating cylinders, *Proc. Roy. Soc.*, **A374**, 299.

Zeldovich, Ya. B. (1983) Structure of the Universe, in: *The Results of Science and Engineering. Astronomy (*ed. R.A. Sunyaev), **22**, VINITI, 4 (in russian).

Zhang, S., Bougher, S.W., and Alexander, M.J. (1996) The impact of gravity waves on the Venus thermosphere and O_2 IR nightglow, *J. Geophys. Res.,* **101**, 23,195.

Zhelazni, S.V., Morgentaler, I.Ch., and Cherendin, D.L. (1973) Models for calculation of friction stresses and intensity of turbulent pulsations for axisymmetrical wake streams, *Rocketry and Cosmonautics*, **11**, 137 (in russian).

Zuev, V.E., Banach, V.A, and Ponasov, V.V. (1988) *Optics of Turbulent Air*, Hydrometeoizdat, Leningrad.

Zurek, R.W., Barnes, J.R., Haberle, R.M. Pollack, J.B., Tillmann, J.E., and Leovy, C.B. (1992) Dynamics of the atmosphere of Mars, in: *Mars* (eds. H.H. Kieffer et al.), The University of Arizona Press, Tucson, Arizona, 835.

SUBJECT INDEX

V

W

Previously published in Astrophysics and Space Science Library book series:

- **Volume 265: Post-AGB Objects as a Phase of Stellar Evolution**
  Editors: R. Szczerba, S.K. Górny
  Hardbound, ISBN 0-7923-7145-3, July 2001
- **Volume 264: The Influence of Binaries on Stellar Population Studies**
  Editor: Dany Vanbeveren
  Hardbound, ISBN 0-7923-7104-6, July 2001
- **Volume 262: Whistler Phenomena**
  Short Impulse Propagation
  Authors: Csaba Ferencz, Orsolya E. Ferencz, Dániel Hamar, János Lichtenberger
  Hardbound, ISBN 0-7923-6995-5, June 2001
- **Volume 261: Collisional Processes in the Solar System**
  Editors: Mikhail Ya. Marov, Hans Rickman
  Hardbound, ISBN 0-7923-6946-7, May 2001
- **Volume 260: Solar Cosmic Rays**
  Author: Leonty I. Miroshnichenko
  Hardbound, ISBN 0-7923-6928-9, May 2001
- **Volume 259: The Dynamic Sun**
  Editors: Arnold Hanslmeier, Mauro Messerotti, Astrid Veronig
  Hardbound, ISBN 0-7923-6915-7, May 2001
- **Volume 258: Electrohydrodynamics in Dusty and Dirty Plasmas**
  Gravito-Electrodynamics and EHD
  Author: Hiroshi Kikuchi
  Hardbound, ISBN 0-7923-6822-3, June 2001
- **Volume 257: Stellar Pulsation - Nonlinear Studies**
  Editors: Mine Takeuti, Dimitar D. Sasselov
  Hardbound, ISBN 0-7923-6818-5, March 2001
- **Volume 256: Organizations and Strategies in Astronomy**
  Author: André Heck
  Hardbound, ISBN 0-7923-6671-9, November 2000
- **Volume 255: The Evolution of the Milky Way**
  Stars versus Clusters
  Editors: Francesca Matteucci, Franco Giovannelli
  Hardbound, ISBN 0-7923-6679-4, January 2001
- **Volume 254: Stellar Astrophysics**
  Editors: K.S. Cheng, Hoi Fung Chau, Kwing Lam Chan, Kam Ching Leung
  Hardbound, ISBN 0-7923-6659-X, November 2000
- **Volume 253: The Chemical Evolution of the Galaxy**
  Author: Francesca Matteucci
  Hardbound, ISBN 0-7923-6552-6, May 2001
- **Volume 252: Optical Detectors for Astronomy II**
  State-of-the-art at the Turn of the Millennium
  Editors: Paola Amico, James W. Beletic
  Hardbound, ISBN 0-7923-6536-4, December 2000
- **Volume 251: Cosmic Plasma Physics**
  Author: Boris V. Somov
  Hardbound, ISBN 0-7923-6512-7, September 2000
- **Volume 250: Information Handling in Astronomy**
  Editor: André Heck
  Hardbound, ISBN 0-7923-6494-5, October 2000
- **Volume 249: The Neutral Upper Atmosphere**
  Author: S.N. Ghosh
  Hardbound, ISBN 0-7923-6434-1, (in production)
- **Volume 247: Large Scale Structure Formation**
  Editors: Reza Mansouri, Robert Brandenberger
  Hardbound, ISBN 0-7923-6411-2, August 2000
- **Volume 246: The Legacy of J.C. Kapteyn**
  Studies on Kapteyn and the Development of Modern Astronomy
  Editors: Piet C. van der Kruit, Klaas van Berkel
  Hardbound, ISBN 0-7923-6393-0, August 2000

- **Volume 245: Waves in Dusty Space Plasmas**
  Author: Frank Verheest
  Hardbound, ISBN 0-7923-6232-2, April 2000
- **Volume 244: The Universe**
  Visions and Perspectives
  Editors: Naresh Dadhich, Ajit Kembhavi
  Hardbound, ISBN 0-7923-6210-1, August 2000
- **Volume 243: Solar Polarization**
  Editors: K.N. Nagendra, Jan Olof Stenflo
  Hardbound, ISBN 0-7923-5814-7, July 1999
- **Volume 242: Cosmic Perspectives in Space Physics**
  Author: Sukumar Biswas
  Hardbound, ISBN 0-7923-5813-9, June 2000
- **Volume 241: Millimeter-Wave Astronomy: Molecular Chemistry & Physics in Space**
  Editors: W.F. Wall, Alberto Carramiñana, Luis Carrasco, P.F. Goldsmith
  Hardbound, ISBN 0-7923-5581-4, May 1999
- **Volume 240: Numerical Astrophysics**
  Editors: Shoken M. Miyama, Kohji Tomisaka, Tomoyuki Hanawa
  Hardbound, ISBN 0-7923-5566-0, March 1999
- **Volume 239: Motions in the Solar Atmosphere**
  Editors: Arnold Hanslmeier, Mauro Messerotti
  Hardbound, ISBN 0-7923-5507-5, February 1999
- **Volume 238: Substorms-4**
  Editors: S. Kokubun, Y. Kamide
  Hardbound, ISBN 0-7923-5465-6, March 1999
- **Volume 237: Post-Hipparcos Cosmic Candles**
  Editors: André Heck, Filippina Caputo
  Hardbound, ISBN 0-7923-5348-X, December 1998
- **Volume 236: Laboratory Astrophysics and Space Research**
  Editors: P. Ehrenfreund, C. Krafft, H. Kochan, V. Pirronello
  Hardbound, ISBN 0-7923-5338-2, December 1998
- **Volume 235: Astrophysical Plasmas and Fluids**
  Author: Vinod Krishan
  Hardbound, ISBN 0-7923-5312-9, January 1999
  Paperback, ISBN 0-7923-5490-7, January 1999
- **Volume 234: Observational Evidence for Black Holes in the Universe**
  Editor: Sandip K. Chakrabarti
  Hardbound, ISBN 0-7923-5298-X, November 1998
- **Volume 233: B[e] Stars**
  Editors: Anne Marie Hubert, Carlos Jaschek
  Hardbound, ISBN 0-7923-5208-4, September 1998
- **Volume 232: The Brightest Binaries**
  Authors: Dany Vanbeveren, W. van Rensbergen, C.W.H. de Loore
  Hardbound, ISBN 0-7923-5155-X, July 1998
- **Volume 231: The Evolving Universe**
  Selected Topics on Large-Scale Structure and on the Properties of Galaxies
  Editor: Donald Hamilton
  Hardbound, ISBN 0-7923-5074-X, July 1998
- **Volume 230: The Impact of Near-Infrared Sky Surveys on Galactic and Extragalactic Astronomy**
  Editor: N. Epchtein
  Hardbound, ISBN 0-7923-5025-1, June 1998
- **Volume 229: Observational Plasma Astrophysics: Five Years of Yohkoh and Beyond**
  Editors: Tetsuya Watanabe, Takeo Kosugi, Alphonse C. Sterling
  Hardbound, ISBN 0-7923-4985-7, March 1998
- **Volume 228: Optical Detectors for Astronomy**
  Editors: James W. Beletic, Paola Amico
  Hardbound, ISBN 0-7923-4925-3, April 1998

- **Volume 227: Solar System Ices**
  Editors: B. Schmitt, C. de Bergh, M. Festou
  Hardbound, ISBN 0-7923-4902-4, January 1998
- **Volume 226: Observational Cosmology with the New Radio Surveys**
  Editors: M.N. Bremer, N. Jackson, I. Pérez-Fournon
  Hardbound, ISBN 0-7923-4885-0, February 1998
- **Volume 225: SCORe'96: Solar Convection and Oscillations and their Relationship**
  Editors: F.P. Pijpers, Jørgen Christensen-Dalsgaard, C.S. Rosenthal
  Hardbound, ISBN 0-7923-4852-4, January 1998
- **Volume 224: Electronic Publishing for Physics and Astronomy**
  Author: André Heck
  Hardbound, ISBN 0-7923-4820-6, September 1997
- **Volume 223: Visual Double Stars: Formation, Dynamics and Evolutionary Tracks**
  Editors: J.A. Docobo, A. Elipe, H. McAlister
  Hardbound, ISBN 0-7923-4793-5, November 1997
- **Volume 222: Remembering Edith Alice Müller**
  Editors: Immo Appenzeller, Yves Chmielewski, Jean-Claude Pecker, Ramiro de la Reza, Gustav Tammann, Patrick A. Wayman
  Hardbound, ISBN 0-7923-4789-7, February 1998
- **Volume 220: The Three Galileos: The Man, The Spacecraft, The Telescope**
  Editors: Cesare Barbieri, Jürgen H. Rahe†, Torrence V. Johnson, Anita M. Sohus
  Hardbound, ISBN 0-7923-4861-3, December 1997
- **Volume 219: The Interstellar Medium in Galaxies**
  Editor: J.M. van der Hulst
  Hardbound, ISBN 0-7923-4676-9, October 1997
- **Volume 218: Astronomical Time Series**
  Editors: Dan Maoz, Amiel Sternberg, Elia M. Leibowitz
  Hardbound, ISBN 0-7923-4706-4, August 1997
- **Volume 217: Nonequilibrium Processes in the Planetary and Cometary Atmospheres: Theory and Applications**
  Authors: Mikhail Ya. Marov, Valery I. Shematovich, Dmitry V. Bisikalo, Jean-Claude Gérard
  Hardbound, ISBN 0-7923-4686-6, September 1997
- **Volume 216: Magnetohydrodynamics in Binary Stars**
  Author: C.G. Campbell
  Hardbound, ISBN 0-7923-4606-8, August 1997
- **Volume 215: Infrared Space Interferometry: Astrophysics & the Study of Earth-like Planets**
  Editors: C. Eiroa, A. Alberdi, Harley A. Thronson Jr., T. de Graauw, C.J. Schalinski
  Hardbound, ISBN 0-7923-4598-3, July 1997
- **Volume 214: White Dwarfs**
  Editors: J. Isern, M. Hernanz, E. García-Berro
  Hardbound, ISBN 0-7923-4585-1, May 1997
- **Volume 213: The Letters and Papers of Jan Hendrik Oort as archived in the University Library, Leiden**
  Author: J.K. Katgert-Merkelijn
  Hardbound, ISBN 0-7923-4542-8, May 1997
- **Volume 212: Wide-Field Spectroscopy**
  Editors: E. Kontizas, M. Kontizas, D.H. Morgan, G.P. Vettolani
  Hardbound, ISBN 0-7923-4518-5, April 1997
- **Volume 211: Gravitation and Cosmology**
  Editors: Sanjeev Dhurandhar, Thanu Padmanabhan
  Hardbound, ISBN 0-7923-4478-2, April 1997
- **Volume 210: The Impact of Large Scale Near-IR Sky Surveys**
  Editors: F. Garzón, N. Epchtein, A. Omont, B. Burton, P. Persi
  Hardbound, ISBN 0-7923-4434-0, February 1997

- **Volume 209: <u>New Extragalactic Perspectives in the New South Africa</u>**
  Editors: David L. Block, J. Mayo Greenberg
  Hardbound, ISBN 0-7923-4223-2, October 1996
- **Volume 208: <u>Cataclysmic Variables and Related Objects</u>**
  Editors: A. Evans, Janet H. Wood
  Hardbound, ISBN 0-7923-4195-3, September 1996
- **Volume 207: <u>The Westerbork Observatory, Continuing Adventure in Radio Astronomy</u>**
  Editors: Ernst Raimond, René Genee
  Hardbound, ISBN 0-7923-4150-3, September 1996
- **Volume 206: <u>Cold Gas at High Redshift</u>**
  Editors: M.N. Bremer, P.P. van der Werf, H.J.A. Röttgering, C.L. Carilli
  Hardbound, ISBN 0-7923-4135-X, August 1996
- **Volume 205: <u>Cataclysmic Variables</u>**
  Editors: A. Bianchini, M. Della Valle, M. Orio
  Hardbound, ISBN 0-7923-3676-3, November 1995
- **Volume 204: <u>Radiation in Astrophysical Plasmas</u>**
  Author: V.V. Zheleznyakov
  Hardbound, ISBN 0-7923-3907-X, February 1996
- **Volume 203: <u>Information & On-Line Data in Astronomy</u>**
  Editors: Daniel Egret, Miguel A. Albrecht
  Hardbound, ISBN 0-7923-3659-3, September 1995
- **Volume 202: <u>The Diffuse Interstellar Bands</u>**
  Editors: A.G.G.M. Tielens, T.P. Snow
  Hardbound, ISBN 0-7923-3629-1, October 1995
- **Volume 201: <u>Modulational Interactions in Plasmas</u>**
  Authors: Sergey V. Vladimirov, Vadim N. Tsytovich, Sergey I. Popel, Fotekh Kh. Khakimov
  Hardbound, ISBN 0-7923-3487-6, June 1995
- **Volume 200: <u>Polarization Spectroscopy of Ionized Gases</u>**
  Authors: Sergei A. Kazantsev, Jean-Claude Henoux
  Hardbound, ISBN 0-7923-3474-4, June 1995
- **Volume 199: <u>The Nature of Solar Prominences</u>**
  Author: Einar Tandberg-Hanssen
  Hardbound, ISBN 0-7923-3374-8, February 1995
- **Volume 198: <u>Magnetic Fields of Celestial Bodies</u>**
  Author: Ye Shi-hui
  Hardbound, ISBN 0-7923-3028-5, July 1994
- **Volume 193: <u>Dusty and Self-Gravitational Plasmas in Space</u>**
  Authors: Pavel Bliokh, Victor Sinitsin, Victoria Yaroshenko
  Hardbound, ISBN 0-7923-3022-6, September 1995
- **Volume 191: <u>Fundamentals of Cosmic Electrodynamics</u>**
  Author: Boris V. Somov
  Hardbound, ISBN 0-7923-2919-8, July 1994
- **Volume 190: <u>Infrared Astronomy with Arrays</u>**
  <u>The Next Generation</u>
  Editor: Ian S. McLean
  Hardbound, ISBN 0-7923-2778-0, April 1994
- **Volume 189: <u>Solar Magnetic Fields</u>**
  <u>Polarized Radiation Diagnostics</u>
  Author: Jan Olof Stenflo
  Hardbound, ISBN 0-7923-2793-4, March 1994
- **Volume 188: <u>The Environment and Evolution of Galaxies</u>**
  Authors: J. Michael Shull, Harley A. Thronson Jr.
  Hardbound, ISBN 0-7923-2541-9, October 1993
  Paperback, ISBN 0-7923-2542-7, October 1993
- **Volume 187: <u>Frontiers of Space and Ground–Based Astronomy</u>**
  <u>The Astrophysics of the 21st Century</u>
  Editors: Willem Wamsteker, Malcolm S. Longair, Yoji Kondo
  Hardbound, ISBN 0-7923-2527-3, August 1994

- **Volume 186: <u>Stellar Jets and Bipolar Outflows</u>**
  Editors: L. Errico, Alberto A. Vittone
  Hardbound, ISBN 0-7923-2521-4, October 1993
- **Volume 185: <u>Stability of Collisionless Stellar Systems Mechanisms for the Dynamical Structure of Galaxies</u>**
  Author: P.L. Palmer
  Hardbound, ISBN 0-7923-2455-2, October 1994
- **Volume 184: <u>Plasma Astrophysics Kinetic Processes in Solar and Stellar Coronae</u>**
  Author: Arnold O. Benz
  Hardbound, ISBN 0-7923-2429-3, September 1993
- **Volume 183: <u>Physics of Solar and Stellar Coronae: G.S. Vaiana Memorial Symposium</u>**
  Editors: Jeffrey L. Linsky, Salvatore Serio
  Hardbound, ISBN 0-7923-2346-7, August 1993
- **Volume 182: <u>Intelligent Information Retrieval: The Case of Astronomy and Related Space Science</u>**
  Editors: André Heck, Fionn Murtagh
  Hardbound, ISBN 0-7923-2295-9, June 1993
- **Volume 181: <u>Extraterrestrial Dust Laboratory Studies of Interplanetary Dust</u>**
  Author: Kazuo Yamakoshi
  Hardbound, ISBN 0-7923-2294-0, February 1995
- **Volume 180: <u>The Center, Bulge, and Disk of the Milky Way</u>**
  Editor: Leo Blitz
  Hardbound, ISBN 0-7923-1913-3, August 1992
- **Volume 179: <u>Structure and Evolution of Single and Binary Stars</u>**
  Authors: C.W.H. de Loore, C. Doom
  Hardbound, ISBN 0-7923-1768-8, May 1992
  Paperback, ISBN 0-7923-1844-7, May 1992
- **Volume 178: <u>Morphological and Physical Classification of Galaxies</u>**
  Editors: G. Longo, M. Capaccioli, G. Busarello
  Hardbound, ISBN 0-7923-1712-2, May 1992
- **Volume 177: <u>The Realm of Interacting Binary Stars</u>**
  Editors: J. Sahade, G.E. McCluskey, Yoji Kondo
  Hardbound, ISBN 0-7923-1675-4, December 1992
- **Volume 176: <u>The Andromeda Galaxy</u>**
  Author: Paul Hodge
  Hardbound, ISBN 0-7923-1654-1, June 1992
- **Volume 175: <u>Astronomical Photometry, A Guide</u>**
  Authors: Christiaan Sterken, J. Manfroid
  Hardbound, ISBN 0-7923-1653-3, April 1992
  Paperback, ISBN 0-7923-1776-9, April 1992
- **Volume 174: <u>Digitised Optical Sky Surveys</u>**
  Editors: Harvey T. MacGillivray, Eve B. Thomson
  Hardbound, ISBN 0-7923-1642-8, March 1992
- **Volume 173: <u>Origin and Evolution of Interplanetary Dust</u>**
  Editors: A.C. Levasseur-Regourd, H. Hasegawa
  Hardbound, ISBN 0-7923-1365-8, March 1992
- **Volume 172: <u>Physical Processes in Solar Flares</u>**
  Author: Boris V. Somov
  Hardbound, ISBN 0-7923-1261-9, December 1991
- **Volume 171: <u>Databases and On-line Data in Astronomy</u>**
  Editors: Miguel A. Albrecht, Daniel Egret
  Hardbound, ISBN 0-7923-1247-3, May 1991
- **Volume 170: <u>Astronomical Masers</u>**
  Author: Moshe Elitzur
  Hardbound, ISBN 0-7923-1216-3, February 1992
  Paperback, ISBN 0-7923-1217-1, February 1992

- **Volume 169: Primordial Nucleosynthesis and Evolution of the Early Universe**
  Editors: Katsuhiko Sato, J. Audouze
  Hardbound, ISBN 0-7923-1193-0, August 1991
- **Volume 168: The Theory of Cosmic Grains**
  Authors: N.C. Wickramasinghe, F. Hoyle
  Hardbound, ISBN 0-7923-1189-2, October 1991
- **Volume 167: Comets in the Post-Halley Era**
  Editors: R.L. Newburn, M. Neugebauer, Jürgen H. Rahe†
  Hardbound Set only of 2 vols., ISBN 0-7923-1164-7, April 1991
  Paperback Set only of 2 vols., ISBN 0-7923-1165-5, April 1991
- **Volume 166: Observatories in Earth Orbit and Beyond**
  Editor: Yoji Kondo
  Hardbound, ISBN 0-7923-1133-7, March 1991
- **Volume 165: Dusty Objects in the Universe**
  Editors: E. Bussoletti, Alberto A. Vittone
  Hardbound, ISBN 0-7923-0863-8, August 1990

Missing volume numbers have not yet been published.
For further information about this book series we refer you to the following web site: www.wkap.nl/prod/s/ASSL